Springer Series in Biophysics 18

Series editor
Boris Martinac

For further volumes:
http://www.springer.com/series/835

Anne H. Delcour
Editor

Electrophysiology of Unconventional Channels and Pores

Springer

Editor
Anne H. Delcour
Department of Biology and Biochemistry
University of Houston
Houston, TX
USA

Series editor
Boris Martinac
University of New South Wales
Victor Chang Cardiac Research Inst.
Darlinghurst
New South Wales
Australia

ISSN 0932-2353 ISSN 1868-2561 (electronic)
Springer Series in Biophysics
ISBN 978-3-319-20148-1 ISBN 978-3-319-20149-8 (eBook)
DOI 10.1007/978-3-319-20149-8

Library of Congress Control Number: 2015954077

Springer Cham Heidelberg New York Dordrecht London

Printed on acid-free paper

Springer International Publishing AG Switzerland is part of Springer Science+Business Media (www.springer.com)

Preface

Electrophysiologists have been fascinated by unconventional channels for a long time. Nature has indeed provided to them a great variety of channels and pores that do not quite fit with the canonical description of ion channels, such as those originally described in excitable cells. When presented with the notion of "ion channel," most of us will mentally refer to the voltage-dependent sodium or potassium channels of neurons, and their relatives, channels usually found in the plasma membrane of eukaryotic cells, often highly selective for ions and tightly gated by specific stimuli, and typically constructed as bundles of α-helical transmembrane segments delineating a narrow pore. We now know that many pore-forming entities do not follow this mold, and the combination of electrophysiology, biochemistry, and structural biology has revealed a vast array of channel architectures and functional phenotypes. A quick look at a typical scientific program of the Biophysical Society meeting will divulge the breath of channel types that are the subject of current research. Besides the more typical plasma membrane channels, here we find channels and pores from intracellular organelles, bacteria, and viruses; membrane active peptides and toxins; translocons and protein-translocation pores; channels spanning two membranes like gap junctions, and even lipid-containing pores. The electrophysiological characterization of many of these is not new and has often preceded the molecular or structural data. For example, the first recordings of gap junction channels and bacterial porins date back to the 1980s, while their structural investigations only appeared in the 1990s. Nowadays, the functional aspects of most of these unconventional pores are examined through a combination of electrophysiology and molecular and structural approaches. These channels appear as complex as their more typical counterparts, and many are extremely dynamic and highly regulated. Some can also be exploited as sensitive devices in nanotechnology.

This book is intended to highlight research on these interesting pores with two main goals: one is to promote awareness to the scientific community of the rich knowledge gathered over the years on these atypical channels and foster an appreciation of the value of studying them; the other is to provide the researchers in these fields with a comprehensive and up-to-date review of the literature on a particular channel type. In compiling this volume, I aimed for a selection of representative pore-forming proteins from a broad base that includes organelles, microbes, and toxins, and thus organized the book in these three sections. Additionally, a few unusual channels, which did not fit into these subdivisions, were also incorporated in a fourth section because of their unique features. Within each section, I was fortunate to enlist experts with a history of excellent and influential contributions to provide a chapter on a well-known member of this eclectic group of unconventional channels. The field is vaster than what is showcased here, and replete with first-rate scientists who are not represented due to space consideration, and to whom I apologize. Hopefully, they will find this volume fulfilling and useful in their own investigations.

Many years ago, Boris Martinac introduced me to the exciting realm of single-channel electrophysiology, and I have valued his support and friendship ever since. How often have we pondered on the abundance of "weird" – unconventional – channels the natural world has to offer, and I thank him for inviting me to edit this volume dedicated to these fascinating molecules.

Houston, TX, USA Anne H. Delcour

Contents

Part I Mitochondria

1 **Function and Regulation of Mitochondrial Voltage-Dependent Anion Channel** 3
Tatiana K. Rostovtseva and Sergey M. Bezrukov

2 **Mitochondrial Protein Import Channels** 33
Richard Wagner, David Schmedt, Patrizia Hanhart, Claudius Walter, Christof Meisinger, and Philipp Bartsch

3 **Electrophysiology of the Mitochondrial Apoptosis-Induced Channel, MAC** 59
Kathleen W. Kinnally and Pablo M. Peixoto

4 **Ceramide Channels** 75
Marco Colombini

Part II Bacteria and Viruses

5 **Bacterial Porins** 101
Vicente M. Aguilella, María Queralt-Martín, and Antonio Alcaraz

6 **Electrophysiology of Bacterial Translocons** 123
Thieng Pham and Anne H. Delcour

7 **Viroporins** 153
Joseph M. Hyser

Part III Toxins and Antimicrobial Peptides

8 **Pore-Forming Colicins: Unusual Ion Channels – Unusually Regulated** 185
Daria Stroukova and Jeremy H. Lakey

9 Anthrax Toxin Protective Antigen Forms an Unusual Channel That Unfolds and Translocates Proteins Across Membranes 209
Bryan A. Krantz

10 Staphylococcal β-barrel Pore-Forming Toxins: Mushrooms That Breach the Greasy Barrier 241
Jack Fredrick Gugel and Liviu Movileanu

11 Properties of Pores Formed by Cholesterol-Dependent Cytolysins and Actinoporins 267
Nejc Rojko, Manuela Zanetti, Gregor Anderluh, and Mauro Dalla Serra

Part IV Other Unconventional Channels

12 Perforins 289
Robert J.C. Gilbert

13 Gap Junction Channels: The Electrical Conduit of the Intercellular World 313
Richard D. Veenstra

14 Amyloid Peptide Channels 343
Rustam Azimov and Bruce L. Kagan

15 From Phototaxis to Biomedical Applications: Investigating the Molecular Mechanism of Channelrhodopsins 361
Ryan Richards and Robert E. Dempski

Index 383

Contributors

Vicente M. Aguilella Laboratory of Molecular Biophysics, Department of Physics, Universitat Jaume I., Castellón, Spain

Antonio Alcaraz Laboratory of Molecular Biophysics, Department of Physics, Universitat Jaume I, Castellón, Spain

Gregor Anderluh Laboratory for Molecular Biology and Nanobiotechnology, National Institute of Chemistry and Department of Biology, Biotechnical Faculty, University of Ljubljana, Ljubljana, Slovenia

Rustam Azimov Department of Psychiatry and Biobehavioral Sciences, Semel Institute for Neuroscience and Human Behavior, David Geffen School of Medicine at UCLA, Los Angeles, CA, USA

Philipp Bartsch MOLIFE Research Center, Jacobs University, Bremen, Germany

Biophysics, Department of Biology/Chemistry, University Osnabrueck, Osnabrueck, Germany

Sergey M. Bezrukov Program in Physical Biology, Eunice Kennedy Shriver National Institute of Child Health and Human Development, National Institutes of Health, Bethesda, MD, USA

Marco Colombini Department of Biology, University of Maryland, College Park, MD, USA

Mauro Dalla Serra Laboratory of Biomolecular Sequence and Structure Analysis for Health, Istituto di Biofisica, Consiglio Nazionale delle Ricerche & Fondazione Bruno Kessler, Trento, Italy

Anne H. Delcour Department of Biology and Biochemistry, University of Houston, Houston, TX, USA

Robert E. Dempski Department of Chemistry and Biochemistry, Worcester Polytechnic Institute, Worcester, MA, USA

Robert J.C. Gilbert Division of Structural Biology, Wellcome Trust Centre for Human Genetics, University of Oxford, Oxford, UK

Jack Fredrick Gugel Department of Physics, Syracuse University, Syracuse, NY, USA

Patrizia Hanhart Biophysics, Department of Biology/Chemistry, University Osnabrueck, Osnabrueck, Germany

Joseph M. Hyser Department of Molecular Virology and Microbiology and Department of Medicine and Alkek Center for Metagenomic and Microbiome Research, Baylor College of Medicine, Houston, TX, USA

Bruce L. Kagan Department of Psychiatry and Biobehavioral Sciences, Semel Institute for Neuroscience and Human Behavior, David Geffen School of Medicine at UCLA, Los Angeles, CA, USA

Kathleen W. Kinnally Department of Basic Sciences, College of Dentistry, New York University, New York, NY, USA

Bryan A. Krantz Department of Microbial Pathogenesis, School of Dentistry, University of Maryland – Baltimore, Baltimore, MD, USA

Jeremy H. Lakey Institute for Cell and Molecular Biosciences, Newcastle University, Newcastle upon Tyne, UK

Christof Meisinger Institute of Biochemistry and Molecular Biology, University Freiburg, Freiburg, Germany

Liviu Movileanu Department of Physics and Structural Biology, Biochemistry, and Biophysics Program and The Syracuse Biomaterials Institute, Syracuse University, Syracuse, NY, USA

Pablo M. Peixoto Department of Natural Sciences, Baruch College and Graduate Center at City University of New York, New York, NY, USA

Thieng Pham Department of Biology and Biochemistry, University of Houston, Houston, TX, USA

María Queralt-Martín Laboratory of Molecular Biophysics, Department of Physics, Universitat Jaume I, Castellón, Spain

Ryan Richards Department of Chemistry and Biochemistry, Worcester Polytechnic Institute, Worcester, MA, USA

Nejc Rojko Laboratory for Molecular Biology and Nanobiotechnology, National Institute of Chemistry, Ljubljana, Slovenia

Tatiana K. Rostovtseva Program in Physical Biology, Eunice Kennedy Shriver National Institute of Child Health and Human Development, National Institutes of Health, Bethesda, MD, USA

David Schmedt Biophysics, Department of Biology/Chemistry, University Osnabrueck, Osnabrueck, Germany

Daria Stroukova Institute for Cell and Molecular Biosciences, Newcastle University, Newcastle upon Tyne, UK

Richard D. Veenstra Department of Pharmacology, SUNY Upstate Medical University, Syracuse, NY, USA

Richard Wagner MOLIFE Research Center, Jacobs University Bremen, Bremen, Germany

Biophysics, Department of Biology/Chemistry, University of Osnabrueck, Osnabrueck, Germany

Claudius Walter Biophysics, Department of Biology/Chemistry, University Osnabrueck, Osnabrueck, Germany

Manuela Zanetti Laboratory of Biomolecular Sequence and Structure Analysis for Health, Istituto di Biofisica, Consiglio Nazionale delle Ricerche & Fondazione Bruno Kessler, Trento, Italy

Part I
Mitochondria

Chapter 1
Function and Regulation of Mitochondrial Voltage-Dependent Anion Channel

Tatiana K. Rostovtseva and Sergey M. Bezrukov

Abstract The voltage-dependent anion channel (VDAC) is the major protein of the mitochondrial outer membrane (MOM). It is now generally accepted that this channel is responsible for most of the metabolite flux in and out of mitochondria. Small ions, adenine nucleotides such as ATP and ADP, and other water soluble mitochondrial respiratory substrates cross the MOM through VDAC. Therefore, any restriction to metabolite exchange through VDAC is able to lead to an essential disturbance of mitochondrial energetic functions and cell metabolism. This review discusses the mechanisms of regulation of these fluxes by VDAC through its ability to "gate" or adopt different conducting states. The focus of the first part is on the biophysical properties of this evolutionary conserved β-barrel channel reconstituted into its "native" lipid environment and especially on the mechanisms of modification of VDAC gating by electrical field, medium pH, and membrane lipid composition. The second part of this review addresses the novel mechanism of VDAC regulation by dimeric tubulin, an abundant cytosolic protein, and the physiological implications of the uncovered VDAC-tubulin interaction. One of the intriguing consequences of VDAC regulation by tubulin is its coupling with the Warburg-type aerobic glycolysis characteristic of many tumor cells where the VDAC-tubulin complex may play a role of "glycolytic switch" moving cells between aerobic glycolysis and oxidative phosphorylation. The discussed biophysical insights of VDAC blockage by tubulin, obtained from experiments at the single-molecule level, could be also important for understanding the molecular mechanisms of functional interactions between water soluble and membrane proteins in general.

Keywords VDAC • Tubulin • Voltage gating • ATP transport • Lipid membranes • Beta-barrel channels

T.K. Rostovtseva (✉) • S.M. Bezrukov
Program in Physical Biology, Eunice Kennedy Shriver National Institute of Child Health and Human Development, National Institutes of Health, Bethesda, MD, USA
e-mail: rostovtt@mail.nih.gov

A.H. Delcour (ed.), *Electrophysiology of Unconventional Channels and Pores*, Springer Series in Biophysics 18, DOI 10.1007/978-3-319-20149-8_1

1.1 Introduction

The latest research points to the exclusive role of mitochondria in human diseases. Cell functioning crucially depends on ATP generated in mitochondria by oxidative phosphorylation. Therefore, cells are extremely vulnerable to mitochondria malfunctioning, and even small reductions in the efficiency of ATP production in some tissues are able to cause pathological symptoms. Mitochondrial dysfunction plays a central role in a broad range of disorders and various forms of cancer. Regulation of the mitochondrial outer membrane (MOM) permeability is critical in normal metabolite exchange between the mitochondria and cytoplasm (thus controlling respiration), and in apoptosis, by releasing apoptogenic factors into the cytosol. The voltage-dependent anion channel (VDAC), the major channel of the MOM, controls a significant portion of the outer membrane function (Colombini 2004; Lemasters and Holmuhamedov 2006; Rostovtseva et al. 2005; Shoshan-Barmatz et al. 2010a; Shoshan-Barmatz and Gincel 2003). As VDAC is being shown to be involved in a wide variety of mitochondria-associated pathologies such as diabetes, hypertension, cardiovascular diseases, various cancers, and neurodegenerative disorders, such as Parkinson's, Alzheimer's, and amyotrophic lateral sclerosis, VDAC is emerging as a promising pharmacological target (Shoshan-Barmatz and Ben-Hail 2012). This large multifunctional channel could be regarded as a conjunction point for a variety of cell signals mediated by its association with various cytosolic proteins. Any restriction to metabolite exchange through VDAC could lead to an essential disturbance of mitochondrial energetic functions and cell metabolism.

VDAC is the most abundant protein in the MOM, and it is now generally accepted that this evolutionary conserved β-barrel channel is responsible for most of the metabolite flux in and out of mitochondria. There are three mammalian VDAC isoforms from three separate genes, VDAC1, VDAC2, and VDAC3 (Messina et al. 2012; Raghavan et al. 2012) with VDAC1 being the most abundant one in most tissues. The uniqueness of VDAC's role in regulation of mitochondrial metabolism mainly arises from its unique location – at the interface between the mitochondria and the cytosol (Colombini 2004) where it serves as a pathway for numerous mitochondrial respiratory substrates. Small ions, such as Ca^{2+}, Na^{+}, Cl^{-}, or OH^{-} as well as water soluble mitochondrial metabolites of a relatively high molecular weight, such as ATP, ADP, pyruvate, succinate, or inorganic phosphate cross the mitochondrial inner membrane through a variety of very specific carriers and exchangers. In contrast, all small and large ions and metabolites can cross the MOM through one channel, VDAC (Maldonado and Lemasters 2014). The weak anionic selectivity of the VDAC pore favors transport of mostly negatively charged mitochondrial metabolites in particular. The key question is how VDAC regulates these fluxes between mitochondria and the cytosol using its conserved ability to "gate" or adopt different conducting states (Colombini 2004; Hodge and Colombini 1997; Lemasters and Holmuhamedov 2006; Lemasters et al. 2012). This review focuses on the biophysical properties of the VDAC channel reconstituted into its "native" lipid environment and discusses mechanisms of VDAC gating as modified by electrical field, medium pH, membrane lipid composition and, most importantly, of VDAC regulation by cytosolic proteins.

1.2 VDAC Is a Pathway for ATP Transport

VDAC is 32 kDa membrane protein which forms a monomeric channel when reconstituted into lipid membranes. The first structural model of VDAC proposed by Colombini and coworkers was developed from a massive number of electrophysiological experiments in combination with point-directed mutagenesis (Blachly-Dyson et al. 1990; Colombini 2004; Song et al. 1998a, b; Thomas et al. 1993). According to this now so-called "functional" model, because it was mostly deduced from functional studies, VDAC is formed by 13 β-strands and one transmembrane α-helix. Later, the 3D structure of VDAC1 was determined using solution nuclear magnetic resonance (NMR) and x-ray crystallographic methods. Structural studies revealed that VDAC1 and VDAC2 form β-barrels composed of 19 β-strands with the α-helical N-terminus located inside the pore (Bayrhuber et al. 2008; Hiller et al. 2008; Schredelseker et al. 2014; Ujwal et al. 2008). The loops that connect the β-strands are exposed to the membrane surfaces where they form channel entrances at the cytosolic and mitochondrial sides of the channel. The internal diameter of the pore is ~2.7 nm and the height is ~3 nm (Ujwal et al. 2008) making the channel somehow embedded into the lipid membrane of ~4–5 nm thickness. Although in both models VDAC adopts β-barrels of similar dimensions and charge distribution, the "native" model of functional VDAC is still the subject of ongoing debates (Colombini 2009; Hiller et al. 2010). The calculated net overall charge of mouse VDAC1 is +3 due to 29 negative Asp and Glu residues and 32 positive Arg and Lys residues (Hiller et al. 2008). Eleven negatively and 15 positively charged residues are located on the β-strands inside the channel lumen while 3 positive and 2 negative charges are located on the N-terminus of VDAC (Ujwal et al. 2008).

Ion channel reconstitution is so far the best available method of direct functional studies of organelle channels, such as mitochondrial VDAC. The experimental setup for channels reconstitution is schematically shown in Fig. 1.1a. In this configuration the compartment to which VDAC is added is defined as the "*cis*" side. When reconstituted into planar lipid membranes, VDAC forms large, 4 nS (in 1 M KCl) channels permeable to non-charged polymers up to a few kDa (Carneiro et al. 2003; Colombini 1989; Gurnev et al. 2011) and to ATP (Rostovtseva and Colombini 1996, 1997). The characteristic and conserved property of VDAC channels reconstituted into lipid bilayers is their ability to move, under applied voltages of >30 mV, from the unique high conducting open state to a variety of low-conducting so called "closed" states (Colombini 2004; Colombini et al. 1996; Rostovtseva et al. 2005) (Fig. 1.2). A typical record of ion current through a single VDAC reconstituted into a planar lipid membrane is shown in Fig. 1.2. The voltage gating is seen as transitions

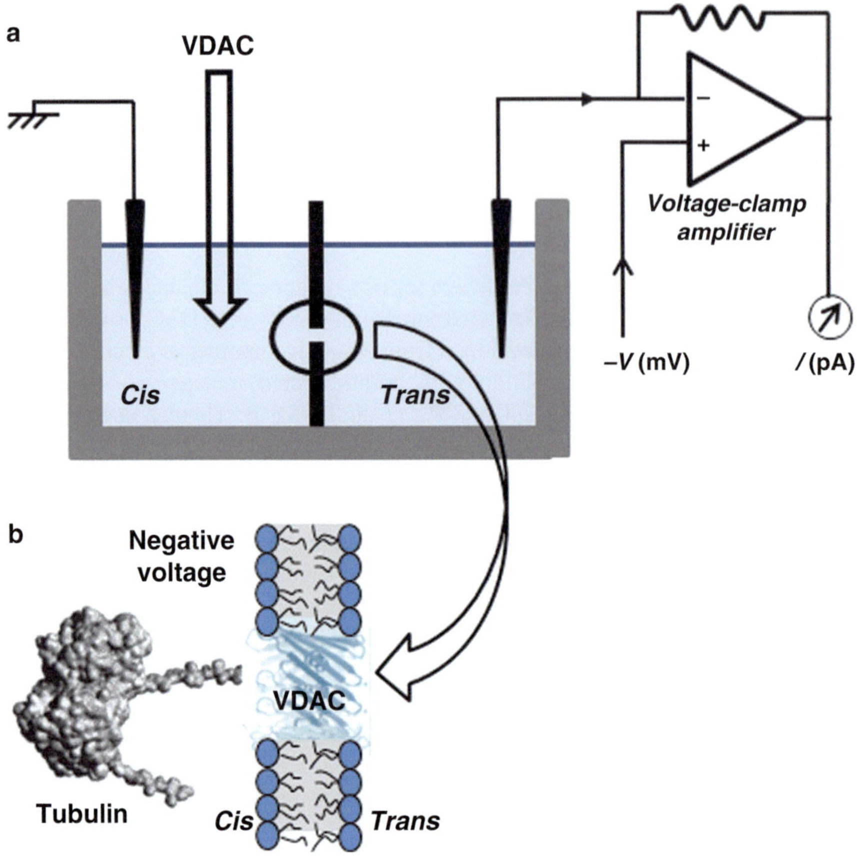

Fig. 1.1 Experimental setup for VDAC reconstitution. (**a**) The experimental chamber consists of two compartments separated by 15-μm-thick Teflon film. The membrane potential is maintained by two Ag/AgCl electrodes with 3 M KCl agarose bridges, a ground electrode in the *cis* compartment and a current-recording electrode in the *trans* compartment connected with the amplifier in the voltage-clamp mode. The potential is defined as positive when it is greater at the *cis* side which is the side of VDAC addition. For more details see (Bezrukov and Vodyanoy 1993; Rostovtseva et al. 2006). A planar bilayer membrane is formed across an aperture of ~70–90 μm diameter in the partition which separates the two compartments by apposition of two lipid monolayers preformed on the buffer solution surfaces. (**b**) The cartoon of a planar lipid bilayer with reconstituted VDAC shows its orientation in the experimental set up. When tubulin is added to the *cis* compartment, the negative voltage applied at the *cis* side induces channel blockage by tubulin

to multiple low conducting states under the ±50 mV applied voltage. Once moved to the closed state, the channel could stay in it for quite a prolonged time, depending on experimental conditions, and the application of low (0 mV) voltage is required to reopen the channel (Fig. 1.2). The VDAC open state is weakly anion selective, characterized by a permeability ratio between Cl^- and K^+ of $P_{Cl^-}/P_{K^+} \cong 2$ (Blachly-Dyson et al. 1990; Colombini et al. 1996), and its closed states are either cation selective or less anion selective than the open state. There is no obvious connection between the

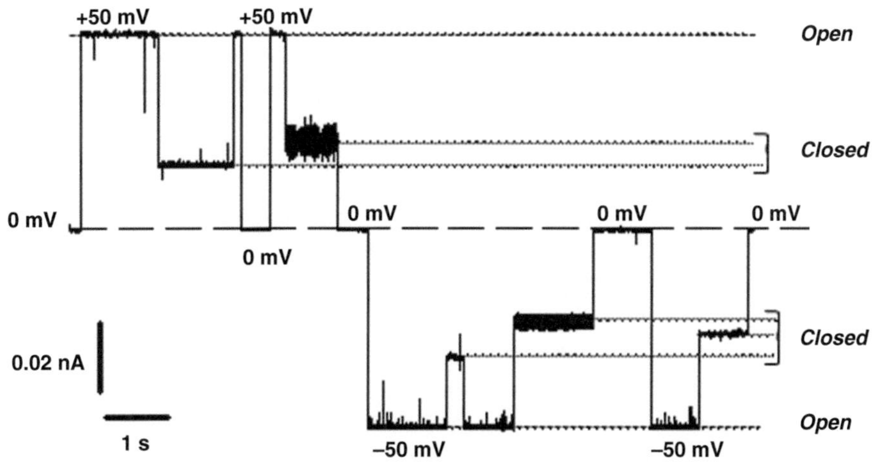

Fig. 1.2 When reconstituted into planar lipid membranes VDAC forms a large ion channel which characteristically gates under the applied voltage. A representative trace of the ion current through a single VDAC channel in a planar lipid membrane bathed by 1 M KCl at pH 7.4. Typical voltage gating of VDAC is seen at the ± 50 mV applied voltage as channel conductance fluctuates between the unique high conducting "open" state and multiple low conducting "closed" states. Application of 0 mV reopens the channel. The *long-dashed line* indicates the zero-current level, the *short-dashed lines* show open states, and *dotted lines* closed states

conductance of a particular closed state and its selectivity (Colombini 1989), which suggests a complex nature of VDAC's closed states.

Since VDAC's main function is to conduct and control fluxes of ATP and ADP, the permeability of the channel's different conducting states to adenine nucleotides is the most relevant physiological question to address. The fluxes of ATP through VDAC were directly measured in experiments with multichannel membranes using the luciferin/luciferase assay for ATP (Rostovtseva and Colombini 1996, 1997). The calculated ATP^{4-} flux through a single channel was found to be ~70 times lower than Cl^{-} flux (Rostovtseva and Colombini 1997). Most importantly, it was shown that ATP does not translocate through VDAC's closed states. Measurements of VDAC single-channel conductance in the presence of ATP, ADP, AMP, NADH, and NADPH revealed weak, in the range of millimolar, binding of adenine nucleotides to the VDAC pore and the ability of VDAC to differentiate between these nucleotides, notwithstanding their very close molecular weights and charge (Rostovtseva and Bezrukov 1998; Rostovtseva et al. 2002a). The effective diffusion coefficient of ATP inside the channel estimated by two independent methods using multichannel membranes and single-channel measurements was found to be ~10 times smaller than the ATP diffusion coefficient in the bulk (Rostovtseva and Colombini 1997; Rostovtseva and Bezrukov 1998). These results suggested the existence of weak binding site(s) for ATP inside the VDAC pore.

Recently, using a powerful computational technique, umbrella-sampling simulations, a few low-affinity binding sites for ATP were found in the open state of VDAC1 (Noskov et al. 2013). The potential of mean force (PMF) for ATP translocation across VDAC1 is shown in Fig. 1.3a. Two ATP binding sites of higher (state 1) and lower (state 3) affinity were found separated by a small barrier (state 2). The most favorable site for ATP is located near the middle of the channel where a constriction zone formed by the VDAC N-terminus with its three basic residues, R15, K12, and K20, stabilizes four negative charges of ATP (Fig. 1.3b). MD simulations showed that the helix-break-helix motif of the N-terminus (residues 6–23) is very stable, while the unstructured N-terminal tail (residues 1–5) is flexible and was found at either entrance of the channel (Noskov et al. 2013). The location of the barrier (state 2) coincides with the constriction zone where the pore radius is

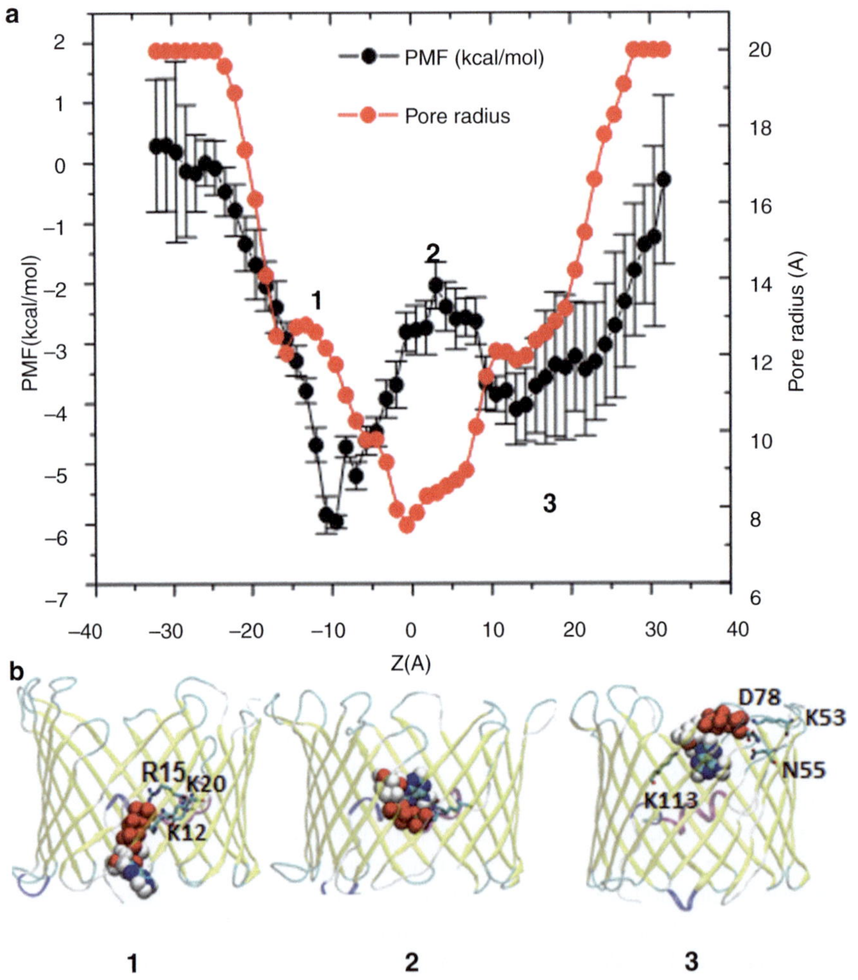

Fig. 1.3 ATP translocates through the open VDAC pore. (**a**) The potential of mean force for ATP in the VDAC1 pore (*left y* axis) and the pore radius (*right y* axis). The higher-affinity ATP-binding site (*1*), the barrier (*2*), and the lower affinity site (*3*) are marked. The Z-axis is along the pore. (**b**) A graphic representation of ATP positions inside the VDAC1 pore in the three states *1*, *2*, and *3* (Adapted from Noskov et al. (2013))

the smallest (Fig. 1.3a). The calculated binding affinity of ATP was in the range of micromolar to millimolar concentration, which agrees with the binding constant obtained in experiments with reconstituted VDAC (Rostovtseva et al. 2002a). The profile of the potential of mean force of ATP translocation along the pore obtained in simulations was used in an analytical approach (Berezhkovskii et al. 2002) to calculate the probability of ATP translocation through the pore. It appeared that in spite of the relatively weak binding affinity, the attractive interaction found by MD simulations increased this probability by ~10 times (Noskov et al. 2013). These counterintuitive – at first glance – results give new insight into the complex permeation mechanism through large β-barrel channels. Going along with these results, by solving the mouse VDAC1 structure in the presence of ATP and combining it with MD simulations, Choudhary and coauthors (Choudhary et al. 2014) also found a weak ATP binding site inside the pore. A Markov state model of permeation used in their calculations gave a prediction that ATP translocates through the open VDAC in 10–50 μs timescale. This result is in a reasonable agreement with the earlier estimations of the characteristic time of ATP diffusion through the pore, obtained from the analysis of the excess of current noise generated in the presence of ATP in the fully open VDAC channel (Rostovtseva and Bezrukov 1998). Using the same approach of current noise analysis on a single channel in the presence of relatively high molecular weight negatively charged metabolites, it was found that VDAC could discriminate between different adenine mono- and di-nucleotides and also between similarly charged molecules, such as ATP and UTP (Rostovtseva et al. 2002a, b). These findings confirm that the electrostatic environment within the channel has been evolutionary selected to favor the passage of adenine nucleotides.

1.3 Mechanism of VDAC Gating

A crucial experimental finding was the demonstration that VDAC's closed state is impermeable for ATP (Rostovtseva and Colombini 1996, 1997). Therefore, VDAC gating could also increase or decrease ATP delivery to the cytosol as well as ADP access to the electron transport chain complexes in the mitochondrial inner membrane, thus directly modulating mitochondrial energy conversion. According to this model, VDAC closure limits mitochondrial oxidative phosphorylation while VDAC opening favors it. Therefore, understanding the precise molecular mechanism of VDAC gating is crucial for our understanding of mechanism(s) of mitochondrial control and modulation of oxidative phosphorylation.

1.3.1 pH-Dependence of VDAC Voltage Gating

The sensitivity of reconstituted VDAC channels to the applied voltage depends on the charged residues that constitute its "voltage sensor". There are two gating processes depending on the sign of the applied voltage (Colombini 1989), which is manifested by a pronounced asymmetry of voltage gating of this otherwise rather symmetrical channel in what regards its other properties (Rostovtseva et al. 2006). The role of charged residues in VDAC gating was investigated by manipulating the pH of the membrane bathing solutions (Bowen et al. 1985; Colombini 1989; Ermishkin and Mirzabekov 1990; Teijido et al. 2014), by chemical modification of amino acid residues (Colombini 1989; Doring and Colombini 1985a, b), and by point-directed mutagenesis (Thomas et al. 1993). Chemical modification of amino groups by succinic anhydride introduced multiple negative charges and resulted in a decrease of the voltage sensitivity (Colombini 1989; Doring and Colombini 1985a, b). An increased pH also reduced the voltage sensitivity of VDAC, which means that higher voltages were required to close the channel (Fig. 1.4b) (Bowen et al. 1985). High pH reduced the overall net positive charge in the pore by titrating the positive charges. Correspondingly, low pH promoted voltage gating, which was manifested by a reduction of the characteristic voltage, V_0, the voltage at which half of the channels are open and half closed (Fig. 1.4b) (Teijido et al. 2014). Low pH neutralizes acidic charges and thus increases the net positive charge involved in gating. The effective gating charge, n, of VDAC increases ~4 times when the pH decreases from 11 to 3 (Fig. 1.4a) (Teijido et al. 2014). Figures 1.4a, b provide a representative combination of the pH effect studied by two independent groups (Bowen et al. 1985; Teijido et al. 2014). At pH >9.5 where positive charges are reduced, gating is decreased, which is shown as a decrease of the gating charge n (Fig. 1.4a) and an increase of V_0 (Fig. 1.4b). At pH <5.5 where acidification leads to a reduction of the net negative charge, gating is increased, which is manifested by an increase of n and a decrease of V_0. Both gating parameters, n and V_0, remain constant in the pH region between 6 and 9.5. The voltage-independent component of the free energy change, nFV_0, is pH independent for the whole range of pH from 3 to 11 (Fig. 1.4c). Two other basic channel parameters, ion selectivity and single-channel conductance also depend on pH with the corresponding increase of both parameters with acidification (Rostovtseva et al. 2000; Teijido et al. 2014). pH titration of negative charges together with His residues titration result in a total increase of the net positive charge inside the pore, which leads to an increase in anion selectivity of the open state. The most dramatic shift in VDAC gating parameters, ion selectivity, and single-channel conductance occurs at pH <5, and, most likely, is due to titration of Asp and Glu residues. Analysis of the pH dependences of the gating and open channel parameters yielded similar pK_a values of ~4.0 (Teijido et al. 2014). All observed effects of pH on VDAC gating were completely reversible, which is in good agreement with the structural CD data (Mannella 1998; Shao et al. 1996).

Another interesting effect of pH on VDAC gating is a pronounced asymmetry in the sensitivity of voltage gating to pH changes with the *cis* side of the channel (see Fig. 1.1a) being highly sensitive to acidification (Teijido et al. 2014). The opposite, *trans* side of the channel was found to be almost insensitive to pH changes. These results were quite unexpected considering a rather symmetrical distribution of charges across the VDAC's β-strands that constitute the channel wall (Ujwal et al. 2008). MD simulations helped to explain these surprising results. A significant asymmetry in the number of stable salt bridges formed by charged residues across

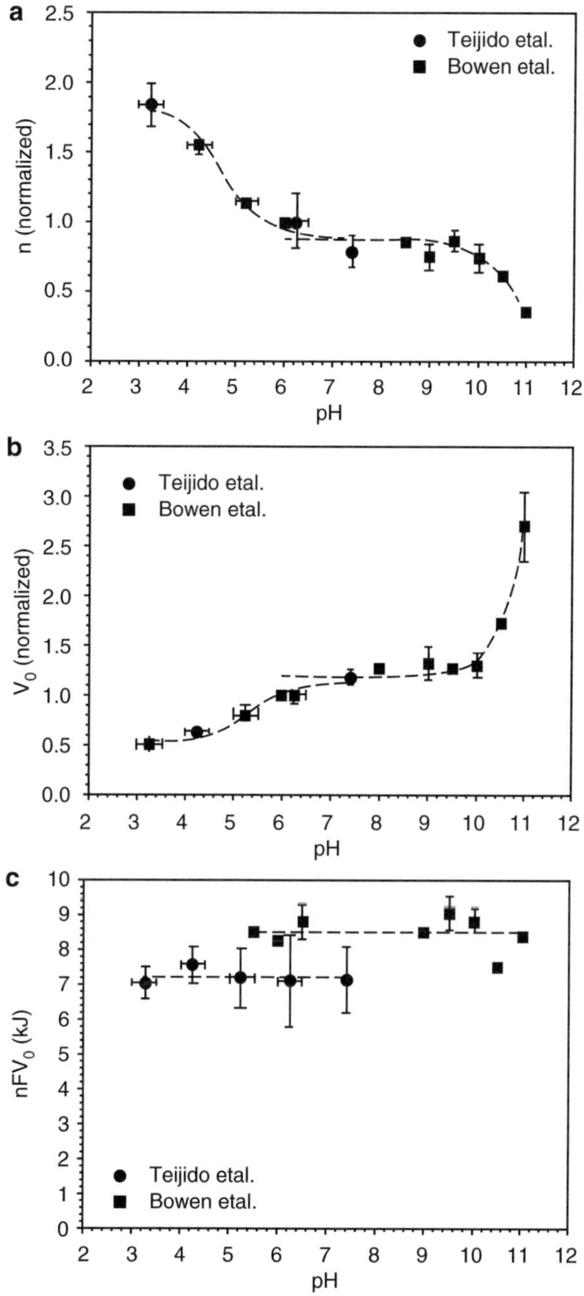

Fig. 1.4 VDAC voltage gating strongly depends on pH. pH dependences of VDAC gating parameters: the effective gating charge, *n* (**a**), and the voltage at which half of the channels are open, V_0 (**b**). Each data point obtained by (Teijido et al. 2014) (*circles*) and by (Bowen et al. 1985) (*squares*) is normalized by division by the corresponding value at pH 6.1. (**c**) The pH dependence of the voltage-independent component of the free energy difference between the open and closed states, nFV_0 (Adapted from Teijido et al. (2014))

the VDAC1 pore between the two sides of the channel was found (Teijido et al. 2014). MD simulations suggested a higher number of salt bridges clustered at the *cis* side of the pore, which are more susceptible to disruption at low pH and, therefore, a pH decrease at the *cis* side leads to the preferential closure of the channel.

The insertion of VDAC into a planar lipid membrane is a spontaneous process; however, the channel inserts directionally and its orientation in the bilayer was established based on the asymmetrical responses of phosphorylated VDAC to tubulin addition (Sheldon et al. 2011) (see below). The data suggested that the VDAC C-terminus is facing the *trans* side of the bilayer (Fig. 1.1). Considering that it was reported (Tomasello et al. 2013) that the C-terminus is not accessible to cleavage by cytosolic caspases and therefore faces the mitochondrial side of the MOM, it was concluded that the *cis* side of reconstituted VDAC in the planar membrane setup (Fig. 1.1) most likely corresponds to the cytosolic channel entrance (Rostovtseva and Bezrukov 2012; Sheldon et al. 2011). Therefore, the pronounced sensitivity of the cytosolic side of VDAC to acidification could be physiologically relevant by responding to the large changes of cytosolic pH during ischemia and reperfusion injury (Lemasters et al. 1996; Murphy and Steenbergen 2008).

1.3.2 Lipid-Dependence of VDAC Voltage Gating

When reconstituted into planar lipid membranes, VDAC exhibits two gating processes: one at the positive and the other at the negative sign of the applied voltage. The gating could be symmetrical or quite asymmetrical with respect to the sign of the applied voltages depending on the salt gradient across the channel (Zizi et al. 1998) or the lipid composition of the planar membranes (Rostovtseva et al. 2006). The asymmetry of voltage gating is especially noticeable even at symmetrical salt conditions when planar membranes have a high content of a non-lamellar lipid, such as phosphatidylethanolamine (PE) or cardiolipin (CL). It was found that in membranes formed from pure dioleoylphosphatidylethanolamine (DOPE) or from a mixture of dioleoylphosphatidylcholine (DOPC) and CL, VDAC closure at *cis-negative* applied voltages is more pronounced (Rostovtseva et al. 2006). Thus, the presence of non-lamellar lipids changes VDAC's conformational equilibrium to promote gating at *cis*-negative voltages, suggesting a coupling between the mechanical pressure in the hydrocarbon lipid region and channel gating. However, it turned out that this coupling takes place not through a change in the gating charge but mostly through a shift in the equilibrium between the open and closed states that could be seen as a shift of $|V_0|$ towards smaller voltages (Rostovtseva et al. 2006). When a non-lamellar lipid, such as PE, which spontaneously forms an inverted hexagonal phase, is forced into the planar bilayer structure, this results in a significant stress of lipid packing in the region of acyl chains (Bezrukov 2000; Cantor 1999; Gruner 1985; Keller et al. 1993; van den Brink-van der Laan et al. 2004). The sensitivity of a membrane protein conformational transition to the changing pressure in the hydrocarbon area of the membrane can be rationalized in a model wherein the transition changes the shape of the protein molecule in a way that either relieves or increases the pressure. In the case of VDAC gating, one of the possibilities is that the outer surface of the β-barrel that is in contact with the hydrocarbon area is shaped as an hourglass in a subset of the closed states at negative applied voltages, but is nearly cylindrical in the open state and the closed ones at positive voltages (Rostovtseva and Bezrukov 2008; Rostovtseva et al. 2006). Importantly, this model assumes a rather flexible β-barrel that could be squeezed and stretched under the applied voltage and, probably, lipid packing stress. This is in contrast with the conventional view of β-barrel channels as quite rigid transmembrane structures.

1.3.3 Models of VDAC Gating

VDAC gating involves large conformational changes leading to a substantial decrease of the channel volume (Zimmerberg and Parsegian 1986) and pore inner dimensions, from ~2.5 to 2.7 nm in the open state to ~1.8 nm in the closed states (Colombini 1989; Colombini et al. 1996) as was found in electrophysiological studies on reconstituted VDAC. Therefore, the transition from the open to the closed states creates both steric and electrostatic barriers for the passage of metabolites due to the decreased pore size and reversed selectivity of the closed states (Hodge and Colombini 1997; Rostovtseva and Colombini 1996). Colombini and co-authors performed an impressive electrophysiological and biochemical analysis of VDAC properties using extensive site-directed mutagenesis and biotin modification approaches (Blachly-Dyson et al. 1990; Colombini et al. 1996; Song et al. 1998b; Thomas et al. 1993). This array of data allowed them to develop a model of VDAC voltage gating which involves translocation of a large positively charged voltage sensor domain towards one of the membrane surfaces, thereby reducing the size of the pore and inverting its selectivity (Colombini et al. 1996; Song et al. 1998a). The molecular identity of the voltage sensor has not been resolved yet because the charged residues affecting voltage gating are spread out through the entire VDAC folding pattern including its N-terminus (Colombini et al. 1996; Song et al. 1998a).

Although there is a general agreement on the N-terminus location inside the VDAC open pore and crucial role in VDAC gating, its position in the closed states is still a subject of discussions. Studies on cells suggest accessibility of the N-terminal region to VDAC antibodies (McDonald et al. 2009; Stanley et al. 1995; Tomasello et al. 2013). Therefore, in several gating models the N-terminal region is considered to move independently out of the channel lumen to the cytoplasm (De Pinto et al. 2003, 2008; Geula et al. 2012; Shoshan-Barmatz et al. 2010b) or to reside on the membrane surface (Guo et al. 1995; Reymann et al. 1995). Some models advanced the idea of a flexible N-terminus moving inside the pore, partially obstructing the ion fluxes and thus providing for VDAC gating (Mertins et al. 2012; Ujwal et al. 2008). However, more recent evidence has questioned this idea (Teijido et al. 2012) and the initially proposed flexibility of the N-terminus. Recent NMR studies by several independent groups reported that the N-terminus was rigid (Eddy et al. 2012; Schneider et al. 2010; Villinger et al. 2010) contrary to the initial thoughts.

The position of the N-terminus is of considerable interest because of its critical role in VDAC gating. Most observations have demonstrated that the N-terminus is required for channel gating (Choudhary et al. 2010; Mannella 1998; Popp et al. 1996). Addressing this question, the N-terminal α-helix of mouse VDAC1 was cross-linked to the β-barrel by using a double cysteine mutations of Leu10Cys on the N-terminus and Ala170Cys on β-strand 11 followed by reconstitution of the mutant VDAC1 into planar lipid membranes (Teijido et al. 2012). The rational of this study was to challenge the available gating models by testing the ability of the N-terminus to move independently under applied voltage and affect gating. Surprisingly, it was found that VDAC with the N-terminal α-helix covalently bound to the channel wall had indistinguishably similar channel properties compared with endogenous VDAC1, suggesting that the N-terminal region is either immobile or does not move independently of the rest of the voltage sensor (Teijido et al. 2012). These data also strongly suggest that the α-helix remains inside the VDAC pore when the channel is open and that during voltage-induced gating the N-terminus either remains inside the pore or moves in a coordinated fashion together with the voltage sensor. Based on these data the scenario of the N-terminus independently undergoing large structural rearrangements to achieve voltage gating seems unlikely.

There is still no clear structural data on VDAC gating, but a new mechanism suggesting a soft β-barrel wall that becomes squeezed and simultaneously elongated upon voltage application was proposed recently (Teijido et al. 2012) and deserves future elaboration. NMR data in combination with MD simulation showed a pronounced dynamics of β-strands 1–6 and 16–19 (Villinger et al. 2010), thus supporting a model of a flexible VDAC β-barrel. The analysis of the B-factor of the VDAC crystal structure showed that it increased measurably in the region of the wall opposite to β-strands 8–15 reinforced by the adjusted N-terminus and that the α-helix has a more rigid structure than the remainder of the pore (Teijido et al. 2012). Deletion of the N-terminal α-helix caused a sharp increase of the overall motion of the VDAC β-barrel and resulted in a pronounced elliptical shape of the β-barrel confirming its intrinsic flexibility (Zachariae et al. 2012). Another piece of evidence of VDAC β-barrel flexibility was obtained on a system of VDAC1 embedded into tethered planar lipid membranes using a spectroelectrochemical approach (Kozuch et al. 2014). Measurements of the electrochemical impedance spectra under applied voltage allowed to associate spectral changes with the structural transition upon gating. The observed spectral changes were interpreted as alterations of the angle of the tilt of the β-strands and could be visualized as expansion of the β-barrel, supporting the above "squeezing" β-barrel gating model.

Taking together all the above data, it seems reasonable to conclude that the β-barrel could undergo partial squeezing with simultaneous elongation along the channel axis, which would lead to a decrease of the pore aperture. Importantly, a new model does not contradict the original model of VDAC gating suggested by Colombini and colleagues where a large part of the β-barrel together with the N-terminus moves under the applied electrical field towards the membrane interface (Colombini 2009; Colombini et al. 1996). However, our intuition does not help in predicting the effect of the squeezing/elongation structural rearrangement on channel selectivity. The crucial test of this novel model would be computation of the ion selectivity of the squeezed (closed) structure and its permeability for ATP. Further evaluations of the MD simulation, and structural and biophysical data are needed to reconcile the reliable experimental findings and provide a better understanding of the structural basis of gating and the role of the N-terminus, in particular.

1.4 VDAC Regulation by Dimeric Tubulin

1.4.1 Tubulin Modulation of VDAC Conductance

Tubulin induces the characteristic reversible blockage of VDAC reconstituted into planar lipid membranes. A typical experiment is shown in Fig. 1.5a where tubulin added to one side of the membrane (*cis* side), in concentrations as low as 1 nM, induced fast time-resolved upward (to zero current) transitions between the high conducting open state and low conducting blocked state. The conductance of the open state in the presence of tubulin remains the same as in control and the conductance of the blocked state is ~0.4 of the open state conductance. Note that this is in contrast to the wide variety of closed states induced by voltage (Fig. 1.2). The characteristic time of the tubulin-blocked state depends on the applied voltage and is in the range of 0.1–100 ms (Rostovtseva et al. 2008), which is much shorter than the seconds to minutes that the channel spends in the closed states induced by voltage (Fig. 1.2). Tubulin blocks VDAC only when a negative potential is applied from the side of tubulin addition. This is illustrated by a typical experiment shown in Fig. 1.5a where tubulin was added asymmetrically to the *cis* side of the channel (see Fig. 1.1b) and induced VDAC blockage when a negative potential of −20 mV (as referred to the *cis* side) was applied (three upper traces in Fig. 1.5a). When the polarity of the applied voltage was reversed to the positive of +20 mV, no blockage events were observed and the channel open state conductance was as steady as in control without tubulin (low trace in Fig. 1.5a).

Tubulin induces VDAC blockage in a concentration-dependent manner. The number of blockage events increases with tubulin bulk concentration (Fig. 1.5a, three upper traces). The on-rate of VDAC-tubulin binding, τ_{on}^{-1}, where τ_{on} is the average time when the channel stays open between closing events, increased linearly at low tubulin concentrations <30 nM and saturated at ~50 nM (Fig. 1.5b). The concentration dependence of the on-rate of VDAC blockage could be described by a simple ligand binding equation with K_d in the range of 10–50 nM depending on experimental conditions such as the membrane lipid composition (Rostovtseva et al. 2012) (see here and in Part 1.4.3 below). The off-rate of VDAC-tubulin binding calculated as τ_{off}^{-1}, where τ_{off} is the average time the channel spends in the blocked state, does not depend on the concentration of tubulin (Rostovtseva et al. 2008). Therefore, at low tubulin concentrations, the blockage could be adequately described by a simple first order reaction where the on-rate linearly increases with tubulin concentration and the off-rate is concentration independent.

VDAC blockage by tubulin is impressively voltage dependent. Figure 1.6a shows current traces of the same single VDAC channel in the presence of 10 nM of tubulin in both compartments at different applied voltages. It could be seen that at 15 mV

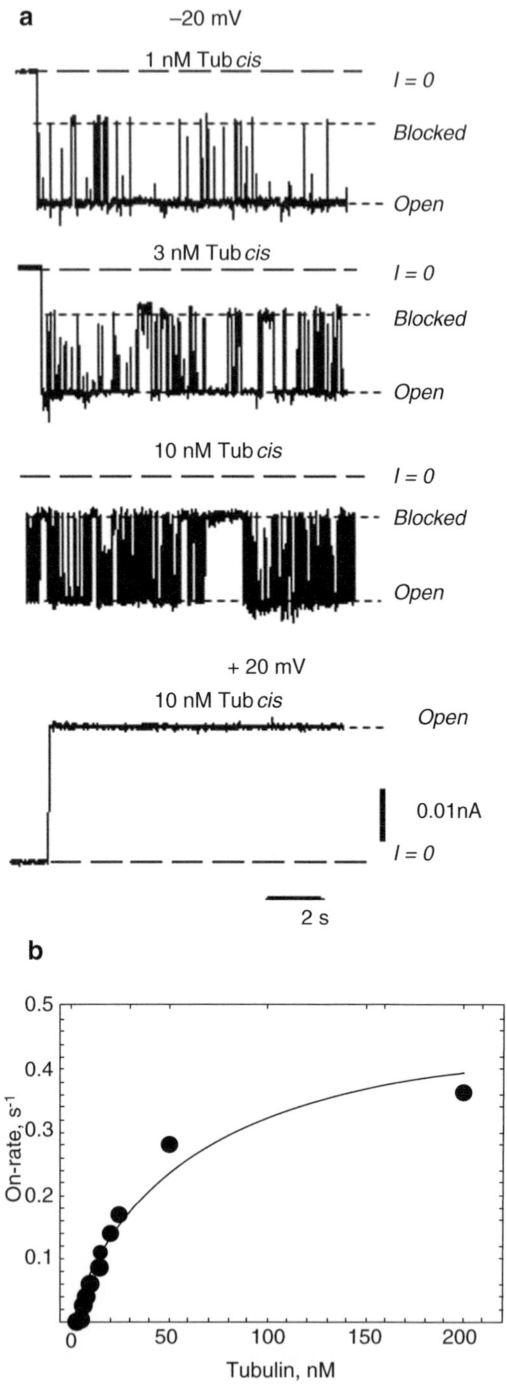

Fig. 1.5 Tubulin induces rapid blockages of VDAC conductance in a concentration dependent manner. (**a**) Representative current records of the same single VDAC channel after sequential additions of 1, 3, and 10 nM of tubulin to the *cis* side of the membrane (see the schematic in Fig. 1.1). The channel conductance fluctuates between the high conducting, "open", and low conducting or "blocked" state. The number of blockage events increases with tubulin concentration. Tubulin induces measurable VDAC blockage only when a negative potential (–20 mV for the *three upper traces*) is applied from the side of tubulin addition (*cis*). At the opposite polarity (+20 mV for the *lower trace*), the channel stays open without any blockage as it occurs in typical control records without tubulin. *Long-dashed lines* indicate the zero-current level, and *short-dashed lines* show open and tubulin-blocked states. (**b**) The concentration dependence of the on-rate of VDAC-tubulin binding. The applied potential was –20 mV. The medium consisted of 1 M KCl buffered with 5 mM HEPES at pH 7.4

there are barely any blockages, at 20 and 25 mV there are plenty of blockages, and at 30 mV the channel is blocked most of the time. (It should be noted that the sign of positive potential refers to the *cis* side, and therefore the negative potential at the *trans* side promotes channel blockage by tubulin added to the *trans* side.) The statistical analysis of the open probability, P_{open}, in the presence of 10 nM of tubulin is shown in Fig. 1.6b. There is a steep dependence of the open probability on the applied voltage with a calculated effective gating charge of 10–13 elementary charges depending on the salt concentration of the membrane-surrounding buffer (Gurnev et al. 2012). This means that the blockage reaction is as responsive to the applied voltage as the most voltage-sensitive channels studied by electrophysiology to date (Swartz 2008).

There is an apparent discrepancy between *in vitro* data, where tubulin could block VDAC at nanomolar concentrations, and the fact that there is up to 20 μM free tubulin in some cells (Gard and Kirschner 1987). Depending on experimental conditions (see below), the on-rate constant of the blockage could be changed by orders of magnitude (Rostovtseva et al. 2012; Sheldon et al. 2011), and at potentials across the MOM close to zero, micromoles of tubulin would be required to block VDAC *in vivo*. The value of the MOM transmembrane potential is a matter of long-standing debates (see discussion in (Colombini 2004; Rostovtseva and Bezrukov 2012)) and usually believed to be close to 0 mV due to the high abundance of VDACs in the MOM. The main source of the potential across the MOM is supposed to be the so called Donnan potential, which arises from the high concentration of VDAC-impermeable charged macromolecules (e.g., cytochrome c of +9 charges) in the mitochondrial intermembrane space and the cytosol (Colombini 2004). The estimates of the MOM transmembrane potential range from 10 mV (Lemeshko 2006) to as high as 46 mV (Porcelli et al. 2005) and even higher (Lemeshko 2014a, b). A recently published model suggests that a VDAC complex with hexokinase could be a generator of the potential across the MOM as high as 50 mV, negative at the cytoplasmic side of the MOM, with the Gibbs free energy of kinase reactions as a driving force (Lemeshko 2014a, b). It should be noted here that a potential, negative from the cytosolic side of the MOM, drives cytosolic tubulin into the VDAC pore.

The heterodimer of α- and β-tubulin is 100 kDa acidic water-soluble protein composed of compactly folded α-helices and β-strands with a well-defined crystal structure (Nogales et al. 1998). Tubulin C-terminal tails (CTTs) are composed of unstructured peptides of 11–15 amino-acids (Westermann and Weber 2003) exposed out of the protein surface. Both α- and β-tubulin CTTs are highly negatively charged and are essentially poly-Glu peptides. The fact that tubulin blocks VDAC only when a negative potential is applied from the side of tubulin addition (Fig. 1.5a) suggests that the negatively charged CTTs are responsible for the channel blockage. Experimental evidence proving this point has been obtained in experiments with tubulin in which CTTs were cleaved off, tubulin-S, which, contrary to tubulin with intact CTTs, did not block VDAC (Rostovtseva et al. 2008). On the other hand, synthetic peptides of α- and β-tubulin CTTs did not block the channel either, up to 10 μM concentration (Rostovtseva et al. 2008), thus suggesting that CTTs should be attached to the main tubulin body to induce the characteristic VDAC blockage. Based on these observations a model of VDAC-tubulin interaction was suggested where the negatively charged tubulin CTT partially blocks the positively charged VDAC pore (Rostovtseva and Bezrukov 2008, 2012; Rostovtseva et al. 2008) (Fig. 1.6c).

The effective gating charge of the VDAC-tubulin blockage equilibrium increases in low salt concentration but the salt dependence is rather weak (Fig. 1.6b). The

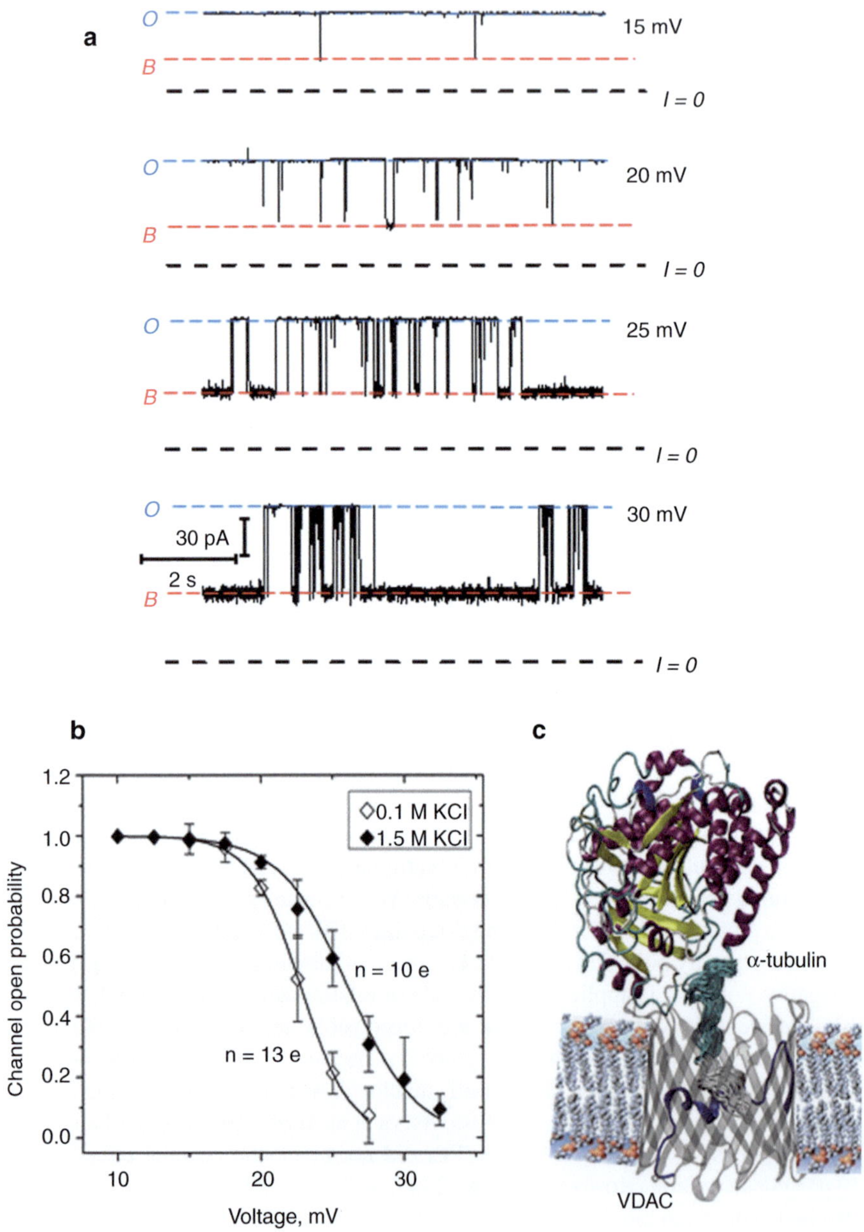

charge of the tubulin tail is not screened even at salt concentrations as high as 1.5 M KCl. This suggests that ion condensation on the tubulin tail in the pore is negligible. The applied electrical field primarily acts on the tubulin tail charges. However, the absolute value of the gating charge of 10–13 elementary charges is too high to be accounted for by the charges on the tubulin tail alone, which are −7 and −10 elementary charges in the α- and β-tubulin CTTs, respectively (Rostovtseva et al. 2008). Surprisingly, the blocked time only weakly increases with salt concentration (Gurnev et al. 2012). This is in contrast with the strong dependence of blocked time on salt concentration found for the interaction of another β-barrel channel, anthrax's protective antigen channel, with positively charged β-cyclodextrins (Nestorovich et al. 2010). This suggests that the Coulomb interactions between the charges on the VDAC interior wall and the tubulin CTT are not predominant, indicating the existence of some kind of short-range interactions between the tubulin CTT and the VDAC pore.

Equilibrium and nonequilibrium MD simulations together with a Roseta protein-protein docking algorithm allowed to further test the structural rearrangements of the VDAC-tubulin complex (Noskov et al. 2013). It was found that the unstructured α-tubulin CTT in the mouse VDAC1 pore decreases its conductance by >40 % and, what is even more important, switches its selectivity from anionic to cationic as was shown in experiments with reconstituted channels. MD simulations showed that the negatively charged residues of the α-tubulin CTT form stable salt bridges with the basic residues of VDAC1 (Fig. 1.6c). Three of the identified basic residues, R15, K12, and K20, are located in the N-terminus of VDAC1 while the other two, K28 and K32, are located on the adjacent β-strands thus creating a kind of positively charged constriction zone around of the N-terminus. These results suggest that the negatively charged CTT of tubulin most of the time does not reach beyond the VDAC N-terminus located in the middle of the pore (Fig. 1.6c). Importantly, MD simulation results support the suggested model (Rostovtseva and Bezrukov 2008, 2012; Rostovtseva et al. 2008) of the VDAC-tubulin interaction where the tubulin CTT partially blocks the channel and reverses its ion selectivity.

Fig. 1.6 VDAC blockage by tubulin is highly voltage dependent. (**a**) Current records from the same single VDAC channel obtained in the presence of 10 nM tubulin in both compartments at different potentials. There are almost no blockage events at 15 mV, but when the voltage is increased to 30 mV, the blocked state dominates. Positive potential induces channel blockage by tubulin added to the *trans* side of the membrane. *Black dashed lines* indicate the zero-current level; *blue dashed lines* show the open (*O*) state, and *red dashed lines* show the tubulin-blocked (*B*) states. (**b**) The VDAC open probability in the presence of tubulin sharply decreases with the applied voltage and is lower in 0.1 M than in 1.5 M KCl. *Solid lines* are drawn according to the Boltzmann equation with the effective gating charges of 13e for 0.1 M KCl and 10e for 1.5 M KCl, respectively (Adapted from Gurnev et al. 2012). (**c**) A model of the restrictive blockage of the VDAC pore by tubulin. One of the unstructured negatively charged CTTs of tubulin partially blocks the channel by entering the positively charged pore. The molecular image of the VDAC1-α-tubulin complex embedded into a lipid bilayer is adapted from (Noskov et al. 2013). The ensemble of complexes has been obtained using protein-protein docking (ROSETTA-docking). Explicit MD simulations were run with the NAMD program for 50 ns in 1 M of KCL and a DOPE lipid bilayer

1.4.2 Probing the Tubulin-Blocked VDAC State

As mentioned above, it was experimentally shown that the voltage-induced VDAC's closed state is essentially impermeable for ATP (Rostovtseva and Colombini 1996, 1997). Correspondingly, it was important to probe the permeability of the tubulin-blocked state for ATP. The tubulin-blocked state is still highly ion-conductive (about 40 % of the open state conductance in 1 M KCl), which suggests that VDAC inhibition by tubulin is limited by the value of this residual conductance. The functional properties of the tubulin-blocked state were assessed using four different experimental approaches (Gurnev et al. 2011).

First, using the method of non-electrolyte polymer partitioning into the ion channels (Bezrukov et al. 1994, 1996; Krasilnikov and Bezrukov 2004; Krasilnikov et al. 1992; Rostovtseva et al. 2002c), the characteristic radius of the VDAC pore blocked by tubulin was estimated to be significantly smaller than the radius of the open state (Gurnev et al. 2011). Based on the characteristic molecular weight of the polymer that separates partitioning from exclusion in VDAC's open and blocked states, the effective cross-sectional area of the channel was estimated as reduced by a factor of two (Gurnev et al. 2011). Second, the small-ion selectivity of the tubulin-blocked state, measured at close to physiological salt concentrations, was shown to be reversed. In particular, in 150 mM vs 50 mM gradient of KCl the selectivity of the channel changes from predominantly anionic in the open state with the permeability ratio of $P_{Cl-}/P_{K+}=4$ to cationic in the tubulin-blocked state with $P_{Cl-}/P_{K+}=0.3$ (Gurnev et al. 2011). Third, ATP partitioning into both the open and tubulin-blocked channel was directly measured using an approach similar to the one described for non-electrolytes partitioning. It was found that addition of ATP to the membrane-bathing solution reduces the conductance of the open-state channel but not the conductance of the tubulin-blocked channel (Gurnev et al. 2011), suggesting that ATP is excluded from the tubulin-blocked state. Fourth, computations of the one-dimensional potential of mean force for the VDAC-tubulin complex demonstrated that tubulin blockage of the pore reduces ATP transport down to negligible (Noskov et al. 2013). The presence of the negatively charged tubulin CTT in the pore creates a barrier for ATP translocation of ~2–3 kcal/mol, due to the strong electrostatic repulsion between ATP and the negative charges on the CTT. Unlike small anions such as Cl^-, the bulkier and more negatively charged ATP molecule cannot overcome the steric and electrostatic barriers created by the CTT inside the pore. Moreover, the tubulin CTT in the pore binds to the same VDAC positive residues that constitute ATP binding sites (Noskov et al. 2013) making them unavailable for binding to ATP.

Altogether these results show that due to the electrostatic and steric barriers induced by tubulin blockage, ATP is excluded from the tubulin-blocked state of VDAC, thus verifying the functional role of the VDAC-tubulin interaction in regulating mitochondrial respiration. More generally, the insights into the nature of VDAC blockage by tubulin, obtained in experiments at a single-molecule level, are important for our understanding of the molecular mechanisms of functional interactions between water soluble and membrane proteins.

1.4.3 *Effect of Membrane Lipid Composition on VDAC-Tubulin Binding*

The initially proposed model of VDAC blockage by tubulin (Fig. 1.6c) considers only interactions between the tubulin CTT and the channel lumen and can be described by a simple bimolecular reaction between VDAC and tubulin (Rostovtseva et al. 2008). However, it turned out that the mechanism of VDAC-tubulin binding is more complex and, quite unexpectedly, strongly depends on the specific composition of the lipid membrane (Rostovtseva et al. 2012). It was found that the blockage varies up to 100-fold depending on the particular lipid used for bilayer formation in reconstitution experiments. For example, it increases with the increasing content of DOPE in a DOPC bilayer (Fig. 1.7). Representative experiments with VDAC single channels reconstituted into membranes formed from pure DOPC, DOPE, or a DOPE/DOPC mixture demonstrate that at the same applied voltage, 10 nM of tubulin induced significantly more blockage events in a DOPE membrane than 60 nM of tubulin induced in pure DOPC or DOPE/DOPC mixtures (Fig. 1.7a). The on-rate constant of VDAC-tubulin binding calculated as $k_{on} = 1/(\tau_{on}[C])$, where τ_{on} is the average time when the channel is open between the consecutive blockage events and $[C]$ is the tubulin concentration in the membrane-surrounding bath, increased gradually with the DOPE content and was ~200 times higher in a pure DOPE membrane than in a pure DOPC membrane (Fig. 1.7b). At the same time, the off-rate of VDAC blockage by tubulin did not depend on the lipid composition (Rostovtseva et al. 2012).

It turned out that the on-rate of VDAC blockage by tubulin depends on both the hydrophobic and polar parts of the phospholipid. The effect of the lipid headgroup charge on the on-rate is more pronounced in physiologically relevant low salt than in high salt concentration (Rostovtseva et al. 2012) suggesting a role of the long-range electrostatic components in the VDAC-tubulin interaction. In 0.1 M KCl, k_{on} was ~100 times higher in neutral diphytanoyl-phosphatidylcholine (DPhPC) membranes than in bilayers containing negatively charged diphytanoyl-phosphatidylserin (DPhPS) (Rostovtseva et al. 2012). Surprisingly, addition of the positively charged synthetic lipid DOTAP to DPhPC also caused a decrease of k_{on}. Another important piece of evidence on the role of the non-electrostatic component in VDAC-tubulin interaction is that the type of lipid hydrocarbon acyl chains also significantly affects the blockage kinetics. When oleoyl chains in DOPC have been replaced with phytanoyl in DPhPC, the on-rate of tubulin blockage increased ~70 times. In addition, the striking similarity between the concentration dependence of the on-rate of VDAC blockage by tubulin (Fig. 1.5b) and tubulin binding to the lipid membranes obtained earlier using ^{125}I-labeled tubulin (Bernier-Valentin et al. 1983), suggests that tubulin binding to the membrane is an important step of the VDAC-tubulin interaction.

This conjecture was verified in independent experiments using confocal fluorescence microscopy of giant unilamellar vesicles (GUVs) in the presence of fluorescently labeled tubulin (Rostovtseva et al. 2012). The confocal image of GUV with bound fluorescently-labeled tubulin is shown in the inset in Fig. 1.8. In

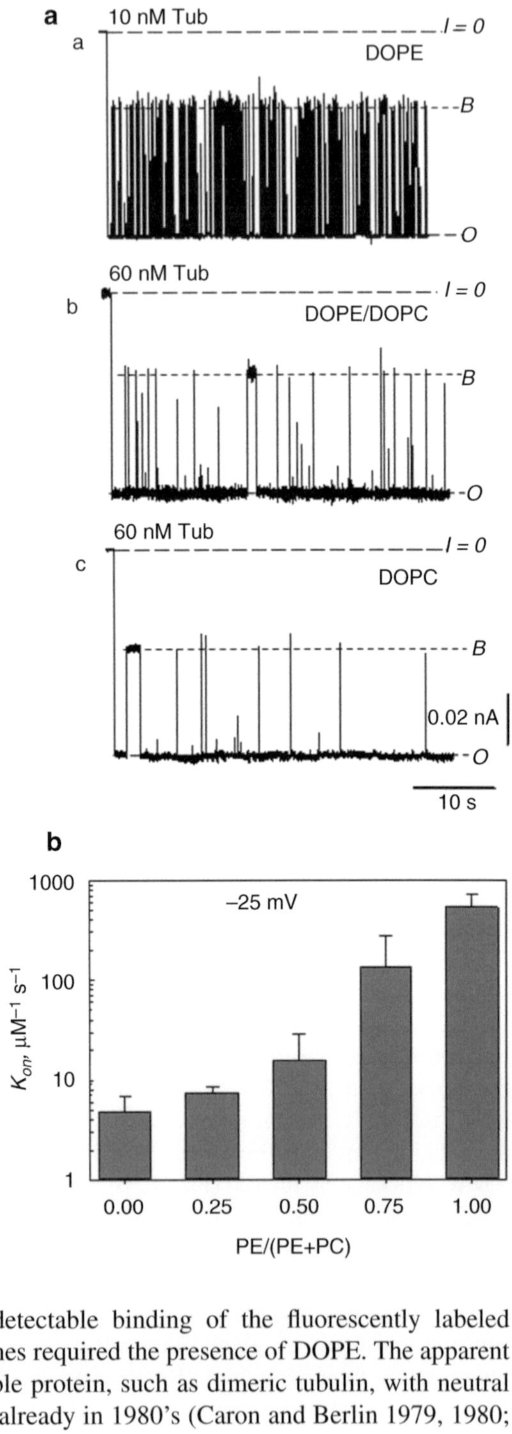

Fig. 1.7 The on-rate of VDAC blockage by tubulin strongly depends on the lipid composition of a planar membrane. (**a**) A comparison of current traces from single VDAC channels reconstituted in planar bilayers of different PC/PE ratios. 10 nM of tubulin effectively blocks VDAC in a pure DOPE membrane (*trace a*) while 60 nM of tubulin blocks VDAC less effectively in a DOPC/DOPE (1:1) mixture (*trace b*) and barely blocks VDAC in a pure DOPC membrane (*trace c*). (**b**) The on-rate of VDAC-tubulin binding increases ~100 times with the increase of PE content from 0 to 100 %. The on-rate is defined as $k_{on} = 1/(\tau_{on}\,[C])$, where τ_{on} is the average time during which the channel is open between consecutive blockages and $[C]$ is bulk tubulin concentration. The applied potential was – 25 mV

accord with our expectations, detectable binding of the fluorescently labeled dimeric tubulin to GUV membranes required the presence of DOPE. The apparent paradox of association of a soluble protein, such as dimeric tubulin, with neutral membranes has been recognized already in 1980's (Caron and Berlin 1979, 1980;

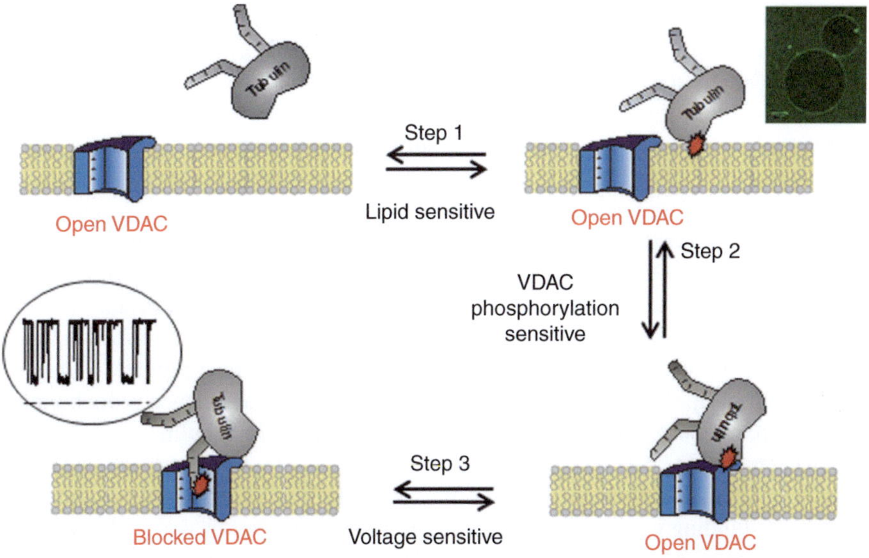

Fig. 1.8 A tentative model of the multistep VDAC blockage by tubulin. *Step 1* is a lipid-sensitive step of globular tubulin binding to the membrane surface with K_d ~10–50 nM, depending on the lipid composition. *Inset* - confocal images of fluorescently-labeled tubulin binding to DOPE-containing GUVs (scale bar is 10 μm). *Step 2* is tubulin binding to the loops forming the channel entrance. This step depends on the VDAC phosphorylation level. *Step 3* is voltage-dependent reversible blockage of the VDAC pore by the tubulin CTT. *Inset* – current trace illustrating tubulin-induced blockage of a single VDAC channel (Adapted from Rostovtseva et al. 2012)

Klausner et al. 1981; Kumar et al. 1981). It was suggested that tubulin-membrane interaction has a hydrophobic component. It was also shown that CTTs of liposome-associated tubulin were accessible to proteolysis (Hargreaves and McLean 1988). Therefore, tubulin CTTs are not directly involved in the association with a membrane and, as our experiments demonstrate, are available for blocking VDAC (Rostovtseva et al. 2008).

The increased association of tubulin with membranes of a specific lipid composition, such as the presence of PE lipids, increases the effective concentration of tubulin on the surface of these membranes, which results in the increased on-rate of VDAC blockage by tubulin. Thus, VDAC-tubulin interaction could be described by a tentative three-step model (Fig. 1.8), where the first step is saturable binding of tubulin to the membrane with the efficiency dependent on the lipid composition. The second step is tubulin binding to the cytosolic loops of VDAC that connect the membrane-spanning β-strands and form the entrance of the channel. This step depends on the level of VDAC posttranslational modifications, especially on phosphorylation (Sheldon et al. 2011). The last step is the voltage-dependent block of the VDAC pore by tubulin CTT. The on-rate of this step depends on the effective concentration of tubulin in the close vicinity of the channel entrance and therefore on tubulin binding to the membrane surface. When the tubulin CTT enters the VDAC pore, the strength of the binding, in terms of its residence time in the pore, does not depend on the lipid composition (Rostovtseva et al. 2012) or VDAC phosphorylation state (Sheldon et al. 2011). Thus, at a constant applied voltage the equilibrium of VDAC-tubulin binding is predominantly defined by the on-rate. Considering a substantial lipid homeostasis in mitochondria during morphological changes such as fusion and fission (Furt and Moreau 2009), under apoptotic stress (Crimi and Esposti 2011), or lipid oxidation by reactive oxygen species produced by mitochondria (Kagan et al. 2004; Pamplona 2008; Paradies et al. 2011), lipids could be potent regulators of the VDAC permeability and therefore MOM permeability *in vivo*.

Overall, these findings provide an example of lipid-controlled protein-protein interactions where the choice of lipid species is able to change the equilibrium binding constant by orders of magnitude. They also suggest a new regulatory role of mitochondrial lipids in the control of MOM permeability, and hence mitochondria respiration, by modulating VDAC sensitivity to blockage by tubulin.

The remaining open question is why tubulin binds more effectively to DOPE than to DOPC membranes. One of the possible explanations is that tubulin-membrane binding is affected by the lipid packing stress. For example, the reduced repulsive forces between DOPE headgroups in comparison with those between DOPC headgroups (Keller et al. 1993) could provide more "flexibility" for PE headgroups, thus making hydrophobic grooves more available for tubulin binding. The answer to this question is a subject of future research.

1.4.4 Physiological Relevance of VDAC Modulation by Tubulin

To summarize, using *in vitro* experiments with reconstituted VDAC, it was shown that depending on the choice of lipid species, VDAC phosphorylation state, and the applied voltage, nanomolar to micromolar concentrations of dimeric tubulin induce functionally important reversible blockage of VDAC. These results suggest a novel mechanism of regulation of MOM permeability where tubulin selectively modulates the fluxes of metabolites between mitochondria and the cytosol. The key question is what is the physiological relevance of this new role for tubulin? Could mitochondria sense the presence of free tubulin in living cells? Experiments with isolated mitochondria (Monge et al. 2008; Rostovtseva et al. 2008) and human hepatoma HepG2 cells (Maldonado et al. 2010, 2011) confirmed that VDAC–tubulin interaction is functionally important for regulating mitochondrial respiration.

Saks and coauthors in experiments with mitochondria isolated from mouse brain or heart demonstrated that dimeric tubulin reduces the isolated mitochondria respiration rate (Guzun et al. 2012; Monge et al. 2008; Rostovtseva et al. 2008), thus confirming that tubulin could control mitochondria respiration by modulating VDAC permeability for adenine nucleotides. There is a dynamic equilibrium exchange in cells between dimeric and polymerized tubulin with tubulin concentration not changing dramatically under physiological conditions. However, microtubule (MT) targeting agents, such as anticancer drugs colchicine and paclitaxel can shift this balance to higher or lower levels of free cytosolic tubulin by promoting or inhibiting MT depolymerization, respectively. It was shown that VDAC blockage by tubulin increases with tubulin concentration (Fig. 1.5). Therefore, one could predict that the increased pool of dimeric tubulin in the cytosol would result in an increase of VDAC blockage and, consequently, in a depletion of cytosolic ADP available for the respiratory chain leading to a decrease of the mitochondrial potential, $\Delta\Psi$, across the inner membrane (Rostovtseva and Bezrukov 2012). Quite remarkably, Maldonado and coauthors (Maldonado et al. 2011) had confirmed this hypothesis in experiments where they treated intact HepG2 human hepatoma cells with MT destabilizing or stabilizing agents, such as colchicine or paclitaxel, respectively, and monitored the mitochondrial membrane potential using potential-sensitive fluorophore, tetramethylrhodamine methylester (TMRM). Treatment of HepG2 cells by colchicine resulted in an increase of the ratio between free and polymerized tubulin in the cytosol measured by immunoblotting and in a loss of the mitochondrial potential assessed by TMRM fluorescence (Maldonado et al. 2011). Accordingly, treatment with paclitaxel resulted in a decrease of free cytosolic tubulin and a rise of $\Delta\Psi$. These effects of MT-targeting drugs on the mitochondrial potential could be sufficiently explained by the modulation of VDAC permeability by free tubulin, thus confirming the results obtained *in vitro* with reconstituted VDAC.

The effect of MT-targeting drugs on mitochondria is well-known (Andre et al. 2000; Esteve et al. 2007; Karbowski et al. 2001; Saetersdal et al. 1990) but, apparently, the mitochondrial potential could be modulated by MT-targeting drugs through a number of pathways other than VDAC-related. Therefore, without a direct genetic evidence of VDAC involvement in the modulation of the mitochondrial potential by MT-targeting drugs the proposed model wouldn't be entirely convincing. Maldonado and coauthors (Maldonado et al. 2013) using siRNA to knockdown individual VDAC isoforms in HepG2 cells unambiguously demonstrated that VDAC contributes to maintaining the mitochondrial membrane potential. They showed that the knockdown of VDAC decreased $\Delta\Psi$ with a magnitude dependent on which of the three VDAC's isoforms was knockdown (Maldonado et al. 2013). Quite unexpectedly, these authors had found that the most significant impact on $\Delta\Psi$ occurred with a knockdown of the VDAC3 isoform, in spite of VDAC3 being a minor isoform among all three (De Pinto et al. 2010; Messina et al. 2012; Raghavan et al. 2012). In HepG2 cells the expression ratio between VDAC1, VDAC2, and VDAC3 isoforms was found to be 0.4: 0.49: 0.11, respectively (Maldonado et al. 2013). The knockdown of VDAC3 resulted in a significant, compared to the knockdown of the other two isoforms, decrease of the total adenine nucleotides and the $NAD(P)H/NAD(P)^+$ ratio, thus conforming a crucial role of VDAC3 in maintaining MOM permeability in cancer cells. The authors offered an interesting interpretation of these surprising results suggesting that in HepG2 cells, VDAC1 and VDAC2 are mostly blocked by tubulin, whereas the less abundant VDAC3 remains open, which explains the prevailing impact of its knockdown on mitochondrial energetics.

The results discussed above unveil a previously unknown mechanism of regulation of mitochondrial energetics, governed by VDAC interaction with tubulin at the mitochondria–cytosol interface. Immediate physiological implications include new insights into the serine/threonine kinase signaling pathways, Ca^{2+} homeostasis, and cytoskeleton/microtubule activity in health and disease, especially in the case of the highly dynamic microtubule network which is characteristic of cancerogenesis and cell proliferation. One of the intriguing implication of VDAC modulation by tubulin is its coupling with the Warburg-type aerobic glycolysis characteristic of many tumor cells (Mathupala et al. 2010), where the VDAC-tubulin complex may play a role of "glycolytic switch" in cells towards aerobic glycolysis or oxidative phosphorylation (Maldonado and Lemasters 2012; Rostovtseva and Bezrukov 2012). These findings may help to identify new mechanisms of mitochondria-associated action of chemotherapeutic MT-targeting drugs via VDAC/tubulin interaction, and also to understand why and how cancer cells preferentially use inefficient glycolysis rather than oxidative phosphorylation (Warburg effect).

Acknowledgements These studies were supported by the Intramural Research Program of the NIH, *Eunice Kennedy Shriver* National Institute of Child Health and Human Development. The authors thank Philip Gurnev, Sergei Noskov, and Oscar Teijido for fruitful discussions.

References

Andre N, Braguer D, Brasseur G, Goncalves A, Lemesle-Meunier D, Guise S, Jordan MA, Briand C (2000) Paclitaxel induces release of cytochrome c from mitochondria isolated from human neuroblastoma cells. Cancer Res 60(19):5349–5353

Bayrhuber M, Meins T, Habeck M, Becker S, Giller K, Villinger S, Vonrhein C, Griesinger C, Zweckstetter M, Zeth K (2008) Structure of the human voltage-dependent anion channel. Proc Natl Acad Sci U S A 105(40):15370–15375

Berezhkovskii AM, Pustovoit MA, Bezrukov SM (2002) Channel-facilitated membrane transport: transit probability and interaction with the channel. J Chem Phys 116(22):9952–9956

Bernier-Valentin F, Aunis D, Rousset B (1983) Evidence for tubulin-binding sites on cellular membranes – plasma-membranes, mitochondrial-membranes, and secretory granule membranes. J Cell Biol 97(1):209–216

Bezrukov SM (2000) Functional consequences of lipid packing stress. Curr. Opin. Colloid Interface Sci. 5:237–43

Bezrukov SM, Vodyanoy I (1993) Probing alamethicin channels with water-soluble polymers. Effect on conductance of channel states. Biophys J 64(1):16–25

Bezrukov SM, Vodyanoy I, Parsegian VA (1994) Counting polymers moving through a single-Ion channel. Nature 370(6487):279–281

Bezrukov SM, Vodyanoy I, Brutyan RA, Kasianowicz JJ (1996) Dynamics and free energy of polymers partitioning into a nanoscale pore. Macromolecules 29(26):8517–8522

Blachly-Dyson E, Peng SZ, Colombini M, Forte M (1990) Selectivity changes in site-directed mutants of the VDAC ion channel – structural implications. Science 247(4947): 1233–1236

Bowen KA, Tam K, Colombini M (1985) Evidence for titratable gating charges controlling the voltage dependence of the outer mitochondrial-membrane channel, Vdac. J Membr Biol 86(1): 51–59

Cantor RS (1999) Lipid composition and the lateral pressure profile in bilayers. Biophys J 76(5): 2625–2639

Carneiro CM, Merzlyak PG, Yuldasheva LN, Silva LG, Thinnes FP, Krasilnikov OV (2003) Probing the volume changes during voltage gating of Porin 31BM channel with nonelectrolyte polymers. Biochim Biophys Acta 1612(2):144–153

Caron JM, Berlin RD (1979) Interaction of microtubule proteins with phospholipid-vesicles. J Cell Biol 81(3):665–671

Caron JM, Berlin RD (1980) Reversible adsorption of microtubule protein to phospholipid-vesicles. J Cell Biol 87(2):A255

Choudhary OP, Ujwal R, Kowallis W, Coalson R, Abramson J, Grabe M (2010) The electrostatics of VDAC: implications for selectivity and gating. J Mol Biol 396(3):580–592

Choudhary OP, Paz A, Adelman JL, Colletier JP, Abramson J, Grabe M (2014) Structure-guided simulations illuminate the mechanism of ATP transport through VDAC1. Nat Struct Mol Biol 21(7):626–632

Colombini M (1989) Voltage gating in the mitochondrial channel, VDAC. J Membr Biol 111(2): 103–111

Colombini M (2004) VDAC: the channel at the interface between mitochondria and the cytosol. Mol Cell Biochem 256(1–2):107–115

Colombini M (2009) The published 3D structure of the VDAC channel: native or not? Trends Biochem Sci 34(8):382–389

Colombini M, Blachly-Dyson E, Forte M (1996) VDAC, a channel in the outer mitochondrial membrane. In: Narahashi T (ed) In ion channels. Plenum Press, New York, pp 169–202

Crimi M, Esposti MD (2011) Apoptosis-induced changes in mitochondrial lipids. Biochim Biophys Acta 1813(4):551–557

De Pinto V, Messina A, Accardi R, Aiello R, Guarino F, Tomasello MF, Tommasino M, Tasco G, Casadio R, Benz R et al (2003) New functions of an old protein: the eukaryotic porin or voltage dependent anion selective channel (VDAC). Ital J Biochem 52(1):17–24

De Pinto V, Reina S, Guarino F, Messina A (2008) Structure of the voltage dependent anion channel: state of the art. J Bioenerg Biomembr 40(3):139–147

De Pinto V, Guarino F, Guarnera A, Messina A, Reina S, Tomasello FM, Palermo V, Mazzoni C (2010) Characterization of human VDAC isoforms: a peculiar function for VDAC3? Biochim Biophys Acta 1797(6–7):1268–1275

Doring C, Colombini M (1985a) The mitochondrial voltage-dependent channel, VDAC, is modified asymmetrically by succinic anhydride. J Membr Biol 83(1–2):87–94

Doring C, Colombini M (1985b) Voltage dependence and ion selectivity of the mitochondrial channel, VDAC, are modified by succinic anhydride. J Membr Biol 83(1–2):81–86

Eddy MT, Ong TC, Clark L, Teijido O, van der Wel PC, Garces R, Wagner G, Rostovtseva TK, Griffin RG (2012) Lipid dynamics and protein-lipid interactions in 2D crystals formed with the beta-barrel integral membrane protein VDAC1. J Am Chem Soc 134(14):6375–6387

Ermishkin LN, Mirzabekov TA (1990) Redistribution of the electric field within the pore contributes to the voltage-dependence of mitochondrial porin channel. Biochim Biophys Acta 1021(2):161–168

Esteve MA, Carre M, Braguer D (2007) Microtubules in apoptosis induction: are they necessary? Curr Cancer Drug Targets 7(8):713–729

Furt F, Moreau P (2009) Importance of lipid metabolism for intracellular and mitochondrial membrane fusion/fission processes. Int J Biochem Cell Biol 41(10):1828–1836

Gard DL, Kirschner MW (1987) Microtubule assembly in cytoplasmic extracts of Xenopus oocytes and eggs. J Cell Biol 105(5):2191–2201

Geula S, Ben-Hail D, Shoshan-Barmatz V (2012) Structure-based analysis of VDAC1: N-terminus location, translocation, channel gating and association with anti-apoptotic proteins. Biochem J 444(3):475–485

Gruner SM (1985) Intrinsic curvature hypothesis for biomembrane lipid composition: a role for nonbilayer lipids. Proc Natl Acad Sci U S A 82(11):3665–3669

Guo XW, Smith PR, Cognon B, D'Arcangelis D, Dolginova E, Mannella CA (1995) Molecular design of the voltage-dependent, anion-selective channel in the mitochondrial outer membrane. J Struct Biol 114(1):41–59

Gurnev PA, Rostovtseva TK, Bezrukov SM (2011) Tubulin-blocked state of VDAC studied by polymer and ATP partitioning. FEBS Lett 585(14):2363–2366
Gurnev PA, Queralt-Martin M, Aguilella VM, Rostovtseva TK, Bezrukov SM (2012) Probing tubulin-blocked state of VDAC by varying membrane surface charge. Biophys J 102(9):2070–2076
Guzun R, Gonzalez-Granillo M, Karu-Varikmaa M, Grichine A, Usson Y, Kaambre T, Guerrero-Roesch K, Kuznetsov A, Schlattner U, Saks V (2012) Regulation of respiration in muscle cells *in vivo* by VDAC through interaction with the cytoskeleton and MtCK within mitochondrial interactosome. Biochim Biophys Acta 1818(6):1545–1554
Hargreaves AJ, McLean WG (1988) The characterization of phospholipids associated with microtubules, purified tubulin and microtubule associated proteins *in vitro*. Int J Biochem 20(10):1133–1138
Hiller S, Garces RG, Malia TJ, Orekhov VY, Colombini M, Wagner G (2008) Solution structure of the integral human membrane protein VDAC-1 in detergent micelles. Science 321(5893):1206–1210
Hiller S, Abramson J, Mannella C, Wagner G, Zeth K (2010) The 3D structures of VDAC represent a native conformation. Trends Biochem Sci 35(9):514–521
Hodge T, Colombini M (1997) Regulation of metabolite flux through voltage-gating of VDAC channels. J Membr Biol 157(3):271–279
Kagan VE, Borisenko GG, Tyurina YY, Tyurin VA, Jiang JF, Potapovich AI, Kini V, Amoscato AA, Fujii Y (2004) Oxidative lipidomics of apoptosis: redox catalytic interactions of cytochrome C with cardiolipin and phosphatidylserine. Free Radical Biol Med 37(12):1963–1985
Karbowski M, Spodnik JH, Teranishi M, Wozniak M, Nishizawa Y, Usukura J, Wakabayashi T (2001) Opposite effects of microtubule-stabilizing and microtubule-destabilizing drugs on biogenesis of mitochondria in mammalian cells. J Cell Sci 114(2):281–291
Keller SL, Bezrukov SM, Gruner SM, Tate MW, Vodyanoy I, Parsegian VA (1993) Probability of alamethicin conductance states varies with nonlamellar tendency of bilayer phospholipids. Biophys J 65(1):23–27
Klausner RD, Kumar N, Weinstein JN, Blumenthal R, Flavin M (1981) Interaction of tubulin with phospholipid-vesicles. 1. Association with vesicles at the phase-transition. J Biol Chem 256(11):5879–5885
Kozuch J, Weichbrodt C, Millo D, Giller K, Becker S, Hildebrandt P, Steinem C (2014) Voltage-dependent structural changes of the membrane-bound anion channel hVDAC1 probed by SEIRA and electrochemical impedance spectroscopy. Phys Chem Chem Phys 16(20):9546–9555
Krasilnikov OV, Bezrukov SM (2004) Polymer partitioning from nonideal solutions into protein voids. Macromolecules 37(7):2650–2657
Krasilnikov OV, Sabirov RZ, Ternovsky VI, Merzliak PG, Muratkhodjaev JN (1992) A simple method for the determination of the pore radius of ion channels in planar lipid bilayer-membranes. FEMS Microbiol Immunol 105(1–3):93–100
Kumar N, Klausner RD, Weinstein JN, Blumenthal R, Flavin M (1981) Interaction of tubulin with phospholipid-vesicles. 2. Physical changes of the protein. J Biol Chem 256(11):5886–5889
Lemasters JJ, Holmuhamedov E (2006) Voltage-dependent anion channel (VDAC) as mitochondrial governator – thinking outside the box. Biochim Biophys Acta 1762(2):181–190
Lemasters JJ, Bond JM, Chacon E, Harper IS, Kaplan SH, Ohata H, Trollinger DR, Herman B, Cascio WE (1996) The pH paradox in ischemia-reperfusion injury to cardiac myocytes. EXS 76:99–114
Lemasters JJ, Holmuhamedov EL, Czerny C, Zhong Z, Maldonado EN (2012) Regulation of mitochondrial function by voltage dependent anion channels in ethanol metabolism and the Warburg effect. Biochim Biophys Acta 1818(6):1536–1544
Lemeshko VV (2006) Theoretical evaluation of a possible nature of the outer membrane potential of mitochondria. Eur Biophys J 36(1):57–66
Lemeshko VV (2014a) VDAC electronics: 1. VDAC-hexo(gluco)kinase generator of the mitochondrial outer membrane potential. Biochim Biophys Acta 1838(5):1362–1371

Lemeshko VV (2014b) VDAC electronics: 2. A new, anaerobic mechanism of generation of the membrane potentials in mitochondria. Biochim Biophys Acta 1838(7):1801–1808
Maldonado EN, Lemasters JJ (2012) Warburg revisited: regulation of mitochondrial metabolism by voltage-dependent anion channels in cancer cells. J Pharmacol Exp Ther 342(3):637–641
Maldonado EN, Lemasters JJ (2014) ATP/ADP ratio, the missed connection between mitochondria and the Warburg effect. Mitochondrion 19(Pt A):78–84
Maldonado EN, Patnaik JR, Lemasters JJ (2010) Free tubulin and cAMP-dependent phosphorylation modulate mitochondrial membrane potential in Hepg2 cells: possible role of VDAC. Biophys J 98(3):735a
Maldonado EN, Patnaik J, Mullins MR, Lemasters JJ (2011) Free tubulin modulates mitochondrial membrane potential in cancer cells. Cancer Res 70(24):10192–10201
Maldonado EN, Sheldon KL, DeHart DN, Patnaik J, Manevich Y, Townsend DM, Bezrukov SM, Rostovtseva TK, Lemasters JJ (2013) Voltage-dependent anion channels modulate mitochondrial metabolism in cancer cells: regulation by free tubulin and erastin. J Biol Chem 288(17): 11920–11929
Mannella CA (1998) Conformational changes in the mitochondrial channel protein, VDAC, and their functional implications. J Struct Biol 121(2):207–218
Mathupala SP, Ko YH, Pedersen PL (2010) The pivotal roles of mitochondria in cancer: Warburg and beyond and encouraging prospects for effective therapies. Biochim Biophys Acta 1797(6–7):1225–1230
McDonald BM, Wydro MM, Lightowlers RN, Lakey JH (2009) Probing the orientation of yeast VDAC1 *in vivo*. FEBS Lett 583(4):739–742
Mertins B, Psakis G, Grosse W, Back KC, Salisowski A, Reiss P, Koert U, Essen LO (2012) Flexibility of the N-terminal mVDAC1 segment controls the channel's gating behavior. PLoS One 7(10):e47938
Messina A, Reina S, Guarino F, De Pinto V (2012) VDAC isoforms in mammals. Biochim Biophys Acta 1818(6):1466–1476
Monge C, Beraud N, Kuznetsov AV, Rostovtseva T, Sackett D, Schlattner U, Vendelin M, Saks VA (2008) Regulation of respiration in brain mitochondria and synaptosomes: restrictions of ADP diffusion in situ, roles of tubulin, and mitochondrial creatine kinase. Mol Cell Biochem 318(1–2):94–1562
Murphy E, Steenbergen C (2008) Mechanisms underlying acute protection from cardiac ischemia-reperfusion injury. Physiol Rev 88(2):581–609
Nestorovich EM, Karginov VA, Bezrukov SM (2010) Polymer partitioning and ion selectivity suggest asymmetrical shape for the membrane pore formed by epsilon toxin. Biophys J 99(3): 782–789
Nogales E, Wolf SG, Downing KH (1998) Structure of the alpha beta tubulin dimer by electron crystallography. Nature 391(6663):199–203
Noskov SY, Rostovtseva TK, Bezrukov SM (2013) ATP transport through VDAC and the VDAC-tubulin complex probed by equilibrium and nonequilibrium MD simulations. Biochemistry 52(51):9246–9256
Pamplona R (2008) Membrane phospholipids, lipoxidative damage and molecular integrity: a causal role in aging and longevity. Biochim Biophys Acta 1777(10):1249–1262
Paradies G, Petrosillo G, Paradies V, Ruggiero FM (2011) Mitochondrial dysfunction in brain aging: role of oxidative stress and cardiolipin. Neurochem Int 58(4):447–457
Popp B, Court DA, Benz R, Neupert W, Lill R (1996) The role of the N and C termini of recombinant Neurospora mitochondrial porin in channel formation and voltage-dependent gating. J Biol Chem 271(23):13593–13599
Porcelli AM, Ghelli A, Zanna C, Pinton P, Rizzuto R, Rugolo M (2005) pH difference across the outer mitochondrial membrane measured with a green fluorescent protein mutant. Biochem Biophys Res Commun 326(4):799–804
Raghavan A, Sheiko T, Graham BH, Craigen WJ (2012) Voltage-dependant anion channels: novel insights into isoform function through genetic models. Biochim Biophys Acta 1818(6): 1477–1485

Reymann S, Florke H, Heiden M, Jakob C, Stadtmuller U, Steinacker P, Lalk VE, Pardowitz I, Thinnes FP (1995) Further evidence for multitopological localization of mammalian porin (VDAC) in the plasmalemma forming part of a chloride channel complex affected in cystic fibrosis and encephalomyopathy. Biochem Mol Med 54(2):75–87

Rostovtseva TK, Bezrukov SM (1998) ATP transport through a single mitochondrial channel, VDAC, studied by current fluctuation analysis. Biophys J 74(5):2365–2373

Rostovtseva TK, Bezrukov SM (2008) VDAC regulation: role of cytosolic proteins and mitochondrial lipids. J Bioenerg Biomembr 40(3):163–170

Rostovtseva TK, Bezrukov SM (2012) VDAC inhibition by tubulin and its physiological implications. Biochim Biophys Acta 1818(6):1526–1535

Rostovtseva T, Colombini M (1996) ATP flux is controlled by a voltage-gated channel from the mitochondrial outer membrane. J Biol Chem 271(45):28006–28008

Rostovtseva T, Colombini M (1997) VDAC channels mediate and gate the flow of ATP: implications for the regulation of mitochondrial function. Biophys J 72(5):1954–1962

Rostovtseva TK, Liu TT, Colombini M, Parsegian VA, Bezrukov SM (2000) Positive cooperativity without domains or subunits in a monomeric membrane channel. Proc Natl Acad Sci U S A 97(14):7819–7822

Rostovtseva TK, Komarov A, Bezrukov SM, Colombini M (2002a) Dynamics of nucleotides in VDAC channels: structure-specific noise generation. Biophys J 82(1):193–205

Rostovtseva TK, Komarov A, Bezrukov SM, Colombini M (2002b) VDAC channels differentiate between natural metabolites and synthetic molecules. J Membr Biol 187(2):147–156

Rostovtseva TK, Nestorovich EM, Bezrukov SM (2002c) Partitioning of differently sized poly(ethylene glycol)s into OmpF porin. Biophys J 82(1):160–169

Rostovtseva TK, Tan WZ, Colombini M (2005) On the role of VDAC in apoptosis: fact and fiction. J Bioenerg Biomembr 37(3):129–142

Rostovtseva TK, Kazemi N, Weinrich M, Bezrukov SM (2006) Voltage gating of VDAC is regulated by nonlamellar lipids of mitochondrial membranes. J Biol Chem 281(49):37496–37506

Rostovtseva TK, Sheldon KL, Hassanzadeh E, Monge C, Saks V, Bezrukov SM, Sackett DL (2008) Tubulin binding blocks mitochondrial voltage-dependent anion channel and regulates respiration. Proc Natl Acad Sci U S A 105(48):18746–18751

Rostovtseva TK, Gurnev PA, Chen MY, Bezrukov SM (2012) Membrane lipid composition regulates tubulin interaction with mitochondrial voltage-dependent anion channel. J Biol Chem 287(35):29589–29598

Saetersdal T, Greve G, Dalen H (1990) Associations between beta-tubulin and mitochondria in adult isolated heart myocytes as shown by immunofluorescence and immunoelectron microscopy. Histochemistry 95(1):1–10

Schneider R, Etzkorn M, Giller K, Daebel V, Eisfeld J, Zweckstetter M, Griesinger C, Becker S, Lange A (2010) The native conformation of the human VDAC1 N terminus. Angew Chem Int Ed Engl 49(10):1882–1885

Schredelseker J, Paz A, Lopez CJ, Altenbach C, Leung CS, Drexler MK, Chen JN, Hubbell WL, Abramson J (2014) High resolution structure and double electron-electron resonance of the zebrafish voltage-dependent anion channel 2 reveal an oligomeric population. J Biol Chem 289(18):12566–12577

Shao L, Kinnally KW, Mannella CA (1996) Circular dichroism studies of the mitochondrial channel, VDAC, from Neurospora crassa. Biophys J 71(2):778–786

Sheldon KL, Maldonado EN, Lemasters JJ, Rostovtseva TK, Bezrukov SM (2011) Phosphorylation of voltage-dependent anion channel by serine/threonine kinases governs its interaction with tubulin. PLoS One 6(10):e25539

Shoshan-Barmatz V, Ben-Hail D (2012) VDAC, a multi-functional mitochondrial protein as a pharmacological target. Mitochondrion 12(1):24–34

Shoshan-Barmatz V, Gincel D (2003) The voltage-dependent anion channel – characterization, modulation, and role in mitochondrial function in cell life and death. Cell Biochem Biophys 39(3):279–292

Shoshan-Barmatz V, De Pinto V, Zweckstetter M, Raviv Z, Keinan N, Arbel N (2010a) VDAC, a multi-functional mitochondrial protein regulating cell life and death. Mol Aspects Med 31(3):227–285
Shoshan-Barmatz V, Keinan N, Abu-Hamad S, Tyomkin D, Aram L (2010b) Apoptosis is regulated by the VDAC1 N-terminal region and by VDAC oligomerization: release of cytochrome c, AIF and Smac/Diablo. Biochim Biophys Acta Bioenerg 1797(6–7):1281–1291
Song J, Midson C, Blachly-Dyson E, Forte M, Colombini M (1998a) The sensor regions of VDAC are translocated from within the membrane to the surface during the gating processes. Biophys J 74(6):2926–2944
Song JM, Midson C, Blachly-Dyson E, Forte M, Colombini M (1998b) The topology of VDAC as probed by biotin modification. J Biol Chem 273(38):24406–24413
Stanley S, Dias JA, Darcangelis D, Mannella CA (1995) Peptide-specific antibodies as probes of the topography of the voltage-gated channel in the mitochondrial outer-membrane of Neurospora-crassa. J Biol Chem 270(28):16694–16700
Swartz KJ (2008) Sensing voltage across lipid membranes. Nature 456(7224):891–897
Teijido O, Ujwal R, Hillerdal CO, Kullman L, Rostovtseva TK, Abramson J (2012) Affixing N-terminal alpha-helix to the wall of the voltage-dependent anion channel does not prevent its voltage gating. J Biol Chem 287(14):11437–11445
Teijido O, Rappaport SM, Chamberlin A, Noskov SY, Aguilella VM, Rostovtseva TK, Bezrukov SM (2014) Acidification asymmetrically affects voltage-dependent anion channel implicating the involvement of salt bridges. J Biol Chem 289(34):23670–23682
Thomas L, Blachly-Dyson E, Colombini M, Forte M (1993) Mapping of residues forming the voltage sensor of the voltage-dependent anion-selective channel. Proc Natl Acad Sci U S A 90(12):5446–5449
Tomasello MF, Guarino F, Reina S, Messina A, De Pinto V (2013) The voltage-dependent anion selective channel 1 (VDAC1) topography in the mitochondrial outer membrane as detected in intact cell. PLoS One 8(12):e81522
Ujwal R, Cascio D, Colletier J-P, Faham S, Zhang J, Toro L, Ping P, Abramson J (2008) The crystal structure of mouse VDAC1 at 2.3 A resolution reveals mechanistic insights into metabolite gating. Proc Natl Acad Sci U S A 105(46):17742–17747
van den Brink-van der Laan E, Killian JA, de Kruijff B (2004) Nonbilayer lipids affect peripheral and integral membrane proteins via changes in the lateral pressure profile. Biochim Biophys Acta 1666(1–2):275–288
Villinger S, Briones R, Giller K, Zachariae U, Lange A, de Groot BL, Griesinger C, Becker S, Zweckstetter M (2010) Functional dynamics in the voltage-dependent anion channel. Proc Natl Acad Sci U S A 107(52):22546–22551
Westermann S, Weber K (2003) Post-translational modifications regulate microtubule function. Nat Rev Mol Cell Biol 4(12):938–947
Zachariae U, Schneider R, Briones R, Gattin Z, Demers JP, Giller K, Maier E, Zweckstetter M, Griesinger C, Becker S et al (2012) beta-Barrel mobility underlies closure of the voltage-dependent anion channel. Structure 20(9):1540–1549
Zimmerberg J, Parsegian VA (1986) Polymer inaccessible volume changes during opening and closing of a voltage-dependent ionic channel. Nature 323(6083):36–39
Zizi M, Byrd C, Boxus R, Colombini M (1998) The voltage-gating process of the voltage-dependent anion channel is sensitive to ion flow. Biophys J 75(2):704–713

Chapter 2
Mitochondrial Protein Import Channels

Richard Wagner, David Schmedt, Patrizia Hanhart, Claudius Walter, Christof Meisinger, and Philipp Bartsch

Abstract Transport of proteins into or across membranes is an essential step of protein biogenesis for about 50 % of the cellular proteins. For all known intracellular protein transport complexes water filled nanopores were shown to be the central protein conducting or membrane insertion unit. Protein conducting nanopores are not just water filled large pores but provide the framework for selectivity and optimization of protein transport energetics, they act as electrostatic traps for preproteins and exert chaperone like functions. They also can adapt to a large variety of polypeptides that are heterogenic substrates with respect to charge, hydrophobicity and flexibility. For unraveling the molecular mechanisms of protein membrane transport or insertion detailed knowledge on the molecular properties of the channel pores are essential. This in-depth knowledge on the channel properties of these protein-conducting nanopores has been gained through electrophysiological techniques and in particular with the planar bilayer technique. The classical ion channel properties (conductance, selectivity and gating properties) of the nanopores with respect to the passage of small anions and cations have been determined which allowed drawing detailed pictures on the molecular properties of these nanopore-channels. Here we summarize the current knowledge on the electrophysiological properties of the mitochondrial protein import channels and provide an example of a detailed analysis of interaction between Tom40 and a presequence peptide.

Keywords Mitochondria • Protein import • Electrophysiology • Protein-import channels

R. Wagner (✉) • P. Bartsch
MOLIFE Research Center, Jacobs University, Bremen,
Campusring 1, Bremen 28759, Germany

Biophysics, Department of Biology/Chemistry, University Osnabrueck,
Osnabrueck, Germany
e-mail: ri.wagner@jacobs-university.de

D. Schmedt • P. Hanhart • C. Walter
Biophysics, Department of Biology/Chemistry, University Osnabrueck,
Osnabrueck, Germany

C. Meisinger
Institute of Biochemistry and Molecular Biology, University Freiburg, Freiburg, Germany

A.H. Delcour (ed.), *Electrophysiology of Unconventional Channels and Pores*,
Springer Series in Biophysics 18, DOI 10.1007/978-3-319-20149-8_2

2.1 Introduction

Mitochondria are not only the powerhouses of eukaryotic cells but play also essential roles in many cell physiological processes as well as in cell pathology. Mitochondria are a central point of apoptosis-regulation, are involved in the calcium signalling of the cells and house numerous essential biosynthetic processes. The traffic of metabolites and ions firmly integrates the mitochondria as essential player into many cellular processes. Mitochondria are divided into four sub-compartments: the outer membrane (OM), the intermembrane space (IMS), the inner membrane (IM) and the matrix. The mitochondrial genome of *Saccharomyces cerevisiae* encodes only eight proteins, which are part of the respiratory chain. The predominant part of the mitochondrial proteins—corresponding to roughly 1000 proteins—is nuclear encoded. These proteins are synthesized in the cytosol and targeted to mitochondria. As shown in Fig. 2.1, five different translocases are known to be responsible for transport either into one of the compartments or for the insertion into one of the membranes (Harbauer et al. 2014). The TOM complex (translocase of the outer membrane) and SAM/TOB complex (sorting and assembly machinery/topogenesis of mitochondrial

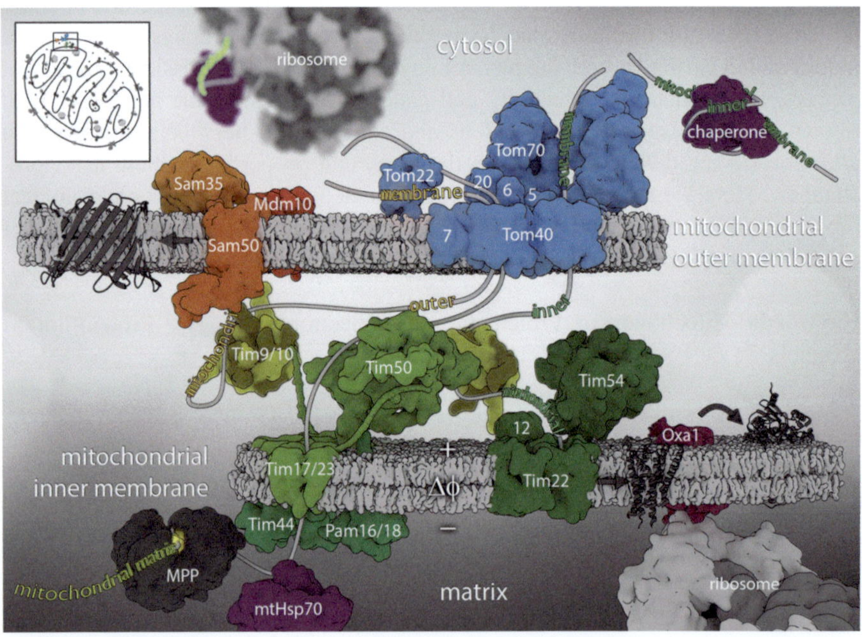

Fig. 2.1 Protein import-machinery of mitochondria (for details see text). *Outer membrane*: The TOM complex (Dekker et al. 1998; Rapaport 2005) and SAM/TOB (Paschen et al. 2003; Wiedemann et al. 2003) complex are located in the outer membrane. *Inner membrane*: The TIM23 complex (Bauer et al. 1996; Truscott et al. 2001), the TIM22 complex (Rehling et al. 2003; 2004; Sirrenberg et al. 1996) and the OXA1 complex (Yi and Dalbey 2005; Krüger et al. 2012) are located in the inner membrane

outer membrane ß-barrel proteins) are located in the outer membrane. The remaining protein-translocases are located in the inner membrane, namely, the TIM23 complex (translocase of the inner membrane or presequence translocase), the TIM22 complex (carrier translocase) and the OXA1 complex (oxidase assembly mutant 1).

2.2 Mitochondrial Targeting

In mitochondria, four different protein import routes and types of targeting sequences direct the various proteins to the correct destination compartment or into the membrane. (i) The more recently discovered ß-signal directs ß-barrel proteins localised in the outer membrane via the SAM complex into this membrane (Kutik et al. 2008). (ii) N-terminal presequences or matrix targeting signals direct the proteins via the TIM23 complex into the matrix the intermembrane space and the inner membrane (Abe et al. 2000; Emanuelsson et al. 2007; Pfanner 2000). In the matrix, the presequence is cleaved by MPP (matrix processing peptidase). Stop-transfer-sequences, containing hydrophobic amino acids are responsible for the insertion of the corresponding proteins by the TIM23 complex into the inner membrane (Glick et al. 1992). (iii) A third class of internal sequence targeting signals direct the hydrophobic inner membrane proteins into the inner membrane ("Carrier-route") (Neupert and Herrmann 2007; Pfanner and Geissler 2001). (iv) Finally some proteins of the IMS (e. g., small Tim proteins) containing a specific cysteine motif (Milenkovic et al. 2009) are imported and refolded with the help of the MIA pathway (mitochondrial intermembrane space import and assembly) (Chacinska et al. 2004; Milenkovic et al. 2007; Sideris and Tokatlidis 2010).

2.3 Protein-Components, Topology and Electrophysiological Properties of the Import Channels

2.3.1 *The TOM Complex*

The central entry gate for the import of nuclear encoded proteins into mitochondria is the TOM complex, a hetero-oligomeric complex of seven proteins. The channel forming subunit Tom40 presumably forms a 19 ß-stranded barrel pore (Lackey et al. 2014; Gessmann et al. 2011; Becker et al. 2005; Hill et al. 1998). Three receptor proteins Tom20, Tom22 and Tom70 (Brix et al. 1997) and additionally Tom5 are involved in the recognition and transfer of the preproteins to Tom40. Current views on the basic mechanism of the mainly unfolded precursor transport through the TOM complex assumes that after initial recognition by the Tom20 receptor, precursors bind to cytosolic domains of Tom22 and Tom5. Subsequently the precursor is guided by a polar slide formed within the pore of Tom40 to an acidic binding-site with high affinity on the trans side of the TOM complex (van Wilpe et al. 1999;

Gessmann et al. 2011; Schatz 1997). An inward increasing gradient of binding affinities (acid/binding chain) of these interactions is proposed to drive the inward-directed movement of precursors. Another driving force could be the membrane potential across the outer membrane, which may reach 30–40 mV (positive at the cytosolic side) (Lemeshko 2006; Porcelli et al. 2005). For presequence-containing proteins, the transport across the outer membrane appears to be tightly coupled to the translocation across or into the inner membrane via the TIM23-machinery through a direct deployment of substrates emerging from the TOM complex (Chacinska et al. 2003; Schatz 1997; Pfanner and Geissler 2001). Cryo-electron microscopy and electrophysiological single channel recordings indicate that channel unit of the active complex is formed by two to three single Tom40 ß-barrel units (Meisinger et al. 2001; Model et al. 2002; Becker et al. 2005).

2.3.1.1 Ion Channel Properties of the Tom40 Pore

The ion channel properties of the Tom40 channel from *Saccharomyces cerevisiae* have been determined from three preparations (i) in *Escherichia coli* heterologously expressed and reconstituted protein, (ii) biochemically purified and reconstituted yeast protein (iii) mitochondrial outer membrane vesicles fused to planar bilayers (Hill et al. 1998; Becker et al. 2005). Similar data were obtained for *Neurospora crassa* Tom40 from the same type of preparations (Künkele et al. 1998a; Ahting et al. 2001).

The Tom40 channel pore from any of the three preparations mentioned above was characterized in detail by planar lipid bilayer measurements (Hill et al. 1998; Becker et al. 2005). The data revealed that the recombinant Tom40 has two conductance states of about $\bar{G}_{max} \cong 400\,\text{pS}$ and $\bar{G}_{sub} \cong 200\,\text{pS}$ in symmetrical 250 mM KCl 10 mM Mops/Tris pH7.5 solution. The $\bar{G}_{max} \cong 400\,\text{pS}$ conductance state corresponds to a complete open$\rightleftharpoons$closed transition of the pore while $\bar{G}_{sub} \cong 200\,\text{pS}$ corresponds to a subconductance state. At high ionic strength, the Tom40 channel displayed a saturating conductance of $G_{sat} \cong 3\,\text{nS}$ (see Fig. 2.2).

Remarkably the Tom40 channel shows a rectifying current voltage relation (Hill et al. 1998; Becker et al. 2005) with an inward/outward current ratio of $r_{rec} = \vec{i} / \overleftarrow{i} \approx 1.2$. At non-saturating concentration of symmetrical 250 mM KCl the pore diameter of $d_{pore} = 1.72\,\text{nm}$ can be calculated from the main conductance state of $\bar{G}_{max} \cong 400\,\text{pS}$ using the ohmic model of Hille (2001) with the improvement of Smart et al. (1997) (for more details see Appendix A). In contrast, the pore geometry deduced from the polymer exclusion method (Krasilnikov et al. 1992; Hill 2001) (see Fig. 2.3) reveals a minimal diameter of $d_{pore}^{min} \cong 1.08\,\text{nm}$ where strong interaction between the inner channel walls and the PEG molecules start to reduce the flow of small ions through the Tom40 channel pore. PEG molecules with a diameter larger than 2.06 nm are completely excluded from the pore (Fig. 2.3). Thus, the fully open channel pore of Tom40 will allow translocation of mainly unfolded proteins.

The reversal potential of the fully open Tom40 channel analyzed under asymmetrical buffer conditions (250 mM: 20 mM KCl) was $V_{rev} = 44 \pm 5\,\text{mV}$, corresponding to a high cation selectivity of $P_{K^+}/P_{Cl^-} \cong 10:1$ (Hill et al. 1998; Hill 2001;

Fig. 2.2 Saturation of conductance in the Tom40 channel (Hill 2001)

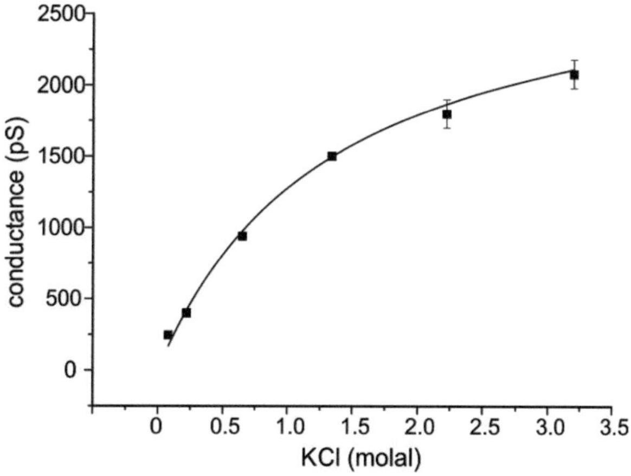

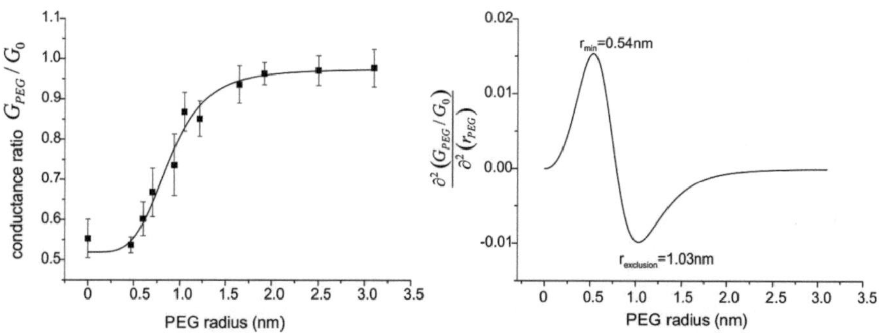

Fig. 2.3 Partitioning of differently sized PEG molecules into the Tom40 channel pore (Data from (Hill 2001)) (*left*) ratio of the *sc*Tom40wt conductance in the presence/absence of differntially sized PEG, (*right*) First derivative of the fitted data from the left depicting r_{min} of the constriction zone and $r_{exclusion}$ of the vestibule

Table 2.1 Relative permeability of the Tom40 channel for different cations (data from (Hill 2001))

Ion	P_{K^+}/P_{X^+}
K^+	1
Na^+	0.7
Cs^+	1.2
Rb^+	1.1
Li^+	0.4
Mg^{2+}	0.27
Ca^{2+}	0.31

Becker et al. 2005) The relative permeability for alkali ions Cs^+ @ $Rb^+ > K^+ > Na^+ > Li^+$ (Table 2.1) follows the weak field site Eisenmann series (Krasne 1973).

This shows that the dipole strength of the negative charges within the Tom40 channel is rather low and that small hydrated cations pass through the channel at rates corresponding to their respective conductance in isotropic solution. It is worth noting that Cu^{2+} ions block the channel completely when forced electrophoretically to enter the pore. This effect is reversible and not related to a redox reaction within the pore (Hill 2001; Becker et al. 2005).

2.3.1.2 Interaction of the Tom40 Channel with Preproteins and Targeting Peptide-Sequences

It has been shown that the specific interaction between the Tom40 channel and preproteins as well as with mitochondrial targeting peptide-sequences can be monitored electrophysiologically. The specific Tom40 substrates modulate the flux of small ions through the Tom40 channel (Hill et al. 1998; Becker et al. 2005; Harsman et al. 2010; Stan et al. 2000; Mahendran et al. 2012). These experiments showed that the Tom40 channel could discriminate between mitochondrial or non-specific peptides (Hill 2001; Becker et al. 2005). Thus the Tom40 channel can be used as a stochastic sensor to investigate the interaction with its substrates and possibly also their translocation through the pore (Harsman et al. 2010; Kasianowicz et al. 2008). A statistical analysis of blocking events (frequency, dwell-time and blocking amplitude) can be used to characterize the peptide channel interactions (Movileanu et al. 2005; Mohammad et al. 2008; Wolfe et al. 2007; Brun et al. 2008). Biological nanopores used as stochastic analyte sensors display a rather voltage independent "silent" open state which can be modulated by partitioning of the analyte into the pore through electrostatic or steric interactions, thus leading to characteristic current fluctuations. The interactions at the binding sites between the pore and the analyte therefore reveal information on both, the pore and the analyte (Kasianowicz et al. 2008). However, with Tom40 the situation is more complicated, since the channel displays rather fast voltage dependent gating (Fig. 2.4a–d). Nevertheless, with aid of the robust mean variance approach (Patlak 1993) which allows rigorous unbiased analysis of single channel recordings it is possible to analyze the interaction of the Tom40 pore with preproteins and short mitochondrial targeting peptides in detail. We performed in detail the analysis of the interactions of Tom40 with CoxIV, a 23 amino acid (–MLSLRQSIRFFKPATRTLSSSRY–) peptide (pI = 12.01) representing the presequence of the cytochrome oxidase c subunit IV (Allison and Schatz 1986).

The rectifying current voltage relation of Tom40 (Hill et al. 1998; Becker et al. 2005) can be used to assign both sites of the channel unambiguously as the high conducting and the low conducting site. As shown previously (Becker et al. 2005) the Tom40 channel displays fast gating at $V_m \geq \pm 80\,\text{mV}$ but mainly between the open and the subconductance state (see Fig. 2.4a). In the presence of CoxIV (cis/trans) more frequent complete closure events were observed (Fig. 2.4b). The frequency of these channel closures the duration of the channel blocking events and the inter-event duration between successive channel blocks contain information on the interaction of the channel pore with the CoxIV peptide. The channel block durations

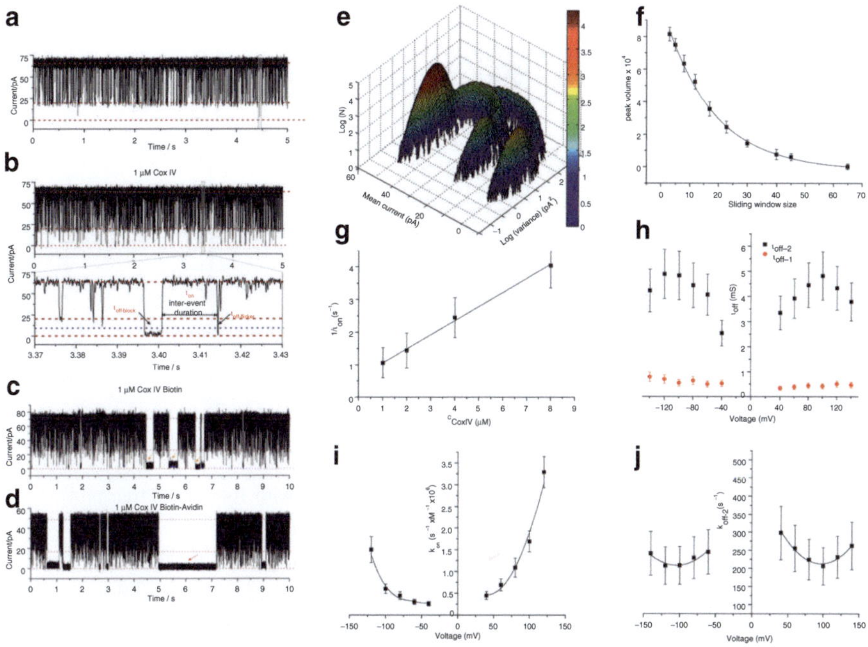

Fig. 2.4 Mean variance analysis of *sc*Tom40wt single channel currents in the presence of symmetrically applied CoxIV (for details and principle of the analysis see Appendix A). (**a**) Single channel recordings from a bilayer containing a single active *sc*Tom40wt in symmetrical (cis/trans) 250 mM KCl, 10 mM MOPS/Tris pH7.5, holding potential $V_m = 80\,\text{mV}$. (**b**) same as (**a**) but with 1 μM CoxIV (cis/trans). (**c**) same as (**a**) but with 1 μM CoxIV-Biotin fusion-protein (cis/trans). **d**) same as (**a**) but with 1 μM CoxIV-Biotin-Avidin. (**e**) Mean variance histogram plot and (**f**) dependence of the mean variance histogram peak volumes on the width of the sliding window (number of data points in the window) of the recording (**b**) (for details see Appendix A). (**g**) Dependence of τ_{on} on the CoxIV concentration at $V_m = 60\,\text{mV}$, data points are averaged with sd from n = 4 experiments (**h**) Voltage dependence of mean dwell times (τ_{off}) where (τ_{off-1}) corresponds to the short flickering events and (τ_{off-2}) to the longer, channel blocking events, data points are averaged with sd from n = 4 experiments, (**i**) voltage dependence of k_{on} (n = 4). (**j**) Voltage dependence of k_{off-2} (n = 4)

are clearly related to the size of the of the CoxIV peptide as shown for the CoxIV-biotin and CoxIV-biotin-avidin fusion peptides (Fig. 2.4c, d). We have analyzed in detail the interaction of CoxIV with single *sc*Tom40wt or *sc*Tom40 S54E mutant channel with respect to the dependence of peptide concentration and the membrane potential in (Schmedt 2010; Hanhart 2011). For this, we used the mean variance analysis of single channel current recordings, which allows an unbiased rigorous analysis of the channel currents in the time and amplitude domains (Patlak 1993) (for details see Appendix A). In short, a sliding window of variable length is used to calculate the mean and the variance yielding the 3-D plot which contained in addition the counts of the binned data points (Fig. 2.4e). From the mean variance histograms at different window sizes the dwell-time of individual channel states can be calculated (Fig. 2.4f). Additionally the current traces are analyzed for the frequency of the transitions mean dwell-times and mean inter-event times (Fig. 2.4g–j). The

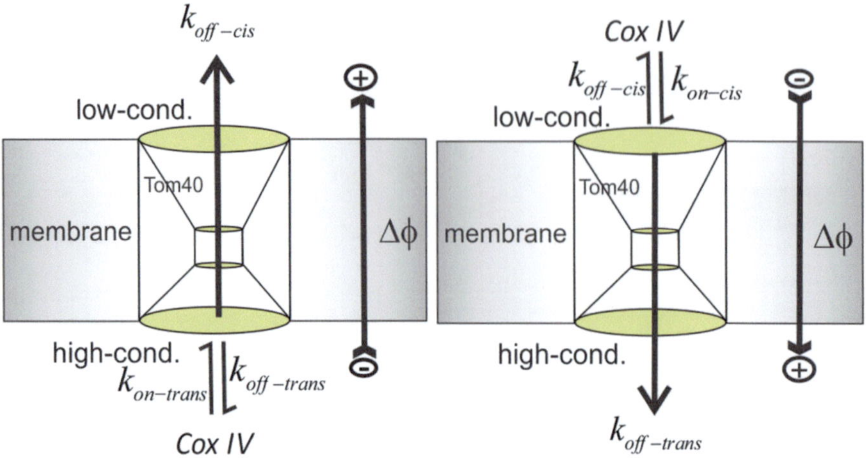

Fig. 2.5 Schematics of voltage dependent interaction of CoxIV with Tom40 and transport

resulting datasets can be interpreted using the Woodhull model (Woodhull 1973) with the extensions of Wolfe et al. (2007) and Mohammand and Movileanu (2008). Basically we observe two classes of peptide induced complete channel closures: fast, non-voltage dependent gating event with $\tau_{\mathrm{off}-1} \approx 0.4-0.8\,\mathrm{ms}$, and 2) slower, voltage dependent events (Fig. 2.4h). The interpretation of these results assumes that the applied membrane potentials drives the peptide electrophoretically into the pore resulting in a channel block. In the voltage range where $\tau_{\mathrm{off}-2}$ progressively increases with increasing voltage, the peptide presumably leaves the pore from the site where it entered. However, in the voltage range where $\tau_{\mathrm{off}-2}$ decreases from $\tau_{\mathrm{off}-2}^{\max}$, the peptide is likely to exit from the opposite site (see Fig. 2.4h) (Movileanu et al. 2005). Assuming a bimolecular reaction between the channel and the peptide we determined from our single channel current data analysis the different on and off rate constants for binding of CoxIV applied to the high (trans) and low (cis) conductance site (Fig. 2.5 (Movileanu et al. 2005)), for both the *sc*Tom40wt and the *sc*Tom40 S54E mutant channel (Table 2.2a and 2.2b).

These results show that the transport of the CoxIV peptide from the high conductance site across the channel occurs at higher rates than from the low conductance site. Moreover, the additional negative charge in the *sc*Tom40 S54E mutant channel increased the transport ratio by three orders of magnitude. It is important to note that these rate constants are extrapolated rate constants at zero membrane potential ($V_{\mathrm{m}} = 0\,\mathrm{mV}$) (Wolfe et al. 2007; Woodhull 1973). Transport of the CoxIV peptide across the Tom40 channel was also inferred from bionic experiments were cis addition of 50 μM CoxIV in symmetrical 10 mM KCl (pH7) buffer introduced a shift of the zero current potential V_{rev} of 3 mV. Taken together our results show that the Tom40 channel has highly asymmetric properties with respect to its affinity for positively charged presequences and rectifying transport rates for the CoxIV peptide. The channel properties of the Tom40 pore thus will allow the transport of short positively charged presequences without additional energy input only due to the electrostatics within the channel. In line with this proposal homology modelling

Table 2.2a Rate constants for the interaction of *sc*Tom40wt channel with CoxIV

Rate constant	CoxIV association from trans (high conductance site)	CoxIV association from cis (low conductance site)
$k_{on}{}^{0}$	$4.3 \pm 0.7 \cdot 10^{3}\,s^{-1}M^{-1}$	$0.5 \pm 0.2 \cdot \leq 10^{3}\,s^{-1}M^{-1}$
$k_{off-trans}{}^{0}$	$409 \pm 60\,s^{-1}$	$\mathbf{16 \pm 2.2\,s^{-1}}$
$k_{off-cis}{}^{0}$	$\mathbf{31 \pm 6.7\,s^{-1}}$	$400 \pm 67\,s^{-1}$

*sc*Tom40wt: $r_{transport} = 31 s^{-1} / 16 s^{-1} = 1.94$; k^0 refers to the extrapolated values of k at $V_m = 0\,mV$

Table 2.2b Rate constants for the interaction of the *sc*Tom40 S54E mutant channel with CoxIV

Rate constant	CoxIV association from trans (high conductance site)	CoxIV association from cis (low conductance site)
$k_{on}{}^{0}$	$14 \pm 0.2 \bullet 10^{3}\,s^{-1}M^{-1}$	$10 \pm 4 \cdot 10^{3}\,s^{-1}M^{-1}$
$k_{off-trans}{}^{0}$	$2639 \pm 64\,s^{-1}$	$\mathbf{0.027 \pm 0.0015\,s^{-1}}$
$k_{off-cis}{}^{0}$	$\mathbf{72 \pm 3.3\,s^{-1}}$	$540 \pm 66\,s^{-1}$

*sc*Tom40 S54E: $r_{transport} = 72\,s^{-1} / 0.27\,s^{-1} = 2.67 \bullet 10^{3}$

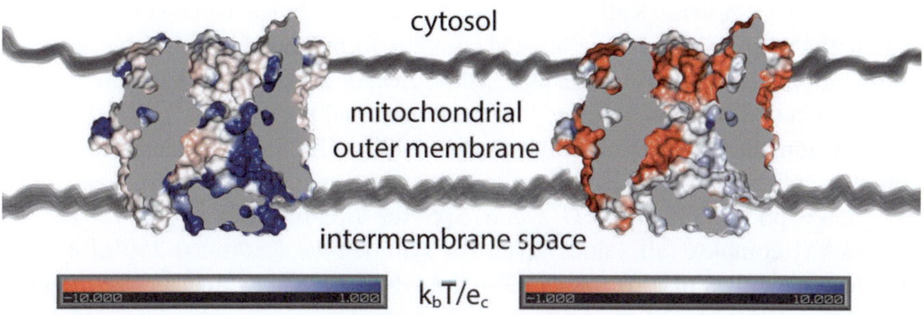

Fig. 2.6 Modeling of the topology of Tom40 and electrostatic surface of charge distribution inside the pore. Charge distribution: *left*: blue positive potential, *right*: red negative potential. Side views with the complexes embedded into a schematic membrane. Modeling was performed as described in the appendix

of the Tom40 structure (Fig. 2.6) shows a steep gradient of the electrostatic surface potential along the z-axis of the pore.

2.3.2 *The SAM Complex*

The SAM complex is responsible for insertion of ß-barrel proteins and membrane proteins with a C-terminal transmembrane segment (C-tail anchored proteins) of the TOM complex (Kozjak et al. 2003; Wiedemann et al. 2003; Paschen et al. 2003). The complex contains three core components: Sam50 (Tob55), forming the pore and the two receptors Sam35 (Tob38) and Sam37 (Mas37) located on the cytosolic side, while Mdm10 an integral membrane protein appears to be only transiently associated with the complex (Meisinger et al. 2007). Sam50 belongs to the Omp85 superfamily (Dolezal et al. 2006); as for other members of the superfamily the pore is formed by a C-terminal presumably ß–barrel and an N-terminal POTRA (polypeptide transport associated) domain (Habib et al. 2007). The recombinant Sam50 and the complex assemble in a pentameric structure, which might provide a central hydrophilic cavity, where the Sam35 protein could be located, for binding of the proteins (Kutik et al. 2008; Paschen et al. 2003). Different models are currently discussed with regards to the mechanism of protein insertion by the SAM complex into the outer mitochondrial membrane (Habib et al. 2007; Chan and Lithgow 2008; Kutik et al. 2008).

2.3.2.1 Ion Channel Properties of the SAM Complex Pore

The electrophysiological properties of the Sam50 channel pore and its interactions with precursor peptides were characterized in detail by planar lipid bilayer measurements from three different preparations: protein expressed heterologously in *E. coli* and reconstituted, as well as purified native complex from *Saccharomyces cerevisiae* and *Homo sapiens* (Kutik et al. 2008; Becker 2008). The isolated recombinant *sc*Sam50, the *hs*Sam50 and the *sc*SAM complex form large channels after incorporation into planar bilayers (Kutik et al. 2008; Becker 2008). For the *sc*Sam50 it has been shown that the pore is formed by the C-terminal part $Sam50_{121–484}$ lacking the POTRA domain (Kutik et al. 2008; Becker 2008). The Sam50 channel displayed complex gating with various subconductance states with the most frequent states $\bar{G}_{main} \cong 450\,pS$ for *sc*Sam50, $\bar{G}_{main} \cong 675\,pS$ for *hs*Sam50 and $\bar{G}_{main} \cong 770\,pS$ for the *sc*SAM complex (all values given for symmetrical (cis/trans) 250 mM KCl, 10 mM MOPS/Tris pH7.5). In addition, the *sc*SAM complex displayed a frequent maximal conductance of the fully open channel of $\bar{G}_{max} \cong 1.1\,nS$ after binding of the ß-signal peptide to the Sam37 subunit while the *hs*Sam50 showed a low frequency of $\bar{G}_{max} \cong 1.3\,nS$ even in the absence of the ß-signal peptide (Kutik et al. 2008; Becker 2008). Frequently occurring lower conductance transitions for the three different Sam50 channel were $\overline{\Delta G_{sub}} \cong 100-300\,pS$ (Becker 2008). With all this precautions in mind, a rough correlation between channel pore size and single channel conductance may be deduced with the modified ohmic approach of Smart et al. ((1997), see also Appendix A). Using the above $\bar{G}_{max} / \bar{G}_{min}$ (with $\bar{G}_{min}$ referring to the lowest conductance state) values and assuming a length of 2 nm for the Sam50 channel constriction zone (Im and Roux 2002) we obtain maximal/minimal pore sizes of $d_{max} \cong 3.2\,nm / d_{min} \cong 0.7\,nm$ for the *sc*SAM pore and $d_{max} \cong 3.6\,nm / d_{min} \cong 0.9\,nm$ for the *hs*Sam50 pore (see Appendix A for details). The analysis of the connectivity of the different conductance states (see also Appendix A) shows that the Sam50 channel of the three different preparations is formed by a single large pore unit which can open to various diameters in an iris-aperture like fashion. The Sam50 channels from the three different preparations revealed a cation selectivity of $P_{K^+} / P_{Cl^-} \approx 6-7:1$ indicating that exposed negative charges are present in the pore. It is noteworthy that the *sc*SAM complex channel resided mainly in the closed state and gating was from closed to different open states, while gating of the *sc*Sam50 and the *hs*Sam50 channels were at $V_m \leq \pm 100\,\text{mV}$ and occurred from the fully open state towards smaller subconductance or closed states.

The Sam50 channel displayed a voltage dependent open probability (Fig. 2.8) decreasing with increasing membrane potential. Similar dependences of P_{open} on the membrane potential have been observed for most of the protein translocation channels (Harsman et al. 2010), although the physiological relevance of this voltage dependent channel closure remain unclear up to now.

2.3.2.2 Interaction of the Sam50 Channel with the Sorting Signal Peptide of Mitochondrial ß-Barrel Proteins

The ß-signal peptides representing the sorting signal of mitochondrial ß-barrel proteins changed the gating behaviour of the SAM complex channel (Kutik et al. 2008; Becker 2008). In the presence of the Tom40 derived ß-peptide (TNDTKIGLGLQFET; Tom40 residues 348–361), the *sc*Sam50 channel displayed voltage and concentration dependent current blocks (Fig. 2.7b) and the relative time of the channel in the closed state increased (Fig. 2.7c).

These results indicate that the ß-signal peptide can enter the pore thereby blocking transiently the channel for small ions similar to what observed for Tom40 and CoxIV (see above). With the synthetic peptide synB2 as control in the same set of experiments, no comparable effect was observed (Becker 2008; Kutik et al. 2008). The peptide synB2 (M L S **R Q Q S Q R Q S R Q Q S Q R Q S R** Y L L) has the same net charge as CoxIV at pH7 but does not form an α-helical structure (Hinnah et al. 2002).

A different effect of the ß-signal peptide on the SAM complex channel was observed. Binding of the ß-signal peptide to the Sam37 subunit in the SAM complex increased not only P_{open} (Fig. 2.8) but also drastically the frequency of gating events with large amplitudes. These events represent mainly transitions from the closed to the fully open channel and reveal the large pore size of $d_{max} \cong 4.4\,\text{nm}$ of the SAM complex channel (see above). This large pore would allow the internal processing and membrane insertion of large even mostly folded proteins (Becker 2008; Kutik et al. 2008).

For import and assembly of the proteins by the SAM complex, no obvious energy source is required. This indicates that increasing affinities of the binding partners on the assembly pathway facilitate unidirectionality to the transport process. In addition, the correct folding and insertion of the proteins into the hydrophobic environment of the membrane is likely to be an energetically favourable process, as the negative free energy of folding of the soluble precursor form of this membrane protein might well be providing the energy for the whole process (Mokranjac and Neupert 2008).

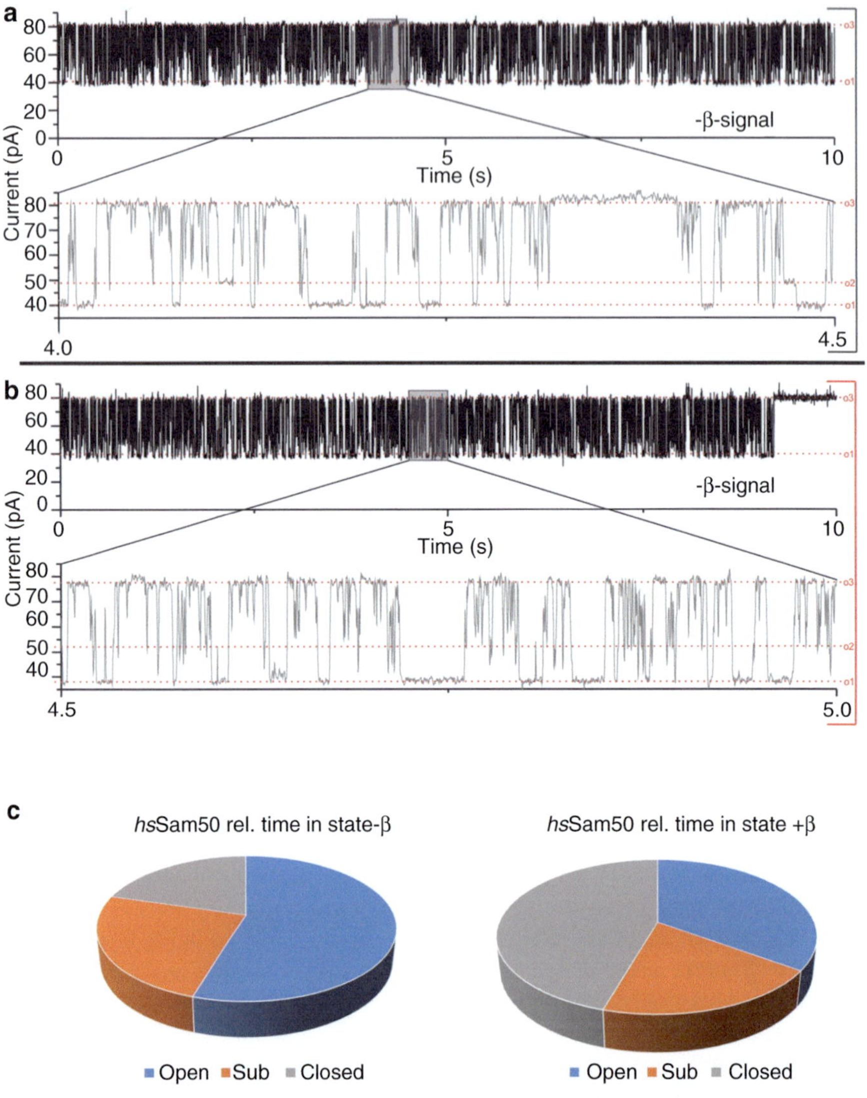

Fig. 2.7 Single channel current traces of reconstituted recombinant *hs*Sam50 in the absence (**a**) and presence (**b**) of 10 μM ß-signal peptide (Becker 2008). (**c**) Relative times that the *hs*Sam50 channel is in the closed state or open states before and after addition of 10 μM ß-signal peptide (Data from a mean variance analysis (see above for details) of a 60 s current recording at $V_m = 80\,\mathrm{mV}$ (Becker 2008))

2.3.3 The TIM23 Complex

Proteins with N-terminal targeting sequences are transported into the matrix via the TIM23 complex. However, when they contain an internal stop transfer signal they also can be released laterally by the TIM23 complex into the inner membrane

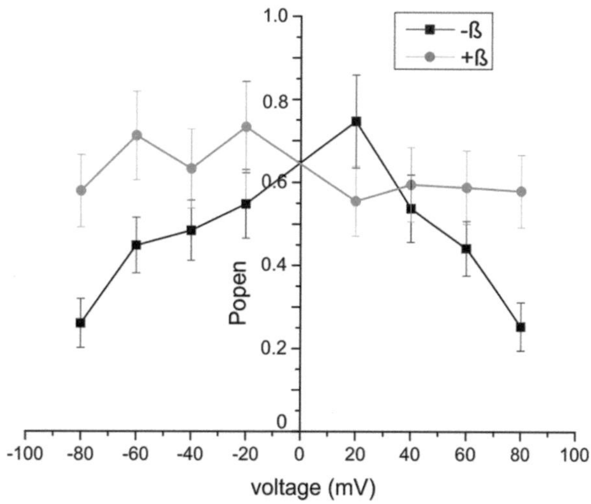

Fig. 2.8 Open probability of the SAM complex in the absence and presence of a ß-signal peptide

(Dudek et al. 2013). The core of the TIM23 complex is formed by three subunits: a peripheral membrane anchored receptor Tim50, as well as two integral membrane proteins Tim23 and Tim17 both and a recently discovered subunit required for the binding of Tim21 to the Tim23 core complex (Gebert et al. 2012). Tim23, a protein with four predicted TM α-helical segments (Alder et al. 2008), is a member of the PRAT-family (protein and amino acid transporter) (Rassow et al. 1999) and forms the pore. The hydrophilic C-terminal domain, which is exposed to the IMS recognises presequences (Bauer et al. 1996; Van der Laan et al. 2007) and mediates the dimerization of Tim23. Although Tim17 is essential for cell viability, its detailed function in the Tim23 complex remains elusive, while the Tim50 receptor has been shown to regulate the gating of the Tim23 pore (Meinecke et al. 2006). Transporting proteins via the TIM23 complex into the inner membrane requires Tim21, an inner membrane protein which also interacts with the TOM complex (Chacinska et al. 2005; Truscott et al. 2001).

2.3.3.1 Channel Properties of the Tim23 Channel

The channel properties of the Tim23 channel have been elucidated in detail (Kovermann 2003; Truscott et al. 2001; Meinecke et al. 2006; Van der Laan et al. 2007). Tim23 displays a voltage and presequence sensitive channel. In contrast to the outer membrane ß-barrel pores Tom40 or Sam50 which close in a voltage dependent manner (Hill et al. 1998; Kutik et al. 2008), the Tim23 channel formed with an α-helical bundle structure (Malhotra et al. 2013; Alder et al. 2008) is clearly activated by the membrane potential and presequences by yet unresolved mechanisms.

The main single-channel conductance of the most frequent transition of the recombinant *sc*Tim23 in symmetrical 250 mM KCl was $\bar{G}_{main} \cong 450\,\text{pS}$. Smaller subconductance states were also present, most prominently $\bar{G}_{sub} = 140\,\text{pS}$. The main conductance states were observed in multiples of 3. Gating of these main conductance states occurred largely independently of each other; however, coupled transitions were also observed such that the Tim23 channel exhibits a complex gating behavior covering a wide range of conductance states, which are combinations of 140 pS and 450 pS and 1.35 nS (Truscott et al. 2001). From the analysis of the connectivity of the different conductance states, it is likely that the functional unit of the Tim23 channel is formed by three mainly independent acting pores. In asymmetric buffer (12.5 fold KCl gradient), the reversal potential of the main conductance state was $E_{\text{rev}} = 49.6\,\text{mV}$, showing that the Tim23 channel is highly selective for cations $P_{\text{K}^+} / P_{\text{Cl}^-} \cong 16:1$. Fast increase of the membrane potential during a voltage step activated the Tim23 channel as shown by the increase in the number of open channels, due to channel voltage dependence. Nevertheless, the Tim23 channels showed a tendency to close during prolonged application of a constant voltage or in response to a slowly increasing voltage ramp. Thus, the Tim23 channel can be activated by a rapid increase in membrane voltage but closes slowly during prolonged exposure to a membrane potential (Truscott et al. 2001; Kovermann 2003). The pore diameter of the Tim23 channel constriction zone calculated by the ohmic model according to Smart et al. (see Appendix A) was $d_{\text{pore}} = 1.8\,\text{nm}$, while the size exclusion experiments revealed a $d_{\text{pore}} = 1.3\,\text{nm}$ and the exclusion limit of $d_{\text{excl}} = 1.8\,\text{nm}$ (Truscott et al. 2001). The Tim23 channel has been shown to interact specifically with known presequence-peptides thereby also changing the voltage sensitivity of the channel. The threshold potential of the Tim23 channel activity of about 120 mV at which peptides start to close the channel (Truscott et al. 2001) may reflect the membrane potential assisted unfolding and transport of the precursor proteins with the matrix import being further assisted by the PAM (presequence translocase associated motor) motor (Bolender et al. 2008). Recently a voltage-coupled conformational change in a transmembrane segment of the Tim23 import channel that may represent the major step in the activation of the voltage-gated Tim23 channel has been discovered (Malhotra et al. 2013).

Two models exist for the proposed pulling mechanism of precursor proteins through the Tim23 pore via *mt*Hsp70, namely the Brownian ratchet model and the power stroke model (Bohnert et al. 2007; Mayer and Bukau 1998; Rehling et al. 2004; Ungermann et al. 1994; von Ahsen et al. 1995). The matrix heat shock protein 70 (mtHsp70) is the central component of the presequence translocase associated motor (PAM) that binds to the unfolded preprotein in transit and drives in cooperation with membrane-associated co-chaperones the preproteins into the matrix (Bolender et al. 2008). Today it is assumed that a combination of both mechanisms describes best the molecular mechanism of Tim23 mediated protein translocation. First passive trapping for loosely folded or spontaneously unfolding proteins has to take place. Then the power stroke by the PAM is needed, in particular in the case of tightly folded proteins (Geissler et al. 2001).

2.3.4 *The TIM22 Complex*

The TIM22 complex is essential for the insertion of carrier proteins with internal targeting sequences into the inner mitochondrial membrane—e. g., the ADP/ATP-carrier (Rehling et al. 2004). This complex consists of four subunits Tim22, Tim54, a peripheral receptor, the subunit Tim18 and the peripheral Tim12 subunit (Adam et al. 1999; Koehler et al. 1998). Tim22, a member of the PRAT-family, with four predicted TM α-helices forms the channel of the complex (Kovermann et al. 2002). The channel properties of the Tim22 cannel have been elucidated in detail (Kovermann et al. 2002; Rehling et al. 2003).

2.3.4.1 Channel Properties of the Tim22 Channel

Similarly to what is described above for Tim23, the Tim22 channel formed by an α-helical bundle structure is activated by the membrane potential and precursor proteins. The Tim22 channel revealed multiple open channel states ranging from $\bar{G}_{min} \cong 65\,\text{pS}$ up to $\bar{G}_{max} \cong 540\,\text{pS}$ and the active unit contains two independently acting pores (Kovermann et al. 2002; Bolender et al. 2008). The Tim22 channel was cation selective with the large conductance states showing a selectivity of $P_{K^+} / P_{Cl^-} \cong 4:1$, while the smaller one revealed a value of $P_{K^+} / P_{Cl^-} \cong 16:1$. The pore diameter of the constriction zone determined by the polymer exclusion method was $d_{pore} \cong 1.8$ nm and the diameter of the channel vestibule $d_{vestibule} \cong 2.3\,\text{nm}$. Calculation of the diameter of the channel constriction zone from $\bar{G}_{max} \cong 540\,\text{pS}$ .using the ohmic model (Smart et al. 1997) with a length of 1 nm for the constriction zone yielded a value of $d_{pore} \cong 1.8$ nm. The Tim22 channel showed specific interactions with carrier derived signal peptides in the nM range but only from one side of the channel. This interaction blocks the current of small ions through the pore. Similar to the Tim23 channel the Tim22 channel was activated by fast changes in the membrane potential with a threshold potential for this activation of about 120 mV (Kovermann et al. 2002; Kovermann 2003). Tim22 forms a twin pore within the TIM22 complex. The twin-pore TIM22 complex inserts precursors into the inner mitochondrial membrane in three steps. (i) The precursor is bound to TIM22 in a $\Delta\psi$ -independent step. (ii) In the first voltage-dependent step, the precursor docks in TIM22 at a $\Delta\psi$ below that needed to effect pore activity, which indicates that $\Delta\psi$ acts directly on the precursor. (iii) In the second voltage-dependent step, a $\Delta\psi \geq 70$ mV and the presence of a signal peptide activate pore gating to complete the membrane-insertion process

Finally, the membrane potential together with an internal signal peptide induces rapid gating transitions in one pore and closing of the other pore thereby driving membrane insertion to completion. Thus, protein insertion is driven by the coordinated action of a twin-pore complex in two voltage-dependent steps (Rehling et al. 2003; Kovermann et al. 2002).

2.3.4.2 The OXA Complex

Only a marginal amount of the mitochondrial proteins are encoded inside the organelle. In *S. cerevisiae*, the mitochondrial genome encodes eight polypeptides, seven of which are integral membrane polypeptides involved in cellular respiration (Stuart 2002). The Oxa1-dependent pathway for the membrane biogenesis of several mitochondrially encoded proteins involves in the matrix the binding of the ribosome to Oxa1, which subsequently targets the nascent proteins to the membrane (Herrmann and Bonnefoy 2004; Hell et al. 2001). The translocation and insertion activity of Oxa1 then promotes the integration of the protein into the bilayer. Oxa1 belongs to a small family of highly conserved proteins acting as insertases with YidC (bacteria) and Alb3 (chloroplasts) as further members (Dalbey and Kuhn 2004). Oxa1 contains five predicted α-helical TM segments with N_{out}–C_{in} topology in which its C-terminal domain interacts with matrix ribosomes (Jia et al. 2003). The main driving forces for co-translational insertion of proteins into the inner membrane appears to be the membrane potential and the input by the ribosomes (Fiumera et al. 2007). The OXA1 complex inserts also hydrophobic nuclear-encoded proteins like Oxa1 itself which are imported posttranslationally. In this case the only driving force seems to be the membrane potential (Herrmann and Bonnefoy 2004; Hell et al. 2001).

Native PAGE experiments suggest that the complex forms a homotetramer (Nargang et al. 2002), whereas cryo-electron microscopy data indicate that the complex is built by a dimer with a pore in the middle (Kohler et al. 2009). The molecular mechanism, by which Oxa1 mediates insertion of membrane spanning segments, entailing the translocation of hydrophilic domains across the inner membrane, is still enigmatic.

2.3.4.3 Channel Properties of the Oxa1 Channel

Oxa1 purified from yeast mitochondria and the recombinant Oxa1 display ion channel activity when reconstituted and incorporated into planar bilayer provided the lipid environment resembles the lipid composition of the inner mitochondrial membrane (Krüger et al. 2012). The basic characteristics of Oxa1 channels purified from yeast and the recombinant channel ($Oxa1_{rec}$) were seemingly identical, with a mean maximal conductance state for $Oxa1_{rec}$ of $\bar{G}_{max} \cong 500\,pS$ and a $\bar{G}_{min} \cong 75\,pS$ at positive V_m and 250 mM KCl (cis/trans). The Oxa1 channel showed a rectifying current voltage relation with several frequent subconductance states and selectivity of $P_{K^+} / P_{Cl^-} \cong 10:1$. Based on the conductance state of $\bar{G}_{max} \cong 500\,pS$, pore diameter of the channel constriction (2 nm length, see Appendix A) was calculated to be $d_{pore} \cong 1.8$ nm using the ohmic channel model (Krüger et al. 2012). This pore size would accommodate polypeptide chains with secondary or unfolded structure, as assessed by a typical a-helix diameter being roughly 0.5 nm (Schwartz and Matouschek 1999) Oxa1 channels are activated by substrate peptides and display two distinct activity states characterized by a high and a low gating frequency. Remarkably, the voltage-dependent open probability showed remarkable differences for the two states. The low frequency gating activity showed steep voltage-dependent channel closure above threshold potentials of $V_m > 100\,\text{mV}$, while the OXA1 complex channel in the high frequency gating state was less sensitive to high voltage-induced channel closure. The overall Oxa1 channel properties, selectivity for cations, activation by physiological peptides, and the dimensions of the pore are similar to those of other protein translocation pores described above, namely Tim23 and Tim22. Importantly, the channel closes in a voltage-dependent manner suggesting that the membrane potential across the inner membrane maintains the channel in a closed state in the absence of substrates. In particular, the substrate and voltage dependence of the Oxa1 channel properties resemble those of the previously extensively characterized Tim22 channel, implying similar inner mitochondrial membrane protein insertion mechanisms might be at work in both pathways. Moreover, it has been shown for Tom40, Tim22, and Tim23 that the specific interaction of these protein translocation pores with signal peptides represent initial translocation steps leading to fast channel gating (flickering) (Rehling et al. 2003). Thus, the high frequency gating state of Oxa1 may be envisaged as a transport initiation state.

2.3.5 ATOM the Bacterial Ancestors of Tom40

Although, Tom40 is conserved in virtually all eukaryotes its evolutionary origin remains unclear since bacterial orthologues have not been identified so far. Recently it was discovered that the parasitic protozoon *Trypanosoma brucei* residing at a basal position at the root of phylogenetic tree of eukaryotes (Cavalier-Smith 2010) lacks a conventional Tom40. Instead, ATOM (archaic translocase of the outer mitochondrial membrane), a protein that shows similarities to both eukaryotic Tom40 and to bacterial protein translocases of the Omp85-family has been discovered (Pusnik et al. 2011; Harsman et al. 2012).

Bioinformatic analysis using the ATOM sequence of several trypanosomatid species revealed similarity to the bacterial Omp85-family protein YtfM/TamA that was recently shown to be involved in protein translocation across the outer membrane of proteobacteria (Pusnik et al. 2011; Selkrig et al. 2012). But also some similarity of ATOM to the canonical Tom40 has been reported (Zarsky et al. 2012). Therefore, ATOM might be an evolutionary intermediate that shares sequence and functional similarity with both bacterial YtfM/TamA and mitochondrial Tom40, respectively.

Using the planar lipid bilayer technique it was shown that ATOM forms a hydrophilic transmembrane pore, which in principle could provide the function of a translocation channel for unfolded proteins (Harsman et al. 2012; Harsman 2012). The functional unit of the ATOM channel contains three independent identical pores. The single pore of the channel displayed a rectifying current voltage relation with $\bar{G}_{max} \cong 400-500\,\text{pS}$ and $\bar{G}_{sub} \cong 100-200\,\text{pS}$ at positive voltages (250 mM KCL, 10 mM MOPS/Tris pH7.5) and a ratio of $G^{+}_{max} / G^{-}_{max} \approx 1.45$ with additional smaller subconductance states. Calculation of the diameter of the channel constriction zone from $\bar{G}_{max} \cong 400\,\text{pS}$ using the ohmic model (Smart et al. 1997, see also Appendix A) with a length of 2 nm for the constriction zone yielded a value of $d_{pore} \cong 2.0$ nm.

While the general characteristics of the pores formed by ATOM are consistent with basic features observed for protein translocases of different origin (mitochondrial/chloroplast), say, Tom40 and Toc75-III, the detailed analysis of single channel properties revealed a pronounced disparity between the channels formed by ATOM and Tom40. As illustrated in Fig. 2.4a, b the gating behavior of recombinant Tom40 single channels was dominated by fast flickering into a prominent subconductance state and full pore closure of the recombinantly produced protein occurred in a single step (Becker et al. 2005; Hill et al. 1998). This contrasts with the tripartite channel closure and low frequency gating that was observed for recombinant ATOM. Certainly, the oligomeric assembly observed for the recombinant proteins reflects the number of β-barrels required for stable integration into the bilayer in the absence of accessory proteins (Naveed et al. 2009). However, it does not necessarily represent the number of pores found *in vivo* in the respective hetero-oligomeric protein translocation complexes. Native TOM complexes purified from mitochondrial outer membranes of *S. cerevisiae* and *N. crassa*, for example, were shown to contain two identical Tom40 pores (Künkele et al. 1998a, b; Becker et al. 2005).

Interestingly, the comparison of electrophysiological characteristics (Harsman et al 2012) however revealed common properties of recombinant ATOM and recombinant Toc75 homologs from plastids and cyanobacteria (Bolter et al. 1998; Hinnah et al. 2002). These are: (i) a similar single channel conductance, (ii) comparable selectivity for cations and most importantly (iii) a homotrimeric assembly, which for recombinant Toc75 was confirmed by cryo-electron microscopy, cross-linking, and other methods (Schleiff et al. 2003; Bredemeier et al. 2007) as well as (iv) a gating behavior that is characterized by infrequent, short gating events between several conductance states. The concordant single channel characteristics of Toc75 and its distant homolog synToc75 from the cyanobacterium *Synechocystis* sp. PCC6803 suggested the development of the chloroplast protein import pore from a preexisting bacterial Omp85-like porin (Bolter et al. 1998). In summary, these common features might indicate that Toc75 as well as ATOM are both derived from bacterial ancestors, which probably were outer membrane porins as predicted earlier (Cavalier-Smith 2006; Herrmann 2003; Gabriel et al. 2001; Gross and Bhattacharya 2009).

In conclusion, ATOM forms a hydrophilic pore, which meets all requirements of a mitochondrial outer membrane protein-translocation channel. The electrophysiological properties of the ATOM channel support the proposal that ATOM may be the evolutionary missing link of the mitochondrial outer membrane translocase and the bacterial Omp85-type protein export machinery.

Acknowledgements *hs*Sam50 preparations for the electrophysiological measurements were kindly provided by Vera Kozjak-Pavlovic, Biocenter, Dep. of Microbiology, University of Wuerzburg, Germany. Excellent technical assistance by Birgit Hemmis, Biophysics, University of Osnabrueck is gratefully acknowledged.

Appendix A

Calculation of Pore Size from Channel Conductance

Even for large channels, it is known that their conductance is a rather poor measure of the pore size. However, with precautions and in a first approximation the pore size may be calculated according to the ohmic model of Hille (2001). As pointed out by Smart et al. (1997), this interpretation is based on the assumption that the pore is a cylindrical conductor having one single characteristic radius. In reality most channel pores have a more complex internal geometry see, for example the hourglass-shaped structure of porins (Cowan et al. 1992) and the effective radius of the pore is also determined by the charges and the electric field within the channel pore (Eisenberg 1996).

With all these precautions in mind, a rough correlation between channel pore size and single channel conductance may be deduced with the approach of Smart et al. (1997) using the following equation and taking the reduced conductivity of the electrolyte solution within the pore into account (Smart et al. 1997):

$$d = \frac{\rho \cdot G}{\pi}\left(\frac{\pi}{2} + \sqrt{\frac{\pi}{2} + \frac{4\pi\ell}{\rho G}}\right)$$

Where: d is the diameter of the pore, G is the conductance (e. g., $G = 1.2\,\mathrm{nS}$, in 250 mM KCl symmetrical solution), l is the length of the constriction zone, assumed to be $\ell_{min} = 0.5$ nm to $\ell_{max} = 2$ nm (following the model of Hille with a short constriction zone flanked by wide vestibules), and ρ is the resistivity of the solution, 49.5 Ω cm at 25 °C for a 250 mM KCl solution, but taking into account the correction factor of Smart et al. (1997) the value is 247.5 Ω cm.

The resulting diameters based on different length of the constriction zone are listed in the following table:

Constriction zone (length l/nm)	Pore-diameter (*d*/nm)
0.5	3.6
1	4.0
1.5	4.4
2	4.70

From high-resolution structure combined with MD calculations the effective length of the constriction zone within OmpF was shown to be 2 nm (Im and Roux 2002) similar to other porins. We therefore use this value when calculating pore diameters based on the ohmic model (Hille 2001; Smart et al. 1997).

Mean Variance Analysis of Single Channel Current Recordings

Reliable quantitative analysis of noteworthy noisy single ion channel recordings in the time and amplitude domain remains a difficult task. Although the desired information is accessible through the commonly applied half-amplitude threshold method in many cases, various limitations that come along with it, for example, signal-to-noise-ratio, clear separation of current fluctuations from gating transitions, sublevel, etc., raised the need for alternatives. In 1993 Patlak developed an alternative way to analyze the time and amplitude domains of complex single ion channel data by introducing a rigorous statistical analysis using mean-variance histograms and a corresponding data evaluation strategy. The main improvement of this concept is to work directly on the raw data in an unbiased statistical manner. Apart from that, it offers a great tool to visualize the complete information of whole recordings in a single graph. We implemented Patlak‘s concept (Patlak 1993) in an MATLAB® based program "Ion Channel Master" (David Schmedt 2010) for single channel data evaluation. Together with a new statistical analysis for idealizing complex current traces—that combines the information extracted from experimental mean-variance histograms with a robust threshold-based transition-scanning algorithm (Anderson and Conder 2011)—this allows for a final verification of the extracted single channel parameters from experimental data. The mean-variance histogram analysis is performed in three steps:

1. **Construction (amplitude domain)**
   - (a) Calculation of mean **<u>and</u>** variance of current within a "sliding window" of N consecutive data points.
   - (b) Assembly of a three-dimensional histogram based on the calculated mean-variance pairs (Fig. 2.4e).
     - (i) Contained information
     - (ii) Defined current levels (low variance regions), gating (high variance regions) between states, changes in conductance
     - (iii) Signal-to-noise ratio, open channel noise, baseline stability
2. **Extraction of channel kinetics (time domain)**
   - (a) Volume (V_{mv}) of low variance regions decreases with increasing size (W) of the "sliding window"
   - (b) Plot of V_{mv} as function of W depends on same time constants as original dwell time probability distribution (Fig. 2.4f).
   - c) Fitting (of Fig. 2.4f) yields dwell times (Fig. 2.4g) and residence times (Fig. 2.4c) for each channel state.
3. **Idealization**
   - (a) The information provided by mean-variance histograms (number of channel states, mean and variance of current in states) are combined with a threshold-based transition-scanning algorithm (Anderson and Conder 2011). To get maximum search efficiency the threshold is set to the lowest possible and

most sensitive level, which depends on the current variance. Between all consecutive transitions, the mean current is calculated and a channel state is assigned according to the state intervals that are extracted from a mean-variance histogram.

^{R}MathWorks, Inc.

Methods for Generation of Structure Cartoons

Template based modeling was performed using: I-TASSER. (http://zhanglab.ccmb.med.umich.edu/I-TASSER).

Electrostatic calculations were performed using APBS Tools 2 in Pymol 1.3.

References

Abe Y, Shodai T, Muto T, Mihara K, Torii H, Nishikawa S, Endo T, Kohda D (2000) Structural basis of presequence recognition by the mitochondrial protein import receptor Tom20. Cell 100(5):551–560

Adam A, Endres M, Sirrenberg C, Lottspeich F, Neupert W, Brunner M (1999) Tim9, a new component of the TIM22 54 translocase in mitochondria. EMBO J 18(2):313–319

Ahting U, Thieffry M, Engelhardt H, Hegerl R, Neupert W, Nussberger S (2001) Tom40, the pore-forming component of the protein-conducting TOM channel in the outer membrane of mitochondria. J Cell Biol 153(6):1151–1160

Alder NN, Sutherland J, Buhring AI, Jensen RE, Johnson AE (2008) Quaternary structure of the mitochondrial TIM23 complex reveals dynamic association between Tim23p and other subunits. Mol Biol Cell 19(1):159–170

Allison DS, Schatz G (1986) Artificial mitochondrial presequences. Proc Natl Acad Sci USA 83:9011–9015

Anderson KB, Conder JA (2011) Discussion of multicyclic Hubbert modeling as a method for forecasting future petroleum production. Energy Fuels 25(4):1578–1584

Bauer MF, Sirrenberg C, Neupert W, Brunner M (1996) Role of Tim23 as voltage sensor and presequence receptor in protein import into mitochondria. Cell 87(1):33–41

Becker L (2008) Electrophysiological characterization of the protein import-pores of the outer mitochondrial membrane. PhD-thesis, Universitaet Osnabrück, FB Biologie/Chemie

Becker L, Bannwarth M, Meisinger C, Hill K, Model K, Krimmer T, Casadio R, Truscott KN, Schulz GE, Pfanner N, Wagner R (2005) Preprotein translocase of the outer mitochondrial membrane: reconstituted Tom40 forms a characteristic TOM pore. J Mol Biol 353(5):1011–1020

Bohnert M, Pfanner N, van der Laan M (2007) A dynamic machinery for import of mitochondrial precursor proteins. FEBS Lett 581(15):2802–2810

Bolender N, Sickmann A, Wagner R, Meisinger C, Pfanner N (2008) Multiple pathways for sorting mitochondrial precursor proteins. EMBO Rep 9(1):42–49

Bolter B, Soll J, Schulz A, Hinnah S, Wagner R (1998) Origin of a chloroplast protein importer. Proc Natl Acad Sci U S A 95(26):15831–15836

Bredemeier R, Schlegel T, Ertel F, Vojta A, Borissenko L, Bohnsack MT, Groll M, von Haeseler A, Schleiff E (2007) Functional and phylogenetic properties of the pore-forming beta-barrel transporters of the Omp85 family. J Biol Chem 282(3):1882–1890. doi:10.1074/jbc.M609598200

Brix J, Dietmeier K, Pfanner N (1997) Differential recognition of preproteins by the purified cytosolic domains of the mitochondrial import receptors Tom20, Tom22, and Tom70. J Biol Chem 272(33):20730–20735

Brun L, Pastoriza-Gallego M, Oukhaled G, Mathé J, Bacri L, Auvray L, Pelta J (2008) Dynamics of polyelectrolyte transport through a protein channel as a function of Applied voltage. Phys Rev Lett 100(15):158302

Cavalier-Smith T (2006) Origin of mitochondria by intracellular enslavement of a photosynthetic purple bacterium. Proc Biol Sci 273(1596):1943–1952

Cavalier-Smith T (2010) Kingdoms Protozoa and Chromista and the eozoan root of the eukaryotic tree. Biol Lett 6(3):342–345

Chacinska A, Rehling P, Guiard B, Frazier AE, Schulze-Specking A, Pfanner N, Voos W, Meisinger C (2003) Mitochondrial translocation contact sites: separation of dynamic and stabilizing elements in formation of a TOM-TIM-preprotein supercomplex. EMBO J 22(20):5370–5381

Chacinska A, Pfannschmidt S, Wiedemann N, Kozjak V, Sanjuan Szklarz LK, Schulze-Specking A, Truscott KN, Guiard B, Meisinger C, Pfanner N (2004) Essential role of Mia40 in import and assembly of mitochondrial intermembrane space proteins. EMBO J 23(19):3735–3746

Chacinska A, Lind M, Frazier AE, Dudek J, Meisinger C, Geissler A, Sickmann A, Meyer HE, Truscott KN, Guiard B, Pfanner N, Rehling P (2005) Mitochondrial presequence translocase: switching between TOM tethering and motor recruitment involves Tim21 and Tim17. Cell 120(6):817–829

Chan NC, Lithgow T (2008) The peripheral membrane subunits of the SAM complex function codependently in mitochondrial outer membrane biogenesis. Mol Biol Cell 19(1):126–136

Cowan SW, Schirmer T, Rummel G, Steiert M, Ghosh R, Pauptit RA, Jansonius JN, Rosenbusch JP (1992) Crystal structures explain functional properties of two E. coli porins. Nature 358(6389):727–733

Dalbey RE, Kuhn A (2004) YidC family members are involved in the membrane insertion, lateral integration, folding, and assembly of membrane proteins. J Cell Biol 166(6):769–774

Dekker PJ, Ryan MT, Brix J, Muller H, Honlinger A, Pfanner N (1998) Preprotein translocase of the outer mitochondrial membrane: molecular dissection and assembly of the general import pore complex. Mol Cell Biol 18(11):6515–6524

Dolezal P, Likic V, Tachezy J, Lithgow T (2006) Evolution of the molecular machines for protein import into mitochondria. Science 313(5785):314–318

Dudek J, Rehling P, van der Laan M (2013) Mitochondrial protein import: Common principles and physiological networks. Biochimica et Biophysica Acta (BBA): Mol Cell Res 1833(2): 274–285

Eisenberg RS (1996) Computing the field in proteins and channels. J Memb Biol 150(1):1–25

Emanuelsson O, Brunak S, von Heijne G, Nielsen H (2007) Locating proteins in the cell using TargetP, SignalP and related tools. Nat Protoc 2(4):953–971

Fiumera HL, Broadley SA, Fox TD (2007) Translocation of mitochondrially synthesized Cox2 domains from the matrix to the intermembrane space. Mol Cell Biol 27(13):4664–4673

Gabriel K, Buchanan SK, Lithgow T (2001) The alpha and the beta: protein translocation across mitochondrial and plastid outer membranes. Trends Biochem Sci 26(1):36–40, doi:S0968-0004(00)01684-4 [pii]

Gebert M, Schrempp SG, Mehnert CS, Heisswolf AK, Oeljeklaus S, Ieva R, Bohnert M, von der Malsburg K, Wiese S, Kleinschroth T, Hunte C, Meyer HE, Haferkamp I, Guiard B, Warscheid B, Pfanner N, van der Laan M (2012) Mgr2 promotes coupling of the mitochondrial presequence translocase to partner complexes. J Cell Biol 197(5):595–604

Geissler A, Rassow J, Pfanner N, Voos W (2001) Mitochondrial import driving forces: enhanced trapping by matrix Hsp70 stimulates translocation and reduces the membrane potential dependence of loosely folded preproteins. Mol Cell Biol 21(20):7097–7104

Gessmann D, Flinner N, Pfannstiel J, Schlosinger A, Schleiff E, Nussberger S, Mirus O (2011) Structural elements of the mitochondrial preprotein-conducting channel Tom40 dissolved by bioinformatics and mass spectrometry. Biochim Biophys Acta 1807(12):1647–1657

Glick BS, Brandt A, Cunningham K, Muller S, Hallberg RL, Schatz G (1992) Cytochromes c1 and b2 are sorted to the intermembrane space of yeast mitochondria by a stop-transfer mechanism. Cell 69(5):809–822
Gross J, Bhattacharya D (2009) Mitochondrial and plastid evolution in eukaryotes: an outsiders' perspective. Nat Rev Genet 10(7):495–505
Habib SJ, Waizenegger T, Niewienda A, Paschen SA, Neupert W, Rapaport D (2007) The N-terminal domain of Tob55 has a receptor-like function in the biogenesis of mitochondrial beta-barrel proteins. J Cell Biol 176(1):77–88
Hanhart P (2011) Effect of charges within the Tom40 pore on the interaction and transport of signal-peptides. BaSc, Universität Osnabrück, FB Biologie/Chemie
Harbauer Angelika B, Zahedi René P, Sickmann A, Pfanner N, Meisinger C (2014) The protein import machinery of mitochondria:—a regulatory hub in metabolism, stress, and disease. Cell Metab 19(3):357–372
Harsman A (2012) Probing protein import machineries of different organisms with the lipid bilayer technique: functional comparison and phylogenetic insights. PhD-thesis, Universität Osnabrück, FB Biologie/Chemie
Harsman A, Kruger V, Bartsch P, Honigmann A, Schmidt O, Rao S, Meisinger C, Wagner R (2010) Protein conducting nanopores. J Phys Condens Matter 22(45):454102
Harsman A, Niemann M, Pusnik M, Schmidt O, Burmann BM, Hiller S, Meisinger C, Schneider A, Wagner R (2012) Bacterial Origin of a Mitochondrial Outer Membrane Protein Translocase: New perspectives from comparative single channel electrophysiology. J Biol Chem 287(37):31437–31445
Hell K, Neupert W, Stuart RA (2001) Oxa1p acts as a general membrane insertion machinery for proteins encoded by mitochondrial DNA. EMBO J 20(6):1281–1288
Herrmann JM (2003) Converting bacteria to organelles: evolution of mitochondrial protein sorting. Trends Microbiol 11(2):74–79
Herrmann JM, Bonnefoy N (2004) Protein export across the inner membrane of mitochondria: the nature of translocated domains determines the dependence on the Oxa1 translocase. J Biol Chem 279(4):2507–2512
Hill K (2001) Tom40, the Proteinimportpore of the outer Mitochondrial Membrane. PhD-thesis, Universitaet Osnabrück, FB. Biologie/Chemie
Hill K, Model K, Ryan MT, Dietmeier K, Martin F, Wagner R, Pfanner N (1998) Tom40 forms the hydrophilic channel of the mitochondrial import pore for preproteins. Nature 395:516–521
Hille B (2001) Ionic channels of excitable membranes, vol 3. Sinauer Ass. Inc., Sunderland
Hinnah SC, Wagner R, Sveshnikova N, Harrer R, Soll J (2002) The chloroplast protein import channel Toc75: pore properties and interaction with transit peptides. Biophys J 83:899–911
Im W, Roux B (2002) Ions and counterions in a biological channel: a molecular dynamics simulation of OmpF Porin from Escherichia coli in an explicit membrane with 1 M KCl aqueous salt solution. J Mol Biol 319(5):1177–1197
Jia L, Dienhart M, Schramp M, McCauley M, Hell K, Stuart RA (2003) Yeast Oxa1 interacts with mitochondrial ribosomes: the importance of the C-terminal region of Oxa1. EMBO J 22(24):6438–6447
Kasianowicz JJ, Robertson JW, Chan ER, Reiner JE, Stanford VM (2008) Nanoscopic porous sensors. Ann Rev Anal Chem (Palo Alto, Calif) 1:737–766
Koehler CM, Jarosch E, Tokatlidis K, Schmid K, Schweyen RJ, Schatz G (1998) Import of mitochondrial carriers mediated by essential proteins of the intermembrane space. Science 279(5349):369–373
Kohler R, Boehringer D, Greber B, Bingel-Erlenmeyer R, Collinson I, Schaffitzel C, Ban N (2009) YidC and Oxa1 form dimeric insertion pores on the translating ribosome. Mol Cell 34(3):344–353
Kovermann P (2003) Electrophysiological chracterization of protein imort channels of the inner mitochondrial membrane. University Osnabrueck, Fac. Biology/Chemistry, Dep. Biophysics

Kovermann P, Truscott KN, Guiard B, Rehling P, Sepuri NB, Muller H, Jensen RE, Wagner R, Pfanner N (2002) Tim22, the essential core of the mitochondrial protein insertion complex, forms a voltage-activated and signal-gated channel. Mol Cell 9(2):363–373
Kozjak V, Wiedemann N, Milenkovic D, Lohaus C, Meyer HE, Guiard B, Meisinger C, Pfanner N (2003) An essential role of Sam50 in the protein sorting and assembly machinery of the mitochondrial outer membrane. J Biol Chem 278(49):48520–48523
Krasilnikov OV, Sabirov RZ, Ternovsky VI, Merzliak PG, Muratkhodjaev JN (1992) A simple method for the determination of the pore radius of ion channels in planar lipid bilayer membranes. FEMS Microbiol Lett 105(1–3):93–100
Krasne SEG (1973) The molecular basis of ion selectivity. Membranes 2:51
Krüger V, Deckers M, Hildenbeutel M, van der Laan M, Hellmers M, Dreker C, Preuss M, Herrmann JM, Rehling P, Wagner R, Meinecke M (2012) The mitochondrial oxidase assembly protein1 (Oxa1) insertase forms a membrane pore in lipid bilayers. J Biol Chem 287(40):33314–33326
Künkele KP, Heins S, Dembowski M, Nargang FE, Benz R, Thieffry M, Walz J, Lill R, Nussberger S, Neupert W (1998a) The preprotein translocation channel of the outer membrane of mitochondria. Cell 93(6):1009–1019
Künkele KP, Juin P, Pompa C, Nargang FE, Henry JP, Neupert W, Lill R, Thieffry M (1998b) The isolated complex of the translocase of the outer membrane of mitochondria. Characterization of the cation-selective and voltage-gated preprotein-conducting pore. J Biol Chem 273(47):31032–31039
Kutik S, Stojanovski D, Becker L, Becker T, Meinecke M, Kruger V, Prinz C, Meisinger C, Guiard B, Wagner R, Pfanner N, Wiedemann N (2008) Dissecting membrane insertion of mitochondrial beta-barrel proteins. Cell 132(6):1011–1024
Lackey SWK, Taylor RD, Go NE, Wong A, Sherman EL, Nargang FE (2014) Evidence supporting the 19 β-strand model for Tom40 from cysteine scanning and protease site accessibility studies. J Biol Chem 289(31):21640–21650
Lemeshko VV (2006) Theoretical evaluation of a possible nature of the outer membrane potential of mitochondria. Eur Biophys J 36(1):57–66
Mahendran Kozhinjampara R, Romero-Ruiz M, Schlösinger A, Winterhalter M, Nussberger S (2012) Protein translocation through Tom40: kinetics of peptide release. Biophys J 102(1):39–47
Malhotra K, Sathappa M, Landin JS, Johnson AE, Alder NN (2013) Structural changes in the mitochondrial Tim23 channel are coupled to the proton-motive force. Nat Struct Mol Biol 20(8):965–972
Mayer MP, Bukau B (1998) Hsp70 chaperone systems: diversity of cellular functions and mechanism of action. Biol Chem 379(3):261–268
Meinecke M, Wagner R, Kovermann P, Guiard B, Mick DU, Hutu DP, Voos W, Truscott KN, Chacinska A, Pfanner N, Rehling P (2006) Tim50 maintains the permeability barrier of the mitochondrial inner membrane. Science 312(5779):1523–1526
Meisinger C, Ryan MT, Hill K, Model K, Lim JH, Sickmann A, Muller H, Meyer HE, Wagner R, Pfanner N (2001) Protein import channel of the outer mitochondrial membrane: a highly stable Tom40-Tom22 core structure differentially interacts with preproteins, small tom proteins, and import receptors. Mol Cell Biol 21(7):2337–2348
Meisinger C, Pfannschmidt S, Rissler M, Milenkovic D, Becker T, Stojanovski D, Youngman MJ, Jensen RE, Chacinska A, Guiard B, Pfanner N, Wiedemann N (2007) The morphology proteins Mdm12/Mmm1 function in the major beta-barrel assembly pathway of mitochondria. EMBO J 26(9):2229–39
Milenkovic D, Gabriel K, Guiard B, Schulze-Specking A, Pfanner N, Chacinska A (2007) Biogenesis of the essential Tim9-Tim10 chaperone complex of mitochondria: site-specific recognition of cysteine residues by the intermembrane space receptor Mia40. J Biol Chem 282(31):22472–22480
Milenkovic D, Ramming T, Muller JM, Wenz LS, Gebert N, Schulze-Specking A, Stojanovski D, Rospert S, Chacinska A (2009) Identification of the signal directing Tim9 and Tim10 into the intermembrane space of mitochondria. Mol Biol Cell 20(10):2530–2539

Model K, Prinz T, Ruiz T, Radermacher M, Krimmer T, Kuhlbrandt W, Pfanner N, Meisinger C (2002) Protein translocase of the outer mitochondrial membrane: role of import receptors in the structural organization of the TOM complex. J Mol Biol 316(3):657–666

Mohammad MM, Movileanu L (2008) Excursion of a single polypeptide into a protein pore: simple physics, but complicated biology. Eur Biophys J 37(6):913–925

Mohammad MM, Prakash S, Matouschek A, Movileanu L (2008) Controlling a single protein in a nanopore through electrostatic traps. J Am Chem Soc 130(12):4081–4088

Mokranjac D, Neupert W (2008) Energetics of protein translocation into mitochondria. Biochim Biophys Acta 1777(7–8):758–762

Movileanu L, Schmittschmitt JP, Scholtz JM, Bayley H (2005) Interactions of peptides with a protein pore. Biophys J 89(2):1030–1045

Nargang FE, Preuss M, Neupert W, Herrmann JM (2002) The Oxa1 protein forms a homooligomeric complex and is an essential part of the mitochondrial export translocase in Neurospora crassa. J Biol Chem 277(15):12846–12853

Naveed H, Jackups R Jr, Liang J (2009) Predicting weakly stable regions, oligomerization state, and protein-protein interfaces in transmembrane domains of outer membrane proteins. Proc Natl Acad Sci U S A 106(31):12735–12740

Neupert W, Herrmann JM (2007) Translocation of proteins into mitochondria. Annu Rev Biochem 76:723–749

Paschen SA, Waizenegger T, Stan T, Preuss M, Cyrklaff M, Hell K, Rapaport D, Neupert W (2003) Evolutionary conservation of biogenesis of beta-barrel membrane proteins. Nature 426(6968):862–866

Patlak JB (1993) Measuring kinetics of complex single ion channel data using mean-variance histograms. Biophys J 65(1):29–42

Pfanner N (2000) Protein sorting: recognizing mitochondrial presequences. Curr Biol 10(11):R412–R415

Pfanner N, Geissler A (2001) Versatility of the mitochondrial protein import machinery. Nat Rev Mol Cell Biol 2(5):339–349

Porcelli AM, Ghelli A, Zanna C, Pinton P, Rizzuto R, Rugolo M (2005) pH difference across the outer mitochondrial membrane measured with a green fluorescent protein mutant. Biochem Biophys Res Commun 326(4):799–804

Pusnik M, Schmidt O, Perry AJ, Oeljeklaus S, Niemann M, Warscheid B, Lithgow T, Meisinger C, Schneider A (2011) Mitochondrial preprotein translocase of trypanosomatids has a bacterial origin. Curr Biol 21(20):1738–1743

Rapaport D (2005) How does the TOM complex mediate insertion of precursor proteins into the mitochondrial outer membrane? J Cell Biol 171(3):419–423

Rassow J, Dekker PJT, van Wilbe S, Meijer M, Soll J (1999) The preprotein translocase of the mitochondrial inner membrane: function and evolution [Review]. J Mol Biol 286(1):105–120

Rehling P, Model K, Brandner K, Kovermann P, Sickmann A, Meyer HE, Kühlbrandt W, Wagner R, Truscott KN, Pfanner N (2003) Protein insertion into the mitochondrial inner membrane by a twin-pore translocase. Science 299(5613):1747–1751

Rehling P, Brandner K, Pfanner N (2004) Mitochondrial import and the twin-pore translocase. Nat Rev Mol Cell Biol 5(7):519–530

Schatz G (1997) Just follow the acid chain [news; comment]. Nature 388(6638):121–122

Schleiff E, Soll J, Kuchler M, Kühlbrandt W, Harrer R (2003) Characterization of the translocon of the outer envelope of chloroplasts. J Cell Biol 160(4):541–551

Schmedt D (2010) Peptid-Translocation through Tom40: dynamic interaction an transport of CoxIV. BaSc, Universität Osnabrück, FB Biologie/Chemie

Schwartz MP, Matouschek A (1999) The dimensions of the protein import channels in the outer and inner mitochondrial membranes. Proc Natl Acad Sci U S A 96(23):13086–13090

Selkrig J, Mosbahi K, Webb CT, Belousoff MJ, Perry AJ, Wells TJ, Morris F, Leyton DL, Totsika M, Phan MD, Celik N, Kelly M, Oates C, Hartland EL, Robins-Browne RM, Ramarathinam

SH, Purcell AW, Schembri MA, Strugnell RA, Henderson IR, Walker D, Lithgow T (2012) Discovery of an archetypal protein transport system in bacterial outer membranes. Nat Struct Mol Biol 19(5):506–10

Sideris DP, Tokatlidis K (2010) Oxidative protein folding in the mitochondrial intermembrane space. Antioxid Redox Signal 13(8):1189–1204

Sirrenberg C, Bauer MF, Guiard B, Neupert W, Brunner M (1996) Import of carrier proteins into the mitochondrial inner membrane mediated by Tim22. Nature 384(6609):582–585

Smart OS, Breed J, Smith GR, Sansom MS (1997) A novel method for structure-based prediction of ion channel conductance properties. Biophys J 72(3):1109–1126

Stan T, Ahting U, Dembowski M, Künkele KP, Nussberger S, Neupert W, Rapaport D (2000) Recognition of preproteins by the isolated TOM complex of mitochondria. EMBO J 19(18): 4895–4902

Stuart R (2002) Insertion of proteins into the inner membrane of mitochondria: the role of the Oxa1 complex. Biochim Biophys Acta 1592(1):79–87

Truscott KN, Kovermann P, Geissler A, Merlin A, Meijer M, Driessen AJ, Rassow J, Pfanner N, Wagner R (2001) A presequence- and voltage-sensitive channel of the mitochondrial preprotein translocase formed by Tim23. Nat Struct Biol 8(12):1074–1082

Ungermann C, Neupert W, Cyr DM (1994) The role of Hsp70 in conferring unidirectionality on protein translocation into mitochondria [see comments]. Science 266(5188):1250–1253

Van der Laan M, Meinecke M, Dudek J, Hutu DP, Lind M, Perschil I, Guiard B, Wagner R, Pfanner N, Rehling P (2007) Motor-free mitochondrial presequence translocase drives membrane integration of preproteins 1 75. Nat Cell Biol 9(10):1152–1159

van Wilpe S, Ryan MT, Hill K, Maarse AC, Meisinger C, Brix J, Dekker PJT, Moczko M, Wagner R, Meijer M, Guiard B, Honlinger A, Pfanner N (1999) Tom22 is a multifunctional organizer of the mitochondrial preprotein translocase [in process citation]. Nature 401(6752):485–489

von Ahsen O, Voos W, Henninger H, Pfanner N (1995) The mitochondrial protein import machinery – role of ATP in dissociation of the hsp70-center-dot-mim44 complex. J Biol Chem 270(50):29848–29853

Wiedemann N, Kozjak V, Chacinska A, Schonfisch B, Rospert S, Ryan MT, Pfanner N, Meisinger C (2003) Machinery for protein sorting and assembly in the mitochondrial outer membrane. Nature 424(6948):565–571

Wolfe AJ, Mohammad MM, Cheley S, Bayley H, Movileanu L (2007) Catalyzing the translocation of polypeptides through attractive interactions. J Am Chem Soc 129(45):14034–14041

Woodhull AM (1973) Ionic blockage of sodium channels in nerve. J Gen Physiol 61:687–708

Yi L, Dalbey RE (2005) Oxa1/Alb3/YidC system for insertion of membrane proteins in mitochondria, chloroplasts and bacteria (review). Mol Membr Biol 22(1–2):101–111

Zarsky V, Tachezy J, Dolezal P (2012) Tom40 is likely common to all mitochondria. Curr Biol 22(12):R479–R481

Chapter 3
Electrophysiology of the Mitochondrial Apoptosis-Induced Channel, MAC

Kathleen W. Kinnally and Pablo M. Peixoto

Abstract MAC is distinct from typical ion channels in its huge pore diameter of 4–6 nm and poor ion selectivity. Although it has been identified during intracellular recordings, MAC is typically studied by patch-clamping mitochondria isolated from apoptotic cells. This mega-channel assembles in the outer membrane of mitochondria containing the pro-apoptotic proteins Bax and/or Bak in response to apoptotic signals. Typically, MAC formation requires assembly of 8–10 molecules of Bax and/or Bak within the outer membrane to achieve a pore size sufficient to release folded 3 nm cytochrome c from the space between the mitochondrial inner and outer membranes. Several compounds have been identified as MAC activators, e.g., BH3-only mimetics, and MAC inhibitors, e.g. iMAC2, and these small molecules are being used to further dissect the apoptotic process. MAC is a novel therapeutic target for a variety of diseases including cancer, infarcts, and neurodegenerative diseases.

Keywords Mitochondrial apoptosis-induced channel (MAC) • Bax • Mitochondria • Megachannel • Bcl2 family • Apoptosis • Inhibitors MAC (iMAC)

K.W. Kinnally (✉)
Department of Basic Sciences, College of Dentistry, New York University, 345 East 24th St., Room 902, New York, NY 10010, USA
e-mail: kck1@nyu.edu

P.M. Peixoto
Department of Natural Sciences, Baruch College and Graduate Center at City University of New York, New York, NY USA
e-mail: Pablo.Peixoto@baruch.cuny.edu

A.H. Delcour (ed.), *Electrophysiology of Unconventional Channels and Pores*, Springer Series in Biophysics 18, DOI 10.1007/978-3-319-20149-8_3

3.1 Introduction

Classical ion channels have long been defined as proteins that open a pathway for an enormous rate of transport of ions across membranes. Ion channels gate in response to voltage, ligand-binding, or mechanical distortion and have a filter that, in general, provides a high level of selectivity determining which ions can pass through. The function of ion channels underlies resting and action potentials as well as muscle contraction and cell signaling. Interestingly, it was previously thought that mitochondria would have no ion channels to compromise the exceptionally low permeability of the inner membrane needed for tightly coupled oxidative phosphorylation. As it turns out, mitochondria harbor primordial channels with some of the largest conductance pores ever reported.

Several mitochondrial channels form part of an incongruent set of channels whose opening essentially breaches cellular compartmentalization; these pores allow passage of larger solutes, including folded proteins, across membranes typically with significantly less selectivity than classical ion channels. Some are gated by voltage, like porins (Mapingire et al. 2013; Schein et al. 1976), but others actually assemble when their function is required (Latorre et al. 1981; Martinez-Caballero et al. 2009). MAC, mitochondrial apoptosis-induced channel, is such a giant channel as it is constructed from the pro-apoptotic proteins, Bax and/or Bak, in response to cell signaling. MAC formation and characterization are the principal subjects of this review.

The commitment step of intrinsic apoptosis corresponds to the permeabilization of the mitochondrial outer membrane (MOM) in higher eukaryotic cells. The mechanism underlying this permeabilization is the assembly of MAC, which then facilitates the release of pro-death factors like cytochrome c from the inter-membrane space (IMS) (Dejean et al. 2005; Guo et al. 2004; Pavlov et al. 2001). While formation of a giant pore was hypothesized to permeabilize the MOM during apoptosis, applications of electrophysiological approaches to demonstrate this pore's existence were difficult.

3.2 Overcoming Electrophysiological Challenges to Recording Mitochondrial Channels

Because of the intracellular location and size of mitochondria, identification of channels by electrophysiology required development of novel approaches. Early on, VDAC (voltage dependent anion-selective channel), or mitochondrial porin, was characterized after protein purification and reconstitution in planar bilayers [reviewed in (Colombini and Mannella 2012)]. Once patch-clamp techniques were developed, several mitochondrial channels were identified in their native membranes (Kinnally et al. 1989; Sorgato et al. 1987; Szabo and Zoratti 1992). As shown in Fig. 3.1a, the application of patch clamp techniques to isolated mitochondria enabled studies of not only MAC but also the mitochondrial protein import

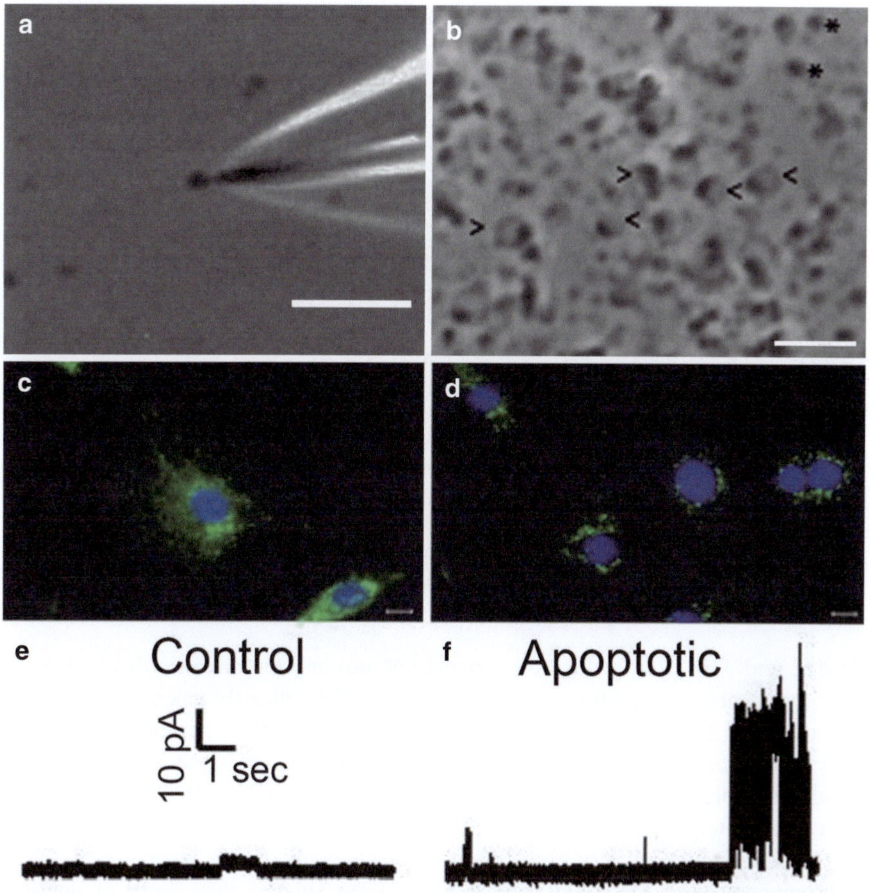

Fig. 3.1 Recording mitochondrial channel activities. (**a**) Phase-contrast image shows high resistance-seal formation between a microelectrode (10–30 MΩ, diameter 0.2–0.5 μm) and a yeast mitochondrion in order to record outer membrane channel activities in their native membranes. (**b**) Phase-contrast image shows isolated mouse heart mitoplasts (<) and mitochondria (*) used to record inner and outer membrane channels, respectively. (**c**, **d**) Serum withdrawal from cultured neural cells (CSM14.1) for 24 h induces apoptosis and cytochrome c release. Fluorescence images of cells that were not (*c*, Control) and were serum-starved for 24 h (*d*, Apoptotic) show this treatment caused loss of cytochrome c (immune-histological stain, *green*). Nuclei were stained with Hoechst (*blue*). (**e**, **f**) Corresponding intra-cellular patch-clamp current traces recorded at 100 mV with the double patch pipette approach are shown below respective images. These recordings revealed high conductance MAC activity in intracellular membranes 24 h after serum withdrawal (**f**) but only low conductance transitions were recorded in control cells (**e**) (Panels **c**–**f** were modified with the kind permission of Springer Science+Business Media from Figure 2 of Martinez-Cabellero et al. (2005)). Scale bar is 5 μm in panels **a**, **b** and 10 μm in panels **c**, **d**

gatekeeper Tom40 which is the subject of Chap. 1. Furthermore, French pressing or osmotic-treatment of isolated mitochondria ruptures the MOM to generate mitoplasts, which reveals the mitochondrial inner membrane (MIM) for patch clamping (Fig. 3.1b). Other giant pores of the MIM, including the protein import channels

Table 3.1 Comparison of single channel characteristics of MAC, Bax, and other mitochondrial channels

	MAC	rBax	Tom40	VDAC
Maximum conductance (nS)	2.5 ± 0.6	0.1–2.0	0.71 ± 0.06	0.68 ± 0.09
Maximum transition size (nS)	1.4 ± 0.4	0.1–0.5	0.34 ± 0.03	0.36 ± 0.04
Ion selectivity	Cation	Cation	Cation	Anion
P_K/P_{Cl}	3.0 ± 0.9	6.8 ± 1.0	3.6 ± 0.8	0.7 ± 0.1
Voltage dependent	No	No	Yes	Yes
Pore diameter[a] (nm)	4.0 ± 0.5	0.8–3.8	2.3 ± 0.1	2.1 ± 0.05

Modified from ©Pavlov et al. (2001). Originally published in *J Cell Biol*

MAC mammalian MAC of apoptotic FL5.12 cells 12 h after Interleukin-3 withdrawal, *rBax* recombinant human Bax

Measurements for all channels were made in 150 mM KCl except ion selectivity, which used a 150/30 mM KCl gradient. Maximum conductance and transition sizes were typically determined from total amplitude histograms (30–60 s of current 2 kHz data at +20 mV) generated with PAT program (Strathclyde Electrophysiological Software, courtesy of J. Dempster, U. of Strathclyde, UK); Maximum transition size was the largest transition observed, whether or not the maximum conductance level was reached; transition sizes were verified by manual examination of current traces and corresponding amplitude histograms

[a]Calculated from maximum conductance in 150 mM KCl assuming 5.5 nm pore length according to Eq. 3.1 (Hille 2001)

TIM22 and TIM23 (translocase inner membrane) and the necrosis death channel PTP (permeability transition pore), were identified by directly patching mitoplasts (Karch et al. 2013; Kinnally et al. 1991; Lohret et al. 1997; Peixoto et al. 2007; Szabo and Zoratti 1992). However, these mitochondrial channels are quite large, hundreds of picoSiemens (pS) at physiological ionic strength as shown in Table 3.1, which presented yet another set of hurtles to overcome.

Characterization of very high conductance channels of tiny organelles requires micropipettes with low resistance and tips even smaller than these organelles. The small-diameter tip is requisite. Otherwise, mitochondria will simply enter the micropipette upon application of the negative pressure typically used to form a seal of gigaohmic resistance. Low-resistance micropipettes are needed to ensure accurate voltage clamp since the resistance of the micropipette should be a fraction of the resistance of the open channel (Hamill et al. 1981). Hence, a balance must be attained between resistance and the size of the micropipette tip opening. Typically, micropipettes used to patch clamp mitochondria are 10–30 MΩ (0.15 M KCl) with tip diameters of 0.2–0.5 μm; the taper is short and stout to minimize access resistance.

Importantly, how can you determine if the mitochondrial channels behave the same in isolated mitochondria as in their normal cellular environment or are regulators lost upon mitochondrial isolation or MOM reconstitution in giant liposomes? The Jonas lab at Yale engineered double-barrel micropipettes to form giga Ω seals for patch clamping mitochondria within cells (Jonas et al. 1999). That is, once the double-barrel micropipette penetrates the plasma membrane to enter the cytoplasm, the patch pipette is extended from within the shielding micropipette to seal directly on a mitochondrion. They applied these novel approaches to determine if a giant channel, presumably MAC, is observed by intracellular patch clamping mitochondria early in apoptosis (Martinez-Caballero et al. 2005). Serum withdrawal from cultured neural cells for 24 h induces apoptosis and cytochrome c release (Fig. 3.1c, d). Importantly, large multiple conductance channels, ostensibly MAC, are recorded from the serum starved (apoptotic) (Fig. 3.1f) but not the normal cells (Fig. 3.1e). This intracellular channel activity is comparable to MAC recorded from mitochondria isolated from apoptotic cells. These findings linked apoptosis with opening of a giant channel in the MOM and justified the use of less technically challenging approaches for MAC characterization.

In part because of the similarity of results found with intracellular recordings, most studies of MAC have been carried out on mitochondria isolated from apoptotic cells of various genetic backgrounds. We have also found MAC reconstitutes with comparable activity when purified MOMs are fused with small liposomes to form giant liposomes, which can be stored in the freezer and are more convenient for patch clamping. This approach relies on dehydration and rehydration of the membranes for reconstitution (Criado and Keller 1987).

3.3 MAC Activity

MAC is a giant channel with a maximum conductance averaging 2–4 nS (in 0.15 M KCl) and multiple conductance levels as shown in Table 3.1 (Pavlov et al. 2001). Unlike Tom40 whose mean open time is ~14 ms (Muro et al. 2003), MAC often remains open for several seconds without transitions (Fig. 3.2). While nS maximum transitions are observed, the most common transition size is ~0.3 nS. The maximum sized transitions are typically 1–2 nS even though the maximum conductance was often much larger. Furthermore, the largest transitions occur between multiple conductance levels, and not necessarily from the fully open or closed states, suggesting the gating of MAC is somewhat atypical. This channel activity is not overtly voltage dependent. As expected for a mega-channel, MAC is only slightly cation selective with a relative permeability of K+ over Cl− of about 3 (Table 3.1).

The pore size of MAC is estimated to be 4–6 nm in diameter. The method of Hille (2001) relies on maximum conductance, which is the reciprocal of the pore resistance, and we assume a pore length of 5.5 nm (Mannella 1981). Hence, pore size is determined after re-arrangement of Eq. 3.1.

$$Rpore = \rho\left(l / \Pi a^2\right) \tag{3.1}$$

where R and ρ are the resistance of the pore and solution, respectively, l is pore length, and a is pore radius. In addition to the resistance of the pore, there are two components of access resistance for current converging on both sides of the pore. The addition of access resistances is equivalent to making the pore longer by Πa/2 (Hille 2001). Hence the total resistance of the channel would be the sum of the resistance of the pore plus access resistances as shown below in Eq. 3.2.

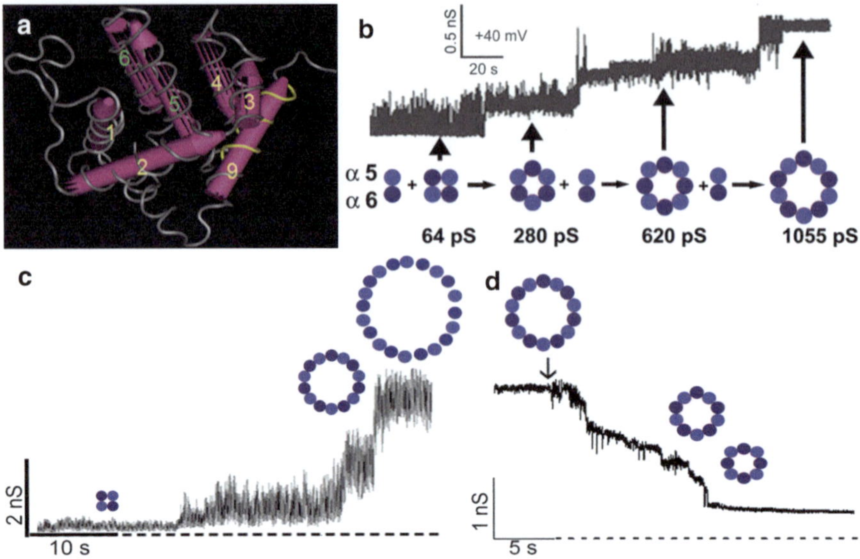

Fig. 3.2 Kinetics of MAC formation (**a**) Structure of Bax with most of the nine helices indicated [(PDB accession code 1F16 (Suzuki et al. 2000)]. Helices 5 and 6 putatively form the staves of the barrel-like pore of MAC. (**b**) Current trace from a liposome patch shows the increase in conductance with time when 380 ng/μl of monomeric BAXΔC and 35 ng/μl tBid are included in the micropipette. The conductance increased in multiple steps of ~0.3 nS until it reached about 1.1 nS. (**c**) Current trace shows putative MAC assembly in a mitochondrion isolated from cultured wild-type fibroblasts and directly patch-clamped using a micropipette whose tip contained 20 nM tBID. Conductance increases stepwise to a maximum of ~4 nS, which would putatively correspond to assembly of ~10 Bax/Bak monomers or 20 helices (Martinez-Caballero et al. 2009). (**d**) Current trace shows MAC recorded in a patch excised from a liposome containing reconstituted mitochondrial outer membranes of apoptotic FL5.12 cells at +20 mV upon perfusion with buffer containing 25 nM iMAC2 (*arrow*). Maximum conductance of ~2 nS would putatively correspond to a minimum of 7 Bax or Bak monomers (~14 helices) and decrease to ~4 monomers. Models near current traces (not drawn to size) show postulated insertion of α−helices (*circles*) that would account for increases in conductance during MAC assembly in panels b and c and possible dissassembly in panel (**d**) (Panels **b**, **c** were modified from research originally published in *J Biol Chem*. Martinez-Caballero et al. (2009). © the American Society for Biochemistry and Molecular Biology; Panel **d** was modified from research originally published in *Biochemical Journal*. Peixoto et al. (2009). © the Biochemical Society)

$$Rchannel = Rpore + Raccess = \left[l + (\Pi a)/2\right]\left(\rho / \Pi a^2\right) \tag{3.2}$$

Both approaches estimate MAC's pore diameter at 4–6 nm suggesting MAC should be large enough to allow passage of folded cytochrome c, which has a diameter of ~3 nm.

If sufficient uncharged molecules enter a pore, displacement of ions will result in a decrease in current flow. Consequently, studies with dextrans (Krasilnikov et al. 1992), polyethylene glycol nano-beads (Martinez-Caballero et al. 2007), and proteins of various molecular weights (Guo et al. 2004) can be used to define pore size. That is, knowledge of the size and shape of molecules that partition into the pore and hence decrease current flow enables estimation of pore size limits. In fact, such studies support the earlier pore sizes interpolated from maximum conductance (Guo et al. 2004). However, the effects of cytochrome c itself on MAC are even more interesting, especially when keeping in mind that MAC's function is to allow permeation of the MOM by this protein.

Cytochrome c is a heme-containing protein normally located in the IMS of mitochondria at a concentration of ~300–700 μM (Gupte and Hackenbrock 1988). This folded protein is released to the cytosol early in apoptosis and becomes an integral part of the caspase-activating complex called the apoptosome. Cytochrome c is positively charged at physiological pH. If cytochrome c passes through MAC in its purported function as a cytochrome c release channel, changes in MAC activity are expected. Cytochrome c (12.4 kDa), but not hemoglobin (64 kDa dimer), induces a voltage dependent transient blockage of MAC, consistent with partitioning of this highly cationic protein into MAC's pore (Guo et al. 2004).

In particular, 100 μM cytochrome c modestly reduces the current flow through MAC. When cytochrome c is introduced in the bath, MAC typically exhibits rectification at pipette negative potentials, because cytochrome c is cationic and is attracted inside the pore at these voltages. These effects of cytochrome c are typically reversible and are accompanied by an increase in noise. In contrast, cytochrome c does not increase noise when added to patches expressing VDAC or Tom40 two other MOM channel activities. Note, in some patches, cytochrome c seems to become "stuck" as large drops in conductance occur that are not always reversible. In summary, the usual effects of cytochrome c on MAC activity include a modest decrease in conductance that is dose- and voltage-dependent, reversible, and accompanied by an increase in noise (Guo et al. 2004). These data are consistent with the release of cytochrome c from the IMS through the pore of MAC.

3.4 The Bcl2 Family of Proteins Is the Rheostat of Intrinsic Apoptosis and MAC Formation

As mentioned earlier, the Bcl2 (B-cell lymphoma 2) family of proteins is central to regulation of the commitment step of intrinsic apoptosis, which is permeabilization of the MOM. This event corresponds to MAC formation and results in release of pro-death factors from the IMS of mitochondria. The Bcl2 family contains both anti- and pro-apoptotic proteins, which interact to suppress each other's action. The anti-apoptotic proteins, e.g. Bcl-2 and Bcl-xL, are pro-survival and their overexpression is linked to cancers. There are two classes of pro-apoptotic proteins, which are the effector (or multi-domain), e.g., Bax and Bak, and the BH3-only proteins, e.g., BID and BAD. BH3-only proteins are activated in response to different cellular stresses and trigger Bax and Bak to form the oligomers that permeabilize the MOM. Over-expression of pro-apoptotic Bcl2 family proteins is associated with cell death and linked to neurodegenerative diseases.

The Bcl2 family of proteins forms a tripartite switch to control apoptosis (Czabotar et al. 2014). The pro-survival proteins sequester both BH3-only and multi-domain pro-apoptotic proteins, thus preventing Bax/Bak oligomerization. Nevertheless, the BH3-only proteins use two mechanisms to initiate apoptosis. They neutralize the pro-survival proteins and directly activate Bax and Bak to form oligomers. This switch is turned on by cellular stresses like DNA damage or lack of growth factors that lead to cleavage (tBID) or dephosphorylation (BAD) of certain BH3-only members. The switch is turned off if a buildup of anti-apoptotic proteins significantly increases the threshold for activation of Bax/Bak by BH3-only proteins.

The proteins of the Bcl2 family are usually either localized at or translocate to mitochondria to perform their apoptotic function. Bak is typically found in its inactive monomeric form inserted in the MOM in complex with VDAC2 (Cheng et al. 2003). Once activated, Bak is released from VDAC2 prior to oligomerization. In contrast, inactive monomers of Bax are usually in the cytosol or loosely bound to the MOM. However, Bax levels in the mitochondria increase dramatically upon activation early in apoptosis (Smaili et al. 2001). Both Bax and Bak undergo conformational changes involving movement of helix 9 when activated by BH3-only proteins that initiate their oligomerization (Gavathiotis et al. 2008).

Interestingly, several Bcl2 family members form ion channels or modify the activity of other channels in artificial membranes. The multi-domain proteins like Bcl-2 and Bax have structures similar to that of diphtheria toxin, which is also known to form ion channels (Antonsson et al. 1997, 2000; Schendel et al. 1998; Schlesinger et al. 1997). As anti-apoptotic proteins prevent mitochondrial permeabilization, it is a bit surprising that Bcl-2 and Bcl-xL form pores in artificial systems (Alavian et al. 2014; Schendel et al. 1997). However, we were unable to attribute any channel activity to Bcl-2 when patch-clamping mitochondria of cells overexpressing this protein (Murphy et al. 2001) suggesting this behavior in bilayers may be limited to reconstituted proteins. Interestingly, overexpression of Bcl-2 modifies the calcium activation of the PTP (permeability transition pore) in the inner membrane. The PTP is another mitochondrial death channel and is linked to necrotic cell death due to ischemia-reperfusion injury in heart (Peixoto et al. 2012; Kinnally and Antonsson 2007; Bernardi and Di Lisa 2015; Karch et al. 2013). More recently, anti-apoptotic Bcl2 family members were found to modify coupling of proton translocation within the ATP synthase, which is a putative pore component of the PTP (Bernardi and Di Lisa 2015; Alavian et al. 2014)).

In contrast to their anti-apoptotic family members, Bax and Bak were expected to form ion channels in artificial systems, in part, because of their putative role in permeabilizing mitochondria (Antignani and Youle 2006). Channel activity similar to MAC is detected in liposomes containing recombinant Bax or Bak that has been mobilized by the activated BH3-only protein tBID (truncated Bid) (Landeta et al. 2011; Martinez-Caballero et al. 2009). Both of these pro-apoptotic proteins form stable channels with long open times and variable pore size. Recombinant Bax and Bak form proteinaceous oligomers, which then induce the formation of lipidic pores capable of releasing large molecules (Gillies et al. 2014; Kuwana et al. 2002). Remarkably, Andrew's group established a battery of fluorescence spectroscopy assays to examine the interactions of Bcl2 family proteins within the membrane while Basanez's group developed assays monitoring the entrance of fluorescently-labeled proteins (~12 kDa cytochrome c and ~100 kDa allophycocyanine) into giant liposomes after addition of activated Bax and Bak (Kale et al. 2014; Schendel et al. 1997). More recent studies using cryo-electron microscopy estimate Bax pore sizes can grow to more than 100 nm in artificial systems (Gillies et al. 2014). Importantly, anti-apoptotic proteins suppress pore formation by these pro-apoptotic proteins (Antonsson et al. 1997).

While artificial systems built with recombinant proteins can go a long way in elucidating mechanism, it is crucial to return to a more native system when validating that mechanism. Interestingly, yeast has made a good model system for Bcl2 family proteins interactions (Gerecova et al. 2013; Priault et al. 2003; Renault et al. 2012). A direct link between MAC and Bax was made through heterologous expression of these proteins in yeast, whose genome does not encode Bcl2 family proteins (Pavlov et al., 2001). MAC activity is recorded from mitochondria isolated from yeast expressing Bax. Finally, co-expression of anti- and pro-apoptotic Bcl2 family members suppresses this yeast MAC activity.

Considerable evidence using several systems indicates Bax and/or Bak are essential components of MAC. Bax antibodies deplete MAC activity from solubilized fractions of MOM isolated from apoptotic cells (Dejean et al. 2005, 2010). These fractions predominantly contain oligomers of Bax. Importantly, genetic ablation of both Bax and Bak or overexpression of Bcl-2 eliminates MAC detection and cytochrome c release (Martinez-Caballero et al. 2009; Pavlov et al. 2001). Interestingly, elimination of both Bax and Bak also modifies the aforementioned PTP channel (Karch et al. 2013). While Bax and Bak remain the principal candidates to form MAC's pore, it is also highly likely that this process is subject to regulation by a variety of additional proteins, which have yet to be identified.

Studies were undertaken to compare the MAC activities of mitochondria isolated from apoptotic cells deficient in one or both of the pro apoptotic proteins Bax and Bak. Patch clamp studies show that MAC does not form and cytochrome c is not released in mitochondria of the double knockout after apoptotic stimulation. Furthermore, the single channel behaviors of MACs containing exclusively Bax or Bak are very similar (Table 3.2) (Martinez-Caballero et al. 2009). Both MAC-Bax and MAC-Bak have a maximum conductance of 2–5 nS and poor ion selectivity (P_K/P_{Cl} ratios of 3–4). Neither channel is voltage dependent. The estimated diameters are 5–6 nm and are purportedly formed by 8–10 molecules of pro-apoptotic protein (see Sect. 3.5).

In summary, the Bcl2 family of proteins regulates apoptosis by controlling the formation of MAC from the multi-domain pro-apoptotic Bax and Bak. Under normal conditions, anti-apoptotic proteins bind to and isolate pro-apoptotic proteins to prevent their interaction. Bax is sequestered as an inactive monomer in the cytosol by anti-apoptotic members while inactive Bak is in the MOM in complex with VDAC2. The pro-apoptotic BH3-only proteins act as guardians monitoring cellular stress levels, e.g. lack of growth factors or DNA damage, and are activated by cleavage (e.g., tBID) or dephosphorylation (e.g., BAD) (Cheng et al. 2001). These activated BH3-only proteins induce conformational changes in Bax and Bak to trigger their oligomerization to form MAC during the commitment step of apoptosis.

Table 3.2 Electrophysiological properties of MAC activity induced by tBid

	Wildtype	Bax KO (MAC-Bak)	Bak KO (MAC-Bax)
Maximum conductance (nS)	3.5 ± 1.8 (n = 7)	3.1 ± 2.0 (n = 8)	3.0 ± 1.9 (n = 8)
Ion selectivity P_K/P_{Cl}	3.1 ± 1.4 (n = 5)	3.7 (n = 1)	3.4 ± 1.2 (n = 4)
Voltage dependent (±50 mV)	No	No	No
Predicted diameter (nm)[a]	6	5.6	5.5
Predicted #helices forming Pore[b]	18	17	17

Modified from research originally published in *J Biol Chem*. Martinez-Caballero et al. (2009).
n number independent determinations, *KO* knockout
[a]Estimated from maximum conductance measured in 150 mM KCl using Eq. 3.1
[b]Based on maximum conductance assuming 1.2 nm diameter α helices are inserted normal to the membrane to form a circular pore with a length of 5.5 nm

3.5 Monitoring MAC Assembly in the Mitochondrial Outer Membrane

MAC assembly can be induced and monitored in *real time* as an increase in current while patch clamping isolated mitochondria with a micropipette containing tBid (Fig. 3.2) (Martinez-Caballero et al. 2009). The process has a time course of minutes and hence is amenable to further scrutiny. Mathematical analysis of the stepwise changes in conductance elicited by the action of tBid on Bax and/or Bak (Fig. 3.2b–d) is consistent with a barrel-stave model of pore assembly (Laver 1994; Sansom et al. 1991). Bax and Bak have nine α-helices (Fig. 3.2). Studies indicate α-helices 5 and 6 of Bax and Bak are the most likely candidates to form the transmembrane domains of MAC (Lovell et al. 2008). Assuming the barrel-staves are the two transmembrane α-helices in Bax and Bak, mature MAC pores with calculated diameters of 5–6 nm would typically be oligomers made of ~9 monomers (Fig. 3.2) (Martinez-Caballero et al. 2009), based on the calculations below (Eq. 3.3).

The number of helices forming MAC is estimated assuming α-helices have a diameter (D) of 1.2 nm and are aligned on center perpendicular to the membrane (Martinez-Caballero et al. 2009). The resulting polygon has a variable number (n) of vertices, where the center of each helix is located at a vertex of the polygon. The pore's cross-sectional area (A) is approximated by a circle whose area is calculated using Eq. 3.3 as more helical staves are added to the polygon in an approach similar to Sansom et al. (1991).

$$A = \pi D^2 \left[\frac{1}{\sqrt{2 - 2\cos\left(\frac{2\pi}{n}\right)}} - \frac{1}{2} \right]^2 \tag{3.3}$$

The diameter of the pore predicted from the conductance can be used to calculate the area A (Eq. 3.2). The number n of helices forming the pore wall can be derived from Eq. 3.3 using the values of A corresponding to an increasingly larger pore.

The reader is referred to the original citations for more details on the mathematical derivations and calculations (Martinez-Caballero et al. 2009). The work established the relationship between the conductance and number of α-helices forming the MAC pore. Furthermore, our findings indicate MAC assembly happens within the native membrane, as opposed to pre-assembly in the cytosol. If MAC were to be assembled prior to insertion into the outer membrane, current transitions of several nS would be expected rather than the stepwise increments we observed. These findings also suggest that at least some of the giant transitions in current observed actually correspond to the addition and subtraction of helices rather than classical gating.

3.6 The Therapeutic Potential of Controlling MAC

Identifying the components of MAC and understanding the mechanisms underlying intrinsic apoptosis have gone a long way in increasing our appreciation of tissue homeostasis and maintaining the adult human cell count at about ten trillion. Nevertheless, the generation and characterization of factors/chemicals that will enable either indirect or direct control of MAC to turn on or turn off cell death processes has inherent value. Specific inhibitors of MAC have the potential to help identify modulators and enable a further functional dissection of apoptosis. Importantly, the ability to turn MAC on in cancer cells or turn MAC off in cells subject to degenerative diseases makes this channel a significant pharmacological target.

Further increasing its potential therapeutic impact, MAC-based cell death has been linked to the phenomena of bystander effects, a collateral spread of apoptosis to nearby cells that is integral to tissue homeostasis and a challenge to anticancer therapies (Frank et al. 2005). We found that MAC assembly, or mimicking MAC assembly by microinjection of cytochrome c, induces a bystander effect in many cell types. For example, microinjection of cytochrome c into a single cell causes a wave of death across an entire *Xenopus laveis* embryo (Peixoto et al. 2011). Death signals may be transmitted through intercellular channels as a bystander effect is suppressed in some cell types, e.g. osteocytes, by the gap junction intercellular communication blocker 18-β glycyrrhetinic (Peixoto et al. 2009b). In contrast extracellular secretion of death signals might underlie bystander effects in breast cancer cell lines (Das et al. 2007). Furthermore, expression of GFP-Bax induces MAC formation and apoptosis in transfected cells as well as a bystander effect in nearby cells where GFP fluorescence is not detectable. Importantly, this MAC-induced bystander effect is Bax/Bak-dependent, as it does not occur in cells lacking both proteins (Peixoto et al. 2011). Hence, developing compounds that can regulate MAC has significant therapeutic applications.

Indirect approaches to activating MAC for chemotherapies currently focus on BH3-only mimetics as these proteins inhibit anti-apoptotic and activate pro-apoptotic proteins. Inhibition of anti-apoptotic proteins lowers the apoptotic threshold to induce death. Novel inhibitors, e.g., ABT267 which inhibits Bcl-2 and Bcl-xL, are proving to be effective chemotherapies with minimal side effects (Tse et al. 2008). Similarly, small molecules are being developed that target the activation of Bax/Bak to turn on the cell death process (Czabotar et al. 2014). It is expected that these molecules would act like tBid in directly activating MAC formation in patch clamp experiments (Fig. 3.2c).

The search for direct blockers of MAC commenced soon after its discovery because of their potential value in treating neurodegenerative diseases such as Parkinson's or Alzheimer's. In our first study, we applied agents known to modify cell death to membrane patches expressing MAC activity. While lidocaine has no effect, the local anesthetic dibucaine induces a dose-dependent fast blockade of MAC with an IC_{50} of 39 μM (Martinez-Caballero et al. 2004). In contrast, the IC_{50} for the β-blocker propranolol and anti-dopaminergic trifluoperazine are 52 and 0.9 μM, respectively; these drugs likely destabilize the open state of MAC. Unlike dibucaine, the effects of propranolol and trifluoperazine on MAC are not reversible. Cyclosporine A is a classical inhibitor of the permeability transition and some types of necrotic cell death. However, this drug has no effect on MAC activity at concentrations as high as 10 μM (Martinez-Caballero et al. 2004).

Antonsson's group took a somewhat different approach to identifying potential MAC inhibitors. They screened compounds that modified the Bax-induced release of carboxyfluorescein from liposomes and referred to them as "bci" or Bax channel inhibitors (Bombrun et al. 2003). Later, they described the ability of these compounds to suppress apoptosis and protect neurons in a model of global brain ischemia (Hetz et al. 2005). We found bci compounds worked equally well in inhibiting MAC activity in patch clamp experiments (Peixoto et al. 2009a).

Several additional compounds, some of which are structurally related, were more recently assessed for their ability to suppress apoptosis by closing MAC and, hence, the pathway for cytochrome c release (Fig. 3.3) (Peixoto et al. 2009a). These compounds are now called iMACs or inhibitors of MAC. Recent derivatives included replacement of a hydroxyl with a fluoride moiety in the propyl linker to produce iMAC2 from iMAC1. Then, iMAC3 was generated by addition of a propylbenzene on the piperazine group. Both iMAC2 and iMAC3 specifically suppressed the onset of apoptotic markers by blocking MAC and suppressing cytochrome c release (Peixoto et al., unpublished results). Thus, iMAC2 and iMAC3 mimicked the behavior of Bax/Bak-double knockouts in which MAC activity, cytochrome c release and apoptosis were eliminated and cell death was delayed. These results confirm the tight link between MAC activity, cytochrome c release and apoptotic death. Furthermore, the lethal dose (LD_{50}) and the effective MAC-blocking dose (IC_{50}) of the iMACs are vastly different, reinforcing their promising therapeutic potential. Several derivatives of these agents have been studied in models of brain injury and neurodegeneration and a novel means of screening for cell death inhibitors in live animals has been described (Pieper et al. 2014). These approaches will be useful in extending the pharmacology of MAC in the future.

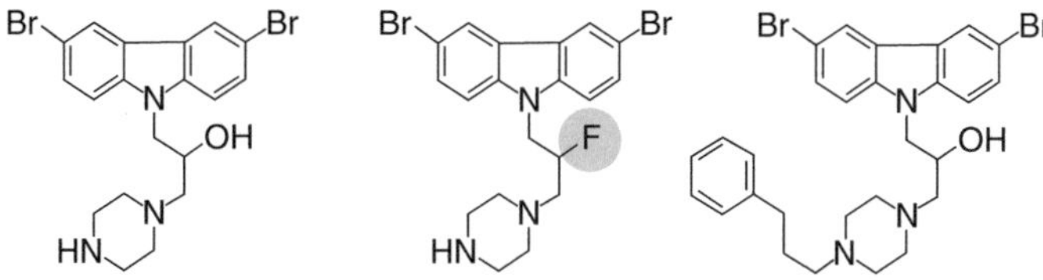

	iMAC1	iMAC2	iMAC3
IC_{50} (nM)	300	28	19
LD_{50} (nM)	50,000	15,000	15,000

Fig. 3.3 Structures of Inhibitors of MAC (iMAC) compounds. The design of iMAC2 and iMAC3 are based on the parent compound iMAC1. iMAC2 has a fluoride instead of a hydroxyl group in the propanol linker while iMAC3 has a propylbenzene on the piperazine group. The IC_{50} determined in patch clamp studies of MAC and the LD_{50} determined in cultured cells are indicated below each compound's structure (This research was in part originally published in *Biochemical Journal*. Peixoto et al. (2009a). © the Biochemical Society)

3.7 Summary

In summary, MAC is a mega-channel that assembles in the outer membrane of mitochondria from the pro-apoptotic proteins Bax and/or Bak in response to apoptotic signals. Typically, MAC must assemble 8–10 molecules of Bax and/or Bak to achieve a pore size sufficient to release folded 3 nm cytochrome c from the space between the mitochondrial outer and inner membranes. This release triggers intrinsic apoptosis and potentially a bystander effect in some cell types. A variety of compounds have become available to activate or block MAC function and these are presently being used to further dissect the apoptotic process. MAC is a novel therapeutic target for a variety of diseases including cancer, infarcts, and neurodegenerative diseases.

Bibliography

Alavian KN, Beutner G, Lazrove E et al (2014) An uncoupling channel within the c-subunit ring of the F1FO ATP synthase is the mitochondrial permeability transition pore. Proc Natl Acad Sci U S A 111(29):10580–10585. doi:10.1073/pnas.1401591111

Antignani A, Youle RJ (2006) How do Bax and Bak lead to permeabilization of the outer mitochondrial membrane? Curr Opin Cell Biol 18(6):685–689. doi:10.1016/j.ceb.2006.10.004

Antonsson B, Conti F, Ciavatta A et al (1997) Inhibition of Bax channel-forming activity by Bcl-2. Science 277(5324):370–372

Antonsson B, Montessuit S, Lauper S et al (2000) Bax oligomerization is required for channel-forming activity in liposomes and to trigger cytochrome c release from mitochondria. Biochem J 345(Pt 2):271–278

Bernardi P, Di Lisa F (2015) The mitochondrial permeability transition pore: Molecular nature and role as a target in cardioprotection. J Mol Cell Cardiol 78C:100–106. doi:10.1016/j.yjmcc.2014.09.023

Bombrun A, Gerber P, Casi G et al (2003) 3,6-dibromocarbazole piperazine derivatives of 2-propanol as first inhibitors of cytochrome c release via Bax channel modulation. J Med Chem 46(21):4365–4368. doi:10.1021/jm034107j

Cheng EH, Wei MC, Weiler S et al (2001) BCL-2, BCL-X(L) sequester BH3 domain-only molecules preventing BAX- and BAK-mediated mitochondrial apoptosis. Mol Cell 8(3):705–711
Cheng EH, Sheiko TV, Fisher JK et al (2003) VDAC2 inhibits BAK activation and mitochondrial apoptosis. Science 301(5632):513–517. doi:10.1126/science.1083995
Colombini M, Mannella CA (2012) VDAC, the early days. Biochim Biophys Acta 1818(6): 1438–1443. doi:10.1016/j.bbamem.2011.11.014
Criado M, Keller BU (1987) A membrane fusion strategy for single-channel recordings of membranes usually non-accessible to patch-clamp pipette electrodes. FEBS Lett 224(1):172–176
Czabotar PE, Lessene G, Strasser A et al (2014) Control of apoptosis by the BCL-2 protein family: implications for physiology and therapy. Nat Rev Mol Cell Biol 15(1):49–63. doi:10.1038/nrm3722
Das S, Nwachukwu JC, Li D et al (2007) The nuclear receptor interacting factor-3 transcriptional coregulator mediates rapid apoptosis in breast cancer cells through direct and bystander-mediated events. Cancer Res 67(4):1775–1782. doi:10.1158/0008-5472.CAN-06-4034
Dejean LM, Martinez-Caballero S, Guo L et al (2005) Oligomeric Bax is a component of the putative cytochrome c release channel MAC, mitochondrial apoptosis-induced channel. Mol Biol Cell 16(5):2424–2432. doi:10.1091/mbc.E04-12-1111
Dejean LM, Ryu SY, Martinez-Caballero S et al (2010) MAC and Bcl-2 family proteins conspire in a deadly plot. Biochim Biophys Acta 1797(6-7):1231–1238. doi:10.1016/j.bbabio.2010.01.007
Frank DK, Szymkowiak B, Josifovska-Chopra O et al (2005) Single-cell microinjection of cytochrome c can result in gap junction-mediated apoptotic cell death of bystander cells in head and neck cancer. Head Neck 27(9):794–800. doi:10.1002/hed.20235
Gavathiotis E, Suzuki M, Davis ML et al (2008) BAX activation is initiated at a novel interaction site. Nature 455(7216):1076–1081. doi:10.1038/nature07396
Gerecova G, Kopanicova J, Jaka P et al (2013) BH3-only proteins Noxa, Bik, Bmf, and Bid activate Bax and Bak indirectly when studied in yeast model. FEMS Yeast Res 13(8):747–754. doi:10.1111/1567-1364.12074
Gillies LA, Du H, Peters B et al (2014) Visual and functional demonstration of growing Bax-induced pores in mitochondrial outer membranes. Mol Biol Cell. doi:10.1091/mbc.E13-11-0638
Guo L, Pietkiewicz D, Pavlov EV et al (2004) Effects of cytochrome c on the mitochondrial apoptosis-induced channel MAC. Am J Physiol Cell Physiol 286(5):C1109–C1117. doi:10.1152/ajpcell.00183.2003
Gupte SS, Hackenbrock CR (1988) Multidimensional diffusion modes and collision frequencies of cytochrome c with its redox partners. J Biol Chem 263(11):5241–5247
Hamill OP, Marty A, Neher E et al (1981) Improved patch-clamp techniques for high-resolution current recording from cells and cell-free membrane patches. Pflugers Arch 391(2):85–100
Hetz C, Vitte PA, Bombrun A et al (2005) Bax channel inhibitors prevent mitochondrion-mediated apoptosis and protect neurons in a model of global brain ischemia. J Biol Chem 280(52):42960–42970. doi:10.1074/jbc.M505843200
Hille B (2001) Ion channels of excitable membranes, 3rd edn. Sinauer Associates, Inc., Sunderland
Jonas EA, Buchanan J, Kaczmarek LK (1999) Prolonged activation of mitochondrial conductances during synaptic transmission. Science 286(5443):1347–1350
Kale J, Chi X, Leber B et al (2014) Examining the molecular mechanism of bcl-2 family proteins at membranes by fluorescence spectroscopy. Methods Enzymol 544:1–23. doi:10.1016/B978-0-12-417158-9.00001-7
Karch J, Kwong JQ, Burr AR et al (2013) Bax and Bak function as the outer membrane component of the mitochondrial permeability pore in regulating necrotic cell death in mice. Elife 2:e00772. doi:10.7554/eLife.00772
Kinnally KW, Antonsson B (2007) A tale of two mitochondrial channels, MAC and PTP, in apoptosis. Apoptosis 12(5):857–868. doi:10.1007/s10495-007-0722-z
Kinnally KW, Campo ML, Tedeschi H (1989) Mitochondrial channel activity studied by patch-clamping mitoplasts. J Bioenerg Biomembr 21(4):497–506
Kinnally KW, Zorov D, Antonenko Y et al (1991) Calcium modulation of mitochondrial inner membrane channel activity. Biochem Biophys Res Commun 176(3):1183–1188

Krasilnikov OV, Sabirov RZ, Ternovsky VI et al (1992) A simple method for the determination of the pore radius of ion channels in planar lipid bilayer membranes. FEMS Microbiol Immunol 5(1–3):93–100

Kuwana T, Mackey MR, Perkins G et al (2002) Bid, Bax, and lipids cooperate to form supramolecular openings in the outer mitochondrial membrane. Cell 111(3):331–342

Landeta O, Landajuela A, Gil D et al (2011) Reconstitution of proapoptotic BAK function in liposomes reveals a dual role for mitochondrial lipids in the BAK-driven membrane permeabilization process. J Biol Chem 286(10):8213–8230. doi:10.1074/jbc.M110.165852

Latorre R, Miller CG, Quay S (1981) Voltage-dependent conductance induced by alamethicin-phospholipid conjugates in lipid bilayers. Biophys J 36(3):803–809. doi:10.1016/S0006-3495(81)84767-4

Laver DR (1994) The barrel-stave model as applied to alamethicin and its analogs reevaluated. Biophys J 66(2 Pt 1):355–359

Lohret TA, Jensen RE, Kinnally KW (1997) Tim23, a protein import component of the mitochondrial inner membrane, is required for normal activity of the multiple conductance channel, MCC. J Cell Biol 137(2):377–386

Lovell JF, Billen LP, Bindner S et al (2008) Membrane binding by tBid initiates an ordered series of events culminating in membrane permeabilization by Bax. Cell 135(6):1074–1084. doi:10.1016/j.cell.2008.11.010

Mannella CA (1981) Structure of the outer mitochondrial membrane: analysis of X-ray diffraction from the plant membrane. Biochim Biophys Acta 645(1):33–40

Mapingire OS, Wager B, Delcour AH (2013) Electrophysiological characterization of bacterial pore-forming proteins in planar lipid bilayers. Methods Mol Biol 966:381–396. doi:10.1007/978-1-62703-245-2_24

Martinez-Caballero S, Dejean LM, Kinnally KW (2004) Some amphiphilic cations block the mitochondrial apoptosis-induced channel, MAC. FEBS Lett 568(1-3):35–38. doi:10.1016/j.febslet.2004.05.006

Martinez-Caballero S, Dejean LM, Jonas EA et al (2005) The role of the mitochondrial apoptosis induced channel MAC in cytochrome c release. J Bioenerg Biomembr 37(3):155–164. doi:10.1007/s10863-005-6570-z

Martinez-Caballero S, Grigoriev SM, Herrmann JM et al (2007) Tim17p regulates the twin pore structure and voltage gating of the mitochondrial protein import complex TIM23. J Biol Chem 282(6):3584–3593. doi:10.1074/jbc.M607551200

Martinez-Caballero S, Dejean LM, Kinnally MS et al (2009) Assembly of the mitochondrial apoptosis-induced channel, MAC. J Biol Chem 284(18):12235–12245. doi:10.1074/jbc.M806610200

Muro C, Grigoriev SM, Pietkiewicz D et al (2003) Comparison of the TIM and TOM channel activities of the mitochondrial protein import complexes. Biophys J 84(5):2981–2989. doi:10.1016/S0006-3495(03)70024-1

Murphy RC, Schneider E, Kinnally KW (2001) Overexpression of Bcl-2 suppresses the calcium activation of a mitochondrial megachannel. FEBS Lett 497(2–3):73–76

Pavlov EV, Priault M, Pietkiewicz D et al (2001) A novel, high conductance channel of mitochondria linked to apoptosis in mammalian cells and Bax expression in yeast. J Cell Biol 155(5):725–731. doi:10.1083/jcb.200107057

Peixoto PM, Grana F, Roy TJ et al (2007) Awaking TIM22, a dynamic ligand-gated channel for protein insertion in the mitochondrial inner membrane. J Biol Chem 282(26):18694–18701. doi:10.1074/jbc.M700775200

Peixoto PM, Ryu SY, Bombrun A et al (2009a) MAC inhibitors suppress mitochondrial apoptosis. Biochem J 423(3):381–387. doi:10.1042/BJ20090664

Peixoto PM, Ryu SY, Pruzansky DP et al (2009b) Mitochondrial apoptosis is amplified through gap junctions. Biochem Biophys Res Commun 390(1):38–43. doi:10.1016/j.bbrc.2009.09.054

Peixoto PM, Lue JK, Ryu SY et al (2011) Mitochondrial apoptosis-induced channel (MAC) function triggers a Bax/Bak-dependent bystander effect. Am J Pathol 178(1):48–54. doi:10.1016/j.ajpath.2010.11.014

Peixoto PM, Dejean LM, Kinnally KW (2012) The therapeutic potential of mitochondrial channels in cancer, ischemia-reperfusion injury, and neurodegeneration. Mitochondrion 12(1):14–23. doi:10.1016/j.mito.2011.03.003

Pieper AA, McKnight SL, Ready JM (2014) P7C3 and an unbiased approach to drug discovery for neurodegenerative diseases. Chem Soc Rev 43(19):6716–6726. doi:10.1039/c3cs60448a

Priault M, Camougrand N, Kinnally KW et al (2003) Yeast as a tool to study Bax/mitochondrial interactions in cell death. FEMS Yeast Res 4(1):15–27

Renault TT, Grandier-Vazeille X, Arokium H et al (2012) The cytosolic domain of human Tom22 modulates human Bax mitochondrial translocation and conformation in yeast. FEBS Lett 586(2):116–121. doi:10.1016/j.febslet.2011.12.003

Sansom MS, Kerr ID, Mellor IR (1991) Ion channels formed by amphipathic helical peptides. A molecular modelling study. Eur Biophys J 20(4):229–240

Schein SJ, Colombini M, Finkelstein A (1976) Reconstitution in planar lipid bilayers of a voltage-dependent anion-selective channel obtained from paramecium mitochondria. J Membr Biol 30(2):99–120

Schendel SL, Xie Z, Montal MO et al (1997) Channel formation by antiapoptotic protein Bcl-2. Proc Natl Acad Sci U S A 94(10):5113–5118

Schendel SL, Montal M, Reed JC (1998) Bcl-2 family proteins as ion-channels. Cell Death Differ 5(5):372–380. doi:10.1038/sj.cdd.4400365

Schlesinger PH, Gross A, Yin XM et al (1997) Comparison of the ion channel characteristics of proapoptotic BAX and antiapoptotic BCL-2. Proc Natl Acad Sci U S A 94(21):11357–11362

Smaili SS, Hsu YT, Sanders KM et al (2001) Bax translocation to mitochondria subsequent to a rapid loss of mitochondrial membrane potential. Cell Death Differ 8(9):909–920. doi:10.1038/sj.cdd.4400889

Sorgato MC, Keller BU, Stuhmer W (1987) Patch-clamping of the inner mitochondrial membrane reveals a voltage-dependent ion channel. Nature 330(6147):498–500. doi:10.1038/330498a0

Suzuki M, Youle RJ, Tjandra N (2000) Structure of Bax: coregulation of dimer formation and intracellular localization. Cell 103(4):645–654

Szabo I, Zoratti M (1992) The mitochondrial megachannel is the permeability transition pore. J Bioenerg Biomembr 24(1):111–117

Tse C, Shoemaker AR, Adickes J et al (2008) ABT-263: a potent and orally bioavailable Bcl-2 family inhibitor. Cancer Res 68(9):3421–3428. doi:10.1158/0008-5472.CAN-07-5836

Chapter 4
Ceramide Channels

Marco Colombini

Abstract Ceramide forms large stable channels capable of translocating proteins across membranes. These channels are dynamic structures that exist in a constant state of assembly and disassembly. Electrophysiological and electron microscopic evidence strongly supports the formation of cylindrical barrel-stave channels with the staves oriented in an anti-parallel fashion. Molecular-dynamic simulations concur that the proposed structure is stable and viable. Ceramide channels are regulated by Bcl-2 family proteins with anti-apoptotic proteins interacting with the hydrophobic regions and thus destabilize the channels and pro-apoptotic proteins acting synergistically with ceramide to permeabilize membranes. Channel formation is not very sensitive to the precise structure of ceramide and will tolerate many chemical changes, probably producing channels with a somewhat different conformation. The ceramide channel itself behaves as an elastic structure capable of being deformed and of spontaneously restoring its circular form.

Keywords Ceramide • Sphingolipid • Apoptosis • Bcl-2 • Bcl-xL • Bax • CED-9 • Planar membrane • Channel • Pore • Mitochondrion • Outer membrane • Lanthanum • Channel dynamics • Channel regulation • Ceramide analogs

4.1 Introduction: Ceramides, Structural and Functional Diversity

The term, ceramide, is used to refer to a large family of waxy (cera in Latin) compounds that contain an amide linkage. The canonical form of ceramide is sphingosine amide linked to a fatty acid (Fig. 4.1). Both of these constituents exist in various altered forms (Pruett et al. 2008; Hannun and Obeid 2011). When the alterations are too drastic the compounds are given different names, for example: lactariamides (Yue et al. 2001), candidamides (Wu et al. 2009), capnoids (Godchaux and Leadbetter 1983). Ceramides serve many functions. The higher molecular weight forms, and those most hydrophobic are found in skin and contribute to the

M. Colombini
Department of Biology, University of Maryland, College Park, MD 20742, USA
e-mail: colombini@umd.edu

A.H. Delcour (ed.), *Electrophysiology of Unconventional Channels and Pores*,
Springer Series in Biophysics 18, DOI 10.1007/978-3-319-20149-8_4

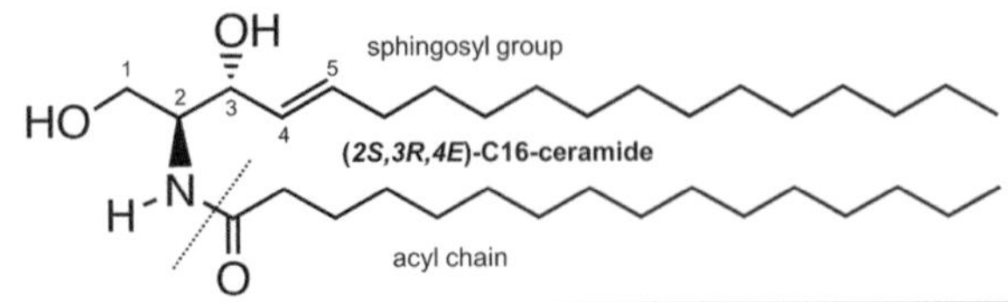

Fig. 4.1 Illustration of the form of ceramide most often studied with regard to the formation of ceramide channels. The sphingosyl group (*upper portion*) is amide linked to a fatty acid, in this case it is palmitic acid (*lower portion*). Carbons 2 and 3 on the sphingosyl group are chiral and there is a trans double bond between carbons 4 and 5

permeability barrier of skin (Meckfessel and Brandt 2014). In cells, ceramides and their derivatives contribute to many essential functions including cell proliferation, differentiation and apoptosis (Zheng et al. 2006). Even this subgroup consists of over 50 different chemical structures (Hannun and Obeid 2008, 2011). The most common structures consist of either sphingosine (containing a trans double bond) or sphinganine (the saturated version) amide linked to saturated fatty acids ranging from palmitate (16 carbons) to lingnocerate (24 carbons). This variety is not produced by accident but determined in large part by the activity of a family of ceramide synthases (Grosch et al. 2012) with different fatty acid specificity. Relatively small differences in the structure of the ceramide can result in profound differences in function. For example the saturated form, dihydroceramide, is not apoptogenic (Bielawska et al. 1993; Obeid et al. 1993) whereas the form containing the trans double bond is apoptogenic (Siskind 2005). Furthermore the former generates channels in phospholipid membranes whereas the latter does not (Siskind and Colombini 2000; Siskind et al. 2006). Clearly, the evolution of the ability to generate a variety of structures in a controlled fashion by investing in a substantial array of proteins highlights the importance of the ceramide family in cellular physiology.

For clarity, in this review the term ceramide will refer to C_{16}-ceramide (Fig. 4.1) where sphingosine is amide linked to the 16 carbon fatty acid, palmitate. More precisely the chemical name is *N*-palmitoyl-D-*erythro*-sphingosine. Please note that there are two chiral centers and that the chiral nature of the molecule is often important. Another frequently used ceramide is C_2-ceramide because of its significant solubility in water due to the short 2-carbon fatty acid moiety, acetate. This mimics some properties of the longer chain-length ceramides but not others.

Another level of complexity in understanding the role of ceramide in cells is its compartmentation within membranes. Like the phosphoglycerolipids, usually just referred to as phospholipids, the virtual insolubility of ceramide in water means that the membrane content of ceramide varies from membrane to membrane in any cell. The synthetic and hydrolytic enzymes also vary in their location. There are specific transport mechanisms to move ceramide from one membrane to another (Hanada 2014). Thus measurements of whole cell ceramide levels are not very useful. As with other lipids, it is likely that ceramide may be asymmetrically distributed between the two membrane leaflets thus injecting some uncertainty even into measurements of mole fraction of ceramide in one membrane type. Finally, more uncertainly exists if one seeks to know how ceramide levels in a given membrane result in a physiological function, such as self-assembly into channels, because the activity of ceramide in a membrane also depends on its interactions with other membrane constituents. This does not only refer to the formation of separate phases due to its higher melting point than that of the typical membrane lipids but also interactions with other lipids.

4.2 Ceramide Permeabilizes Membranes by Forming Organized Structures

It has long been known that elevating ceramide levels in the mitochondrial outer membrane (MOM) results in protein release from the mitochondrial intermembrane space to the cytosol (Ghafourifar et al. 1999; Di Paola et al. 2000; Birbes et al. 2001; Siskind et al. 2002, 2006), although the nature of this release pathway is still debated. That this is a bidirectional flow was demonstrated (Siskind et al. 2002) by recording the rate of oxidation of exogenously-added reduced cytochrome c. The latter must traverse the MOM to be oxidized by cytochrome oxidase on the outer surface of the inner membrane but then be released into the medium so that all the reduced cytochrome c can be oxidized. These findings are consistent with either the formation of a channel in the MOM or the physical damage of that membrane. Evidence for the former conclusion was reported (Siskind et al. 2002) by observing that the permeabilization of the MOM could be reversed by the removal of short-chain ceramide (C_2-ceramide). This finding was consistent with the discovery that ceramide can permeabilize phospholipid membranes in the absence of any proteins (Siskind and Colombini 2000; Siskind et al. 2003; Montes et al. 2002; Pajewski et al. 2005; Stiban et al. 2006). Thus, not only is the permeabilization of the MOM probably the result of the formation of channels but these channels need not involve the use of proteins because ceramide alone can form such channels. That is not to say that proteins are not involved because, indeed, there is evidence to the contrary (Siskind et al. 2008; Ganesan et al. 2010).

There is agreement among research groups that the pathways formed by ceramide in phospholipid membranes are large (Siskind et al. 2002, 2003; Montes et al. 2002; Samanta et al. 2011) and thus require the cooperation of many ceramide molecules. However, there is no general agreement on the physical nature of these pathways because different groups use vastly different mole fractions of ceramide in the phospholipid membrane. When unphysiologically high mole fractions of ceramide are used to make liposomal membranes containing ceramide, under these conditions ceramide forms separate lipid phases (Montes et al. 2002). These workers have proposed that the interface between these lipid phases may result in defects that allow polar molecules to cross the membrane. If such defects do exist, their structure is unclear. In any case, these structures are totally unrelated to the ceramide channels that are discussed in this review. Physiological levels of ceramide in the MOM are at mole fractions of 0.5 % or less (Garcia-Ruiz et al.

1997; Rodriguez-Lafrasse et al. 2002; Birbes et al. 2005) even when elevated by the induction of apoptosis. These levels are below those that form separate phases because at low mole fractions entropy wins out. It is at these levels that ceramide can form highly organized and stable cylindrical channels capable of allowing proteins to cross membranes (Siskind et al. 2006; Samanta et al. 2011). These ceramide channels interact with and are regulated by Bcl-2 family proteins in ways consistent with their playing important roles in the early, decision-making stages of apoptosis (Siskind et al. 2008, 2010; Ganesan et al. 2010; Perera et al. 2012b).

4.3 Columns and Barrels: A Working Model of the Channel

Like the typical phospholipids, ceramide molecules are amphipathic with a polar head, two hydroxyl groups and the amide linkage, and two long aliphatic chains (Fig. 4.1). Naturally the polar head would be expected to both line the inner wall of a membrane channel and act as an organizing center because of its hydrogen bonding capacity. The first proposed structure (Siskind and Colombini 2000) took advantage of the organizing ability of the amide linkage, responsible for the secondary structure of proteins, to produce ceramide columns that would be the staves of a barrel-like channel. Such a structure would produce a polar lining that would hydrogen-bond with the water in the channel and would be of variable size depending on the number of columns incorporated into the structure. The structure could grow and shrink by adding or removing columns. This dynamic structure of variable size is in agreement with experimental observations (Siskind and Colombini 2000; Siskind et al. 2003). Columns consisting of six ceramide monomers would span the 3.5–4 nm hydrophobic portion of the membrane but would generate rings of exposed hydrophobic ceramide tails at each end of the channel. These would need to be covered by a distortion of the phospholipid bilayer. Of course such a distortion would generate local stress on the phospholipids and this stress would need to be balanced by an equal and opposite stress in the ceramide channel. Molecular dynamic simulations (Anishkin et al. 2006) of a small ceramide channel that started with straight columns quickly transformed into columns with a positive curvature in an orientation perpendicular to the plane of the membrane. Naturally the channel would have a negative curvature in the plane of the membrane. Thus a structure with dual curvature forms and is stable indefinitely in the simulation.

Figure 4.2 illustrates the columns, how they could be arranged in the channel, and how they may interface with the phospholipid bilayer to form a continuous polar surface capable of polar interactions with water. The ceramide columns are asymmetric; having a vectorial orientation, both structurally and from the point of view of charge separation. Like the alpha helix, the amide linkages are aligned to form a strong dipole. Initially the columns were arranged in a parallel fashion but this was changed to an antiparallel structure to be in harmony with the results of electrophysiological experiments (Siskind et al. 2003).

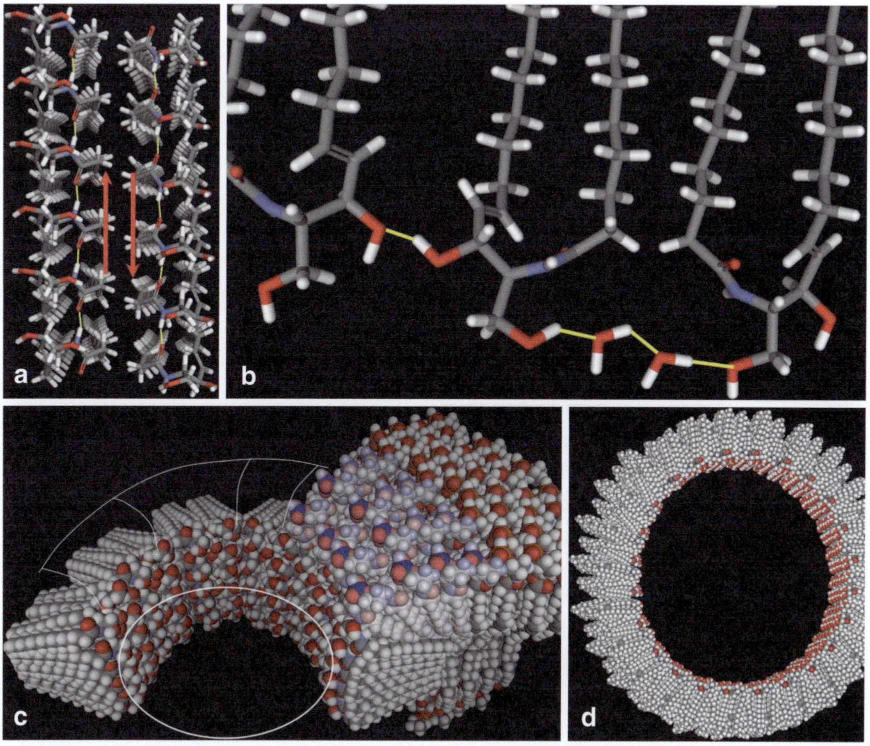

Fig. 4.2 Four views of various aspects of a model of a ceramide channel. (**a**) Two anti-parallel ceramide columns are illustrated. Ceramide columns are formed by six ceramide molecules held together by hydrogen bonding (*yellow*) between adjacent amide linkages. The antiparallel arrangement results in favorable dipole-dipole interactions (*red arrows*) between adjacent columns. (**b**) Hydrogen bonds (*yellow*) connect adjacent columns at the ceramide-water interface, either directly (*left*) or via bridging water molecules (*right*). (**c**) Columns form the wall of the channel and interface with the phospholipid bilayer. The lighter colored phospholipids are shown as curving to meet the channel and cover the aliphatic chains of ceramide. The columns take on a positive curvature, sharing the interfacial strain. (**d**) *Top view* of a channel forming a 10 nm pore, a typical size. The wall consists of 48 columns arranged in an antiparallel fashion

Although ceramide channels vary widely in terms of size, the typical ceramide channel is composed of hundreds of monomers. These are held together by hydrogen bonds and thus would be expected to be in dynamic equilibrium with ceramide monomers or ceramide aggregates in the membrane. Although no evidence exists, it is expected that ceramide molecules in a leaflet of a membrane would hydrogen bond and form columns and these would be in dynamic equilibrium, growing and shrinking in length and perhaps associating with other columns. Groups of columns could then spontaneously form a channel or they could add to an existing channel. Siskind et al. (2003) observed conductance changes consistent with groups of columns adding to and leaving a channel as a unit. Certainly the high degree of insolubility of the medium (C_{16}-, C_{18}-ceramide) and long-chain ceramides (C_{24}-ceramide) makes it very unlikely that ceramide monomers would leave the membrane for the aqueous phase. Thus any gain or loss of columns would come from or return to leaflets of the membrane. The short chain ceramide, C2-ceramide, is somewhat water soluble and there is evidence that it can leave the membrane, binding to proteins in solution. This property was exploited to demonstrate that ceramide channels can disassemble and thus restore the impermeability of the MOM (Siskind et al. 2002).

4.4 Electrophysiological and Electronmicroscopic Evidence for the Structure of the Channel

The rapid dispersal of ceramide dissolved in DMSO or isopropanol into an aqueous phase containing a membrane results in some of the ceramide entering the membrane. If enough ceramide enters the membrane it forms discrete conductance increments that are characteristic of channels. These are best observed when formed in a planar phospholipid membrane under voltage clamp conditions. The conductance increments vary greatly in size and in the first publication on ceramide channels (Siskind and Colombini 2000) these increments were interpreted to be individual channels. This is the standard interpretation. However, subsequent research demonstrated that the conductance increments and decrements represented changes in the permeability (most probably size) of one channel-forming structure (Siskind et al. 2003). Figure 4.3 shows the electrophysiological recording of a typical experiment. Following the dispersal of the ceramide, there is usually a lag phase followed by conductance increments that represent the growth of a single channel. The left axis shows both the measured conductance and the calculated diameter of the channel assuming a right cylinder. The conductance typically reaches a steady value that varies from experiment to experiment as shown in the inset.

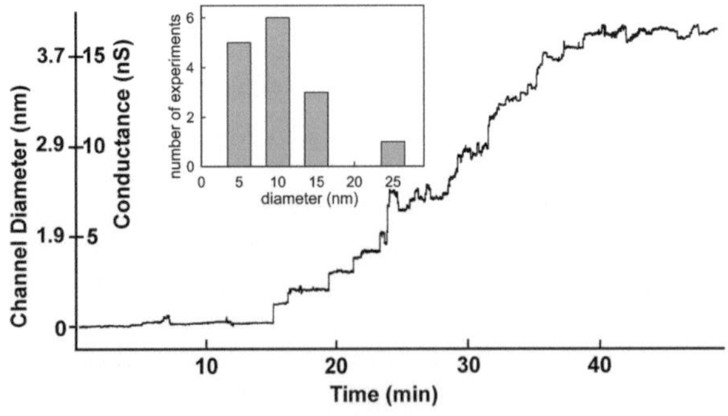

Fig. 4.3 The growth of a ceramide channel in a planar phospholipid membrane. A typical trace of conductance increments in a membrane following the dispersal of ceramide to the aqueous compartment. The calculated channel diameter is also indicated on the y-axis. Ceramide was added 40 min before the start of the record. The *inset* illustrates the distribution of channel sizes from 15 experiments performed under similar conditions (The figure was revised from Samanta et al. (2011))

The lag phase followed by rapid growth is consistent with the formation of a single channel but could also be the result of some cooperative insertion process such as the auto-catalyzed insertion process displayed by VDAC channels (Xu and Colombini 1996, 1997). However, there is far more compelling evidence that a single channel forms rather than a population (Siskind et al. 2003). At times one observes multiple conductance increments followed by a sudden total loss of conductance. This indicates the growth of a channel in steps followed by its rapid collapse. More evidence came from selectivity measurements. The overall selectivity of the membrane following ceramide-induced conductance increments, declined as the conductance increased. If the system consisted of a population of channels, the addition of more channels would not cause the selectivity to decrease. However, when a channel grows in size, the interaction of the permeating ions with the wall of the channel would decrease resulting in a progressively reduced influence of the charged nature of the wall on ion present in the channel and ion movement (i.e. ion selectivity). By far, the most compelling evidence (Siskind et al. 2003) that a single channel formed in the planar membrane is the demonstration of stochastic behavior. The addition of La^{+3} to the aqueous phase next to a ceramide-channel-containing membrane resulted in the total loss of conductance. However the channel disassembly did not have the properties expected of a population of channels (exponential decay) but rather showed a variable delay, during which the conductance remained essentially unchanged, followed by rapid disassembly. The delay varied from experiment to experiment in a way expected for a single energy barrier experienced by a single channel. Pooling many experiments together resulted in the generation of an effective population of channels and this pool showed the expected exponential decay with a time constant of 17 s. If each experiment consisted of a population of channels, pooling experiments would not result in a different outcome.

Thus these results provide compelling evidence that ceramide addition to a planar membrane results in the formation of a single large aqueous pathway in the phospholipid membrane. The membrane that was used had a large surface area: around 8000 μm^2. Compared to the estimated size of the typical ceramide channel, 10 nm in diameter, this is enormous. Why would ceramide addition not result in the formation of multiple channels? It seems that the nucleation of the first "protochannel" is an unfavorable process. Sometimes one must wait for hours for a channel to form. Once formed it grows rapidly. The likelihood of a second forming may be very low. It is also possible that there is a propensity for the second, channel, if it forms, to fuse with the first.

Direct visual evidence of the formation of large channels was obtained by using negative stain electron microscopy (Samanta et al. 2011). Phospholipid vesicles were treated with ceramide with the same method used for planar membranes. Simply generating liposomes with a phospholipid/ceramide mixture was not used because the structures formed by ceramide self-assembly may be path dependent and the sudden delivery of ceramide may reflect the physiological situation that leads to channel formation. This method resulted in channels in the liposomes of the expected size and shape. Figure 4.4a shows four electron micrographs of ceramide channels formed in phospholipid membranes. Control membranes showed no such structures. Negative stain fills both the holes in the channel and the space around the dried and flattened vesicles. This amorphous material scatters electrons in proportion to the number of osmium atoms present and thus the intensity of the stain is proportional to the thickness of the stain. Note that the density of the stain in Fig. 4.4b within the channel is about the same as the density at the edges of the vesicle attesting to the conclusion that the stain indeed penetrates right through the membrane. An analysis of the shape of the stain is consistent with the channel being a right cylinder because the rounding is attributable to the resolution of the technique. Higher resolution methods would be necessary to gain more insight into the shape of the pore. As to the size distribution of the channels observed by electron microscopy, these were very comparable to the distribution of channel sizes calculated from the conductance achieved in planar membranes in many separate experiments. Each experiment yielded one estimate of the channel diameter calculated from the conductance reached a steady level. The most frequently observed ceramide channel had a diameter of 10 nm, certainly large enough to allow the release of all mitochondrial intermembrane space proteins.

The information gap between the structural model of the channel and the experimental evidence for a large channel was closed somewhat by the discovery of remarkable patters in the magnitude of the conductance decrements either occurring spontaneously or induced by the addition of La^{3+} (Siskind et al. 2003). Whereas

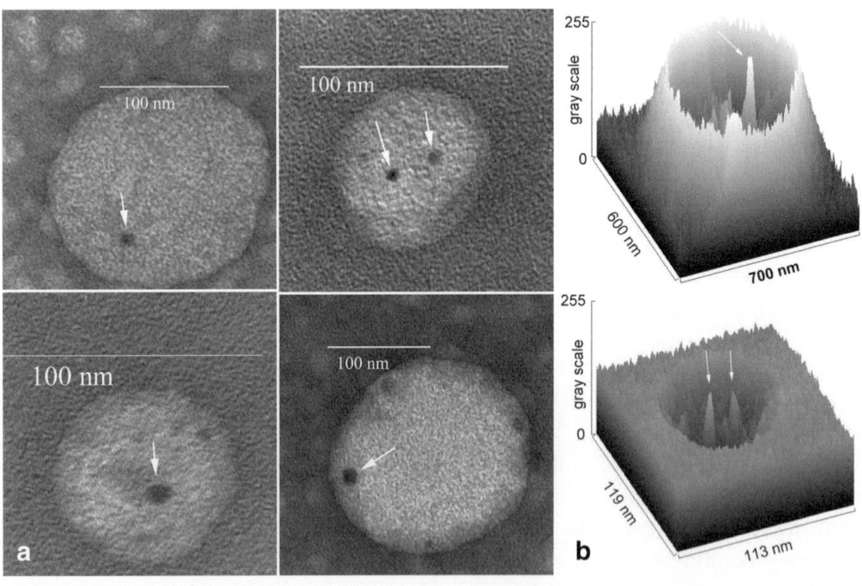

Fig. 4.4 (**a**) Electron micrographs of liposomes treated with C_{16}-ceramide followed by osmium tetroxide fixation and negatively-staining using uranyl acetate. The *arrows* point to the ceramide channels. (**b**) Densitometry scans of two of these electron micrographs showing the intensity of the negative stain. The vesicles form a footprint in the stain. The *arrows* point to the stain filling the channels (The figure was revised from Samanta et al. (2011))

decrements showed the expected decrease in frequency[1] with increase in decrement size, the large decrements showed a large preference for being multiples of a fundamental unit. Beginning with 12 nS decrements, there was a strong preference for decrements that were a multiple of 4 nS (Fig. 4.5). This pattern could be explained if the channels were essentially right cylinders and the loss of conductance was due to a reduction in the circumference of the channel by multiples of a fundamental structure. Calculations of the size of the segment of a circle that would explain the results were made based on the standard equation:

$$G = \left(\kappa_{sp}\pi \bullet r^2\right) / \left(L + 0.5\pi \bullet r\right)$$

where G is the conductance of the channel, κ_{sp} is the specific conductivity of the solution (0.112 Scm^{-1} for 1.0 M KCl), L is the length of the channel and r is the radius of the aqueous pore. Thus the loss of a segment of the circumference would result in a change in conductance and from this change the arc of the circumference can be calculated. For very large channels L <<0.5 πr and so:

$$\text{segment of circumference} = \frac{\pi}{k_{sp}}(\Delta G)$$

This works out to 1.12 nm for a change in conductance of 4.0 nS. For the typical 10 nm channel, calculations without the approximation yield 1.3 nm. This corresponds to a pair of ceramide columns Fig. 4.2a in the 48-column model illustrated in Fig. 4.2d. The loss of an even number of columns could be explained if the columns were arranged in an antiparallel fashion, held together by dipole-dipole interactions. The hydrogen bonding between adjacent ceramides in each column results in the combination of electric dipoles forming one large overall dipole on one side of the ceramide column. This would have a favorable interaction with the opposite dipole of the adjacent column oriented in an opposite direction. Thus they might be expected to dissociate as a unit. Clearly the data do not exclude the dissociation of an odd number of columns but it is the remarkable preference for conductance changes corresponding to an even number of columns that strongly supports both the existence of columns and their antiparallel orientation. Support for preference for an antiparallel orientation was obtained from simulated annealing studies performed by Andriy Anishkin (personal communication 2008).

[1] In general one expects that small changes require less energy and thus are more frequent than large changes. Molecular dynamic simulations show that a variety of structural changes can take place that change the conductance. In the simulation one does not observe the loss of a strand because the time of simulation is too short but one observes strands become partially disconnected from adjacent strands. In the experiments one sees conductances increasing and decreasing spontaneously. Thus the small changes are likely due to these types of structural fluctuations combined with thermal fluctuations in the overall structure. They probably include the loss of single, pairs of strands, and other multiples since the multiplicity of 4 nS pattern becomes clear at 12 nS, an event that corresponds to the loss of 6 strand.

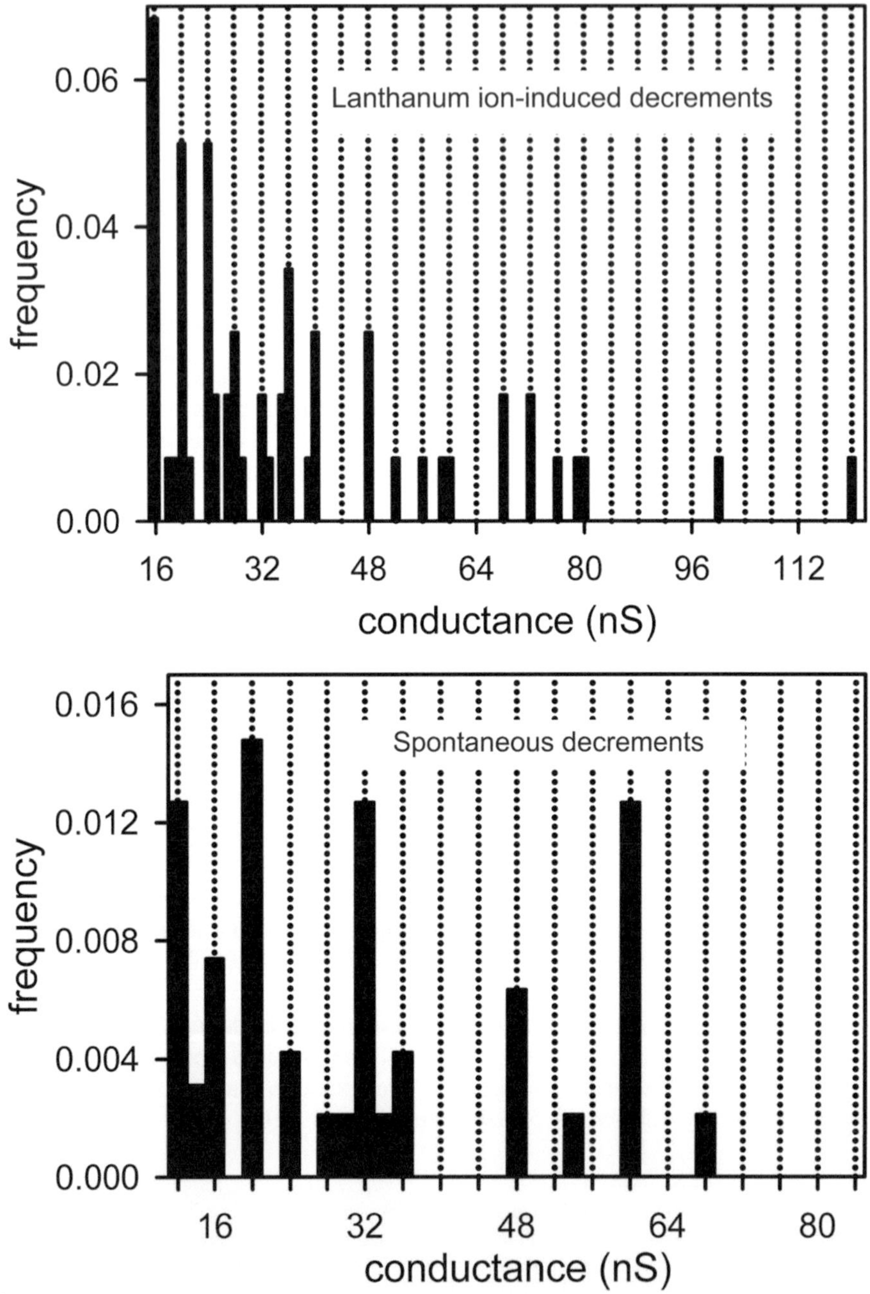

Fig. 4.5 The large conductance decrements from ceramide channel disassembly are more likely to be multiples of 4 nS. The figure is composed of pooled data from many experiments. The *upper panel* shows the decrements observed following the addition of $LaCl_3$ whereas the *lower* contains only spontaneous decrements that occur sporadically. The *dotted lines* are at multiples of 4 nS. In the *upper panel* the bin size was 1 nS whereas in the lower it was 2 nS (The data were taken from Siskind et al. (2003))

4.5 Regulation by Bcl-xL

The regulation of ceramide channels can occur in a variety of ways. The most obvious would be in the control of the steady state level of ceramide in a membrane. Ceramide can be produced by the de novo pathway or the salvage pathway (sphingomyelinase cleavage of sphingomyelin). Ceramide can also be used, and thus consumed, to produce more complex lipids such as sphingomyelin and glycosylated sphingolipids. Clearly multiple enzymatic activities could influence steady-state levels of ceramide in a particular membrane and thus change the propensity for the formation of ceramide channels and indeed, in a variety of systems, ceramide levels increase early in the apoptotic process to levels more than sufficient to result in ceramide channel generation (Siskind et al. 2006). However, it remains to be determined which of these enzymatic activities might be part of a genuine regulatory system. In addition to influencing the steady state level of ceramide, some proteins actually affect the ability of ceramide channels to form and/or their stability. The ones intensely studied are members of the Bcl-2 family. This is the protein family that regulates the onset of apoptosis, including the likelihood of the release of mitochondrial intermembrane space proteins into the cytosol.

Anti-apoptotic proteins, Bcl-2, Bcl-xL, and CED-9, inhibit the onset of apoptosis by inhibiting the release of the mitochondrial intermembrane space proteins that initiate the execution phase of apoptosis. These same proteins have also been shown (Siskind et al. 2008) to inhibit the formation of ceramide channels in the MOM (Fig. 4.6a, b). This implicates ceramide channels as a pathway responsible for said protein release otherwise why would all these proteins inhibit ceramide channel formation? Most of the research has focused on Bcl-xL and its ability to inhibit ceramide channel formation in the outer membrane of isolated rat liver mitochondria (Fig. 4.6a) but essentially the same results were observed when Bcl-xL was added to mitochondria isolated from the yeast, *S. cerevisiae*. Similar abilities observed on ceramide channels formed in the defined system of the planar phospholipid membrane clearly indicated a direct cause and effect relationship. CED-9, the anti-apoptotic protein from *C. elegans*, worked just as well as the mammalian ortholog, Bcl-xL, on both of these experimental systems (Fig. 4.6b), indicating that the ability of anti-apoptotic proteins to inhibit ceramide channels was conserved evolutionarily. The other important mammalian anti-apoptotic protein, Bcl-2, was tested by expressing it in yeast as it lacks the Bcl-2 family of proteins. Nevertheless Bcl-2 localized to the MOM of yeast mitochondria rendered these mitochondria, when isolated, resistant to ceramide permeabilization in that higher levels of ceramide were needed to achieve the same level of permeabilization.

The most straight-forward way by which Bcl-xL could affect the stability of ceramide channels would be to physically bind to the channels. If that were so then there would need to be a region of the Bcl-xL protein that would recognize a complementary region of the ceramide channel. By using ceramide analogs and ceramides of varying chain length, Perera and co-workers demonstrated that Bcl-xL preferentially binds to the hydrophobic tails. They found optimal inhibitory potency when the fatty acyl chain was 16, 18 and 20 carbons. Bcl-xL was much less effective

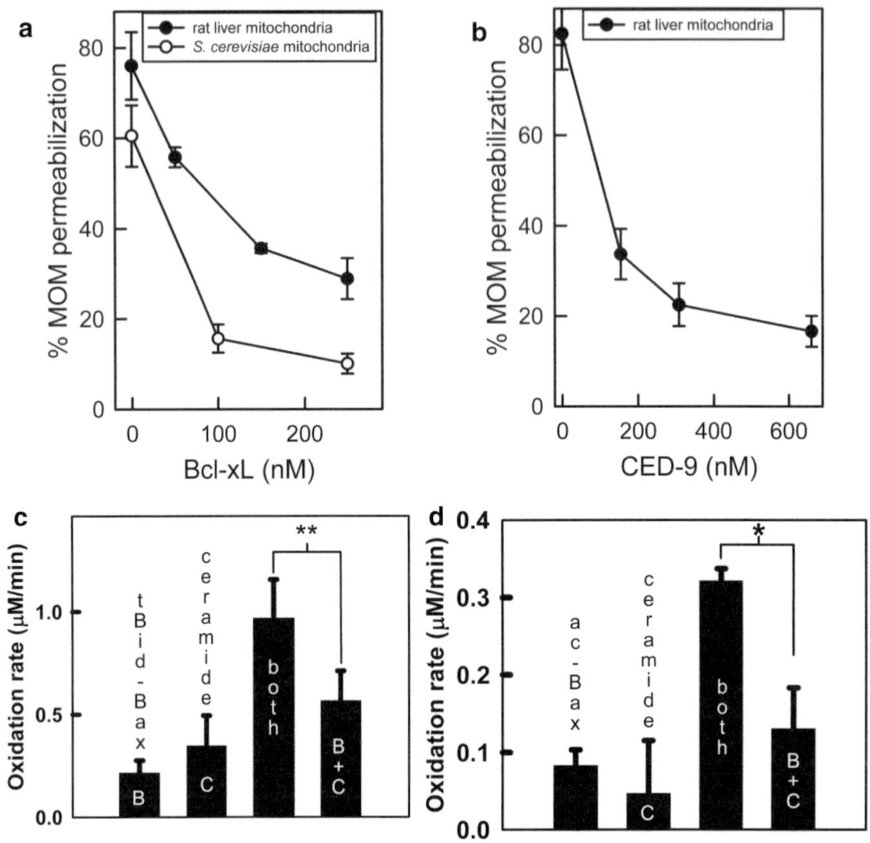

Fig. 4.6 Bcl-2 family proteins regulate ceramide channels. Bcl-xL (**a**) and CED-9 (**a**) inhibit ceramide-induced permeabilization of the MOM. t-Bid plus Bax (**a**) and octyl glucoside-activated Bax (**a**) act synergistically with ceramide to permeabilize the MOM. Without activation Bax had no effect. Statistics: *, 95 %; **, 99 % confidence (The research illustrated in panels **a**, **b** was originally published in the *Journal of Biological Chemistry*. Siskind et al. (2008) © the American Society for Biochemistry and Molecular Biology. The research illustrated in panels **b**, **c** was published in Ganesan et al. (2010))

on channels formed by ceramides with shorter and longer chains. Modification of a chain by the introduction of a double bond also reduced the effectiveness of Bcl-xL. Modifications of the polar regions were much better tolerated. This is very different from what was observed with Bax and thus is not due to a generic physical phenomenon but rather to the way Bcl-xL binds to a ceramide channel.

Identification of the site on Bcl-xL that binds to the ceramide channel was achieved both by the use of Bcl-xL binding drugs and by site-directed mutagenesis. BH3 mimetic drugs developed by Abbott Laboratories (Oltersdorf et al. 2005) bind to the hydrophobic groove of Bcl-xL. Two of these, ABT-263 and ABT-737, were tested (Perera et al. 2012b) for their ability to interfere with Bcl-xL to inhibit ceramide channel formation. These and a third compound all were effective, indicating that Bcl-xL binds to ceramide channels using its hydrophobic groove. This is consistent with the inhibitory action of Bcl-xL being sensitive to changes in the hydrophobic portions of ceramide. The use of point mutations in the hydrophobic groove demonstrated the sensitivity of Bcl-xL's inhibitory ability to changes in the charge, polarity and apolar bulk of sites in the groove (Chang et al. 2015). The binding of individual ceramide molecules to Bcl-xL and associated simulations further support the hydrophobic groove as the binding site. Thus this same region on Bcl-xL binds to ceramide presumably on ceramide channels and to Bax, preventing Bax channel formation.

In the *in vitro* experiments showing the ability of Bcl-xL to inhibit the formation of ceramide channels, there was a large molar excess of ceramide over Bcl-xL. Thus Bcl-xL is not acting by merely sequestering ceramide monomers. The specificity for the aliphatic chains means that Bcl-xL should only be able to bind to ceramide if these chains are accessible. In an aqueous environment, these chains would be buried into the hydrophobic milieu of a membrane or a micelle. However, at the end of a ceramide channel, there is presumably stress at the ceramide-phospholipid interface. Bcl-xL anchored to the membrane may find it energetically easy to displace the phospholipids covering ceramide's hydrocarbon chains. Other ceramides in the membrane would have their hydrocarbon chains buried deep into the membrane's core and these could be inaccessible to Bcl-xL because of the high energy barrier to exposing these chains.

How does Bcl-xL inhibit ceramide channels? Hill plots of dose response curves where varying amounts of Bcl-xL were added to mitochondria treated with a constant amount of ceramide, had a Hill coefficient of 1 (Siskind et al. 2008). Thus 1 Bcl-xL molecule is sufficient to destabilize one channel. How one relatively small protein can destabilize a large channel is not clear. The working hypothesis (Perera et al. 2012b) is that the binding of the Bcl-xL to one end of one column of a channel that may consist of 50 columns will result in a change in the positive curvature of that column. That change will result in stress to adjacent columns, whose curvature was not altered, through the hydrogen-bonded network that holds the channel together. That stress is propagated throughout the entire structure through this network in much the same way that allosteric interactions are propagated in a protein. There is as yet no evidence to support this hypothesis.

4.6 Synergistic Permeabilization by Ceramide and Bax

Bax exists in the cytosol as an inactive soluble globular protein. *In vivo* it is activated and targeted to insert into the MOM by another protein called Bid but only when Bid itself is activated by being truncated into what is called t-Bid (truncated Bid). Alternatively, Bax can be activated by being treated with the detergent, octyl glucoside. Both activated forms permeabilize the MOM and phospholipid membranes to proteins without the need for the addition of ceramide. The natural presence of ceramide in the MOM cannot be excluded as contributing to the "Bax only" permeabilization. Regardless, in a planar membrane system lacking ceramide, Bax can produce channels (Antonsson et al. 1997; Schlesinger et al. 1997). As for ceramide, the presence of Bax or BAK is not required for ceramide channel formation in either the MOM or planar membranes. Nevertheless, doses of Bax or ceramide that each result in weak permeabilization of the MOM, when combined (Fig. 4.6c, d) resulted in a level of permeabilization that exceeded that observed with either agent alone (Ganesan et al. 2010). Thus it was concluded that activated Bax and ceramide act synergistically. This synergy was not seen at doses when each agent produces substantial permeabilization of the MOM. Lee and coworkers (2011) also reported compelling evidence that ceramide and Bax act synergistically to permeabilize the MOM. They provided evidence for the formation of ceramide-rich microarrays that favored Bax oligomerization resulting in outer membrane permeabilization. Pastorino et al. (1999) reported synergy but interpreted their results in terms of the mitochondrial permeability transition.

In experiments with isolated mitochondria, Bax seemed to favor ceramide channel formation by having a higher apparent affinity for channels of larger size (Ganesan et al. 2010), thus favoring the growth of ceramide channels. Experiments showed a decrease in the half-maximal concentration of Bax needed to enhance the ceramide permeability as the experiment was performed at higher doses of ceramide (and thus higher ceramide-only MOM permeability). It was speculated that activated Bax acts as a molecular scaffold, driving the ceramide channel to a specific radius of curvature. This interpretation also explained why the Bax enhancement of the ceramide-induced MOM permeabilization was reduced as the ceramide level was increased. Indeed at high enough ceramide-only permeabilization, no enhancement by Bax was observed. Presumably the appropriate "best-fit" size of the ceramide channel had already been achieved. This did not represent the maximal possible MOM permeability as the addition of more ceramide resulted in a higher level of MOM permeability. Thus it appeared that an optimal size of ceramide was achieved. These results and associated interpretation predict that Bax should reduce the ceramide-induced permeabilization if ceramide channels larger than the optimum size were formed. This prediction was not tested.

The presence of synergistic permeabilization by Bax and ceramide was used as a diagnostic tool to determine what features of ceramide are required for this synergy to take place. Ceramide analogs that are capable of forming channels were tested for their ability to achieve synergistic permeabilization of the MOM with Bax. The elimination of either one of the hydrocarbon chains did not affect the synergy and this is in sharp contrast with the observations with Bcl-xL. However, synergy was lost when the chirality of carbon 2 on the sphingosyl moiety was changed or when the amide nitrogen was methylated. L-ceramide, which is as capable of forming ceramide channels as the natural stereoisomer, D-ceramide, is incapable of permeabilizing the MOM synergistically with Bax. Other changes in the polar region had a smaller or no effect. These results indicate that Bax binds to a different portion of the ceramide channel than Bcl-xL and perhaps this is the reason why these proteins of the same family have opposite effects on ceramide channels. Regardless of the mechanism, it must be noted that these two related Bcl-2 family proteins influence the structure/stability of ceramide channels in a way that matches the role of these proteins in controlling the onset of apoptosis. A further difference between the actions of Bcl-xL and Bax comes from dose-response experiments. Whereas only a 1:1 interaction between Bcl-xL and ceramide channels is necessary to inhibit channel formation (Siskind et al. 2008), many Bax molecules are involved in the synergy.

4.7 Channel Formation by Ceramide Analogs

The unexpected ability of a lipid, ceramide, to form large stable channels in membranes, raised the question: what is it about the structure of ceramide that results in the self-assembly of hundreds of ceramide molecules resulting in channel formation? Ceramide has several structural features that could contribute to the stability of a channel. If all these features were necessary to achieve a stable channel structure then changing any of these would interfere with channel formation. The hydrocarbon chains should be necessary to anchor ceramide to the membrane by hydrophobic interactions but the largely saturated chains could also be important in the packing of and orientation of each ceramide molecule making up the channel. The *trans* double bond should add to the ordering of the hydrocarbon chains, reducing the entropic loss inherent in the formation of an organized structure. The location of the double bond relative to the hydroxyl on carbon 3, makes the latter an allylic hydroxyl, changing its hydrogen-bonding characteristics. This could be important in forming a stable channel. The two hydroxyl groups and the amide linkage form hydrogen bonds and thus should provide both mechanical strength and a degree of rigidity. Ceramide has two chiral centers and a planar amide linkage, all of which limit possible conformational structures and intra or intermolecular interactions. Each of these features was altered and the ability of the resulting structures to form channels, assessed (Perera et al. 2012a). Examples of the dose dependence of MOM permeabilization for some analogs is shown in Fig. 4.7.

The ability of ceramide to form channels was not significantly altered by shortening either one of the two hydrocarbon chains. Any changes in ability to permeabilize the MOM could be attributed to differences in the ability to deliver ceramide to the target membrane. The efficiency of ceramide dispersal in water from the isopropanol solution and the propensity to aggregate prior to reaching the target membranes are all expected to change with alterations in chain length. Thus certainly only one chain is needed not the two found in natural ceramide. The presence of two chains is important in other functions of ceramide, such as interaction with Bcl-xL. The existence of a panel of ceramide synthases (Grosch et al. 2012) that have different preferences for the synthesis of ceramides with difference acyl chains, demonstrates the importance of chain length in ceramide function in cells. However, this is not necessary for channel formation.

The ceramide desaturase forms a trans double bond specifically between carbon 4 and 5 of the sphingosyl moiety but that specific location is not important for channel formation. A cis double bond is also acceptable even though this should be more disordering. The double bond can also be replaced with a third hydroxyl, as in

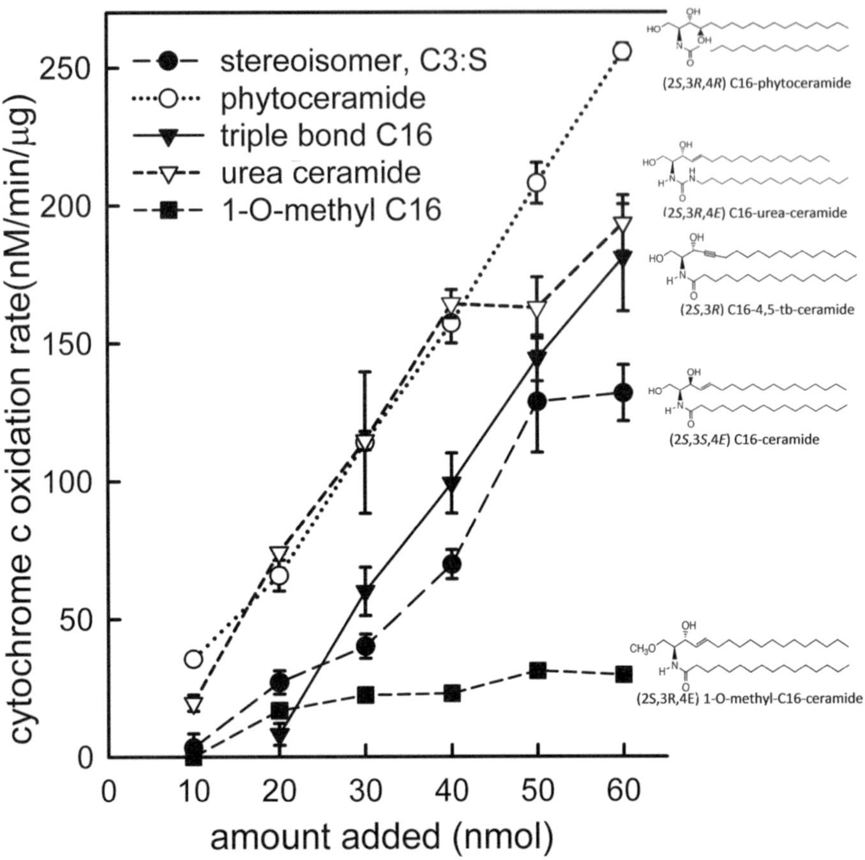

Fig. 4.7 Ceramide analogs permeabilize the MOM of rat liver mitochondria. The structures of the analogs used are shown on the *right side* (These are unpublished data collected in association with the publication of Perera et al. (2012a))

phytoceramide, and channel formation takes place quite easily (Perera et al. 2012a). However, inserting a cis double bond half way down the acyl chain did destabilize the channel. All this raises the question of why the double bond is needed. Dihydroceramide differs from ceramide only in lacking the double bond and that is sufficient to prevent channel formation. Perhaps this failure to form channels is due to an inability to reach a critical concentration of free lipid in the membrane that can assemble into channels. Dihydroceramide may form aggregates in the membrane that reduce the free dihydroceramide concentration in the membrane. Alternatively, the packing of dihydroceramide in the form of a channel may be inherently unstable. Indeed, mixtures of ceramide and dihydroceramide are less effective in making channels as opposed to ceramide alone. As little as one part in ten of dihydroceramide reduces the ability of ceramide to form channels (Stiban et al. 2006).

Methylation of the polar residues would reduce hydrogen bonding and is expected to destabilize ceramide channels (Fig. 4.2). Methylation of the C3-hydroxyl resulted in no significant change in channel-forming ability whereas methylation of the C1-hydroxyl resulted in failure to measurably permeabilize the MOM to cytochrome *c*. However, the C1 methylated analog did result in adenylate kinase release, indicating the formation of transient, unstable channels. When the C1 hydroxyl analog was tested on planar phospholipid membranes, small short-lived conductances were observed as opposed to the large channels characteristic of ceramide (Perera et al. 2012a). This difference between the modification of the two hydroxyls mirrors changes observed when the stereochemistry was changed at the chiral centers. Conversion of carbon 3 from R to S, a modification that would change the orientation of the C3-hydroxyl, had no effect on channel formation. However, an additional change at carbon 2 (from S to R) resulted is a large drop in channel-forming ability. The latter change would have altered the orientation of the C1-hydroxyl. These all point to the importance of the C1-hydroxyl to channels stability. Methylation at a third site, the amide nitrogen, resulted in a marked reduction in the ability to permeabilize the MOM and the stability of channels in planar phospholipid membranes. Yet, the formation of any channels at all is inconsistent with the channel structure illustrated in Fig. 4.2. The amide linkage is critical to the stability of the ceramide columns and the inability of the amine nitrogen to hydrogen bond should be devastating. However, molecular dynamic simulations using this analog resulted in a structure in which the carbonyl group of the amide linkage hydrogen bonds with one of the hydroxyls (supplemental figure to Perera et al. 2012a). This altered structure seems stable enough and illustrates how different structures can still form channels.

Two different attempts to strengthen the hydrogen-bonded network believed to hold the channel together succeeded in increasing the potency for permeabilization of the MOM. Urea ceramide has an extra amide nitrogen potentially increasing the number of hydrogen bonds. The diene has a second double bond conjugated with the first and thus should further change the electronegativity of the allylic hydroxyl. Both decreased the amount of analog needed to permeabilize the MOM. Thus, rather than being disruptive, presumably the structure of the channel adapted to the new molecular properties.

Overall, channel formation is tolerant of many changes in the structure of ceramide and thus the constraints on the structure probably lie in the ability to interact with regulating proteins, in other functions of ceramide, and in the importance of keeping ceramide localized. For instance, if ceramide has only one long acyl chain it is likely to hop from one membrane to another and its localization to specific membranes is important for cellular function.

4.8 Synergy and Antagonism Between Ceramide and Related Lipids

Ceramide channels are dynamic structures, constantly assembling and disassembling. The virtual water insolubility of ceramide means that this dynamic process takes place almost in two dimensions. Lipids monomers or lipid aggregates dissolved in the phospholipid membranes collide with the channel and can insert into it if they produce a new structure that is reasonably stable. Similarly fragments of the channel can come off and dissolve into the membrane bilayer. If the colliding molecule cannot intercalate into the channel structure in such a way as to form a low-energy interaction, the intercalation attempt will fail. Thus one might expect that molecules quite different from ceramide (e.g. the phospholipids) would not intercalate into the channel and, in fact ceramide channels are stable despite the presence of large amounts of these lipids. This was confirmed with molecular dynamic simulations (Anishkin et al. 2006). Molecules similar to ceramide might or might not intercalate depending on the details of the interaction. In addition to the favorable or unfavorable interactions, which would contribute to the enthalpy changes, entropic effects would generally favor intercalation. Entropy driven intercalations that have a weak or unfavorable enthalpic component could lead to channel destabilization.

Ceramide channel destabilization was observed with both the addition of sphingosine and dihydroceramide. Sphingosine by itself is capable of forming small channels but when added to a ceramide channel it disassembles that channel (Elrick et al. 2006). Each, alone, forms structures with some order but the combination may form poorly organized structures that result in instability. Dihydroceramide does not form channels and merely destabilizes ceramide channels. In both cases, the destabilization could arise from the considerations just discussed or merely by the possibility that each of these agents interacts with ceramide to form structures that are non-conducting. In any case it seems more likely that similar structures would display this antagonistic behavior.

Indeed, suppression of MOM permeabilization was observed when certain analogs were mixed with ceramide prior to addition to the mitochondrial suspension. This antagonism was seen with phytoceramide and the analog methylated at the amide nitrogen. Both of these analogs are modified in a fairly major way. Synergy was observed when analogs differed from ceramide by just the length of the acyl chain. Perhaps mixing ceramide with different chain length forms channels that are more stable than those with the same chain length because the surface curvature varies in different parts of the channel. At the interface with the phospholipid bilayer, ceramide molecules with shorter chains might pack with less strain. Some mixtures showed neither synergy nor antagonism presumably because the ceramide and its analog were essentially interchangeable in the structure. Surprisingly this cooperativity was found to be associated with the amount of ceramide or analog that was able to insert into the MOM indicating that insertion of ceramide into the MOM may involve some autocatalytic events that requite intimate lipid-lipid interactions.

It is important to emphasize that ceramide has many roles in cells and that the evolutionary constraints on its structure must be many and varied. From the studies with the analogs one can conclude that a fairly large set of structures would form channels capable of releasing proteins from mitochondria and thus initiating the execution phase of apoptosis. Thus the biophysical constraints on forming channels are not very stringent. Of this set, some would form channels capable of being controlled by Bcl-xL or Bax. A much smaller set would be controlled by both. Of course evolutionary selection also acts on the proteins and their binding specificity. An additional constraint that needs further study is the specificity for ceramide channel formation in the MOM but not in the plasma membrane. If such specificity were not in place, ceramide formed in the plasma membrane by the hydrolysis of sphingomyelin could result in channel formation and cell death. Using erythrocytes as model plasma membranes, the addition of ceramide to these does not result in channel formation and therefore cell lysis (Siskind et al. 2002). Even channels small enough to allow the translocation of NaCl would result in lysis and thus even small channels were not formed. By measuring the amount of ceramide that had been delivered to the erythrocyte membrane, Siskind and coworkers (2006) demonstrated that a molar ratio of ceramide to membrane lipids that was 50-fold greater than that needed to permeabilize the MOM still did not cause detectable erythrocyte lysis. Yet ceramide channels form in planar phospholipid membranes and liposomes. Whatever the molecular mechanism, there must be evolutionary selection pressure against ceramide channel formation on the plasma membrane. There could be evolutionary constraints on the molecular form of what we call ceramide or on other components on the membranes in question that either favor or inhibit ceramide channel formation.

4.9 Elastic Properties of Ceramide Channels

Membrane channels are usually thought of as rigid structures. The discrete changes in conductance upon channel formation and gating support this view. In addition, a rigid structure helps to account, at least in part, for the selectivity displayed by some channels. However, one knows well that at least thermal energy will result in motions of channels as it does with all structures but more evident as the size of objects decrease. Beyond thermal energy, there are structural changes arising from mechanical stresses generated by neighboring objects or processes. The structural changes experienced by a channel of course depend on its elastic properties and any conformations to which the channel has access. Thus depending on the channel's mechanical properties and the environmental conditions, structural changes could occur thus making the channel a less than rigid structure.

The size of the ceramide channel and thinness of its wall should make it quite sensitive to forces such as the lateral pressure of the membrane. Ceramide channels are typically 10 nm in diameter, at least twice the membrane thickness, but can be much larger. As the radius of curvature of the wall of the channel increases, its susceptibility to membrane tension increases. Increases in membrane lateral pressure following the addition of La^{3+} is believed to be the mechanism by which La^{3+} causes complete ceramide channel disassembly. Studies on the effects of lanthanides on stretch-sensitive channels show that these trivalents bind to negatively-charged lipids resulting in pressure changes within the membrane that inhibit the ability of these channels to open following membrane stretch (Ermakov et al. 2010). The La^{3+}-induced complete disassembly of ceramide channels was observed when such channels were formed in solvent-free planar membranes (i.e. made from phospholipid monolayers). When these channels were formed in decane-containing phospholipid

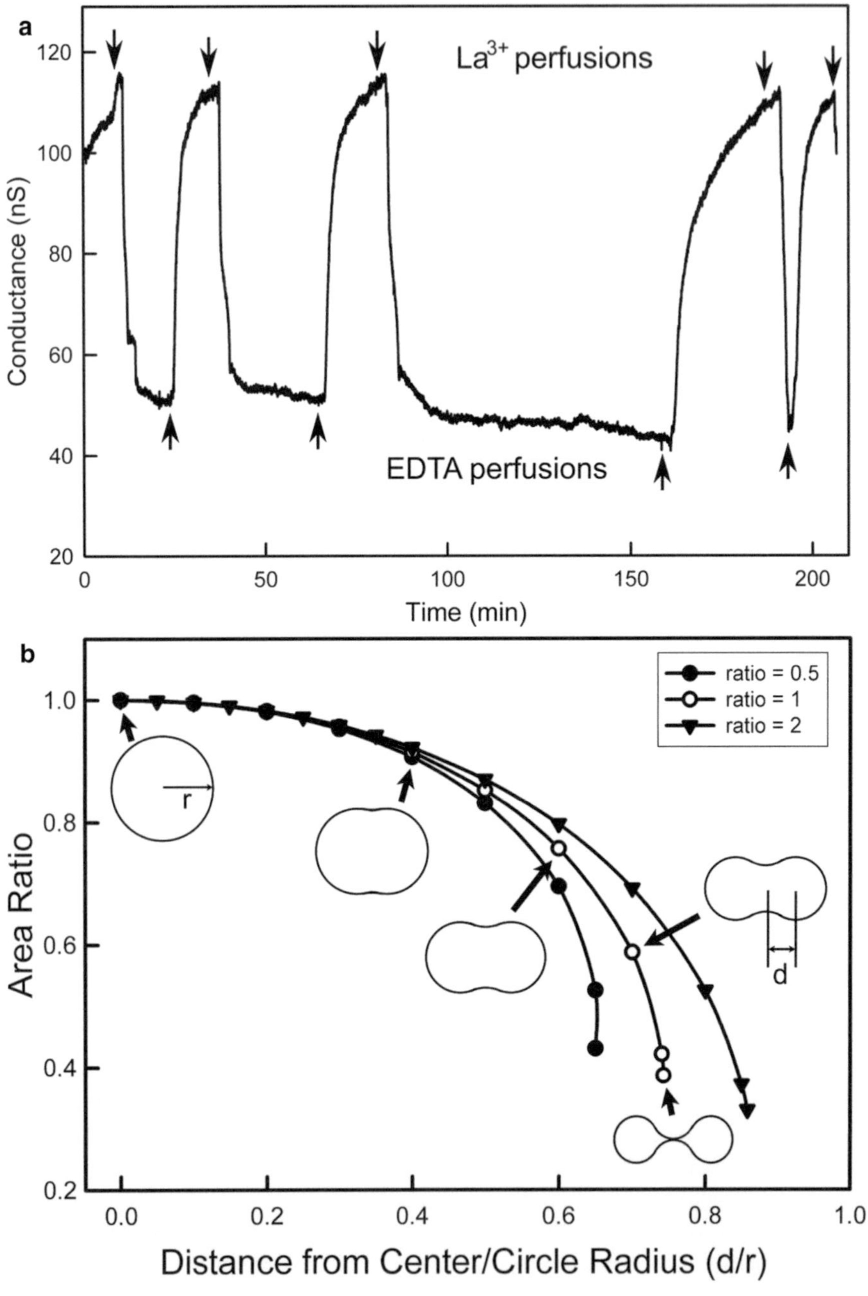
a
La3+ perfusions
EDTA perfusions
Conductance (nS)
Time (min)
b
ratio = 0.5
ratio = 1
ratio = 2
r
d
Area Ratio
Distance from Center/Circle Radius (d/r)

membranes, La^{3+} caused reversible changes in conductance (Fig. 4.8a) attributed to elastic changes in the channels (Shao et al. 2012). The original conductance was restored when La^{3+} was chelated by perfusion with EDTA-containing buffer. The conductance drop following La^{3+} addition and its reversal with EDTA could be repeated many times on the same membrane by means of a microfluidic perfusion system. Note that reversal became slower if the channel was kept in a squeezed conformation for a longer time indicating some mechanical adaptation took place. An alternative interpretation that the channel partially disassembled and reassembled to the same degree at each perfusion, seems highly unlikely. Furthermore, theoretical calculations that quantitatively modeled the hypothesized structural change in the channel, conversion of the channel cross-section from a circle to a biconcave disk, accounted for the observed conductance change (Fig. 4.8b). Illustrated is the ratio of the cross-sectional area of the biconcave structures to that of the unmodified circular structure. The decrease of this ratio should reflect the drop in conductance induced by La^{3+}. The point at which the two concave area touch (the very end of the curves) has a drop in cross-sectional area that matches closely the drop in conductance. This theory also accounts quantitatively for the linear relationship between the initial conductance and the final La^{3+}-induced conductance for a number of different experiments with different starting conductances. That's because the ratio of the areas is constant regardless of the radius of the original channel. Thus the ceramide channels are acting as elastic structures, responding to changes in lateral pressure by channel distortion forming a stressed state that relaxes when the pressure is reduced.

Fig. 4.8 Ceramide channel shows elastic properties. (**a**) Cyclic changes in ceramide channel conductance formed in decane-containing planar phospholipid membranes. Perfusion with La^{3+} (50 μM)-containing medium (beginning at *down arrows* and ending at *up arrows*) resulted in a conductance drop to a constant level. Perfusions with EDTA (50 μM)-containing medium (*up arrow* until *down arrow*) resulted in return to the original conductance. (**b**) Theoretical calculations of cross-sectional area changes (expressed as a ratio of the area of the modified version to that of a perfect circle) resulting from the formation of a biconcave structure with positive and negative radii of curvature. The perimeter of the structure was maintained constant as the axial ratio was increased while maintaining the absolute value of the ratio of the two curvatures (negative/positive) at 0.5, 1 and 2 as shown. Calculations were performed at the plotted points. The cross-sections shown are the results of calculations and are to scale relative to each other. The "distance from center" referred to on the label of the x-axis is the distance, "d", from the center of the structure to the center of the radius of positive curvature, as illustrated (The figure was revised from Shao et al. (2012))

4.10 Conclusion

Our view of what membrane channels are, what they can do, and how they can be controlled needs to expand to include a greater diversity of structures and functions. Some channels regulate the flow of small ions and have properties consistent with their function. Other channels control the flow of metabolites and thus have a larger pore size and a selectivity mechanism appropriate to that function (Komarov et al. 2005). Ceramide channels allow the flow of proteins across membranes and thus their properties are also appropriate to that function. These channels need to form rapidly from existing molecular precursors so as to begin the execution phase of apoptosis as soon as that decision is made. They need to be regulated so that they do not form at inappropriate times and inappropriate locations. Ceramide channels fulfill all these needs. Much work is needed to probe the structure of these channels and the mechanisms by which they are controlled. From a practical aspect, these are new targets for influencing the apoptotic process that is central to much cellular physiology and pathology.

Acknowledgments Support from the National Science Foundation (MCB-1023008) is gratefully acknowledged.

References

Anishkin A, Sukharev S, Colombini M (2006) Searching for the molecular arrangement of transmembrane ceramide channels. Biophys J 90:2414–2426

Antonsson B, Conti F, Ciavatta A et al (1997) Inhibition of Bax Channel-Forming Activity by Bcl-2. Science 277:370–372

Bielawska A, Crane HM, Liotta D et al (1993) Selectivity of ceramide-mediated biology. Lack of activity of erythro-dihydroceramide. J Biol Chem 268:26226–26232

Birbes H, El Bawab S, Hannun YA et al (2001) Selective hydrolysis of a mitochondrial pool of sphingomyelin induces apoptosis. FASEB J 15:2669–2679

Birbes H, Luberto C, Hsu YT et al (2005) A mitochondrial pool of sphingomyelin is involved in TNFalpha-induced Bax translocation to mitochondria. Biochem J 386:445–451

Chang K-T, Anishkin A, Patwardhan GA et al (2015) Ceramide channels: destabilization by Bcl-xL and role in apoptosis. Biochim Biophys Acta 1848:2374–2384

Di Paola M, Cocco T, Lorusso M (2000) Ceramide interaction with the respiratory chain of heart mitochondria. Biochemistry 39:6660–6668

Elrick MJ, Fluss S, Colombini M (2006) Sphingosine, a product of ceramide hydrolysis by ceramidase, disassembles ceramide channels. Biophys J 91:1749–1756

Ermakov YA, Kamaraju K, Sengupta K, Sukharev S (2010) Gadolinium ions block mechanosensitive channels by altering the p and lateral pressure of anionic lipids. Biophys J 98:1018–1027

Ganesan V, Perera MN, Colombini D et al (2010) Ceramide and activated Bax act synergistically to permeabilize the mitochondrial outer membrane. Apoptosis 15:553–562

Garcia-Ruiz C, Colell A, Mari M et al (1997) Direct effect of ceramide on the mitochondrial electron transport chain leads to generation of reactive oxygen species. Role of mitochondrial glutathione. J Biol Chem 272:11369–11377

Ghafourifar P, Klein SD, Schucht O et al (1999) Ceramide induces cytochrome *c* release from isolated mitochondria. Importance of mitochondrial redox state. J Biol Chem 274: 6080–6084

Godchaux W 3rd, Leadbetter ER (1983) Unusual sulfonolipids are characteristic of the Cytophaga-Flexibacter group. J Bacteriol 153:1238–1246
Grosch S, Schiffmann S, Geisslinger G (2012) Chain length-specific properties of ceramides. Prog Lipid Res 51:50–62
Hanada K (2014) Co-evolution of sphingomyelin and the ceramide transport protein CERT. Biochim Biophys Acta 1841:704–719
Hannun YA, Obeid LM (2008) Principles of bioactive lipid signaling: lessons from sphingolipids. Nat. Rev Mol Cell Biol 9:139–150
Hannun YA, Obeid LM (2011) Many ceramides. J Biol Chem 286:27855–27862
Komarov AG, Deng D, Craigen WJ et al (2005) New insights into the mechanism of permeation through large channels. Biophys J 89:3950–3959
Lee H, Rotolo JA, Mesicek J et al (2011) Mitochondrial ceramide-rich macrodomains functionalize Bax upon irradiation. PLoS One 6(6):e19783
Meckfessel MH, Brandt S (2014) The structure, function, and importance of ceramides in skin and their use as therapeutic agents in skin-care products. J Am Acad Dermatol 71:177–184
Montes LR, Ruiz-Arguello MB, Goni FM et al (2002) Membrane restructuring via ceramide results in enhanced solute efflux. J Biol Chem 277:11788–11794
Obeid LM, Linardic CM, Karolak LA et al (1993) Programmed cell death induced by ceramide. Science 259:1769–1771
Oltersdorf T, Elmore SW, Shoemaker AR et al (2005) An inhibitor of Bcl-2 family proteins induces regression of solid tumours. Nature 435:677–681
Pajewski R, Djedovic N, Harder E et al (2005) Pore formation in and enlargement of phospholipid liposomes by synthetic models of ceramides and sphingomyelin. Bioorgan Med Chem 13:29–37
Pastorino JG, Tafani M, Rothman RJ et al (1999) Functional consequences of the sustained or transient activation by Bax of the mitochondrial permeability transition pore. J Biol Chem 274:31734–31739
Perera MN, Ganesan V, Siskind LJ et al (2012a) Ceramide channels: influence of molecular structure on channel formation in membranes. Biochim Biophys Acta 1818:1291–1301
Perera MN, Lin SH, Peterson YK et al (2012b) BAX, Bcl-xL exert their regulation on different sites of the ceramide channel. Biochem J 445:81–91
Pruett ST, Bushnev A, Hagedorn K et al (2008) Biodiversity of sphingoid bases ("sphingosines") and related amino alcohols. J Lipid Res 49:1621–1639
Rodriguez-Lafrasse C, Alphonse G, Aloy MT et al (2002) Increasing endogenous ceramide using inhibitors of sphingolipid metabolism maximizes ionizing radiation-induced mitochondrial injury and apoptotic cell killing. Int J Cancer 101:589–598
Samanta S, Stiban J, Maugel TK et al (2011) Visualization of Ceramide Channels by Transmission Electron Microscopy. Biochim Biophys Acta 1808:1196–1201
Schlesinger PH, Gross A, Yin XM et al (1997) Comparison of the ion channel characteristics of proapoptotic BAX and antiapoptotic BCL-2. Proc Natl Acad Sci U S A 94:11357–11362
Shao C, Sun B, DeVoe DL, et al (2012) Dynamics of ceramide channels detected using a microfluidic system. PLoS One 7(9):e43513
Siskind LJ (2005) Mitochondrial ceramide and the induction of apoptosis. J Bioenerg Biomembr 37:143–53
Siskind LJ, Colombini M (2000) The lipids C_2- and C_{16}-ceramide form large stable channels: implications for apoptosis. J Biol Chem 275:38640–38644
Siskind LJ, Kolesnick RN, Colombini M (2002) Ceramide channels increase the permeability of the mitochondrial outer membrane to small proteins. J Biol Chem 277:26796–26803
Siskind LJ, Davoody A, Lewin N et al (2003) Enlargement and contracture of C_2-ceramide channels. Biophys J 85:1560–1575
Siskind LJ, Kolesnick RN, Colombini M (2006) Ceramide forms channels in mitochondrial outer membranes at physiologically relevant concentrations. Mitochondrion 6:118–125
Siskind LJ, Feinstein L, Yu T et al (2008) Anti-apoptotic Bcl-2 family proteins disassemble ceramide channels. J Biol Chem 283:6622–6630

Siskind LJ, Mullen TD, Rosales KR et al (2010) The Bcl-2 protein BAK is required for long-chain ceramide generation during apoptosis. J Biol Chem 285:11818–11826
Stiban J, Fistere D Jr, Colombini M (2006) Dihydroceramide hinders ceramide channel formation: implications on apoptosis. Apoptosis 11:773–780
Wu ZP, Chen Y, Xia B et al (2009) Two novel ceramides with a phytosphingolipid and a tertiary amide structure from Zephyranthes candida. Lipids 44:63–70
Xu X, Colombini M (1996) Self-catalyzed insertion of proteins into phospholipid membranes. J Biol Chem 271:23675–23682
Xu X, Colombini M (1997) Auto-directed insertion: pre-inserted VDAC channels greatly shorten the delay to the insertion of new channels. Biophys J 72:2129–2136
Yue JM, Fan CQ, Xu J et al (2001) Novel ceramides from the fungus Lactarium volemus. J Nat Prod 64:1246–1248
Zheng WJ, Kollmeyer J, Symolon H et al (2006) Ceramides and other bioactive sphingolipid backbones in health and disease: lipidomic analysis, metabolism and roles in membrane structure, dynamics, signaling and autophagy. Biochim Biophys Acta 1758:1864–1884

Part II
Bacteria and Viruses

Chapter 5
Bacterial Porins

Vicente M. Aguilella, María Queralt-Martín, and Antonio Alcaraz

Abstract Bacterial porins are the most abundant proteins in the outer membrane of gram-negative bacteria. The term porin was originally introduced to denote proteins that form nonspecific channels but it evolved to include other substrate-specific pore-forming proteins produced by bacteria. These channels display interesting effects that go beyond passive diffusion and their electrophysiological characterization may provide new insights to learn about ion transport regulation in many other protein channels. Aside from their common structural motifs they share a number of other features, the most relevant of which are their multiionic character, their weak selectivity and the key role of long range electrostatic interactions in their regulation of ion permeation. Here, we review recent advances in the molecular basis of the interaction of general diffusion bacterial porins with small inorganic ions. Single channel current and reversal potential measurements obtained from planar bilayer electrophysiology are analyzed either using continuum electrodiffusion models or with the aid of simple statistical mechanics concepts.

Keywords Ion channel • Porin • Planar bilayer • Ion transport • Modulation • Membrane protein • Lipid • β-barrel structure • Cell membrane permeability • Electrostatics • Bioelectrochemistry

5.1 Introduction

The outer membrane (OM) of Gram-negative bacteria hosts a number of channel-forming proteins called porins and designed for exchanging nutrients and other charged or neutral solutes (usually of molecular weight below 1 kDa) over the OM. Depending on the way OM proteins mediate the transport of metabolites and ions across the membrane, three types of channels can be distinguished: porins (general and substrate-specific porins), substrate specific transporters, and active transporters (Nikaido 2003). Although the term porin was originally introduced to

---

V.M. Aguilella (✉) • M. Queralt-Martín • A. Alcaraz
Laboratory of Molecular Biophysics, Department of Physics, Universitat Jaume I, 12080 Castellón, Spain
e-mail: aguilell@uji.es

A.H. Delcour (ed.), *Electrophysiology of Unconventional Channels and Pores*, Springer Series in Biophysics 18, DOI 10.1007/978-3-319-20149-8_5

refer to proteins that form general or nonspecific channels, it is also currently used to denote a few substrate-specific pore-forming proteins (Nikaido 2003; Galdiero et al. 2007). In this chapter the word porin will denote one of the members of the General Bacterial Porin Family according to the Transport Classification Database (TCDB) (Saier et al. 2014). This superfamily comprises a variety of channel forming proteins, usually homotrimeric, which share an antiparallel β-barrel motif with 16 or 18 strands, according to the available atomic resolution 3D structures. Porins made of 16 strands are called general or non-specific porins and allow the diffusion of hydrophilic molecules, with no particular solute specificity, despite some cationic or anionic selectivity; whereas 18 strands porins are substrate specific porins (Schulz 2002). Representative examples of general porins are OmpF and OmpC from *Escherichia coli* (Nikaido 2003), PorA from *Neisseria meningitidis* (Song et al. 1999; Cervera et al. 2008), Omp32 from *Comamonas acidovorans* (Zeth and Thein 2010) and OmpK36 from *Klebsiella pneumoniae* (Dutzler et al. 1999), to mention just a few. Within the group of substrate specific porins we find the sugar specific LamB from *Escherichia coli* (Benz et al. 1986; Schirmer et al. 1995; Van Gelder et al. 2000; Kullman et al. 2002) and the sucrose-specific porin ScrY from *Salmonella typhimurium* (Charbit 2003; Forst et al. 1998) among others. For an extensive survey the reader is referred to the TCDB (Saier et al. 2014) and other structure and function studies (Koebnik et al. 2000; Galdiero et al. 2007; Fairman et al. 2011).

General porins were the first bacterial pores to be investigated by electrophysiology (Delcour 2013). The planar bilayer technique was already used in the early studies by Benz et al. (1978) and Schindler and Rosenbusch (1978). Three general porins, the major porin from *Rhodobacter capsulatus* (Weiss et al. 1991) and OmpF and PhoE from *Escherichia coli* (Cowan et al. 1992) were also the first ones whose 3D crystal structure was obtained at atomic resolution. Structural studies revealed that each trimer consists of three channels with β-strands connected by several loops and a constriction at the barrel center formed by a long loop L3 folded back into the pore lumen at half the height of the channel. Many additional porin structures have been determined since then. As a result, the simple classification between 16-strand unspecific and 18-strand specific porins became challenged by other OM protein structures (e.g. 14-stranded monomeric) that hardly fit into one of the two categories both structurally and functionally (Schulz 2002; Zeth and Thein 2010). Figure 5.1 shows the 3D structure of three of these porins.

Notwithstanding, general diffusion porins continue to be the most extensively studied porins, partly because of the homology shown between porins from different bacteria and also because they are by far the most abundant in OMs. Furthermore, the relevance of general porins to the gram-negative bacteria resistance to beta-lactam antibiotics is well established (Nikaido 1989; Pagès et al. 2008; Delcour 2009; Galdiero et al. 2012; Page 2012) and a great number of studies on the affinity of antibiotics to porins have been reported either from a computational approach (Ceccarelli and Ruggerone 2008; Ceccarelli 2009) or based on electrophysiological measurements (Nestorovich et al. 2002).

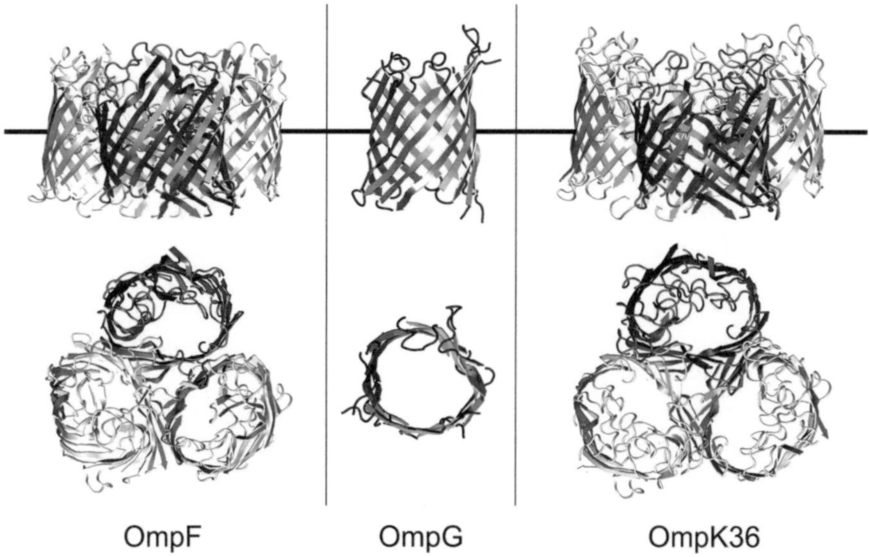

Fig. 5.1 Three dimensional structures of three different bacterial porins: OmpF and OmpG from *Escherichia coli* and OmpK36 from *Klebsiella pneumonia*

Different methodologies have been applied to characterization of porins: from liposome swelling assays and antibiotic flux measurements in live cells up to patch-clamp or planar bilayer electrophysiology (Delcour 2003). However, the last one has been the technique of choice for a detailed functional characterization of their transport and selective properties in a variety of experimental conditions.

Reviewing succinctly the large amount of electrophysiological studies performed on porins is a formidable task beyond the space limitations of this chapter. Our goal here is more modest but so much or even more interesting. We aim to show that general porins display interesting effects that go beyond passive diffusion and that their electrophysiological characterization may provide new insights to learn about ion transport regulation in many other protein channels, in particular those with a similar beta-barrel structure even in prokaryotes (Zeth and Thein 2010). Single channel current and reversal potential measurements in a representative nonspecific porin as is, for instance, OmpF from *Escherichia coli* reveal a wealth of information about the way porins interact with small inorganic ions, something that is potentially useful to understand the permeation of other charged solutes like antibiotic molecules. Furthermore, because of the lack of ion especificity of OmpF and its homologous porins, manipulating the charge of the protein itself, the host lipid and that of the permeating ions, allows us to understand the molecular basis of porin selectivity, their pH-regulated conductance, the channel response to multivalent ions, the electrostatic effects of a charged membrane and other related phenomena. This knowledge may be used in porins of unknown 3D atomic structure to predict structural features from electrophysiological measurements.

Last but not least, in recent years biological channels that are well characterized functionally and structurally have inspired synthetic nanopores with novel properties like rectification, and promising applications as solute detection and counting, nucleotide sequencing, etc. The alpha-hemolysin toxin from *Staphylococcus aureus* is the best example of those channels. We will show here that, even without chemical modification or mutation of aminoacid residues, the matrix porin of *Escherichia coli*, OmpF, exhibits some rectification properties that can be translated into similar size nanopores.

Otherwise noticed, in all experiments reported in this chapter, the transmembrane potential is defined as positive when it is greater at the *cis* side (the side of the protein addition). Furthermore, the measured single channel conductance is always that of the whole channel rather than a monomeric conductance (in the case of trimeric channels).

5.2 Electrostatic Properties Regulate Electrodiffusion

The specific function of a particular channel arises from the competition between two antagonistic features: efficient discrimination between charged species and ability to transport ions and charged solutes as fast as possible. The pore size is the factor that usually tips the balance in favour of one or the other. Bacterial porins are wide channels that allow simultaneous transport of water molecules, solvated ions and even small solutes (metabolites and antibiotics) at the price of displaying no particular substrate specificity and only a weak selectivity between anions and cations. Narrow channels usually conduct one type of ion and exclude not only ions of opposite charge but also other ions with the same charge. The steric constraints arising from the close ion-residue interactions generate a peculiar transport mechanism in a single-file fashion.

Ion selectivity is not an inherent channel feature but inevitably comprises both the protein channel and the electrolyte flowing through it. Thus, the discrimination between ions produced by the channel could be remarkably different depending not only on the ion intrinsic properties (size, solvation, diffusivity in water, electrical charge, shielding effects, etc.) but also on the conditions of the bathing solution (concentration, acidity, temperature, etc.) (Alcaraz et al. 2009). The leading contributions to ion selectivity in large channels like most bacterial porins, are: electrostatic exclusion due to the interaction between permeating ions and channel ionizable residues, diffusional effects due to the differences between cation and anion mobilities, and short range non-coulombic interactions.

Ion channel characterization is usually performed under physiological conditions. This involves experiments performed at moderate gradients of KCl solutions buffered at neutral pH and with salts of multivalent cations present in micromolar or millimolar concentrations. At such conditions specific binding of ions is usually irrelevant, and diffusion potentials are also negligible because K^+ and Cl^- have almost equal bulk mobilities (Alcaraz et al. 2009). This allows one to reduce the description to electrostatic exclusion only and to interpret selectivity exclusively in terms of the effective channel charge. Thus, the reversal potential measurements in OmpF porin with KCl performed at a constant concentration ratio c_{cis}/c_{trans} (Fig. 5.2a) but at different concentrations reveal that the reversal potential is a function of absolute concentrations. The decrease in the selectivity with the increasing salt concentration is then a natural consequence of the screening of the channel charges.

The correlation between the ionization state of protein residues and the channel discrimination is even more evident in selectivity experiments at different pH, as seen in the pH dependence of the K^+ transport number (Fig. 5.2b). The successive protonation and deprotonation of acidic and basic residues explains how this cation-selective channel at basic and neutral pH turns into anion-selective in an acidic environment (Nestorovich et al. 2003; Aguilella-Arzo et al. 2007). Interestingly, the correlation between charge and selectivity is not limited to OmpF. This porin was compared to its structural homolog OmpC in different experimental conditions in the studies by Biró et al. (2010) and Pezeshki et al. (2009). Both porins show weak cation selectivity, being OmpC slightly smaller than OmpF. The small differences between OmpC and OmpF lie in the number and role of ionizable residues, leading to a different density of counterions interacting with the protein, as shown in Fig. 5.3a–d. An additional negatively charged residue and the less-exposed positive charges explain the larger cation selectivity of OmpC compared to OmpF.

Figure 5.3a–d show also that both OmpF and OmpC pores display similar permeation pathways, with positive and negative ions following screwlike pathways. The trajectories of positive and negative ions do not intersect, as shown in Fig. 5.3f

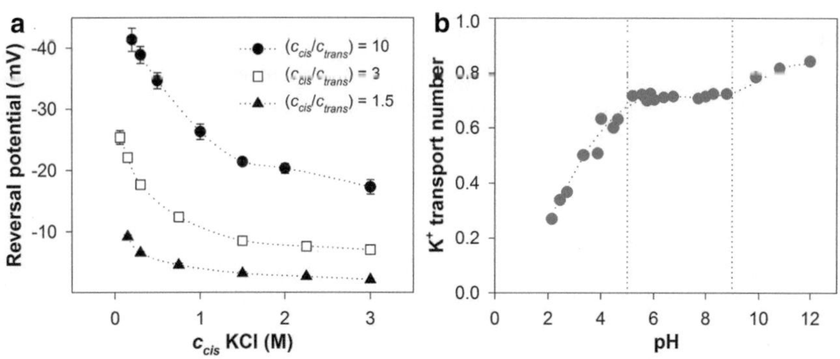

Fig. 5.2 (**a**) Reversal potential measured at three constant *cis/trans* KCl concentration ratios but different absolute concentrations. The channel exhibits cationic selectivity: higher salt concentration at the *cis* side generates negative reversal potential. The absolute value of this potential decreases as salt concentration goes up (Adapted from Alcaraz et al. (2004), Copyright 2004, with permission from Elsevier), (**b**) pH-dependent ion selectivity of a fully open single channel, given as potassium transport number. Dotted lines are shown to guide the eye (Adapted from Nestorovich et al. (2003), Copyright 2003, with permission from Elsevier)

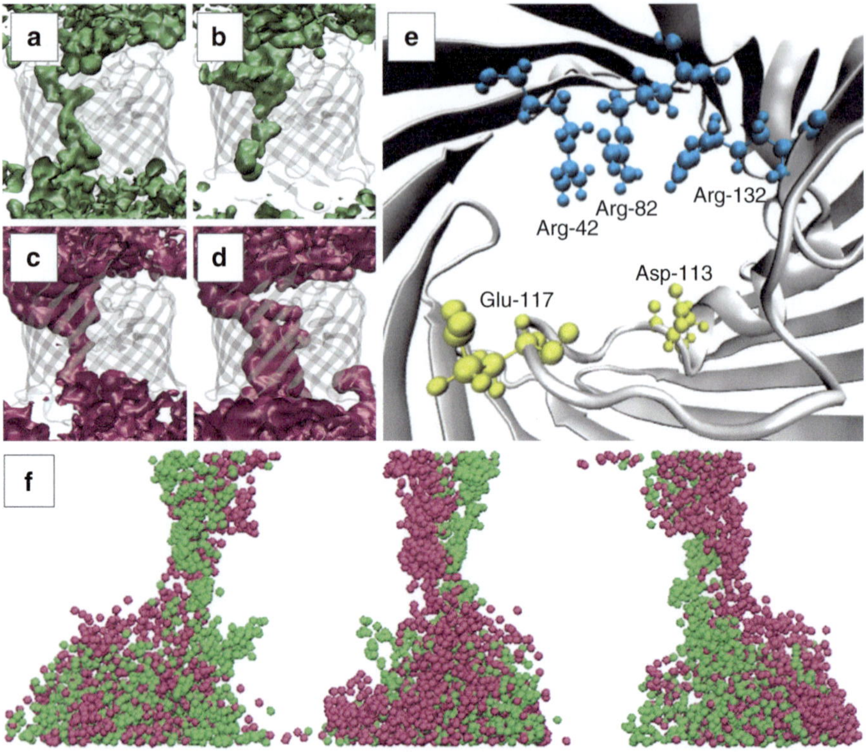

Fig. 5.3 Isosurfaces of ion density obtained from Molecular Dynamics (MD) simulations with 0.5 M KCl at 22 °C. (**a**) Cl^- density in OmpF, (**b**) Cl^- density in OmpC, (**c**) K^+ density in OmpF, (**d**) K^+ density in OmpC. All isosurfaces represent the same density value (Adapted from Biró et al. (2010), Copyright 2010, with permission from Elsevier), (**e**) View of OmpF eyelet with two acidic residues (D113, E117) facing a cluster of three positive arginines (R42, R82, R132) (Reproduced from Aguilella et al. (2011) with permission from The Royal Society of Chemistry), (**f**) K^+ (*magenta*) and Cl^- (*green*) MD ion pathways of OmpF channel showing left-handed screw-like fashion trajectories without intersecting (Reprinted from Im and Roux (2002a), Copyright (2002), with permission from Elsevier)

for OmpF (Im and Roux 2002a). This arises from a conserved feature of general diffusion porins, the existence of a peculiar charge distribution at the pore eyelet, containing charged residues at opposite sides, shown in detail for OmpF in Fig. 5.3e.

The pore constriction of OmpC is very similar to that of OmpF with more that 50 % of the pore lining residues being conserved (Baslé et al. 2006). Several computational studies have reported that the electric field generated by the charged residues of the constriction is strong enough to orient water molecules as well as other dipolar molecules (Karshikoff et al. 1994; Schirmer and Phale 1999; Robertson and Tieleman 2002; Im and Roux 2002a, b). In addition, this charge configuration serves to enhance ion permeability of either cations or anions, dependent on the type of porin.

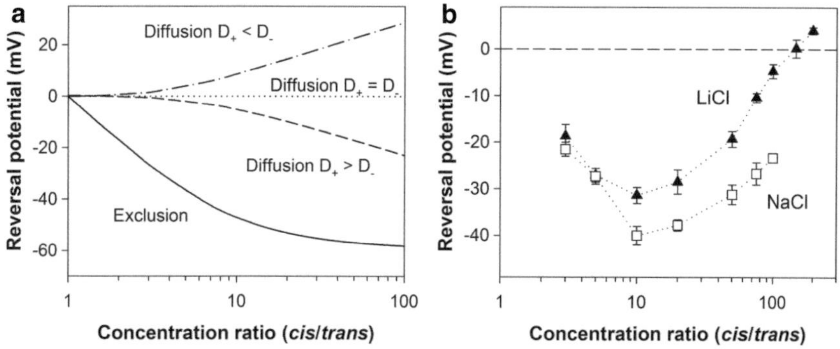

Fig. 5.4 (**a**) Theoretical calculations of exclusion and diffusion contributions to reversal potential. *Solid line*: total Donnan potential drop (exclusion contribution). *Dashed and dot lines*: Diffusion potential across the channel in three different cases of relative cation and anion mobilities as labeled ($D_+=0.5D_-$; $D_+=D_-$; $D_+=2D_-$), (**b**) OmpF reversal potential measured in LiCl (*triangles*) and NaCl (*squares*) at pH 6. The salt concentration is 0.03 M on the *trans* side, and the concentration on the *cis* side varies from 0.09 to 6 M (LiCl) or 3 M (NaCl). The dashed line marks the limit between cationic (negative reversal potential) and anionic (positive reversal potential) channel selectivity under these experimental conditions (Adapted with permission from López et al. (2009). Copyright (2009) American Chemical Society)

5.3 Not Only Electrostatic Exclusion: The Role of Diffusion and Ion Binding

The previous section showed that the way in which bacterial porins regulate the ion transport under physiological conditions can be understood in terms of partitioning. However, in reversal potential experiments with concentrated solutions at large concentration ratios (50–100) and electrolytes such as chloride salts of alkaline or multivalent cations, the description of selectivity only in terms of ion accumulation/depletion could be an oversimplification because the contributions of specific binding and diffusion potentials can be substantial (Alcaraz et al. 2009). Those results challenge the classical view of the ion selectivity as an intrinsic feature of the channel and open new possibilities to tune the transport properties of the pores both in terms of current asymmetry and selectivity (Queralt-Martín et al. 2013a).

The role of diffusion in selectivity experiments can be studied theoretically by comparing its relative contribution to the partitioning described by Donnan equilibrium (López et al. 2009). As shown in Fig. 5.4a for a negatively charged channel like OmpC or OmpF, the diffusion potential calculated under Planck's electroneutrality assumption (López et al. 2009) has the same sign as the Donnan potential for $D_+>D_-$, but the opposite for $D_+<D_-$.

Taking into account that channel characterization is usually carried out in solutions like NaCl, LiCl, $MgCl_2$, $CaCl_2$ or $LaCl_3$ (in all of them $D_+<D_-$) we observe that exclusion and diffusion play antagonistic roles. At sufficiently high concentration ratio the diffusion dominates the overall selectivity and even could reverse the canonical

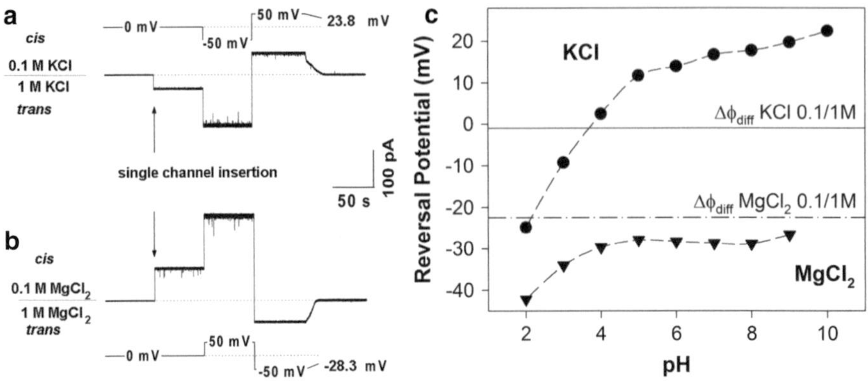

Fig. 5.5 Selectivity inversion by divalent cations. (**a**) Current traces of spontaneous channel insertion at 10-fold gradients of KCl, (**b**) A similar experiment with $MgCl_2$. The potentials that should be applied in order to zero currents in the two cases are of opposite signs, (**c**) Reversal potential of OmpF channel in salts of KCl and $MgCl_2$. The lines represent the diffusion potential values ($\Delta\phi_{diff}$) arising from the difference between cation and anion diffusivities for concentration ratios of 0.1/1 M in KCl (*solid line*) and in $MgCl_2$ (*dashed-dotted line*) (Adapted from Aguilella et al. (2011) with permission from The Royal Society of Chemistry)

channel selectivity reported under moderate gradients and physiological concentrations. The experimental realization of this situation is shown in Fig. 5.4b for OmpF.

Furthermore, the channel selectivity can be dramatically altered when a short-range interaction between permeant ions and protein groups outweighs the Coulombic attraction or repulsion. A remarkable example is the selectivity of bacterial porins in presence of multivalent cations. The moderate cationic selectivity of the OmpF channel in salts of monovalent ions at neutral pH turns into anionic selectivity in salts of divalent and trivalent cations. Figure 5.5 shows sample current traces from reversal potential measurements performed under a 10-fold concentration gradient of KCl and $MgCl_2$ and the corresponding reversal potential values as a function of pH.

In salts of multivalent cations the connection between charge and selectivity is not so straightforward as mentioned in Sect. 5.2 for monovalent ions. The large difference between cation and anion diffusivities in salts of multivalent cations yields a noteworthy diffusion potential. The diffusion potential shown in Fig. 5.5c (dashed-dotted line) would be the measured reversal potential in a neutral ideal channel devoid of any electrostatic interaction and just filled with a $MgCl_2$ solution. This means that the selectivity inversion may simply arise from the counterbalance between the diffusional contribution and the channel electrostatic preference for cations.

5.4 Correlation Between Structure and Function

There is ample experimental evidence that the charge state of ionizable residues of general diffusion porins determine channel transport properties to a large extent. However, contrary to what is usual in other channel families, it is not the charge state

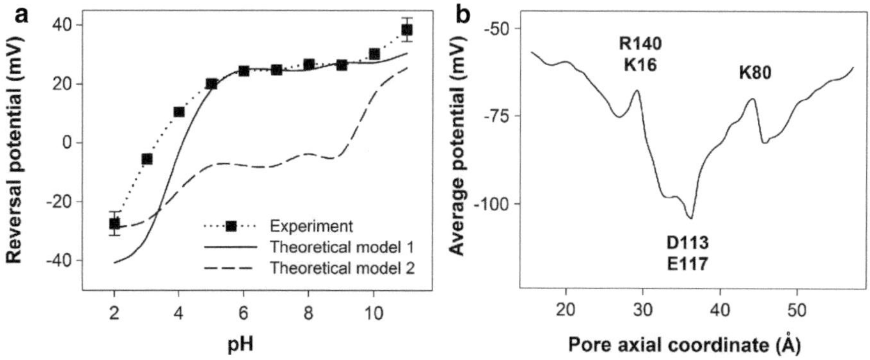

Fig. 5.6 (**a**) Experimental (*squares*) and theoretical predictions (*solid and dashed lines*) of the reversal potential for KCl $c_{cis} = 0.1$ M | $c_{trans} = 1$ M as a function of the pH. Good agreement is found between the theoretical model 1 (PNP model using the same conditions than in experiments and including in the calculations the ionization state of all the channel residues, *solid line*) and experiments but for extreme pH conditions. The reversal potential predicted exclusively by the residues at the channel constriction (theoretical model 2, *dashed line*) is also shown. In this case, the model predicts an anion selective channel for most of the pH range, (**b**) Model calculations of the electric potential profile along the pore longitudinal axis for an OmpF channel in KCl solution. The potential at each axial position is the averaged value over the solvent-accessible region of every channel cross-section. Labels with the main ionizable residues to which the principal peaks and wells may be ascribed are included (Adapted from Aguilella-Arzo et al. (2007), Copyright 2007, with permission from Elsevier)

of the residues at the channel constriction what determines the channel selectivity but the overall net charge of the protein, which also includes the charge of residues that are not solvent accessible, so called "buried". This implies that the concept of selectivity filter, crucial in highly selective (narrow) channels is less meaningful in wide multiionic channels. In line with this fact, studies of mutants (Phale et al. 2001) show no clear correlation between charge alterations introduced by mutations and changes in channel selectivity. In addition, the net charge of ionizable residues along the lumen of the OmpF pore is positive at neutral pH (Alcaraz et al. 2004), as well as the net charge of the constriction, in apparent contradiction with the channel cation selectivity: see in Fig. 5.6a (dashed line) the reversal potential calculated considering only those eyelet residues (Aguilella-Arzo et al. 2007). Both facts are in apparent contradiction with the channel cation selectivity. The involvement of residues far from the eyelet was also confirmed in recent studies of OmpF mutants (Vrouenraets and Miedema 2010). However, continuum electrostatics calculations using as input the atomic structure of OmpF (Cowan et al. 1992) yield a negative effective net charge and an electric potential profile along the pore with a characteristic negative potential well as shown in Fig. 5.6b (Aguilella-Arzo et al. 2007).

By averaging the potential over the pore crossection, the fine electrostatic details are blurred (with the exception of the relevant ionizable residues of the constriction, labeled in Fig. 5.6b) but it becomes clear that the channel cationic selectivity is consistent with the negative potential distribution. In addition to accounting properly for all protein charges, simulations and continuum electrostatic calculations in porins face the problem of an adequate representation of the dielectric properties of the protein and the solvent. Electrostatic calculations of the electric field in the bacterial porin OmpF reported values that are high enough to orient water molecule dipoles and consequently decrease the water dielectric constant up to 50 % at the channel constriction (Aguilella-Arzo et al. 2009). As for the protein charges, since their ionization state (specially that of buried charges) is critically dependent on the surrounding dielectric environment, the protein dielectric constant becomes a kind of hidden parameter of MD simulations that may change the initial charge distribution throughout the channel. This may be also relevant to other channels where long range electrostatic interactions control ion transport.

The distribution of ionizable residues in many bacterial porins is not homogeneous but asymmetric. This means that the channel opening facing the extracellular side may exclude or accumulate ions in a different way than the side at the intermembrane space. However, this asymmetrical channel charge distribution cannot be inferred from the usual current–voltage measurements in porins because its effect is not high enough to produce current rectification. Reversal potential measurements are, though, more sensitive to asymmetries in the effective channel fixed charge density. Selectivity measurements in both directions of the salt gradient give measurable differences in reversal potential (Alcaraz et al. 2004). When the 3D structure of a channel is not known, these experiments may help to predict an asymmetrical charge distribution provided that the channel insertion in the lipid bilayer is directional (Cervera et al. 2008).

The change in the channel effective charge induced by titration of the residues along the pore is always accompanied by variations in channel conductance. The addition of protons would normally foreshadow an increase in conductance but there is no such general trend among porins. On the contrary, exhaustive studies have shown that as the pH is lowered, the channel conductance in bacterial porins can remain constant, as is the case of OmpT from *Vibrio cholerae* (Duret et al. 2007) or drastically decrease, as happens in other porins like OmpF.

Likewise, the effect of lipid charge on bacterial porins does not show a unique trend, but it has been reported that channel conductance is different when channels are reconstituted in charged membranes and in neutral membranes; depending on the cationic or anionic channel selectivity an increase or a decrease of ionic conductance is observed. In most porins the change in conductance is only measurable at low electrolyte concentrations and solutions above physiological concentration screen the lipid charges up to the point at which channel currents in charged and neutral lipids become almost indistinguishable. As lipid charged headgroups surrounding the pore entrance are responsible for the alteration of local ion concentrations, the smaller the channel mouth the higher the effect of the charged lipid (Aguilella et al. 2014).

5.5 Current Rectification and Asymmetric Blocking

Unlike specific channels like inward-rectifying potassium channels (Hille 2001), bacterial porins do not naturally exhibit a rectifying behavior. However, such an asymmetry can be achieved in some particular conditions. Figure 5.7a shows current rectification observed in the OmpF porin when the extracellular and periplasmic sides of the channel face different pH solutions. In this case, the *I-V* curves display a clear non-linearity with higher currents at negative voltages. This remarkable feature obeys to the particularities of the reversible titration of the channel ionizable residues. As mentioned before, ion conduction in bacterial porins is mainly influenced by electrostatic interactions between the permeating ions and the titratable residues within the pore (Nestorovich et al. 2003; Nikaido 2003). In this scenario, the ionization state of these residues determines the measured ion current. Under the pH conditions of Fig. 5.7a, the acidic residues at the *cis* side are titrated in such a way that the corresponding channel mouth becomes positively charged, accumulating negative ions. On the other side, basic residues at the *trans* side are neutralized, yielding a negatively charged channel end, which accumulates positive ions. This particular ion distribution induces current rectification (Alcaraz et al. 2006). The ratio of rectification has been reported to decrease as salt concentration increases because of the screening of the channel fixed charges by mobile ions. Indeed, at high salt concentrations, the difference in titration at the channel mouths is attenuated and current becomes more similar under both polarities (Alcaraz et al. 2006).

The remarkable effect of an asymmetric pH on OmpF may help to understand the role of porins located in the OM in protecting bacteria against acidic media (Bearson et al. 1997). Furthermore, it opens the door to the design of new rectifying devices based on synthetic nanopores. In this sense, bacterial porins have become a valuable candidate for applications in nanotechnology, as will be shown in Sect. 5.7.

Besides a pH-asymmetry of the external solutions, there are other strategies to obtain current rectification in bacterial porins, as is protein engineering. The design of protein mutations may lead to engineer a bacterial porin with intrinsic diode characteristics that can rectify without the need of asymmetric external conditions. Figure 5.7b shows the current asymmetry obtained with a particular residue modification of the bacterial porin OmpF. In that study, the existence of two spatially

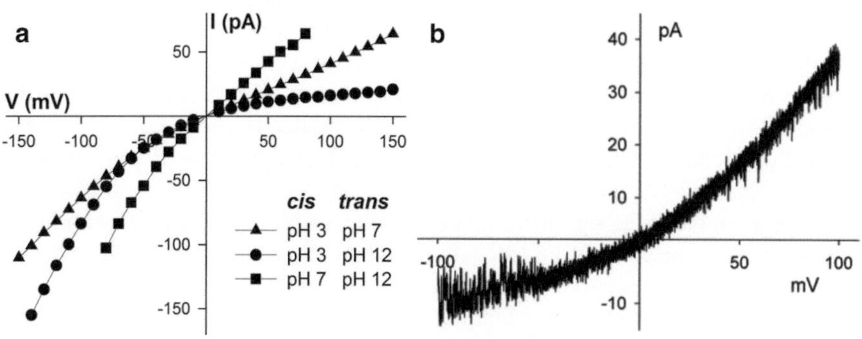

Fig. 5.7 (**a**) Current–voltage curves of OmpF channel in 0.1 M KCl under several asymmetric pH conditions. Only when the pH is asymmetric, the channel shows noticeable rectification (Adapted with permission from Alcaraz et al. (2006). Copyright (2006) American Chemical Society), (**b**) OmpF-RRR mutant current profile recorded in 0.1 M NaCl, pH 7.4 (in these experiments, the protein was added at the *trans* side of the membrane, the side set to ground) (Adapted with permission from Miedema et al. (2007). Copyright (2007) American Chemical Society)

separated areas where either cations or anions accumulate was obtained with site directed mutagenesis (Miedema et al. 2007). Indeed, particular negative residues located at the extracellular vestibule of the protein were replaced with arginines (mutant OmpF-RRR), changing the net charge by 6*e* and inducing a strong anion accumulation in that area. This particular channel arrangement induces a redistribution of ions, causing the existence of a zone depleted of charge carriers only for a particular current direction. This situation, in turn, yields current rectification, which may be tuned by changing the selected residues to be replaced. Similar asymmetric I-V curves to that shown in Fig. 5.7b were obtained by substituting the residues of the central and extracellular zones of the protein (so-called first and second selectivity filter in the study) with amino acids of different charges (Miedema et al. 2007). It was also proposed to introduce a third selectivity filter in the periplasmic vestibule of the channel. However, the refolding efficiency of some of the mutated proteins is still the most important difficulty in protein engineering.

Besides pH or protein modification, other factors can induce current asymmetry in bacterial porins. For instance, multivalent ions are able to modify the ionization state of channel residues or to directly block the pore, thus inducing a significant effect on channel current that is not observed with monovalent ions. Figure 5.8a, b show that salts of multivalent cations may shape in different ways the current in OmpF. While the interaction between a divalent cation, Mg^{2+}, and the channel is not strong enough to produce any current asymmetry (Verdiá-Báguena et al. 2012) a trivalent cation, La^{3+}, does induce a voltage-dependent current blockage, thus yielding non-linear *I-V* curves (Queralt-Martín et al. 2013b).

In contrast to the rectification induced by asymmetric pH explained above, in this case $LaCl_3$ is added symmetrically, which indicates that the mechanism by which La^{3+} ions induce current rectification in OmpF is clearly different from an asymmetric ionization of channel residues. On the contrary, it points to a particular channel blocking by La^{3+} ions.

The investigation of La^{3+} partial current inhibition in OmpF was addressed in two ways. First, the *I-V* curve itself was analyzed by means of a blocking model, as shown in Fig. 5.8b. The solid lines represent the fit to a blocking model (Queralt-Martín et al. 2013b) that includes the Hill formalism (Hill 1910) and the Woodhull model (Woodhull 1973). These theoretical approaches give information about the blocking through different parameters as the Hill coefficient, the electrical distance, and the dissociation constant (Queralt-Martín et al. 2013b). In addition, the study was complemented by performing experiments with channel mutants. Figure 5.8c shows an example of the significant changes in the rectification induced by La^{3+} ions in a selected OmpF mutant, D113R/E117R, where the addition of $LaCl_3$ seems to have no effect in the *I-V* curves. The residues D113 and E117 were modified because they have been reported to be a binding site for monovalent and divalent cations (Kobayashi and Nakae 1985; Suenaga et al. 1998; Danelon et al. 2003; Nestorovich et al. 2003; Yamashita et al. 2008; López et al. 2009). This study allowed not only to identify which residues are more likely to be responsible for the blocking, but also demonstrated that the asymmetric effect of La^{3+} in OmpF is not due to the position of the interacting site, which is actually located at the center of the pore, as shown in Fig. 5.8d.

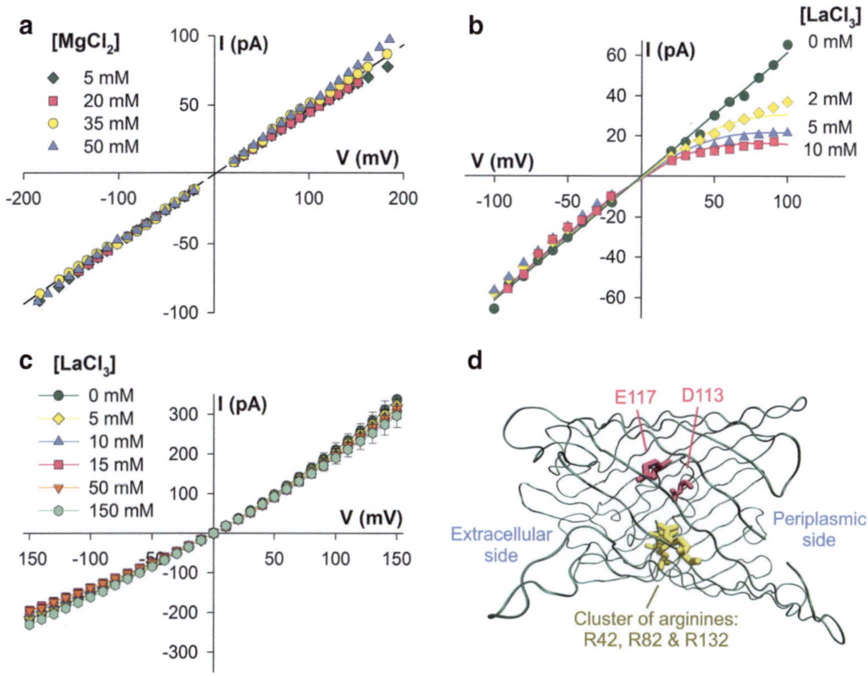

Fig. 5.8 (**a**) Single – channel current-voltage curves in 0.1 M KCl symmetrical solution at pH 6 after the addition of millimolar concentrations of $MgCl_2$ (Adapted with permission from Verdiá-Báguena et al. (2012). Copyright (2012) American Chemical Society), (**b**) Rectification induced by millimolar concentrations of $LaCl_3$ in 0.1 M KCl pH 6 solution. Solid lines correspond to the fit of the experimental data to a modified Woodhull model, (**c**) Single – channel current-voltage curves of the mutant OmpF-D113R/E117R in 1 M KCl symmetrical solution at pH 6 after the addition of millimolar concentrations of $LaCl_3$, (**d**) Lateral view of the OmpF channel obtained from the crystal structure (PDB code: 2OMF). Two acidic residues (D113 and E117) and three basic residues (R42, R82, and R132) of the constriction are highlighted (Adapted with permission from Queralt-Martín et al. (2013b). Copyright (2013) American Chemical Society)

Altogether, the results shown in this section highlight different strategies used to obtain current rectification in bacterial porins, a property that is not naturally exhibited by these systems. These different approaches are based in a deep knowledge of the channel under study, which has been obtained after many years of electrophysiological investigations of the ion transport in bacterial porins.

5.6 pH-Mediated Current Inhibition and Ion Competition

The sensitivity of bacterial porins to pH seen in their selectivity and current rectification displays other interesting facets, particularly when salt concentration is high. The reduction of the channel conductance observed at acidic pH in some bacterial porins could seem counter-intuitive because the addition of protons would normally be expected to increase the conductance. Several explanations have been proposed. In some cases (OmpC, PhoE), a current decrease is observed due to the existence of new channel conformations with reduced pore aperture that dominate at low pH (Todt et al. 1992). In other cases (OmpU), the gating is the responsible for the current decrease (Duret et al. 2007).

However, in other bacterial porins more subtle mechanisms are responsible for the conductance decrease at low pH, as happens with OmpF. In this case, the partial current inhibition under moderate acidic conditions is consistent with the titration of some negative residues and the subsequent reduction of the net negative charge of the channel. However, the further current decrease at very low pH, (where pH reduction increases the net positive charge and the proton concentration may contribute significantly to the amount of mobile charge carriers), goes beyond pure electrostatics. It has been explained by invoking the existence of a competitive interaction between protons and cations at the channel eyelet (Alcaraz et al. 2012).

This subtle mechanism is revealed, first, by a conductance decrease accompanied by a remarkable increase in the low-frequency current spectral density, which accounts for the current noise (Fig. 5.9a, b). Interestingly, the noise peak and the steepest conductance decrease occur around the same pH. Figure 5.9a, b also show that the effect of salt concentration on the conductance drop by pH is the opposite of the well-known salt screening effect (Alcaraz et al. 2004). In this case, increasing salt concentration enhances the interactions responsible for the conductance decrease. This indicates some kind of interaction between the two types of ions involved, protons and cations.

Whether this interaction is competitive or non-competitive can be clarified by double reciprocal plots of conductance versus salt concentration (Brelidze and Magleby 2004). In the case of OmpF, this kind of plot revealed a competitive interaction between protons and cations at intermediate pH, that turned into non–competitive at low pH (Alcaraz et al. 2012). The residues involved in this interaction were localized by means of current measurements with mutants. Figure 5.9c shows that the pH titration of the OmpF conductance differs considerably between the wild type and mutant forms in which one (D113C) or two (D113C/E117C, labeled as CC) of the key acidic residues of the channel constriction were replaced by neutral cysteines.

Identifying the particular residues involved in an interaction opens the door to the development of simplified molecular models that evaluate the microscopic configurations of the system. These formalisms, usually based on statistical thermodynamics, probe different quantities that can be compared with those measured in the experiments. For instance, Fig. 5.9d (top) shows the fraction of unprotonated residues (α) of OmpF channel as a function of pH. It was calculated using the grand canonical ensemble and considering the acidic residues D113 and E117 as the key sites for the interaction between protons and cations. There is a clear correspondence between the sigmoidal curve for α in this figure and that for the measured conductance shown in Fig. 5.9c (circles). This correspondence links directly the protonation of those residues at the eyelet with the conductance decrease.

Figure 5.9d (bottom) shows the configurational entropy obtained from the same molecular model. Its shape clearly resembles the pattern of the current spectral density (Fig. 5.9b), with a peak around pH 2.5. But the computation of the entropy

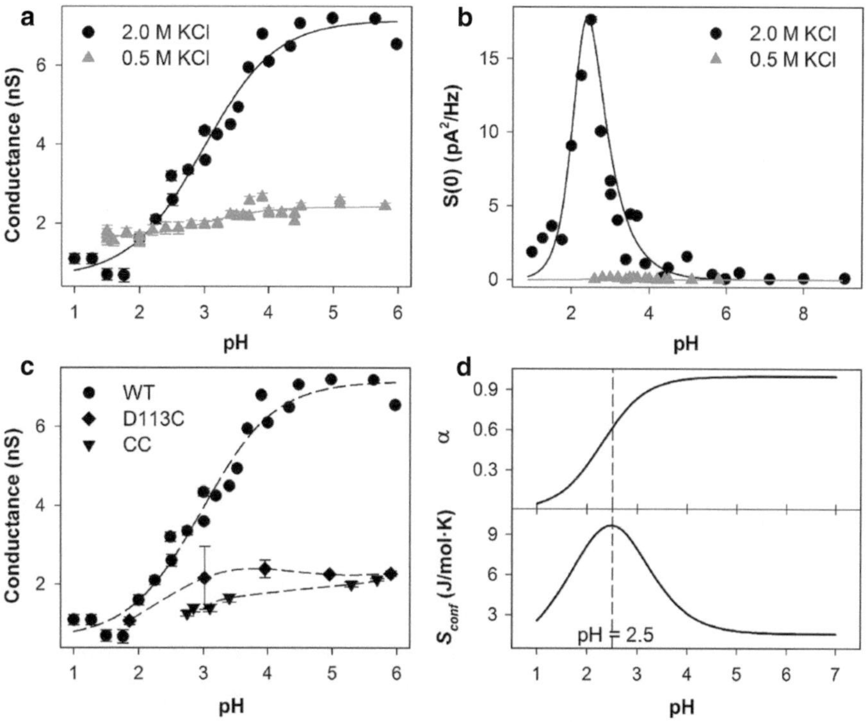

Fig. 5.9 (**a**) OmpF single−channel conductance in 2 M and 0.5 M KCl solutions, (**b**) Low−frequency spectral density of current fluctuations S(0) at the same conditions. The observed decrease of conductance is accompanied by an increase of S(0) that exhibits a sharp peak around pH 2.5 in 2 M KCl. The solid lines correspond to a theoretical fit using the Hill formalism, (**c**) Single−channel conductance as a function of pH for wild type OmpF (WT) and mutants D113C and D113C/E117C (CC) measured in 2 M KCl solution. Dashed lines are shown to guide the eye (Adapted from Alcaraz et al. (2012), Copyright 2012, with permission from Elsevier), (**d**) Fraction α of unprotonated acid residues (*top*) and configurational entropy S_{conf} (*bottom*) versus pH, obtained from a statistical thermodynamics model using $c = 1$ M and pKc = 1.7 (Adapted from Alcaraz et al. (2014) with permission from The Royal Society of Chemistry)

goes beyond its comparison with other quantities because it can also be obtained through electrophysiology by means of temperature-dependent current measurements (Chimerel et al. 2008).

Temperature-controlled experiments allow analyzing the different contributions to the free energy of ion conduction as a function of different parameters as cation type, salt concentration or solution acidity. This is possible because the ion electrodiffusion through a porin can be related to the Eyring's approach or Kramers theory (Hänggi and Borkovec 1990), providing a theoretical framework to obtain the activation enthalpies and entropies of the system (Hajjar et al. 2010; Cheneke et al. 2012; Alcaraz et al. 2014). The decomposition of free energy changes into the enthalpic and entropic terms is useful to understand essential functions at the single ion channel level. Furthermore, this approach opens the door to discuss a possible entropy-enthalpy compensation, an effect already reported in many protein-ligand associations that has become a matter of debate for its implications in ligand engineering (Sharp 2001; Fenley et al. 2012; Chodera and Mobley 2013; Martin and Clements 2013). Temperature-dependent electrophysiology is also used to study the facilitated permeation of antibiotics through bacterial porins (Mahendran et al. 2009). In this case, the mode of antibiotic permeation is determined through the analysis of current fluctuations at different temperatures.

5.7 Future Prospects: Applications in Nanotechnology

The detailed information that electrophysiology has provided over the last years about the functional properties of bacterial porins makes these systems valuable candidates for applications in nanotechnology. The fact that bacterial porins can be easily inserted in many different types of lipid environments is a crucial feature to produce robust synthetic scaffolds necessary for different technological applications (Alcaraz et al. 2006). In this sense, nanotechnology can benefit enormously from the knowledge of the transport mechanisms in nanoscaled biological systems, as is the case of bacterial porins, by mimicking and designing devices inspired by nature's nanopores (Majd et al. 2010).

Previous studies have taken advantage of the current modulation of different bacterial porins to design different nanodevices. For instance, the rectifying properties of OmpF under an asymmetric pH configuration were used to pump potassium ions against an external concentration gradient. Figure 5.10 shows a characteristic experiment that promotes cation pumping with OmpF. Salt concentration of the solutions is kept constant (*cis* side is more concentrated than *trans* side). A symmetric square voltage wave is used as the time-dependent input signal (blue solid lines). Under the appropriate pH conditions, an asymmetrical output current is obtained (dark grey solid lines). At low applied voltages, the current is always positive because ions flow from the concentrated solution towards the diluted one. Increasing the magnitude of the square voltage wave leads to a shift of the output current toward negative values, indicating a switch in the flowing directions of mobile ions (Queralt-Martín et al. 2013a).

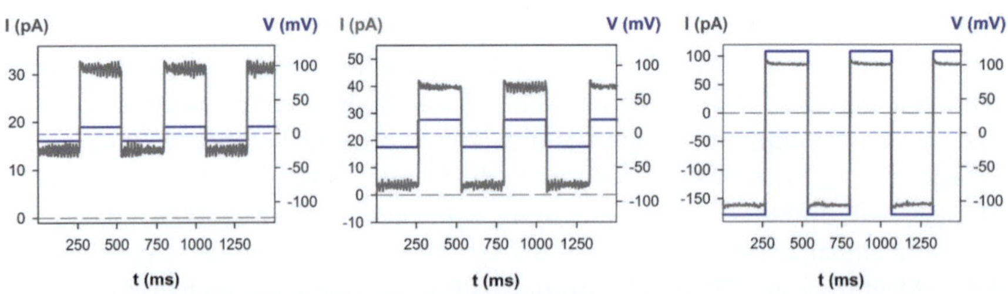

Fig. 5.10 Single channel output current traces (*dark grey solid lines*) obtained upon application of a square voltage wave (*blue solid lines*) of amplitude $V_{max} = \pm 10$, ± 20, and ± 120 mV. The *horizontal grey and blue dashed lines* correspond to 0 pA and 0 mV, respectively. Experiments were made in 0.25 M KCl pH 3 *cis* || 0.05 M KCl pH 12 *trans*

The evaluation of a putative pumping of potassium ions is made through the calculation of the average current. Because the OmpF channel, as most bacterial porins, is only partially selective (Nestorovich et al. 2003; Alcaraz et al. 2004), the existence of a non-zero average current does not necessarily mean that there is pumping of potassium ions (Siwy and Fuliński 2002). The fact that the individual contribution of cations and anions to the overall current changes with voltage calls for a more detailed analysis to confirm the existence of ion pumping. The use of mean-field continuum models based on the three-dimensional structure of the channel –which is available for many bacterial porins– arises as an useful tool for the evaluation of ion contributions (Queralt-Martín et al. 2013a).

Engineered bacterial porins have also shown potential for use as stochastic sensors based on single molecule detection. Although these kind of studies have mostly focused on staphylococcal α-hemolysin (Bayley and Cremer 2001), monomeric and trimeric bacterial porins have arisen as a potential alternative (Reiß and Koert 2013). Specifically, the monomeric porin OmpG of *Escherichia coli* is an attractive choice (Chen et al. 2008) because of its robust β-barrel structure with a single subunit.

Figure 5.11 reports a potential use of OmpG as a sensor for the detection of ADP. The monomeric porin OmpG was mutated to avoid the characteristic gating of the wild-type form. This gating is characterized by rapid spontaneous switches between open and closed states and would obviously interfere with the application

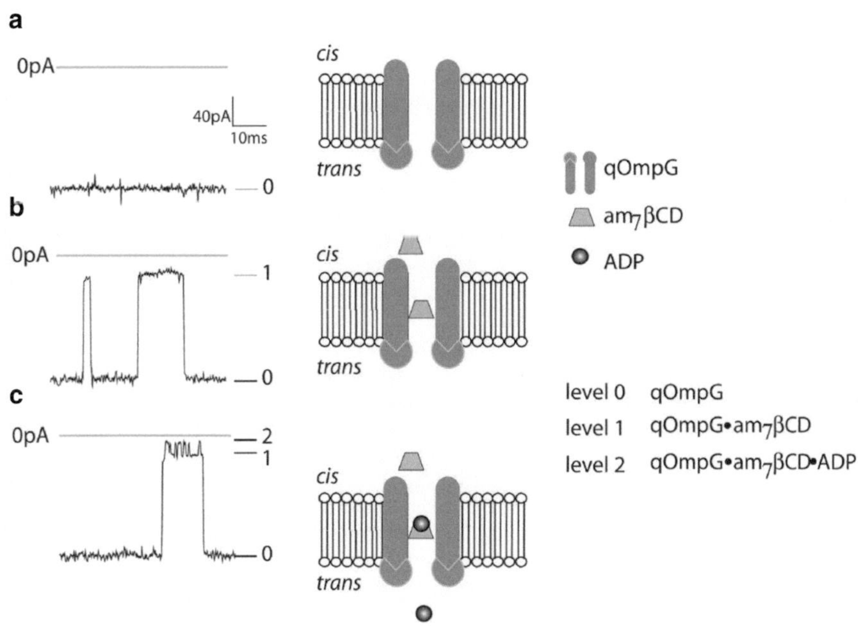

Fig. 5.11 Detection of ADP with the qOmpG · $am_7\beta CD$ complex at −100 mV in 10 mM Tris · HCl (pH 8.5)/1 M KCl. (**a**) Current through a single qOmpG pore, −125 pA (*level 0*), (**b**) $am_7\beta CD$ binding to qOmpG produces transient partial blockades of the channel, −20 pA (*level 1*), (**c**) ADP binding to $am_7\beta CD$ produces additional blockades, −4 pA (*level 2*). In this figure, a positive potential indicates a higher potential in the trans chamber (Adapted with permission from Chen et al. (2008). Copyright (2008) National Academy of Sciences, U.S.A.)

of the pore as a biosensor. The mutated form (qOmpG), free of gating (Fig. 5.11a), was fitted with heptakis-(6-deoxy-6-amino)-β-cyclodextrin ($am_7\beta CD$). $am_7\beta CD$ is a cyclodextrin, a molecular adapter used in stochastic sensing to detect a wide variety of organic molecules (Gu and Bayley 2000). The interaction between qOmpG and $am_7\beta CD$ induced transient partial blockades of the channel current (Fig. 5.11b). The addition of ADP to the complex qOmpG · $am_7\beta CD$ caused a further current block on top of the $am_7\beta CD$ binding events (Fig. 5.11c), which actually allowed the identification and quantification of ADP (Chen et al. 2008).

In summary, the examples shown in this section reveal not only the potential of bacterial porins for use in nanotechnology, but also highlight that electrophysiology provides the basis for the future progress on the design of nature's inspired nanodevices.

Acknowledgments We wish to acknowledge support from the Spanish Ministry of Economy and Competitiveness (MINECO Project FIS2013-40473-P), Generalitat Valenciana (Prometeo 2012/069), and Fundació Caixa Castelló-Bancaixa (Project No. P1-1B2012-03)

References

Aguilella VM, Queralt-Martín M, Aguilella-Arzo M et al (2011) Insights on the permeability of wide protein channels: measurement and interpretation of ion selectivity. Integr Biol 3: 159–172

Aguilella VM, Verdiá-Báguena C, Alcaraz A (2014) Lipid charge regulation of non-specific biological ion channels. Phys Chem Chem Phys 16:3881–3893

Aguilella-Arzo M, García-Celma JJ, Cervera J et al (2007) Electrostatic properties and macroscopic electrodiffusion in OmpF porin and mutants. Bioelectrochemistry 70:320–327

Aguilella-Arzo M, Andrio A, Aguilella VM, Alcaraz A (2009) Dielectric saturation of water in a membrane protein channel. Phys Chem Chem Phys 11:358–365

Alcaraz A, Nestorovich EM, Aguilella-Arzo M et al (2004) Salting out the ionic selectivity of a wide channel: the asymmetry of OmpF. Biophys J 87:943–957

Alcaraz A, Ramírez P, García-Giménez E et al (2006) A pH-tunable nanofluidic diode: electrochemical rectification in a reconstituted single ion channel. J Phys Chem B 110:21205–21209

Alcaraz A, Nestorovich EM, López ML et al (2009) Diffusion, exclusion, and specific binding in a large channel: a study of OmpF selectivity inversion. Biophys J 96:56–66

Alcaraz A, Queralt-Martín M, García-Giménez E, Aguilella VM (2012) Increased salt concentration promotes competitive block of OmpF channel by protons. Biochim Biophys Acta 1818:2777–2782

Alcaraz A, Queralt-Martín M, Verdiá-Báguena C et al (2014) Entropy-enthalpy compensation at the single protein level: pH sensing in the bacterial channel OmpF. Nanoscale 6:15210–15215

Baslé A, Rummel G, Storici P et al (2006) Crystal structure of osmoporin OmpC from E. coli at 2.0 A. J Mol Biol 362:933–942

Bayley H, Cremer PS (2001) Stochastic sensors inspired by biology. Nature 413:226–230

Bearson S, Bearson B, Foster JW (1997) Acid stress responses in enterobacteria. FEMS Microbiol Lett 147:173–180

Benz R, Janko K, Boos W, Läuger P (1978) Formation of large, ion-permeable membrane channels by the matrix protein (porin) of Escherichia coli. Biochim Biophys Acta 511:305–319

Benz R, Schmid A, Nakae T, Vos-Scheperkeuter GH (1986) Pore formation by LamB of Escherichia coli in lipid bilayer membranes. J Bacteriol 165:978–986

Biró I, Pezeshki S, Weingart H et al (2010) Comparing the temperature-dependent conductance of the two structurally similar E. coli porins OmpC and OmpF. Biophys J 98:1830–1839

Brelidze TI, Magleby KL (2004) Protons block BK channels by competitive inhibition with K+ and contribute to the limits of unitary currents at high voltages. J Gen Physiol 123:305–319

Ceccarelli M (2009) Simulating transport properties through bacterial channels. Front Biosci (Landmark Ed) 14:3222–3238

Ceccarelli M, Ruggerone P (2008) Physical insights into permeation of and resistance to antibiotics in bacteria. Curr Drug Targets 9:779–788

Cervera J, Komarov AG, Aguilella VM (2008) Rectification properties and pH-dependent selectivity of meningococcal class 1 porin. Biophys J 94:1194–1202

Charbit A (2003) Maltodextrin transport through LamB. Front Biosci 8:265–274

Chen M, Khalid S, Sansom MSP, Bayley H (2008) Outer membrane protein G: engineering a quiet pore for biosensing. Proc Natl Acad Sci U S A 105:6272–6277

Cheneke BR, Indic M, van den Berg B, Movileanu L (2012) An outer membrane protein undergoes enthalpy- and entropy-driven transitions. Biochemistry 51:5348–5358

Chimerel C, Movileanu L, Pezeshki S et al (2008) Transport at the nanoscale: temperature dependence of ion conductance. Eur Biophys J 38:121–125

Chodera JD, Mobley DL (2013) Entropy-enthalpy compensation: role and ramifications in biomolecular ligand recognition and design. Annu Rev Biophys 42:121–142

Cowan SW, Schirmer T, Rummel G, Steiert M, Ghosh R et al (1992) Crystal structures explain functional properties of two E. coli porins. Nature 358:727–733

Danelon C, Suenaga A, Winterhalter M, Yamato I (2003) Molecular origin of the cation selectivity in OmpF porin: single channel conductances vs. free energy calculation. Biophys Chem 104:591–603

Delcour AH (2003) Solute uptake through general porins. Front Biosci 8:1055–1071

Delcour AH (2009) Outer membrane permeability and antibiotic resistance. Biochim Biophys Acta 1794:808–816

Delcour AH (2013) Electrophysiology of bacteria. Annu Rev Microbiol 67:179–197

Duret G, Simonet V, Delcour AH (2007) Modulation of Vibrio cholerae porin function by acidic pH. Channels (Austin) 1:70–79

Dutzler R, Rummel G, Albertí S, Hernández-Allés S, Phale P, Rosenbusch J, Benedí V, Schirmer T (1999) Crystal structure and functional characterization of OmpK36, the osmoporin of Klebsiella pneumoniae. Structure 7:425–434

Fairman JW, Noinaj N, Buchanan SK (2011) The Structural biology of β-barrel membrane proteins: a summary of recent reports. Curr Opin Struct Biol 21:523–531

Fenley AT, Muddana HS, Gilson MK (2012) Entropy-enthalpy transduction caused by conformational shifts can obscure the forces driving protein-ligand binding. Proc Natl Acad Sci U S A 109:20006–20011

Forst D, Welte W, Wacker T, Diederichs K (1998) Structure of the sucrose-specific porin ScrY from Salmonella typhimurium and its complex with sucrose. Nat Struct Biol 5:37–46

Galdiero S, Galdiero M, Pedone C (2007) β-Barrel membrane bacterial proteins: structure, function, assembly and interaction with lipids. Curr Protein Pept Sci 8:63–82

Galdiero S, Falanga A, Cantisani M, Tarallo R, Della Pepa ME, D'Oriano V, Galdiero M (2012) Microbe-host interactions: structure and role of Gram-negative bacterial porins. Curr Protein Pept Sci 13:843–854

Gu LQ, Bayley H (2000) Interaction of the noncovalent molecular adapter, beta-cyclodextrin, with the staphylococcal alpha-hemolysin pore. Biophys J 79:1967–1975

Hajjar E, Mahendran KR, Kumar A et al (2010) Bridging timescales and length scales: from macroscopic flux to the molecular mechanism of antibiotic diffusion through porins. Biophys J 98:569–575

Hänggi P, Borkovec M (1990) Reaction-rate theory: fifty years after Kramers. Rev Mod Phys 62:251–341

Hill AV (1910) The possible effects of the aggregation of the molecules of haemoglobin on its dissociation curves. J Physiol 40:i–vii

Hille B (2001) Ion channels of excitable membranes, 3rd edn. Sinauer, Sunderland, p 814
Im W, Roux B (2002a) Ion permeation and selectivity of OmpF porin: a theoretical study based on molecular dynamics, Brownian dynamics, and continuum electrodiffusion theory. J Mol Biol 322:851–869
Im W, Roux B (2002b) Ions and counterions in a biological channel: a molecular dynamics simulation of OmpF porin from Escherichia coli in an explicit membrane with 1 M KCl aqueous salt solution. J Mol Biol 319:1177–1197
Karshikoff A, Spassov V, Cowan SW, Ladenstein R, Schirmer T (1994) Electrostatic properties of two porin channels from Escherichia coli. J Mol Biol 240:372–384
Kobayashi Y, Nakae T (1985) The mechanism of ion selectivity of OmpF-porin pores of Escherichia coli. Eur J Biochem 151:231–236
Koebnik R, Locher KP, Van Gelder P (2000) Structure and function of bacterial outer membrane proteins. Mol Microbiol 37:239–253
Kullman L, Winterhalter M, Bezrukov SM (2002) Transport of maltodextrins through maltoporin: a single-channel study. Biophys J 82:803–812
López ML, Aguilella-Arzo M, Aguilella VM, Alcaraz A (2009) Ion selectivity of a biological channel at high concentration ratio: insights on small ion diffusion and binding. J Phys Chem B 113:8745–8751
Mahendran KR, Chimerel C, Mach T, Winterhalter M (2009) Antibiotic translocation through membrane channels: temperature-dependent ion current fluctuation for catching the fast events. Eur Biophys J 38:1141–1145
Majd S, Yusko EC, Billeh YN et al (2010) Applications of biological pores in nanomedicine, sensing, and nanoelectronics. Curr Opin Biotechnol 21:439–476
Martin SF, Clements JH (2013) Correlating structure and energetics in protein-ligand interactions: paradigms and paradoxes. Annu Rev Biochem 82:267–293
Miedema H, Vrouenraets M, Wierenga J et al (2007) A biological porin engineered into a molecular, nanofluidic diode. Nano Lett 7:2886–2891
Nestorovich EM, Danelon C, Winterhalter M, Bezrukov SM (2002) Designed to penetrate: time-resolved interaction of single antibiotic molecules with bacterial pores. Proc Natl Acad Sci U S A 99:9789–9794
Nestorovich EM, Rostovtseva TK, Bezrukov SM (2003) Residue ionization and ion transport through OmpF channels. Biophys J 85:3718–3729
Nikaido H (1989) Outer membrane barrier as a mechanism of antimicrobial resistance. Antimicrob Agents Chemother 33:1831–1836
Nikaido H (2003) Molecular basis of bacterial outer membrane permeability revisited. Microbiol Mol Biol Rev 67:593–656
Page MG (2012) The role of the outer membrane of Gram-negative bacteria in antibiotic resistance: Ajax' shield or Achilles' heel? Handb Exp Pharmacol 211:67–86
Pagès JM, James CE, Winterhalter M (2008) The porin and the permeating antibiotic: a selective diffusion barrier in Gram-negative bacteria. Nat Rev Microbiol 6:893–903
Pezeshki S, Chimerel C, Bessonov AN, Winterhalter M, Kleinekathöfer U (2009) Understanding ion conductance on a molecular level: an all-atom modeling of the bacterial porin OmpF. Biophys J 97:1898–1906
Phale PS, Philippsen A, Widmer C et al (2001) Role of charged residues at the OmpF porin channel constriction probed by mutagenesis and simulation. Biochemistry 40:6319–6325
Queralt-Martín M, García-Giménez E, Aguilella VM et al (2013a) Electrical pumping of potassium ions against an external concentration gradient in a biological ion channel. Appl Phys Lett 103:043707
Queralt-Martín M, Verdiá-Báguena C, Aguilella VM, Alcaraz A (2013b) Electrostatic interactions drive the nonsteric directional block of OmpF channel by La(3+). Langmuir 29:15320–15327
Reiß P, Koert U (2013) Ion-channels: goals for function-oriented synthesis. Acc Chem Res 46:2773–2780

Robertson KM, Tieleman DP (2002) Orientation and interactions of dipolar molecules during transport through OmpF porin. FEBS Lett 528:53–57
Saier MH, Reddy VS, Tamang DG, Vastermark A (2014) The transporter classification database. Nucleic Acids Res 4:251–258
Schindler H, Rosenbusch JP (1978) Matrix protein from Escherichia coliouter membranes forms voltage-controlled channels in lipid bilayers. Proc Natl Acad Sci U S A 75:3751–3755
Schirmer T, Phale PS (1999) Brownian dynamics simulation of ion flow through porin channels. J Mol Biol 294:1159–1167
Schirmer T, Keller TA, Wang YF, Rosenbusch JP (1995) Structural basis for sugar translocation through maltoporin channels at 3.1 Å resolution. Science 267:512–514
Schulz GE (2002) The structure of bacterial outer membrane proteins. Biochim Biophys Acta 1565:308–317
Sharp K (2001) Entropy-enthalpy compensation: fact or artifact? Protein Sci 10:661–667
Siwy ZS, Fuliński A (2002) Fabrication of a synthetic nanopore ion pump. Phys Rev Lett 89:198103
Song JM, Minetti CASA, Blake MS, Colombini M (1999) Meningococcal PorA/C1, a channel that combines high conductance and high selectivity. Biophys J 76:804–813
Suenaga A, Komeiji Y, Uebayasi M et al (1998) Computational observation of an ion permeation through a channel protein. Biosci Rep 18:39–48
Todt JC, Rocque WJ, McGroarty EJ (1992) Effects of pH on bacterial porin function. Biochemistry 31:10471–10478
Van Gelder P, Dumas F, Rosenbusch JP, Winterhalter M (2000) Oriented channels reveal asymmetric energy barriers for sugar translocation through maltoporin of Escherichia coli. Eur J Biochem 267:79–84
Verdiá-Báguena C, Queralt-Martín M, Aguilella VM et al (2012) Protein ion channels as molecular ratchets. Switchable current modulation in outer membrane protein F porin induced by millimolar La 3+ ions. J Phys Chem C 116:6537–6542
Vrouenraets M, Miedema H (2010) The ionization state of D37 in E. coli porin OmpF and the nature of conductance fluctuations in D37 mutants. Eur Biophys J 39:1563–1571
Weiss MS, Kreusch A, Schiltz E, Nestel U, Welte W, Weckesser J, Schulz GE (1991) The structure of porin from Rhodobacter capsulatus at 1.8 Å resolution. FEBS Lett 280:379–382
Woodhull AM (1973) Ionic blockage of sodium channels in nerve. J Gen Physiol 61:687–708
Yamashita E, Zhalnina MV, Zakharov SD et al (2008) Crystal structures of the OmpF porin: function in a colicin translocon. EMBO J 27:2171–2180
Zeth K, Thein M (2010) Porins in prokaryotes and eukaryotes: common themes and variation. Biochem J 431:13–22

Chapter 6
Electrophysiology of Bacterial Translocons

Thieng Pham and Anne H. Delcour

Abstract Protein-translocating channels, or translocons, are ubiquitous membrane proteins found in all cell types as well as intracellular organelles. In bacteria, they often constitute one of the components of molecular machines that participate in membrane biogenesis, assembly of surface exposed appendages, and the secretion of protein substrates across the cell envelope. Their pore-forming properties make them directly amenable to electrophysiology, even though ionic movement is not their natural function. Several bacterial translocons have been investigated by patch-clamp or planar lipid bilayer electrophysiology. These studies have revealed that bacterial translocons are often dynamic channels, capable of sampling a large conformational landscape, and thus displaying multiplicity of conductance states. They typically form large pores, but are often gated by internal plug elements that limit the flow of ions and molecules in the resting state. Combined with structural data, electrophysiological studies have allowed the discovery of structure-function relationships that shed light on functional mechanisms. In some cases, the channel properties have been investigated in the presence of natural substrates to understand the molecular steps involved in translocation itself. This review highlights the electrophysiological investigations of SecYEG, BamA, TolC, secretins, ushers, type III secretion translocons, autotransporters and two-partner secretion systems.

Keywords Secretion • Bacteria • Membrane biogenesis • Patch clamp • Planar lipid bilayer • Translocation • Virulence

6.1 Introduction

Gram-negative bacteria perform numerous functions in order to aid in their survival, including the import of a multitude of necessary nutrients. However, within the host environment, pathogens often export various toxic effectors and virulence factors. These secreted proteins typically help to subjugate host cell machinery to support bacterial growth and expansion. Some of these are released in the external medium,

T. Pham • A.H. Delcour (✉)
Department of Biology and Biochemistry, 369 Science & Research Building 2,
University of Houston, Houston, TX 77204-5001, USA
e-mail: adelcour@uh.edu

A.H. Delcour (ed.), *Electrophysiology of Unconventional Channels and Pores*,
Springer Series in Biophysics 18, DOI 10.1007/978-3-319-20149-8_6

either by crossing the whole bacterial envelope in a single step or by traversing each of the inner and outer membranes through individual pathways. Others are directly injected into host cells via complex secretion machineries spanning the bacterial envelope and the host plasma membrane. In all cases, the daunting task for a water soluble protein to pass through membrane barriers is accomplished by macromolecular assemblies containing a protein-translocation component, or translocon.

Seven types of secretion systems have been identified in Gram-negative bacteria (Holland 2010), and include at least one translocon. In addition, the transport of protein substrates through the bacterial plasma membrane, together with the membrane insertion machineries that participate in the biogenesis of resident inner or outer membrane proteins, also involve translocons. Translocons vary not only in location, but also in structure and functional mechanisms. Outer membrane (OM) translocons are β-barrel proteins, as most OM proteins. Those assembled in the bacterial inner membrane or the plasma membrane of host cells are α-helical. Most of these machines are gated in some way, i.e. they are not present in a constitutively open state. Such a state would expose the bacterial cell to the influx and efflux of various unwanted molecules that would prove detrimental to survival. Some translocons transport folded polypeptides, while others require unfolded proteins to pass through their narrower diameters.

In all cases, however, translocons contain a hydrophilic pore, and therefore are amenable to electrophysiological techniques, such as planar lipid bilayer and patch clamp. Such techniques have, in the past, established evidence and confirmed that these secreting machines are pore-forming proteins. At present, they are commonly used to analyze the behaviors and kinetics of these channels and associate them with their structures. Thus, electrophysiology plays a key role in determining the structure-function relationship of various translocons. In addition, the transport of protein substrates can also be monitored as a kinetic modification in electrophysiological traces, and therefore electrophysiology may enable a single molecule analysis of translocation. This chapter will summarize the various findings uncovered by the use of electrophysiological techniques on translocons studied in this manner to date.

6.2 TolC: The Channel Tunnel

6.2.1 General Introduction

The presence of antibiotics and other harmful chemicals represent a prevailing danger for the continued survival of bacteria. As a defensive mechanism, Gram-negative bacteria such as *Escherichia coli* and *Pseudomonas aeruginosa* employ drug efflux systems to pump noxious material out of the cell (Du et al. 2015; Nikaido 2009; Zgurskaya et al. 2015). These macromolecular assemblies contain a component that resides in the outer membrane and extends into the periplasmic space. An example of such a component is TolC (Zgurskaya et al. 2011). Synthesized in the cytoplasm, the nascent protein is targeted to the general secretory pathway for export into the periplasm. Here, because of its peculiar trimeric structure and high α-helical content (see below), it folds in a unique way without the aid of known periplasmic chaperones. It is then believed to insert into the outer membrane via the Beta-barrel Assembly Machinery complex (BAM), but more concrete evidence for this is still largely undiscovered (Zgurskaya et al. 2011). TolC can be recruited by a large number of various transporters in the inner membrane for drug efflux or the secretion of large proteins (Zgurskaya et al. 2011). One of the most studied examples is its association with two envelope proteins, the periplasmic AcrA and the inner membrane residing AcrB, to form a complex pump that generates a conduit spanning from the inner membrane through the periplasmic space to the outer membrane (Nikaido 2009). As such, it can be used by the bacterium specifically for the efflux of a wide variety of antibiotics.

Initial crystallographic diffraction data identified TolC as an unusual trimeric protein (Koronakis et al. 2000). Each protomer within this oligomer is composed of 4 β-sheets from which 2 pairs of coiled-coil α-helices extend. Together the protomers entwine to create a trimer that exists as a 12-stranded 40 Å-long β-barrel anchored in the outer membrane, with six pairs of coiled-coil α-helices (2 pairs from each protomer) protruding into the periplasm to form a 100 Å long tunnel. Because of this elongated structure, TolC was coined a "channel-tunnel" (Koronakis et al. 2000). The crystal structure of the TolC trimer also indicated that because the protomers twist around each other, in particular in the coiled-coil region, the entrance to the TolC channel tunnel is constricted and hence closed (Koronakis et al. 2000). The modeled structure of the tripartite complex of TolC, AcrA and AcrB shows that the assembly of the three components in a 3:6:3 stochiometry results in the formation of a complex spanning the whole envelope (Zgurskaya et al. 2011). A recent cyro-EM study provides a pseudo atomic model of the whole isolated drug efflux pump (with an additional accessory component, AcrZ), suggesting the formation of a continuous internal conduit through the TolC, AcrA and AcrB oligomers, and thus proposing that the TolC channel must open upon docking with AcrA (Du et al. 2014).

6.2.2 Channel Properties

Electrophysiological analysis has been rather critical in elucidating structure-function relationship in order to provide insight into the molecular mechanism behind TolC channel opening. Initial studies used purified TolC reconstituted into planar lipid bilayers in order to determine the effects of varying membrane potentials on channel properties (Andersen et al. 2002a). Such experiments with bilayers containing multiple inserted channels identified polarity dependent behavior indicating asymmetry consistent with its structure. This was illustrated by an increased channel noise at negative membrane potentials relative to positive potentials. Single channel experiments further cemented this observation while providing evidence that purified TolC inserts itself into the lipid bilayer in the same orientation each time. It was also noted that while no voltage dependent opening or closing of TolC was seen, higher resolution single channel recordings identified the presence of three conductance substates as well as cation selectivity. Cation selectivity was presumed to be due to the highly electronegative aspartate ring at the periplasmic entrance to the TolC tunnel domain. Interestingly, conductance varied depending on positive or negative membrane potentials. In particular, negative membrane potentials relative to the *cis* side resulted in a larger conductance presumably because of increased ion flow from the β-barrel to the periplasm. Experiments utilizing different pHs also provided some insight into the opening mechanism in TolC. Low pH experiments with reconstituted TolC showed a much smaller conductance level. This is believed to be due to the protonation of the periplasmic ring of aspartate residues at lower pHs. Consequently, this would limit ion flow and thus conductance through the channel tunnel. Importantly, these initial studies established a conductance of ~80 pS in 1 M KCl, a value that seems too small to represent a pore able to accommodate large protein substrates, but consistent with a "closed" state of the channel. Based on the X-ray structure, these electrophysiological results suggested that the periplasmic entrance may be constricting for channel conduction and therefore a critical factor underlying the mechanism for TolC channel gating. In particular, it was hypothesized that the periplasmic entrance to the tunnel domain must somehow untwist in an iris-like fashion in order to allow the export of large proteins or efflux of drugs.

Insight into the molecular mechanism of opening was provided by later studies seeking to test this hypothesis (Andersen et al. 2002b). Close inspection of the X-ray structure and modeling of the open state revealed that the "closed" state of the channel might be maintained by two intra-monomeric hydrogen bonds, an inter-monomeric salt bridge and an inter-monomeric hydrogen bond, which altogether keep the coiled-coil helices in close proximity. The authors introduced mutations of the residues involved in these interactions to disrupt the links between the coils. Single mutations had relatively marginal effects. However the combination of some mutations resulted in a profound enhancement of the conductance, indicative of an enlarged pore opening. In particular the combined disruption of the inter-monomeric links and one of the intra-monomeric hydrogen bonds produced variants with conductances of 500–1000 pS. The channel with the highest conductance also showed a higher propensity at substate gating at negative potentials, indicating a somewhat destabilized structure, with increased flexibility of the α-helices. Altogether, these results support the notion that opening of the channel might occur via the weakening of a ring of links in the coiled-coil domain that keeps the α-helices of this region in close proximity. The open state, though, would need to be maintained in the fully assembled and functional machine, perhaps through interactions of the TolC α-helices with AcrB.

6.2.3 *The Trojan Horse: Parasitization of TolC by Colicins*

Electrophysiological analyses also proved crucial in showing how this translocon can be exploited naturally. Within environments of limited resources, bacterial warfare can be an efficient means by which to eliminate competition in order to ensure continued survival. *Escherichia coli* cells indeed produce and employ an arsenal of antibacterial toxins, known as colicins, to target sensitive strains in the environment (Cascales et al. 2007; Riley and Wertz 2002). Colicin cytotoxicity ranges from pore-forming to nuclease activity. In order to overcome the permeability barrier of the outer membrane, colicins hijack separate outer membrane proteins to serve as receptors and translocators (Jakes and Cramer 2012). In an example of this so-called "fishing pole" mechanism of colicin translocation through the OM, BtuB, the normal transporter for the import of vitamin B12 into the periplasm, is used as a receptor by colicin E1 (ColE1), which is then handed over to TolC for translocation through the outer membrane (Zakharov et al. 2004, 2012).

In elegant experiments, TolC was indeed confirmed as the passageway for ColE1 in planar lipid bilayers. Purified TolC was added to the *cis* side of a bilayer chamber for insertion into the membrane, and ColE1 to the opposite side. At negative membrane potentials, which would pull ColE1 into the channel towards the *cis* side of the chamber, a marked decrease in channel conductance was observed, suggesting that ColE1 was occluding TolC (Zakharov et al. 2004). Positive membrane potentials, however, were ineffective. Perfusion of the *trans* side of the bilayer chamber restored the full conductance, indicating that the association of ColE1 with TolC was reversible. On the other hand, other colicins failed to occlude TolC. Later studies demonstrated that occlusion of TolC could also occur with an α-helical peptidic segment of the colicin translocation domain (T-domain) (Zakharov et al. 2012). Thus, electrophysiology was essential in demonstrating that the import of ColE1 into the periplasm actually occurs through TolC, although the exact mechanism for this process still remains enigmatic. The extracellular entrance of TolC is large enough to accommodate the α-helical T domain of ColE1, but navigation through the whole length of the channel tunnel may require unfolding.

6.3 Type III Secretion System: Secreted Translocons

6.3.1 General Introduction

Various pathogenic Gram-negative bacterial species such as *Pseudomonas spp.* or *Yersinia spp.* are able to manipulate the host cell machinery through numerous effector proteins directly injected into the host cell via a delivery system known as the Type III Secretion System (T3SS) (Burkinshaw and Strynadka 2014; Cornelis 2010; Kosarewicz et al. 2012; Mattei et al. 2011).

The "injectisome", composed of over 20 different proteins, represents the secreting component of the T3SS that assembles in the bacterial envelope. It includes the basal body, the proteinaceous needle terminated by a tip complex, and the translocon (Burkinshaw and Strynadka 2014; Mattei et al. 2011). The basal body is made up of ring-like structures spanning the inner and outer membranes and the periplasm in-between, and an inner rod connecting with the hollow needle. The needle extends outward and was recently confirmed to serve as a conduit for secreted effector proteins (Radics et al. 2014). Because the effective diameter within the needle is only approximately 2 nm (Cordes et al. 2003), secreted effectors are thought to traverse the needle in a mostly unfolded, molten globule, conformation (Dohlich et al. 2014; Faudry et al. 2007). An ATPase is embedded within the inner membrane located at the floor of the basal body facing the cytoplasmic interior, and may serve at enabling the entry of T3SS effectors into the injectisome as they are delivered at the base of the apparatus by dedicated chaperones. The first proteins to be secreted are the components of the translocon that assembles in the eukaryotic host plasma membrane. The translocon contains a hydrophilic protein thought to form an assembly platform at the tip of the needle to aid in the oligomerization and membrane insertion of two hydrophobic translocator proteins. These two proteins are thought to form a pore within the eukaryotic cell membrane, for the delivery of secreted effector proteins into the host cell.

6.3.2 *Electrophysiology of the Yersinia enterocolitica T3SS Translocon*

Yersinia enterocolitica, which causes gastrointestinal symptoms in humans, along with its more infamous cousin, *Yersinia pestis*, the causative agent of plague, and *Yersinia pseudotuberculosis*, harbors a plasmid-encoded T3SS that acts as a major virulence factor (Cornelis 2002; Cornelis and Wolf-Watz 1997). Its role is to disarm immune cells, in particular neutrophils and macrophages, by delivering effector proteins into their cytosol. These T3SS effectors disrupt the cytoskeleton and intracellular signaling pathways, thereby promoting survival in the host by blocking phagocytosis and secretion of pro-inflammatory mediators, among other effects. For many years, the Ysc-Yop system of *Yersiniae* was the best understood T3SS (Cornelis 2002; Cornelis and Wolf-Watz 1997). The *Yersiniae* translocon is composed of the hydrophilic platform protein LcrV and the hydrophobic, pore-forming proteins YopB and YopD, which are thought to form a mixed oligomer (Montagner et al. 2011).

The first electrophysiological experiments that support the formation of a channel by YopB and YopD were performed by fusing with a planar lipid bilayer proteoliposomes which had been preincubated with bacteria induced to release the Yop proteins in the extracellular medium (Tardy et al. 1999). Using a strain overexpressing YopB, YopD and LcrV, the authors provided evidence of channel activity upon voltage application across the membrane, and attributed it to the YopBD oligomer. At physiological salt concentrations, the single channel conductance was approximately 105 pS. Experiments in asymmetric conditions revealed that the channel had essentially no ionic selectivity. Upon impairing the hydrophobic nature of YopD via truncation, the kinetics of the channel in bilayer differed drastically. The channel displayed a predominantly closed configuration with occasional transitions to conductance states of various size and dwell times. This suggested that YopD is critical to assembling a stable channel conformation. On the other hand, YopD alone was unable to form a channel at all, providing evidence for the absolute requirement of YopB in pore formation. These electrophysiological studies support the function of YopB and YopD in forming a channel that may perhaps be responsible for the import of secreted effectors into the host cell cytosol. However, although the presence of YopB and YopD in the liposome membrane was confirmed by SDS-PAGE, validation of these findings would be strengthened by using purified proteins. This was not performed with the Yop system but with the homologous proteins from the T3SS of *Pseudomonas aeruginosa*.

6.3.3 *The Translocon of Pseudomonas aeruginosa*

Similarly to *Y. enterocolitica*, *P. aeruginosa* uses a T3SS to break down host immune defenses. In fact, the YopB and YopD homologues of *P. aeruginosa*, PopB and PopD, have been shown to lyse red blood cells as well as disrupt the membrane integrity of macrophages (Goure et al. 2004; Mueller et al. 2008). Dye leakage experiments and other *in vitro* assays demonstrated that PopBD was able to form pores within membranes, even in the absence of the LcrV homologue, PcrV (Faudry et al. 2006). However, electrophysiological experiments helped provide insight into the kinetics of pore formation *in vitro* (Wager et al. 2013). PopB and PopD were purified in guanidine HCl, which mimicked the unfolded conformation seen *in vivo* during transit through the needle (Faudry et al. 2007). Dilution of an equimolar mixture of PopB and PopD into a planar lipid bilayer chamber allowed the proteins to renature for insertion into the membrane. Upon voltage application, bursts of openings to various conductance states were observed with a delay that was taken to represent time for insertion and/or activation of PopB/PopD oligomers. The use of PopB alone resulted in a longer time of activation necessary to reach the characteristic bursts in conductance associated with the PopB and PopD mixture, suggesting that PopD might facilitate PopB insertion and/or oligomerization. Experiments with PopD only, on the other hand, yielded very little to no channel activity at similar protein concentrations. The kinetic signature of the PopB/PopD mixture is illustrated in Fig. 6.1a. Current fluctuations of various amplitudes were observed. Many are extremely transient and seen as spikes, but are interspersed with square-top events defining genuine channel openings. With time, the channel activity typically ramped up due to the continuous insertions of pores, leading to extremely large transitions, as seen at the end of the top trace of Fig. 6.1a, and culminating with the rupture of the bilayer. Interestingly, PopB/PopD displayed a very distinct voltage polarity-dependent behavior, as illustrated in Fig. 6.1b. With PopB/PopD added to the *cis* side of the membrane, channel activity was robust at positive membrane potentials, but inactivated upon application of negative voltages. The reverse was true when PopB/PopD was added to the *trans* side. This polarity dependence may

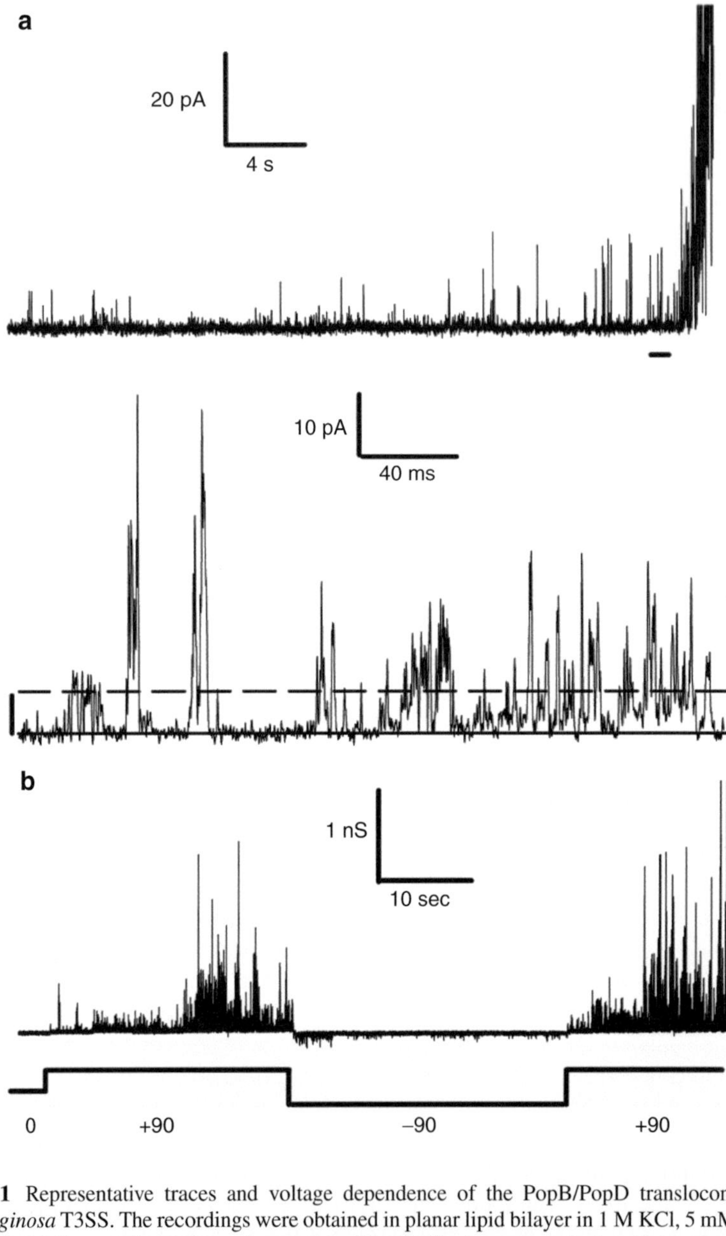

Fig. 6.1 Representative traces and voltage dependence of the PopB/PopD translocon of the *P. aeruginosa* T3SS. The recordings were obtained in planar lipid bilayer in 1 M KCl, 5 mM Hepes (pH 7.2). (**a**) Current fluctuations observed at +90 mV. The lower trace is the segment marked by a *horizontal bar* in the upper trace and shown on an expanded time scale, to highlight the transitions of various conductance amidst square top events with a conductance of 57 pS (marked by the *dashed line*). (**b**) The trace shows the reversible inactivation of the PopB/PopD translocon at negative voltages. The voltage protocol is shown below the recording. In both panels, channel openings are seen as upward deflections at positive voltages, and downward deflections at negative voltages. (Reproduced from Wager et al. (2013))

have important implication *in vivo*, as it would favor the activity of newly released PopB and PopD as they approach a polarized eukaryotic membrane, whose natural membrane potential is relatively more positive on the extracellular side than on the intracellular side.

In order to provide further insight into the process of PopB/PopD pore formation, the properties of the PopB/PopD translocon were investigated in bilayers of different lipid compositions (Wager et al. 2013). The insertion of the translocon into the membrane was found to require higher membrane potentials with thicker bilayers. In addition, the mild cation-selectivity of the pore was further decreased in neutral bilayers. These findings suggested that the nature of the lipids surrounding the pore influences pore properties, and led to the proposal that the PopB/PopD translocon might form a toroidal pore, akin to some antimicrobial peptides. In toroidal pores, the channel wall is lined by both protein residues and lipid headgroups. An example is provided with the bee toxin melittin (Yang et al. 2001), which showed in planar lipid bilayer experiments astonishing similarities with the PopB/PopD translocon, such as bursts of current fluctuations of various sizes and the same polarity dependence of activity.

6.4 The Secretin Family

6.4.1 General Introduction

The term "secretin" encompasses a broad family of outer membrane proteins that are present across many species of bacteria, and form channels for the translocation of proteinaceous effectors and substrates. In fact, the type II secretion system (T2SS), type III secretion system (T3SS), type IV pili system (T4PS), as well as the phage assembly system all utilize secretins as one of their core components (Korotkov et al. 2011). Many of these systems constitute virulence factors for bacterial pathogens. *In vivo,* targeting and insertion of the channel into the outer membrane often require interaction with an accessory lipoprotein known as the pilotin. Secretins are oligomeric in nature, with 12–15 subunits involved in the quaternary structure, depending on the secretin type. The ensuing ring-like structures, nicely demonstrated in electron microscopy (Bitter et al. 1998; Collins et al. 2003; Crago and Koronakis 1998; Koster et al. 1997; Nouwen et al. 1999), range from 50 to 100 Å, and therefore are large enough for the transport of folded or partially folded protein substrates. It is assumed that the complex might be gated in some fashion to prevent leakage of periplasmic content or entry of unwanted material. Indeed, in all the secretins investigated to date, plug-like constrictions are observed as localized densities within the channel (Kowal et al. 2013; Tosi et al. 2014). Each monomer is organized in domains with specific functions. The C-terminal domain, which is conserved across members of the family and thus defines the secretin domain, is embedded in the outer membrane, and participates in the formation of the membrane spanning channel. The very last 50 or so residues of the C-terminal domain interact with the pilotin. The N-terminal domain is periplasmic and, in the multimeric structure, defines a large cavity connected to the outer membrane channel (Tosi et al. 2014). This region interacts with other components of the secretion machinery, and the secreted proteins (Korotkov et al. 2011). Because of their obvious pore-forming architecture, a few secretins have been analyzed by electrophysiology, namely XcpQ of *P. aeruginosa* (Brok et al. 1999), PulD of *Klebsiella oxytoca* (Nouwen et al. 1999), and YscC of *Yersinia enterocolitica* (Burghout et al. 2004).

6.4.2 PulD, a T2SS Secretin

K. pneumoniae uses the secretin PulD of the T2SS machine for the secretion of pullulanase, a cell-surface exposed outer membrane lipoprotein that degrades pullulan, a form of starch (d'Enfert et al. 1987, 1989). PulD from its close relative *K. oxytoca* was the first secretin investigated by electrophysiology (Nouwen et al. 1999). Purified together with its pilotin PulS, it was incorporated into liposomes. These proteoliposomes were then fused with an artificial membrane in a planar lipid bilayer setup. The experiments revealed that PulD is indeed a channel capable of conducting ions. The channel was predominantly closed at a low membrane potential of +20 mV, but holding the voltage at +80 mV increased the propensity for the channel to open, indicating that it may be voltage sensitive. Negative membrane potentials also yielded different channel kinetics compared to positive ones, as opening transitions were better defined and regular in occurrence. It is interesting that opening transitions were observed, despite the presence of an apparent plug at the center of the channel, as revealed by cryo-EM in the same study. The conductance of these events is however ~300 pS, presumably too small to originate from an unplugged channel. It is possible that these transitions represent small fluctuations in the ionic pathway, possibly due to small movements of the plug. Recent dye leakage experiments actually suggest that, in its resting, non-protein translocating state, PulD might exist in a leaky conformation of size comparable to the open state of general diffusion porins (Disconzi et al. 2014). The authors suggest that the discrepancy with earlier studies may stem from different membrane insertion protocols, in particular with respect to detergents.

6.4.3 YscC, a T3SS Secretin

The secretin YscC from *Y. enterocolitica* functions as the outer membrane component of the T3SS basal body (Koster et al. 1997). Purified YscC was initially reported to form a homooligomer containing 13 subunits (Burghout et al. 2004), but a more recent study rather suggests a dodecameric quaternary structure, as in PulD (Kowal et al. 2013). Near the bacterial cell surface, the YscC oligomer forms a large opening of 70 Å in diameter to enclose the proteinaceous needle that forms the extracellular part of the T3SS (Kowal et al. 2013). Similarly to PulD, the YscC secretin has a large domain protruding into the periplasm to connect with the inner membrane rings that form the rest of the basal body.

The channel conducting property of the YscC secretin was also established via planar lipid bilayer experiments demonstrating the usefulness of electrophysiological analysis as a means to identify pore-forming proteins (Burghout et al. 2004). Insertions of purified YscC revealed two different levels of conductance at 2.6 nS and 3.0 nS (in 1 M KCl), and occasional ~0.5 nS transitions shortly after a 2.6 nS insertion, suggesting that the YscC secretin might be able to undergo a conformational change that allows it to increase from 2.6 nS to its maximal open state at 3.0 nS. Surprisingly, the channel appeared to be more selective for chloride than potassium. Unlike PulD, YscC was found to be in a permanently open state. Despite increasing the membrane potential to voltages upwards of ±200 mV, no closing transitions were observed. The conductance remained linear in voltage-ramps unlike PulD, which appeared to be voltage sensitive. YscC's lack of voltage dependence as well as its 100 % open probability indicates that the T3SS system may be constitutively open *in vivo* to allow a continuous conduit for effectors to reach the hollow needle that will deliver them into the host target cell.

6.5 Autotransporters and Two Partner Secretion Systems

6.5.1 General Introduction

Gram-negative bacteria also use dedicated secretion systems for the cell surface display of proteases, adhesins, and other enzymes. Two such protein export machineries are autotransporters and two partner secretion systems (Leo et al. 2012; van Ulsen et al. 2014). An autotransporter is composed of (1) a N-terminal signal peptide, (2) a functional passenger domain, and (3) a translocator domain. The N-terminal signal peptide targets the protein to the general secretory pathway of the plasma membrane, then is cleaved once it emerges in the periplasmic space. Chaperones and the BAM complex are involved in the insertion of the C-terminal translocator domain into the outer membrane. There, this domain behaves as a conduit for the translocation of its passenger domain into the extracellular environment. In a similar pathway, the two-partner secretion system, the passenger and translocator functions are found on two individual proteins, the passenger exoprotein commonly referred to as TpsA, and the transporter protein, TpsB, that serves as the translocon for its partner. TpsA and TpsB each have an N-terminal signal peptide sequence that targets them to the general secretory pathway for export into the periplasm. Upon cleavage of the TpsA N-terminal signal peptide, a targeting sequence, called the TPS domain, is revealed to allow the TpsA to recognize its cognate TpsB in the outer membrane. The similarities between these two systems lend support to the idea that they may have evolved and diverged from a common ancestral secretory pathway. Because TpsBs and the translocator domains of autotransporters are thought to provide a passageway for protein substrates, they have been the focus of electrophysiological investigations designed to demonstrate their pore-forming capabilities.

6.5.2 The Autotransporters NalP and Hbp

Neisseria meningitidis, the causative agent of meningitis, utilizes the autotransporter NalP to secrete a protease which processes other autotransporters and virulence factors at the cell surface (van Ulsen et al. 2014). High resolution crystal structure data of the translocator domain of NalP (Oomen et al. 2004) indicate that it is a 12-stranded β-barrel embedded within the outer membrane and protruding slightly into the extracellular space. The interior core of the channel is very hydrophilic but is plugged by a highly charged N-terminal α-helix extending outward from the base of the barrel, which is thought to provide stability to the barrel (Roussel-Jazede et al. 2011). This α-helix represents the remnant of the cleaved passenger domain after its translocation through the translocator domain and proteolysis by other neighboring cell surface exposed NalPs (van Ulsen et al. 2014). Because the translocator domain of NalP is predicted to be a channel for the translocation of its passenger domain, its pore activity was measured *in vitro* via electrophysiological techniques. Purified NalP translocator domain containing the alpha helical plug was reconstituted into planar lipid bilayers (Oomen et al. 2004; Roussel-Jazede et al. 2011). Membrane potential clamped to +100 mV revealed large openings of conductance ~1.3 nS, with additional small conductance steps of ~150 pS (in 1 M KCl). The authors postulated that the large conductance, corresponding to an estimated internal lumen diameter of 8.4 Å, represents the channel without the inserted α-helix, while the small conductance would correspond to an obstructed channel with small water-filled conduits at the interface of the helix and the barrel wall. The displacement of the α-helix was attributed to the possible effect of salt and detergent disrupting the interactions between the α-helix and barrel wall, but this proposal was never confirmed. Unfortunately, purified NalP translocator domain lacking the α-helix (NalP TD Δα) did not show stable openings, similar to those postulated to represent NalP with a displaced α-helix. On the contrary, the recordings from the NalP TD Δα showed frequent variable fluctuations, indicating an unstable channel (Roussel-Jazede et al. 2011).

The hemoglobin protease Hbp is an autotransporter from a virulent strain of *Escherichia coli*. Here, the passenger domain cleaves human hemoglobin to release heme, which can be used by the bacteria to scavenge iron. Hbp belongs to the family of SPATE (serine protease autotransporters of *Enterobacteriaceae*) autotransporters, which undergo a spontaneous autoproteolytic reaction within the β-domain itself to release the passenger domain (van Ulsen et al. 2014). The structure of an engineered non-cleavable Hbp mutant shows that the structure of the Hbp translocator domain is very similar that of NalP (Tajima et al. 2010). The channel properties of Hbp translocator domain were compared to those of NalP (Roussel-Jazede et al. 2011). Here, the Hbp translocator domain was purified from inclusion bodies and was found to have undergone spontaneous cleavage at the site of release of the passenger domain. For this reason, the α-helix appears to be stably lodged within the channel lumen, and only small openings of ~0.2 nS were observed. This was in contrast with another Hbp construct, with just one additional N-terminal amino acid ahead of the cleavage side, which showed only the large conductance steps, as in NalP. It is interesting that, whether the cleavage releasing the passenger domain is intramolecular, as in the case of Hbp, or intermolecular, as for NalP (van Ulsen et al. 2014), the remaining α-helix is likely to efficiently block the channel lumen after translocation, thus preventing leakage of periplasmic components.

6.5.3 FHA and FhaC: A Two-Partner Secretion System

The bacterium responsible for whooping cough is *Bordetella pertussis*. In the initial stages of infection, *B. pertussis* adheres to the ciliated and non-ciliated epithelial cells of the upper respiratory tract (Locht et al. 1993). The infection then proceeds to the lungs which may constitute a bacterial reservoir. A major virulence factor of this bacterium is the filamentous haemagglutinin adhesin FHA, which is secreted by a two-partner secretion system and represents the TpsA exotoxin in this pathway (van Ulsen et al. 2014). FHA is actually derived from a precursor FhaB (Coutte et al. 2001), which reaches the cell surface by translocating in an unfolded conformation through its cognate TpsB, FhaC (Baud et al. 2014). Upon exit from FhaC, FHA then folds and is proteolytically cleaved to be released in the extracellular milieu (Coutte et al. 2001), although a significant amount of the protein also remains associated with the bacterium. The presence of a carbohydrate binding site on FHA allows it to bind to sulphated polysaccahrides on the ciliated epithelium and macrophages, while binding to non-ciliated cells involves a heparin binding site (Locht et al. 1993; Mattoo et al. 2001). FHA also appears to play a major role in the formation of *B. pertussis* biofilms in the respiratory tract (Serra et al. 2011).

The high-resolution crystal structure of FhaC represented a major advance in the understanding of the two-partner secretions system (Clantin et al. 2007), and confirmed the predicted domain organization of the protein, which is now known to be a hallmark of the Omp85-TpsB superfamily (Jacob-Dubuisson et al. 2009). The translocation domain is a single 16-stranded β-barrel found at the C-terminus. Two salient features of the β-domain are the fact that it is plugged by the N-terminal α-helix H1, and further obstructed by the presence of a large extracellular loop L6, which folds as a hairpin back inside the β-barrel. Modeling of the FhaC structure with the H1 α-helix removed reveals that the pore delineated by the β-barrel itself has an 8 Å diameter. Two N-terminal POTRA (polypeptide transport associated) domains, canonical features of the members of this family, form a so-called "periplasmic module", and have been shown to recognize the FHA TPS domain (targeting sequence) in an unfolded conformation (Jacob-Dubuisson et al. 2009).

FhaC is one of the first secretion translocons to be investigated by electrophysiology (Jacob-Dubuisson et al. 1999). In planar lipid bilayers, the conductance generated by FhaC in 1 M KCl was approximately 1.2 nS. This conductance is similar to that of NalP, and as for this protein, is assumed to represent a configuration where the internal alpha helix is dislodged from the pore (Clantin et al. 2007). Figures 6.2a, b show representative traces obtained in planar lipid bilayers and patch-clamp, respectively, illustrating that FhaC behaves as a predominantly closed channel with frequent well-resolved openings at both negative and positive membrane potentials. At negative voltages, the kinetics revealed the existence of a sub-conductance state. A number of mutated constructs of FhaC have been examined to establish structure-function relationships and gain insight into the mechanism of FhaC. In particular, it was clearly established that the C-terminal 350 amino acids of the protein are responsible for channel activity, although these mutants lacking the first 200 N-terminal residues displayed a noisy kinetic signature, suggesting some destabilization of the protein (Meli et al. 2006). This domain organization was later validated by the crystal structure (Clantin et al. 2007). The deletion of the inwardly folded loop L6 has drastic functional consequences: it abolishes secretion of FHA, impacts the channel conductance and increases antibiotic sensitivity. This later phenotype, together with the intra-barrel location of the loop, suggests a role for L6 in constricting the channel, and thus its possible displacement upon substrate translocation.

Recently, further insight into the mechanism of channel opening was obtained in an innovative study that combines EPR spectroscopy and electrophysiology to probe the conformational dynamics of the protein (Guerin et al. 2014). The authors reported that FhaC can adopt different conformations, even in the absence of substrates, and in particular that the H1 helix can spontaneously slide out of the pore, towards the periplasmic side, where it can interact with the POTRA2 domain. Additional experiments suggested that the conformation with H1 displaced into the periplasm is prevalent during substrate secretion. Interestingly, in the course of this study, the authors purified a FhaC variant which contained a sub-population of proteins with an intermolecular disulfide bond linking the extracellular tip of the H1 α-helix to the extracellular loop L7. This variant yielded two types of channel recordings, presumably originating from the two protein populations found in the sample: some had the typical FhaC activity and other displayed very little activity, but could be converted to the regular pattern upon application of a reducing agent during the recording. For this reason, this latter activity was attributed to the population of channels with the intramolecular disulfide bond. These results suggest that hampering the displacement of the H1 helix towards the cytoplasm essentially maintains the channel in a constricted conformation and that the sliding of the H1 helix in and out of the pore might correlate with the dynamics of channel activity.

6.5.4 Other Two-Partner Secretion Systems

Electrophysiological experiments have also confirmed the pore-forming activity of other TpsB proteins, such as HMW1B (Duret et al. 2008), EtpB (Meli et al. 2009) and ShlB (Konninger et al. 1999). Figure 6.2c illustrates the electrophysiological

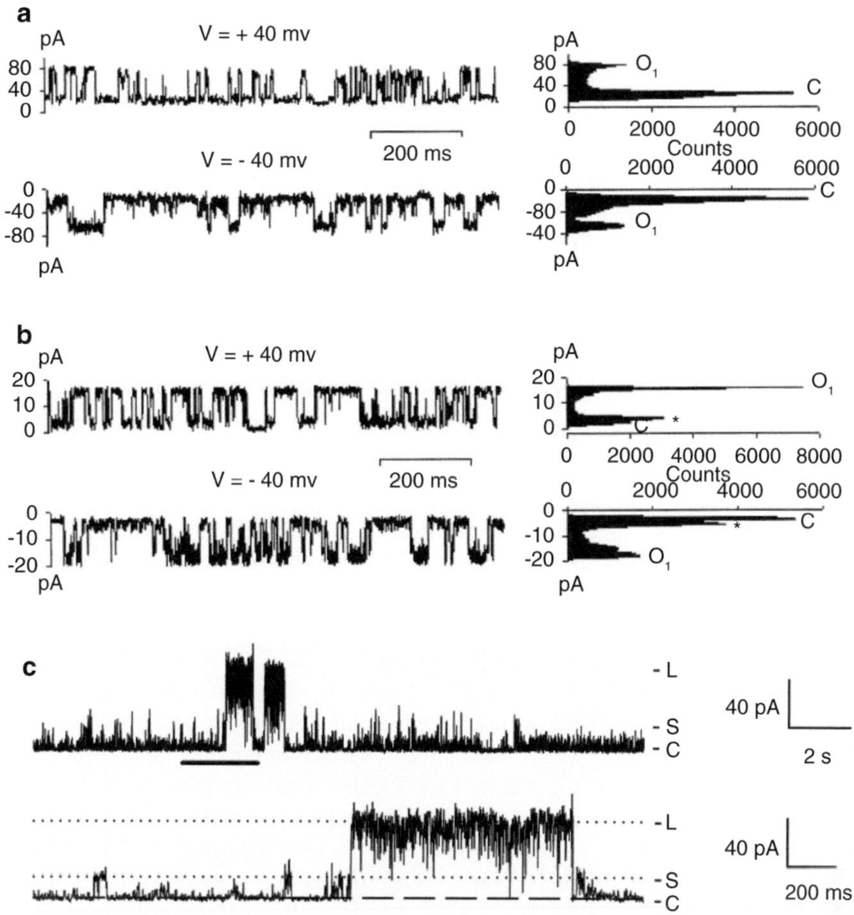

Fig. 6.2 Representative traces from the TpsBs FhaC and HMW1B. (**a**) Activity of FhaC in an azolectin planar lipid bilayer in 1 M KCl, 10 mM Hepes (pH 7.2) at the voltages indicated. The *right panels* show the associated amplitude histograms with a conductance of the open level O1 of 1200 pS. (**b**) Activity of reconstituted FhaC in azolectin liposomes recorded by patch clamp in 200 mM KCl, 10 mM Hepes (pH 7.2) at the indicated voltages. The *right panels* show the associated amplitude histograms with a conductance of the open level O1 of 300 pS. A small substate of 30–50 pS is denoted by the *asterisks* (Reproduced from Jacob Dubuisson et al. (1999) with permission). (**c**) Activity of HMW1B in a diphytanoyl phosphatidylcholine planar lipid bilayer in 1 M KCl, 10 mM Hepes (pH 7.2) at +50 mV. "C" denotes the closed state. Two open states are represented by "L" and "S", with the largest one ("L") having a conductance of ~1400 pS. The bottom trace shows at an expanded time scale the segment of the top trace highlighted by *horizontal bar* (This research was originally published in the *Journal of Biological Chemistry*. Duret et al. (2008). © the American Society for Biochemistry and Molecular Biology)

signature of HMW1B, a TpsB from *Haemophilus influenzae* used for the secretion of an adhesin (Duret et al. 2008). In these planar lipid bilayer experiments, two levels of current are also clearly observed. The largest one represents a conductance of ~1400 pS, a value similar to that of FhaC (1200 pS) (Jacob-Dubuisson et al. 1999) (both in 1 M KCl). The smaller conductance is about 1/4th of the larger one.

EtpB also behaves as a pore with clearly defined multiple conductance states (Meli et al. 2009). Interestingly, the smallest one is also 1/4th of the largest one, but an additional, intermediate substate, was also revealed in this case. The molecular basis for these substates is unclear. The HMW1B protein does exist as a tetramer, and it is possible that the different sub-conductance states represent the monomeric and oligomeric form. But, FhaC crystallizes as a monomer, and is believed to function as such. Sub-conductance states are frequent in large protein-translocating pores, probably due to some inherent flexibility of the pore constriction itself and the ability of the protein to sample a large number of conformational states.

6.6 The Usher PapC of the Chaperone-Usher System

6.6.1 General Introduction

While many bacterial species utilize translocons in order to release toxic effectors into target host cells or the extracellular milieu, translocation machines can also be used for the assembly of pili at the cell surface. For example, uropathogenic *E. coli* utilizes the chaperone-usher pathway for the biogenesis of Type 1 and P pili (Geibel and Waksman 2014; Kline et al. 2010; Thanassi et al. 2012; Waksman and Hultgren 2009). Type 1 pili, assembled by the *fim* system, are used in bladder colonization, while P pili, of the *pap* system, are involved in infection of the kidney, leading to pyelonephritis. These pili extend from the cell surface to provide adherence to the host cell epithelium lining the walls of the kidney or urinary tract. This adhesion represents an initial critical step in colonization, and animal studies have shown that pili are essential virulence factors in the disease process (Crepin et al. 2012). Type 1 pili also trigger the internalization of bacteria by the bladder epithelial cells (Martinez et al. 2000), and are required for the formation of intracellular biofilm-like communities in these cells (Wright et al. 2007).

6.6.2 Pilus Biogenesis and Structural Features of the Usher

The *fim* and *pap* systems share extensive structural and functional homologies, and the molecular steps taking part in the biogenesis of the type 1 and P pili have been extensively reviewed (Busch and Waksman 2012; Geibel and Waksman 2014; Kline et al. 2010; Thanassi et al. 2012; Waksman and Hultgren 2009). The pili themselves are composite appendages made up of several different types of subunits assembled in a precise stoichiometry and order. The terminal subunit plays the role of adhesin and binds to glycolipids or glycoproteins in the host cell membrane. In the Pap system, the secretion machinery comprises a periplasmic chaperone, PapD, and an outer membrane component, the usher PapC. The biogenesis of the pilus begins in the cytoplasm where all the individual subunits of the pilus and the translocation machinery are synthesized. Each of these subunits contains an N-terminally located signal peptide sequence that targets them for secretion through the Sec Translocon into the periplasmic space. Once in the periplasm, the individual pilus subunits associate with the chaperone PapD. This association serves three purposes: (1) the PapD chaperone helps to prevent the spontaneous aggregation of pili subunits or assembly of the pilus in the periplasm; (2) binding to the chaperone induces the correct folding of the pili subunits into an immunoglobulin-like fold of antiparallel β-sheets; and (3) PapD targets the subunits to the base of the PapC usher for translocation and assembly into a functional pilus.

Our understanding of the biogenesis process has benefited from major advances in the determination of the three-dimensional structures of the ushers of the *fim* and *pap* systems, FimD and PapC, respectively (Ford et al. 2010; Phan et al. 2011; Remaut et al. 2008). The usher is composed of a twin β-barrel imbedded within the outer membrane, and globular N- and C-terminal domains facing the periplasm. Each monomeric β-barrel comprises 24 β strands, and a globular plug domain that gates the channel shut. The plug domain extends from two linker regions located at the base of the barrel wall and connecting with β-strands 6 and 7. Additional structural features of importance are a β5-6 hairpin which constrains the plug domain in place, and an α-helix residing externally to the β5-6 hairpin and stabilizing it. A remarkable snapshot of the translocation process has been provided by the X-ray crystal structure of the complex of the usher, the chaperone and the first pilus subunit of the *fim* system (Phan et al. 2011). The data reveal that, during translocation, the plug domain is displaced from the channel lumen towards the periplasmic side, where it makes extensive contact with the usher N-terminal domain. However, the exact molecular mechanism behind this complex event has yet to be elucidated.

6.6.3 Electrophysiological Characteristics of the PapC Usher

Planar lipid bilayer and patch clamp techniques have been used for the analysis of the PapC usher. PapC exhibits a characteristic kinetic signature at relatively low membrane potentials (<100 mV), which indicates that the usher is a mostly closed, but still extremely dynamic channel. As seen in the trace of Fig. 6.3a, the wild-type (WT) channel exhibits frequent transient openings of small, but varying conductance (50–600 pS in 1 M KCl). These current fluctuations likely originate from ion movements through water-filled conduits that exist at the barrel-plug interface (Mapingire et al. 2009). Interruptions of the current might be due to thermal motion of the plug within the channel lumen or spontaneous conformational changes of the barrel or the periplasmic domains leading to transient occlusions of these small pathways. Initial studies of domain mutants also demonstrated that removal of the plug domain generates a pore with an extremely large conductance (~3 nS in 1 M KCl, for the monomer), which, surprisingly, still displays a highly dynamic behavior (Mapingire et al. 2009). The channel tends to spend more time in the open state, but closures remain very frequent. Even in the wild-type channel, occasional spontaneous large transitions of a conductance of approximately 3 nS (in 1 M KCl) do occur and are believed to be short plug displacement events (Fig. 6.3a) (Farabella et al. 2014; Mapingire et al. 2009), which become more frequent as voltage increases. Much higher voltages (>130 mV) appear to trigger prolonged displacement of the plug, as shown in Fig. 6.3b. This trace was obtained in patch clamp experiments of PapC reconstituted in liposomes. In this case, the plug displacement events are always observed at negative voltages, indicating a polarity dependence of the phenomenon and the reproducible orientation of the protein. Interestingly, the kinetic signature of these events is reproducibly characterized by a transient sojourn of the current at the level corresponding to opening of the dimer, followed by a quick relaxation to current levels oscillating around the monomeric current. A straightforward explanation would be that the plug quickly relocates inside one of the monomers. However, this pattern is also observed in plugless channels at high voltages (Pham et al. in preparation). Possible explanations are that perhaps the N-terminal or C-terminal globular domains could occlude the open barrel resulting in relaxation of the conductance, or that the displacement of the plug leads to some amount of barrel collapse.

Plug displacement being a necessary event for pilus biogenesis, it is of prime importance to identify the molecular determinants involved in this process. Interestingly, the kinetic signature of the "plugless" channel can also be attained at times by mutant channels lacking the β5-6 hairpin. In addition, mutants lacking the α-helix or where interactions of the α-helix and the plug domain have been compromised appear to have an increased pore permeability (Volkan et al. 2013). Altogether, these results suggest that the α-helix and the β5-6 hairpin may play a role in keeping the plug in place. Therefore, it has been proposed that an allosteric mechanism mediated by a network of residues linking these domains to the plug domain might underlie the conformational changes necessary for plug displacement. By using a hybrid approach combining sequence conservation and coevolution analysis in combination with molecular dynamics of domain mutants, 5 connected communities were identified (Farabella et al. 2014). 'Hot spots' within each community were

Fig. 6.3 Electrophysiological analysis of the PapC usher and allosteric network controlling channel gating. (**a**) Planar lipid bilayer recording of WT PapC in 1 M KCl at +50 mV, illustrating occasional spontaneous short-lived monomeric plug displacement events (marked by "M") amongst smaller conductance transitions. (**b**) Patch-clamp recording of WT PapC in 150 mM KCl at −160 mV to illustrate the prolonged displacement of the plug observed at high voltages. Note that the current reaches the dimeric level (marked by "D") for a short period of time, but then relaxes to lower values. (**c**) Cartoon summarizing the electrophysiological and cell physiological observations obtained in various mutants in the putative allosteric network. "Large opening" and "absence of large opening" means that plug displacement was observed more often or less often than in WT, respectively; "transient mixed" means that occasional plug displacement events were seen as often as in WT (Reproduced from Farabella et al. (2014))

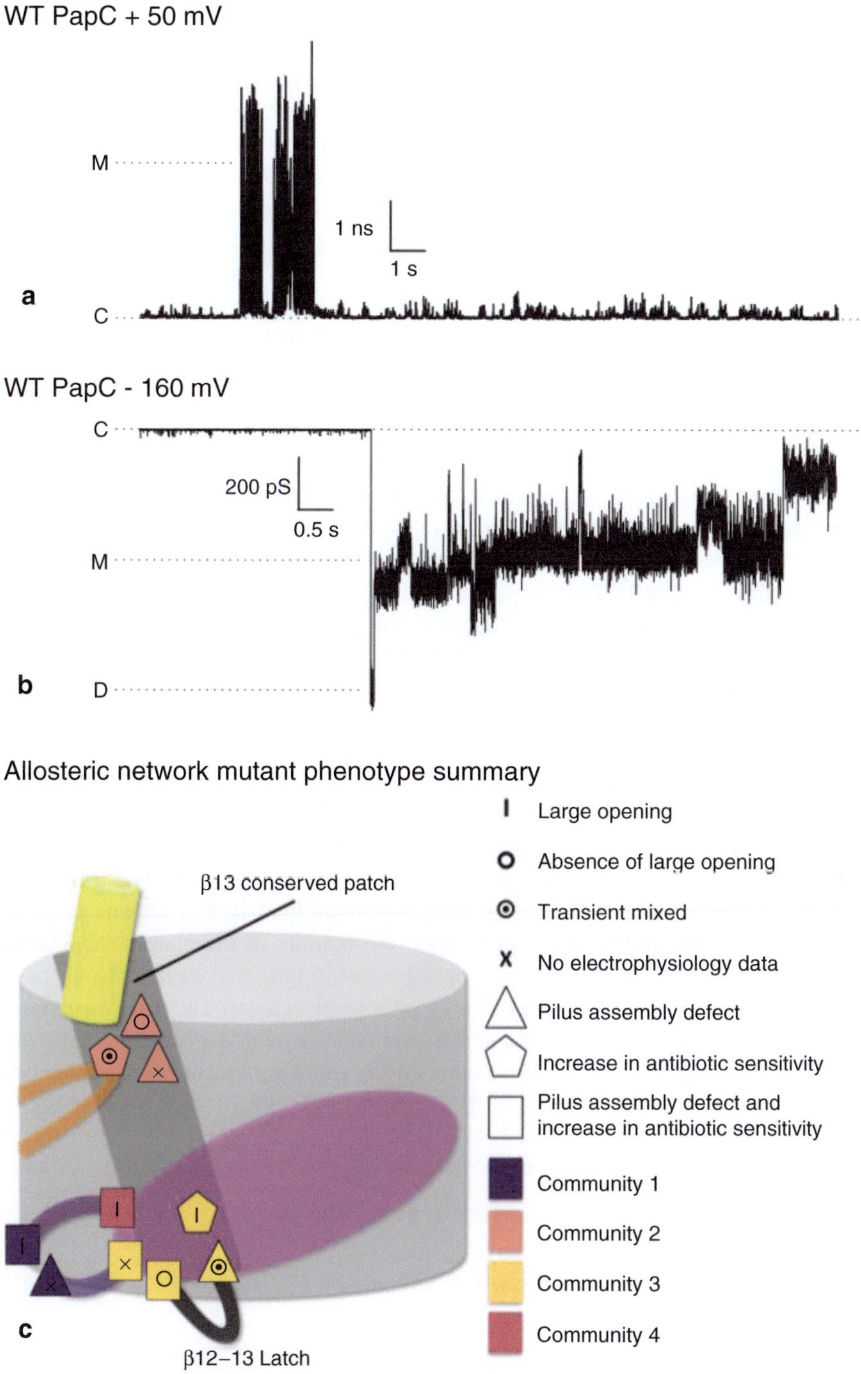
WT PapC + 50 mV
M
1 ns
1 s
a
C
WT PapC - 160 mV
C
200 pS
0.5 s
M
b
D
Allosteric network mutant phenotype summary
β13 conserved patch
β12–13 Latch
c
Large opening
Absence of large opening
Transient mixed
No electrophysiology data
Pilus assembly defect
Increase in antibiotic sensitivity
Pilus assembly defect and increase in antibiotic sensitivity
Community 1
Community 2
Community 3
Community 4

then defined as residues having strong interactions with each other, which are weakened in the absence of the α-helix and/or the β-hairpin. The list of potentially critical residues was then further refined through analysis of cellular phenotypes and electrophysiological properties of alanine substituted mutants. Mutations of residues in four of the five communities were found to have an impact on pilus biogenesis and antibiotic sensitivity (Fig. 6.3c). These communities are located at the base of the plug linkers (C1); between the β-hairpin and the conserved region at the base of the α-helix (C2); between β12-13 loop and the 1st plug linker (C3); between the β-hairpin, 2nd plug linker and the plug domain (C4). Mutations of residues in the 5th community on the tip of the plug domain did not alter these cellular phenotypes. The electrophysiological experiments showed that all of the mutants were capable of plug displacement, as seen on a representative WT trace (Fig. 6.3a), but the frequency of these events varied from mutant to mutant. Mutants with increased antibiotic sensitivity exhibited high propensities for plug displacement relative to WT, in accordance with a more "leaky" usher phenotype conferring increased antibiotic entry. In contrast, mutants impaired in their ability to form functional pili displayed very low propensities for plug displacement, as expected. A summary of the electrophysiological and cellular phenotypes is provided in Fig. 6.3c. The results of this study suggest that an allosteric pathway connecting residues close to the extracellular and the periplasmic sides of the usher might control the gating of the pore. This pathway includes conserved elements on the 13th β strand (β13 patch), the linkers to the plug domain and the periplasmic loop linking strands 12 and 13 (β12-13 latch) (Fig. 6.3c). Interestingly the β12-13 latch was shown to be in different conformations in the open and closed states (Phan et al. 2011).

Electrostatic interactions also exist at the interface of the plug domain with the α-helix, β5-6 hairpin, and barrel wall, and may participate in maintaining the plug in place. Indeed multiple alanine substitutions at selected residues effectively destabilize the channel such that plug displacement events similar to the one illustrated in Fig. 3b can occur at lower membrane potentials than in WT (Pham et al, in preparation). Presumably, in the WT usher, the binding of the chaperone-subunit complex may lead to conformational changes resulting in the weakening of these electrostatic interactions, and thus prompt plug displacement. The high membrane potential used in electrophysiological experiments would trigger local intrinsic charge movements that essentially mimic or bring about the conformational changes that normally result from binding of the chaperone-subunit complex to the usher.

6.7 Proteins Involved in Membrane Biogenesis

6.7.1 SecYEG and the General Secretory Pathway

Many proteins of Gram-negative bacteria destined to the plasma membrane or to compartments exterior to the plasma membrane utilize the general secretory pathway to exit the cytoplasm and gain access to these locations (Driessen and Nouwen 2008). This pathway uses the translocon SecYEG, where a transmembrane conduit for protein translocation is provided by the SecY subunit. This translocon is homologous to the Sec61 complex of Eukaryotes, also involved in nascent polypeptide translocation across the membrane of the endoplasmic reticulum (Pohlschroder et al. 1997). Targeting to the Sec translocon typically requires N-terminal signal sequences on pre-proteins, that are often cleaved after passage across the membrane. In co-translational translocation, utilized primarily for insertion of nascent membrane proteins, the pre-protein first associates with the signal recognition particle (SRP) and the ribosome. This complex is then recognized by the SRP-receptor FtsY before being delivered to the Sec translocon. For the vast majority of secreted proteins, post-translational translocation is used. Here, pre-proteins associate with SecA which targets them to SecYEG and powers translocation through rounds of ATP hydrolysis coupled to threading of the unfolded polypeptide through the channel (Chatzi et al. 2014). The process of moving a polypeptide across the plasma membrane without loss of the proton-motive force is an amazing feat that is still not completely understood.

The X-ray structure was determined for the homologous complex SecYEβ from the archeon *Methanococcus jannaschii* (Van den Berg et al. 2004) and found to superimpose well on the cryo-EM structure of an active *E. coli* SecYEG complex (Breyton et al. 2002). SecY, which contains 10 transmembrane (TM) helices, is organized like a clamshell, with the N- and C-terminal domains, made up of 5 TM helices each, acting as two halves surrounding a central pore. In the closed state observed in this structure, the pore is obstructed by a plug formed by a distorted helix connecting the first two TM segments of SecY. A lateral gate at the interface of the membrane lipids and SecY is formed by TM 2 and 7 which come in close proximity, and to which pre-protein signal sequences are known to bind. It is proposed that the binding of a signal sequence at the lateral gate triggers a conformational change of the clamshell with a concomitant displacement of the plug and widening of the pore. Indeed, the structure of the *Thermotoga maritima* SecYEG complexed with SecA and presumably in a pre-open state, reveals movements of the lateral gate helices and partial removal of the plug (Zimmer et al. 2008). A ring of isoleucine residues constricts the channel to its narrowest diameter of 4 Å, large enough to accommodate a polypeptide chain in its unfolded conformation. However, as long as some opening of the lateral gate is permitted, the pore can accommodate bulky moieties attached to translocation peptides with a diameter extending up to 24 Å (Bonardi et al. 2011). In fact, recent data suggest that the hydrophobic part of the signal sequence forms a helix within the lateral gate and effectively becomes part of the channel wall (Park et al. 2014).

Planar lipid bilayer experiments detect no ion conductance in the presence of purified wildtype Sec complex, but current deflections are readily observed when complexes with SecY mutants lacking the plug helix are used (Saparov et al. 2007). Single channel conductance of the mutant carrying the largest deletion (from residues 60 to 74) was estimated at ~940 pS (presumably in 1 M KCl), with an estimated pore diameter of 7.3 Å. These results support a model whereby the plug helix indeed functionally seals the channel shut. They also suggest that the hydrophobic pore ring alone is not sufficient for preventing ion conduction, at least in the absence of translocating peptides. However, it was suggested that the hydrophobic ring might form a gasket that seals against a translocating peptide, allowing for the maintenance of the proton gradient during protein translocation (Park and Rapoport 2011).

The first electrophysiological demonstration that signal sequences open the pore was provided by Simon and Blobel over two decades ago (Simon and Blobel 1992). Here inner membrane vesicles were fused with planar lipid bilayers. The addition of the LamB signal peptide led to an increase in macroscopic currents. With low peptide concentration, the appearance of individual channels could be detected, but only when the signal peptide was applied from the presumed cytoplasmic side of the channel. These initial observations were made on rather crude membrane preparations, but similar results have been obtained more recently with purified SecYEG complexes, where addition of proOmpA signal peptide led to channel openings of conductance (~220 pS) similar to those reported by Simon and Blobel (Knyazev et al. 2014). Surprisingly increases in membrane potential to physiological values (~130 mV) drove these channels to closures in step-wise fashion. In the presence of a stalled translocation intermediate, the voltage-driven closure of the channel leaves a residual conductance of ~6 pS, a value that, according to the authors, would ensure the preservation of the proton-motive force across the bacterial membrane. Thus, perhaps it is not the pore ring that limits ion flow during polypeptide translocation but rather a conformational change in SecYEG itself. The nature of the voltage sensor and the resulting conformational changes remain, however, unresolved.

6.7.2 BamA and Outer Membrane Biogenesis

Whereas SecYEG functions in the passage of proteins from the cell cytosol into the periplasmic space, BamA (a.k.a Omp85 or YaeT), in association with other proteins of the BAM (β-barrel Assembly Machinery) complex, functions as the main machinery for insertion of proteins into the outer membrane (Selkrig et al. 2014). This protein plays such a vital role for the cell that its identification in this process was not accomplished until relatively recently with the use of conditional mutants (Wu et al. 2005). BamA/Omp85 itself is a conserved protein with a widespread distribution among prokaryotes and homologues in mitochondria as well as chloroplasts (Gentle et al. 2005). Members of the Omp85 family are characterized by the presence of 1–5 N-terminal POTRA (Polypeptide Translocation Associated) domains, and a C-terminal β-barrel domain inserted in the membrane (Jacob-Dubuisson et al. 2009). These POTRA domains may function as receptors for outer membrane proteins (OMPs) that use the BAM complex for insertion into the outer membrane, or to assist in folding of the substrates, or even in the maintenance of the complex as a whole (Selkrig et al. 2014). Two structures of BamA have been solved by X-ray crystallography (Noinaj et al. 2013), and suggest possible conformational changes that may occur during the insertion of substrates into the membrane. Based on these results, the authors propose two possible mechanisms: (1) large OMPs might actually fold by β-augmentation using exposed BamA β-strands at a lateral gate to initiate the formation of a barrel which would eventually "bud off" the BamA template, and (2) smaller OMPs might reach the BAM complex already in a β-barrel configuration which would interact with the POTRA domains and use the local membrane destabilization provided by BamA to directly insert in the membrane. More recently, the same group has expanded the first model by including a substrate exit pore located above the lateral gate and whose presence was revealed by molecular dynamics simulations (Noinaj et al. 2014). The authors propose that this exit pore may provide a conduit for the extracellular domains and loops of the substrate to gain access to the cell surface, while transmembrane β-strands exit within the membrane through the lateral gate. The X-ray structure of BamA shows that its extracellular loops form a dome capping the barrel opening. This observation was in apparent contradiction with the clear pore-forming properties of BamA revealed by electrophysiology (see below). However, the presence of a lateral substrate exit pore might explain these conduction properties. Alternatively, the extracellular loops of the dome might be loose or dynamic enough to still allow ion flux.

Electrophysiological recordings were first reported in 2006 in two complementary studies from two distinct groups, using the nomenclature of Omp85 (Robert et al. 2006) and YaeT (Stegmeier and Andersen 2006). Both reports document that the channels display various conductances, but agree that the dominant conductance is in the order of 400–500 pS in 1 M KCl. The channel appears cation-selective, with a reported P_K/P_{Cl} of 5.7, and inactivated by acidic pH (Stegmeier and Andersen 2006). As was found for TPS translocons which also belong to the Omp85 superfamily (see above), the Omp85 proteins from various sources are constructed with a two-domain architecture, with the C-terminal domain forming the pore (Bredemeier et al. 2007; Stegmeier and Andersen 2006). There are several reports that suggest that the protein might indeed function as a folding platform for nascent β-barrels that are ultimately released laterally in the membrane plane (model 1 above). This proposal was first suggested by Stegmeier and Andersen (2006) based on the observation of multiple conductance levels, in particular when the C-terminal domain was studied in isolation. The appearance of multiple conductance levels hints at a dynamic and perhaps relatively plastic β-barrel structure, which may be linked to its physiological function of laterally opening up toward the membrane environment through a gap in the barrel. Indeed, Estrada Mallarino et al. calculated the so-called "barrel-closure energies" reflecting the interactions between β-strand #1 and #16 of the *Thermophilus thermus* Omp85 homolog (TtOmp85), and found them to be significantly weaker than the corresponding ones from members of the porin family (Estrada Mallarino et al. 2015). In the same study, the authors report that events of larger conductance appear when TtOmp85 is investigated in presence of peptides originating from a natural substrate OMP. The authors interpret these larger conductance events as representing pores with widened diameters due to β-augmentation by the substrates. In essence, they claim to have caught assembly intermediates made up of compound barrels including the 16 TtOmp85 strands and a few strands of the substrate OMP. Interestingly, these larger events are only seen if peptides corresponding to the C-terminal or last two β-strands of the substrate OMP are used, but not with N-terminal peptide mimics. These observations agree with prior results showing that larger currents are observed when bilayers containing multiple *E. coli*

Omp85 channels are exposed to denatured PhoE or even to a peptide corresponding to the last 10 C-terminal residues of PhoE (Robert et al. 2006). In this paper, the authors interpret the increased in current to opening of the channel, but since the experiments were performed on bilayers containing multiple channels, it is actually not possible to know if the increased current is truly due increase open probability or increase conductance of a single pore. The latter interpretation would support the model of β-augmentation proposed in more recent studies.

6.8 Conclusions

The application of electrophysiology to the study of bacterial translocons has been fruitful and has revealed some surprising properties of these complex machines. Foremost, far from behaving as gaping holes for the transport of large protein substrates, translocons have proven to be rather dynamic channels. In a way, this is not completely unexpected, as these large proteins are bound to undergo thermally driven conformational changes. This behavior makes their electrophysiological characterization challenging at times, as they rarely display current levels of single conductance. Transitions are often observed between multiple open states, and traces can be noisy due to frequent fast current fluctuations. Still, the combination of structural and electrophysiological approaches has established that translocons are often gated by internal elements that act as plugs or constricting devices, most likely to limit the potentially damaging influx or loss of material in the resting state. Future work should be directed at refining these studies by mapping more precisely the residues that participate in the gating mechanisms. A few studies have investigated the interaction of the channels with their substrates, but this is an area that one would hope could be expanded, despite the technical difficulty of such experiments. Indeed, ultimately one would like to gain a molecular understanding of the transport process through these translocons, in a similar way as it has been done for other channels, such as mitochondrial import channels or anthrax toxin. Finally, an area that has been under-explored so far is that of modulation or inhibition of the pore function. As many of these translocons participate in the release or biogenesis of virulence factors, the discovery and characterization of inhibitory molecules would represent an important step towards the design of new therapeutic strategies.

References

Andersen C, Hughes C, Koronakis V (2002a) Electrophysiological behavior of the TolC channel-tunnel in planar lipid bilayers. J Membr Biol 185(1):83–92. doi:10.1007/s00232-001-0113-2

Andersen C, Koronakis E, Bokma E, Eswaran J, Humphreys D, Hughes C, Koronakis V (2002b) Transition to the open state of the TolC periplasmic tunnel entrance. Proc Natl Acad Sci U S A 99(17):11103–11108. doi:10.1073/pnas.162039399

Baud C, Guerin J, Petit E, Lesne E, Dupre E, Locht C, Jacob-Dubuisson F (2014) Translocation path of a substrate protein through its Omp85 transporter. Nat Commun 5:5271. doi:10.1038/ncomms6271

Bitter W, Koster M, Latijnhouwers M, de Cock H, Tommassen J (1998) Formation of oligomeric rings by XcpQ and PilQ, which are involved in protein transport across the outer membrane of Pseudomonas aeruginosa. Mol Microbiol 27(1):209–219

Bonardi F, Halza E, Walko M, Du Plessis F, Nouwen N, Feringa BL, Driessen AJ (2011) Probing the SecYEG translocation pore size with preproteins conjugated with sizable rigid spherical molecules. Proc Natl Acad Sci U S A 108(19):7775–7780. doi:10.1073/pnas.1101705108

Bredemeier R, Schlegel T, Ertel F, Vojta A, Borissenko L, Bohnsack MT, Groll M, von Haeseler A, Schleiff E (2007) Functional and phylogenetic properties of the pore-forming beta-barrel transporters of the Omp85 family. J Biol Chem 282(3):1882–1890. doi:10.1074/jbc.M609598200

Breyton C, Haase W, Rapoport TA, Kuhlbrandt W, Collinson I (2002) Three-dimensional structure of the bacterial protein-translocation complex SecYEG. Nature 418(6898):662–665. doi:10.1038/nature00827

Brok R, Van Gelder P, Winterhalter M, Ziese U, Koster AJ, de Cock H, Koster M, Tommassen J, Bitter W (1999) The C-terminal domain of the Pseudomonas secretin XcpQ forms oligomeric rings with pore activity. J Mol Biol 294(5):1169–1179. doi:10.1006/jmbi.1999.3340

Burghout P, van Boxtel R, Van Gelder P, Ringler P, Muller SA, Tommassen J, Koster M (2004) Structure and electrophysiological properties of the YscC secretin from the type III secretion system of Yersinia enterocolitica. J Bacteriol 186(14):4645–4654. doi:10.1128/JB.186.14.4645-4654.2004

Burkinshaw BJ, Strynadka NC (2014) Assembly and structure of the T3SS. Biochim Biophys Acta 1843(8):1649–1663. doi:10.1016/j.bbamcr.2014.01.035

Busch A, Waksman G (2012) Chaperone-usher pathways: diversity and pilus assembly mechanism. Philos Trans R Soc Lond B Biol Sci 367(1592):1112–1122. doi:10.1098/rstb.2011.0206

Cascales E, Buchanan SK, Duche D, Kleanthous C, Lloubes R, Postle K, Riley M, Slatin S, Cavard D (2007) Colicin biology. Microbiol Mol Biol Rev 71(1):158–229. doi:10.1128/MMBR.00036-06

Chatzi KE, Sardis MF, Economou A, Karamanou S (2014) SecA-mediated targeting and translocation of secretory proteins. Biochim Biophys Acta 1843(8):1466–1474. doi:10.1016/j.bbamcr.2014.02.014

Clantin B, Delattre AS, Rucktooa P, Saint N, Meli AC, Locht C, Jacob-Dubuisson F, Villeret V (2007) Structure of the membrane protein FhaC: a member of the Omp85-TpsB transporter superfamily. Science 317(5840):957–961. doi:10.1126/science.1143860

Collins RF, Ford RC, Kitmitto A, Olsen RO, Tonjum T, Derrick JP (2003) Three-dimensional structure of the Neisseria meningitidis secretin PilQ determined from negative-stain transmission electron microscopy. J Bacteriol 185(8):2611–2617

Cordes FS, Komoriya K, Larquet E, Yang S, Egelman EH, Blocker A, Lea SM (2003) Helical structure of the needle of the type III secretion system of Shigella flexneri. J Biol Chem 278(19):17103–17107. doi:10.1074/jbc.M300091200

Cornelis GR (2002) The Yersinia Ysc-Yop 'type III' weaponry. Nat Rev Mol Cell Biol 3(10):742–752. doi:10.1038/nrm932

Cornelis GR (2010) The type III secretion injectisome, a complex nanomachine for intracellular 'toxin' delivery. Biol Chem 391(7):745–751. doi:10.1515/BC.2010.079

Cornelis GR, Wolf-Watz H (1997) The Yersinia Yop virulon: a bacterial system for subverting eukaryotic cells. Mol Microbiol 23(5):861–867

Coutte L, Antoine R, Drobecq H, Locht C, Jacob-Dubuisson F (2001) Subtilisin-like autotransporter serves as maturation protease in a bacterial secretion pathway. EMBO J 20(18):5040–5048. doi:10.1093/emboj/20.18.5040

Crago AM, Koronakis V (1998) Salmonella InvG forms a ring-like multimer that requires the InvH lipoprotein for outer membrane localization. Mol Microbiol 30(1):47–56

Crepin S, Houle S, Charbonneau ME, Mourez M, Harel J, Dozois CM (2012) Decreased expression of type 1 fimbriae by a pst mutant of uropathogenic Escherichia coli reduces urinary tract infection. Infect Immun 80(8):2802–2815. doi:10.1128/IAI.00162-12 d'Enfert C, Ryter A, Pugsley AP (1987) Cloning and expression in Escherichia coli of the Klebsiella pneumoniae genes for production, surface localization and secretion of the lipoprotein pullulanase. EMBO J 6(11):3531–3538

d'Enfert C, Reyss I, Wandersman C, Pugsley AP (1989) Protein secretion by gram-negative bacteria. Characterization of two membrane proteins required for pullulanase secretion by Escherichia coli K-12. J Biol Chem 264(29):17462–17468

Disconzi E, Guilvout I, Chami M, Masi M, Huysmans GH, Pugsley AP, Bayan N (2014) Bacterial secretins form constitutively open pores akin to general porins. J Bacteriol 196(1):121–128. doi:10.1128/JB.00750-13

Dohlich K, Zumsteg AB, Goosmann C, Kolbe M (2014) A substrate-fusion protein is trapped inside the type III secretion system channel in Shigella flexneri. PLoS Pathog 10(1), e1003881. doi:10.1371/journal.ppat.1003881

Driessen AJ, Nouwen N (2008) Protein translocation across the bacterial cytoplasmic membrane. Annu Rev Biochem 77:643–667. doi:10.1146/annurev.biochem.77.061606.160747

Du D, Wang Z, James NR, Voss JE, Klimont E, Ohene-Agyei T, Venter H, Chiu W, Luisi BF (2014) Structure of the AcrAB-TolC multidrug efflux pump. Nature 509(7501):512–515. doi:10.1038/nature13205

Du D, van Veen HW, Luisi BF (2015) Assembly and operation of bacterial tripartite multidrug efflux pumps. Trends Microbiol. doi:10.1016/j.tim.2015.01.010

Duret G, Szymanski M, Choi KJ, Yeo HJ, Delcour AH (2008) The TpsB translocator HMW1B of haemophilus influenzae forms a large conductance channel. J Biol Chem 283(23):15771–15778. doi:10.1074/jbc.M708970200

Estrada Mallarino L, Fan E, Odermatt M, Müller M, Lin M, Liang J, Heinzelmann M, Fritsche F, Apell HJ, Welte W (2015) TtOmp85, a β-barrel assembly protein, functions by barrel augmentation. Biochemistry 54(3):844–852. doi:10.1021/bi5011305.

Farabella I, Pham T, Henderson NS, Geibel S, Phan G, Thanassi DG, Delcour AH, Waksman G, Topf M (2014) Allosteric signalling in the outer membrane translocation domain of PapC usher. Elife 3. doi:10.7554/eLife.03532

Faudry E, Vernier G, Neumann E, Forge V, Attree I (2006) Synergistic pore formation by type III toxin translocators of Pseudomonas aeruginosa. Biochemistry 45(26):8117–8123. doi:10.1021/bi060452+

Faudry E, Job V, Dessen A, Attree I, Forge V (2007) Type III secretion system translocator has a molten globule conformation both in its free and chaperone-bound forms. FEBS J 274(14):3601–3610. doi:10.1111/j.1742-4658.2007.05893.x

Ford B, Rego AT, Ragan TJ, Pinkner J, Dodson K, Driscoll PC, Hultgren S, Waksman G (2010) Structural homology between the C-terminal domain of the PapC usher and its plug. J Bacteriol 192(7):1824–1831. doi:10.1128/JB.01677-09

Geibel S, Waksman G (2014) The molecular dissection of the chaperone-usher pathway. Biochim Biophys Acta 1843(8):1559–1567. doi:10.1016/j.bbamcr.2013.09.023

Gentle IE, Burri L, Lithgow T (2005) Molecular architecture and function of the Omp85 family of proteins. Mol Microbiol 58(5):1216–1225. doi:10.1111/j.1365-2958.2005.04906.x

Goure J, Pastor A, Faudry E, Chabert J, Dessen A, Attree I (2004) The V antigen of Pseudomonas aeruginosa is required for assembly of the functional PopB/PopD translocation pore in host cell membranes. Infect Immun 72(8):4741–4750. doi:10.1128/IAI.72.8.4741-4750.2004

Guerin J, Baud C, Touati N, Saint N, Willery E, Locht C, Vezin H, Jacob-Dubuisson F (2014) Conformational dynamics of protein transporter FhaC: large-scale motions of plug helix. Mol Microbiol 92(6):1164–1176. doi:10.1111/mmi.12585

Holland IB (2010) The extraordinary diversity of bacterial protein secretion mechanisms. Methods in Molecular Biology 619:1–20

Jacob-Dubuisson F, El-Hamel C, Saint N, Guedin S, Willery E, Molle G, Locht C (1999) Channel formation by FhaC, the outer membrane protein involved in the secretion of the Bordetella pertussis filamentous hemagglutinin. J Biol Chem 274(53):37731–37735

Jacob-Dubuisson F, Villeret V, Clantin B, Delattre AS, Saint N (2009) First structural insights into the TpsB/Omp85 superfamily. Biol Chem 390(8):675–684. doi:10.1515/BC.2009.099

Jakes KS, Cramer WA (2012) Border crossings: colicins and transporters. Annu Rev Genet 46:209–231. doi:10.1146/annurev-genet-110711-155427

Kline KA, Dodson KW, Caparon MG, Hultgren SJ (2010) A tale of two pili: assembly and function of pili in bacteria. Trends Microbiol 18(5):224–232. doi:10.1016/j.tim.2010.03.002

Knyazev DG, Winter L, Bauer BW, Siligan C, Pohl P (2014) Ion conductivity of the bacterial translocation channel SecYEG engaged in translocation. J Biol Chem 289(35):24611–24616. doi:10.1074/jbc.M114.588491

Konninger UW, Hobbie S, Benz R, Braun V (1999) The haemolysin-secreting ShlB protein of the outer membrane of Serratia marcescens: determination of surface-exposed residues and formation of ion-permeable pores by ShlB mutants in artificial lipid bilayer membranes. Mol Microbiol 32(6):1212–1225

Koronakis V, Sharff A, Koronakis E, Luisi B, Hughes C (2000) Crystal structure of the bacterial membrane protein TolC central to multidrug efflux and protein export. Nature 405(6789):914–919. doi:10.1038/35016007

Korotkov KV, Gonen T, Hol WG (2011) Secretins: dynamic channels for protein transport across membranes. Trends Biochem Sci 36(8):433–443. doi:10.1016/j.tibs.2011.04.002

Kosarewicz A, Konigsmaier L, Marlovits TC (2012) The blueprint of the type-3 injectisome. Philos Trans R Soc Lond B Biol Sci 367(1592):1140–1154. doi:10.1098/rstb.2011.0205

Koster M, Bitter W, de Cock H, Allaoui A, Cornelis GR, Tommassen J (1997) The outer membrane component, YscC, of the Yop secretion machinery of Yersinia enterocolitica forms a ring-shaped multimeric complex. Mol Microbiol 26(4):789–797

Kowal J, Chami M, Ringler P, Muller SA, Kudryashev M, Castano-Diez D, Amstutz M, Cornelis GR, Stahlberg H, Engel A (2013) Structure of the dodecameric Yersinia enterocolitica secretin YscC and its trypsin-resistant core. Structure 21(12):2152–2161. doi:10.1016/j.str.2013.09.012

Leo JC, Grin I, Linke D (2012) Type V secretion: mechanism(s) of autotransport through the bacterial outer membrane. Philos Trans R Soc Lond B Biol Sci 367(1592):1088–1101. doi:10.1098/rstb.2011.0208

Locht C, Bertin P, Menozzi FD, Renauld G (1993) The filamentous haemagglutinin, a multifaceted adhesion produced by virulent Bordetella spp. Mol Microbiol 9(4):653–660

Mapingire OS, Henderson NS, Duret G, Thanassi DG, Delcour AH (2009) Modulating effects of the plug, helix, and N- and C-terminal domains on channel properties of the PapC usher. J Biol Chem 284(52):36324–36333. doi:10.1074/jbc.M109.055798

Martinez JJ, Mulvey MA, Schilling JD, Pinkner JS, Hultgren SJ (2000) Type 1 pilus-mediated bacterial invasion of bladder epithelial cells. EMBO J 19(12):2803–2812. doi:10.1093/emboj/19.12.2803

Mattei PJ, Faudry E, Job V, Izore T, Attree I, Dessen A (2011) Membrane targeting and pore formation by the type III secretion system translocon. FEBS J 278(3):414–426. doi:10.1111/j.1742-4658.2010.07974.x

Mattoo S, Foreman-Wykert AK, Cotter PA, Miller JF (2001) Mechanisms of Bordetella pathogenesis. Front Biosci 6:E168–E186

Meli AC, Hodak H, Clantin B, Locht C, Molle G, Jacob-Dubuisson F, Saint N (2006) Channel properties of TpsB transporter FhaC point to two functional domains with a C-terminal protein-conducting pore. J Biol Chem 281(1):158–166. doi:10.1074/jbc.M508524200

Meli AC, Kondratova M, Molle V, Coquet L, Kajava AV, Saint N (2009) EtpB is a pore-forming outer membrane protein showing TpsB protein features involved in the two-partner secretion system. J Membr Biol 230(3):143–154. doi:10.1007/s00232-009-9195-z

Montagner C, Arquint C, Cornelis GR (2011) Translocators YopB and YopD from Yersinia enterocolitica form a multimeric integral membrane complex in eukaryotic cell membranes. J Bacteriol 193(24):6923–6928. doi:10.1128/JB.05555-11

Mueller CA, Broz P, Cornelis GR (2008) The type III secretion system tip complex and translocon. Mol Microbiol 68(5):1085–1095. doi:10.1111/j.1365-2958.2008.06237.x

Nikaido H (2009) Multidrug resistance in bacteria. Annu Rev Biochem 78:119–146. doi:10.1146/annurev.biochem.78.082907.145923

Noinaj N, Kuszak AJ, Gumbart JC, Lukacik P, Chang H, Easley NC, Lithgow T, Buchanan SK (2013) Structural insight into the biogenesis of beta-barrel membrane proteins. Nature 501(7467):385–390. doi:10.1038/nature12521

Noinaj N, Kuszak AJ, Balusek C, Gumbart JC, Buchanan SK (2014) Lateral opening and exit pore formation are required for BamA function. Structure 22(7):1055–1062. doi:10.1016/j.str.2014.05.008

Nouwen N, Ranson N, Saibil H, Wolpensinger B, Engel A, Ghazi A, Pugsley AP (1999) Secretin PulD: association with pilot PulS, structure, and ion-conducting channel formation. Proc Natl Acad Sci U S A 96(14):8173–8177

Oomen CJ, van Ulsen P, van Gelder P, Feijen M, Tommassen J, Gros P (2004) Structure of the translocator domain of a bacterial autotransporter. EMBO J 23(6):1257–1266. doi:10.1038/sj.emboj.7600148

Park E, Rapoport TA (2011) Preserving the membrane barrier for small molecules during bacterial protein translocation. Nature 473(7346):239–242. doi:10.1038/nature10014

Park E, Menetret JF, Gumbart JC, Ludtke SJ, Li W, Whynot A, Rapoport TA, Akey CW (2014) Structure of the SecY channel during initiation of protein translocation. Nature 506(7486):102–106. doi:10.1038/nature12720

Phan G, Remaut H, Wang T, Allen WJ, Pirker KF, Lebedev A, Henderson NS, Geibel S, Volkan E, Yan J, Kunze MB, Pinkner JS, Ford B, Kay CW, Li H, Hultgren SJ, Thanassi DG, Waksman G (2011) Crystal structure of the FimD usher bound to its cognate FimC-FimH substrate. Nature 474(7349):49–53. doi:10.1038/nature10109

Pohlschroder M, Prinz WA, Hartmann E, Beckwith J (1997) Protein translocation in the three domains of life: variations on a theme. Cell 91(5):563–566

Radics J, Konigsmaier L, Marlovits TC (2014) Structure of a pathogenic type 3 secretion system in action. Nat Struct Mol Biol 21(1):82–87. doi:10.1038/nsmb.2722

Remaut H, Tang C, Henderson NS, Pinkner JS, Wang T, Hultgren SJ, Thanassi DG, Waksman G, Li H (2008) Fiber formation across the bacterial outer membrane by the chaperone/usher pathway. Cell 133(4):640–652. doi:10.1016/j.cell.2008.03.033

Riley MA, Wertz JE (2002) Bacteriocins: evolution, ecology, and application. Annu Rev Microbiol 56:117–137. doi:10.1146/annurev.micro.56.012302.161024

Robert V, Volokhina EB, Senf F, Bos MP, Van Gelder P, Tommassen J (2006) Assembly factor Omp85 recognizes its outer membrane protein substrates by a species-specific C-terminal motif. PLoS Biol 4(11), e377. doi:10.1371/journal.pbio.0040377

Roussel-Jazede V, Van Gelder P, Sijbrandi R, Rutten L, Otto BR, Luirink J, Gros P, Tommassen J, Van Ulsen P (2011) Channel properties of the translocator domain of the autotransporter Hbp of Escherichia coli. Mol Membr Biol 28(3):158–170. doi:10.3109/09687688.2010.550328

Saparov SM, Erlandson K, Cannon K, Schaletzky J, Schulman S, Rapoport TA, Pohl P (2007) Determining the conductance of the SecY protein translocation channel for small molecules. Mol Cell 26(4):501–509. doi:10.1016/j.molcel.2007.03.022

Selkrig J, Leyton DL, Webb CT, Lithgow T (2014) Assembly of beta-barrel proteins into bacterial outer membranes. Biochim Biophys Acta 1843(8):1542–1550. doi:10.1016/j.bbamcr.2013.10.009

Serra DO, Conover MS, Arnal L, Sloan GP, Rodriguez ME, Yantorno OM, Deora R (2011) FHA-mediated cell-substrate and cell-cell adhesions are critical for Bordetella pertussis biofilm formation on abiotic surfaces and in the mouse nose and the trachea. PLoS One 6(12), e28811. doi:10.1371/journal.pone.0028811

Simon SM, Blobel G (1992) Signal peptides open protein-conducting channels in E. coli. Cell 69(4):677–684

Stegmeier JF, Andersen C (2006) Characterization of pores formed by YaeT (Omp85) from Escherichia coli. J Biochem 140(2):275–283. doi:10.1093/jb/mvj147

Tajima N, Kawai F, Park SY, Tame JR (2010) A novel intein-like autoproteolytic mechanism in autotransporter proteins. J Mol Biol 402(4):645–656. doi:10.1016/j.jmb.2010.06.068

Tardy F, Homble F, Neyt C, Wattiez R, Cornelis GR, Ruysschaert JM, Cabiaux V (1999) Yersinia enterocolitica type III secretion-translocation system: channel formation by secreted Yops. EMBO J 18(23):6793–6799. doi:10.1093/emboj/18.23.6793

Thanassi DG, Bliska JB, Christie PJ (2012) Surface organelles assembled by secretion systems of Gram-negative bacteria: diversity in structure and function. FEMS Microbiol Rev 36(6):1046–1082. doi:10.1111/j.1574-6976.2012.00342.x

Tosi T, Estrozi LF, Job V, Guilvout I, Pugsley AP, Schoehn G, Dessen A (2014) Structural similarity of secretins from type II and type III secretion systems. Structure 22(9):1348–1355. doi:10.1016/j.str.2014.07.005

Van den Berg B, Clemons WM Jr, Collinson I, Modis Y, Hartmann E, Harrison SC, Rapoport TA (2004) X-ray structure of a protein-conducting channel. Nature 427(6969):36–44. doi:10.1038/nature02218

van Ulsen P, Rahman S, Jong WS, Daleke-Schermerhorn MH, Luirink J (2014) Type V secretion: from biogenesis to biotechnology. Biochim Biophys Acta 1843(8):1592–1611. doi:10.1016/j.bbamcr.2013.11.006

Volkan E, Kalas V, Pinkner JS, Dodson KW, Henderson NS, Pham T, Waksman G, Delcour AH, Thanassi DG, Hultgren SJ (2013) Molecular basis of usher pore gating in Escherichia coli pilus biogenesis. Proc Natl Acad Sci U S A 110(51):20741–20746. doi:10.1073/pnas.1320528110

Wager B, Faudry E, Wills T, Attree I, Delcour AH (2013) Current fluctuation analysis of the PopB and PopD translocon components of the Pseudomonas aeruginosa type III secretion system. Biophys J 104(7):1445–1455. doi:10.1016/j.bpj.2013.02.018

Waksman G, Hultgren SJ (2009) Structural biology of the chaperone-usher pathway of pilus biogenesis. Nat Rev Microbiol 7(11):765–774. doi:10.1038/nrmicro2220

Wright KJ, Seed PC, Hultgren SJ (2007) Development of intracellular bacterial communities of uropathogenic Escherichia coli depends on type 1 pili. Cell Microbiol 9(9):2230–2241. doi:10.1111/j.1462-5822.2007.00952.x

Wu T, Malinverni J, Ruiz N, Kim S, Silhavy TJ, Kahne D (2005) Identification of a multicomponent complex required for outer membrane biogenesis in Escherichia coli. Cell 121(2):235–245. doi:10.1016/j.cell.2005.02.015

Yang L, Harroun TA, Weiss TM, Ding L, Huang HW (2001) Barrel-stave model or toroidal model? A case study on melittin pores. Biophys J 81(3):1475–1485. doi:10.1016/S0006-3495(01)75802-X

Zakharov SD, Eroukova VY, Rokitskaya TI, Zhalnina MV, Sharma O, Loll PJ, Zgurskaya HI, Antonenko YN, Cramer WA (2004) Colicin occlusion of OmpF and TolC channels: outer membrane translocons for colicin import. Biophys J 87(6):3901–3911. doi:10.1529/biophysj.104.046151

Zakharov SD, Sharma O, Zhalnina M, Yamashita E, Cramer WA (2012) Pathways of colicin import: utilization of BtuB, OmpF porin and the TolC drug-export protein. Biochem Soc Trans 40(6):1463–1468. doi:10.1042/BST20120211

Zgurskaya HI, Krishnamoorthy G, Ntreh A, Lu S (2011) Mechanism and function of the outer membrane channel TolC in multidrug resistance and physiology of enterobacteria. Front Microbiol 2:189. doi:10.3389/fmicb.2011.00189

Zgurskaya HI, Weeks JW, Ntreh AT, Nickels LM, Wolloscheck D (2015) Mechanism of coupling drug transport reactions located in two different membranes. Front Microbiol 6:100. doi:10.3389/fmicb.2015.00100

Zimmer J, Nam Y, Rapoport TA (2008) Structure of a complex of the ATPase SecA and the protein-translocation channel. Nature 455(7215):936–943. doi:10.1038/nature07335

Chapter 7
Viroporins

Joseph M. Hyser

Abstract Virus encoded ion channels, termed viroporins, are expressed by a diverse set of viruses and have been found to target nearly every host cell membrane and compartment, including endocytic/exocytic vesicles, ER, mitochondria, Golgi, and the plasma membrane. Viroporins are generally very small (<100 amino acids) integral membrane proteins that share common structure motifs (conserved cluster of basic residues adjacent to an amphipathic alpha-helix) but only limited sequence homology between viruses. Ion channel activity of viroporins is either required for replication or greatly enhances replication and pathogenesis. Channel characteristics have been investigated using standard electrophysiological techniques, including planar lipid bilayer, liposome patch clamp or whole-cell voltage clamp. In general, viroporins are voltage-independent non-specific monovalent cation channels, with the exception of the influenza A virus M2 channel that forms a highly specific proton channel due to a conserved HXXXW motif. Viroporin channel currents range between highly variable ('burst-like') fluctuations to well resolved unitary ('square-top') transitions, and emerging data indicates the quality of channel activity is influenced by many factors, including viroporin synthesis/solubilization, the lipid environment and the ionic composition of the buffers, as well as intrinsic differences between the viroporins themselves. Compounds that block viroporin channel activity are effective antiviral drugs both *in vitro* and *in vivo*. Surprisingly distinct viroporins are inhibited by the same compounds (e.g., amantadines and amiloride derivatives), despite wide sequence divergence, raising the possibility of broadly acting antiviral drugs that target viroporins. Electrophysiology of viroporins will continue to play a critical role in elucidating the functional roles viroporins play in pathogenesis and to develop new drugs to combat viroporin-encoding pathogens.

Keywords Viral ion channel • Amphipathic alpha-helix • Proton channel • Non-selective cation channels • Planar lipid bilayer • *Xenopus* oocytes

---

J.M. Hyser
Department of Molecular Virology and Microbiology, Department of Medicine,
Alkek Center for Metagenomic and Microbiome Research, Baylor College of Medicine,
One Baylor Plaza, MS: BCM385, Houston, TX 77030-3411, USA
e-mail: joseph.hyser@bcm.edu

A.H. Delcour (ed.), *Electrophysiology of Unconventional Channels and Pores*,
Springer Series in Biophysics 18, DOI 10.1007/978-3-319-20149-8_7

7.1 Introduction

Viruses are obligate cellular parasites that require host cells to support their replication cycle, which utilizes virus-encoded proteins and nucleic acids to exploit cellular processes needed from progeny virus assembly. Many viruses have evolved their own ion channels to help subvert cellular ionic gradients to their advantage. These virus-encoded ion channels are called viroporins. Research on viroporins over the last 22 years has demonstrated that most (if not all) viroporins are *bona fide* ion channels, in that they can display unitary single channel conductance states. However, identification and characterization of new candidate viroporins is challenging due to the lack of sequence homology between viroporins, the wide array of small molecules that can be conducted through these proteins when expressed in exogenous systems such as *E. coli*, and differences in ion channel characteristics (Lamb and Pinto 1997; Gonzalez and Carrasco 2003; Kelly et al. 2003). This chapter will summarize the different approaches used to identify viroporins and to demonstrate *bona fide* ion channel activity, with a particular emphasis on how differences in experimental details may explain differences in ion channel characteristics. The goal is to provide a guide for current and future electrophysiology analysis of viroporins and a more rational approach to understanding the electrophysiology of these truly unconventional ion channels.

7.1.1 Viroporins

Viroporins are a diverse class of small, hydrophobic virus encoded membrane associated proteins that oligomerize, particularly in lipid membranes or lipid-like detergents, such that an aqueous pore is formed through the membrane by an amphipathic α-helix within the viroporin domain (Gonzalez and Carrasco 2003). While the initial concept of viroporins focused on proteins that induced cell lysis or permeability to small molecule compounds, such as hygromycin B, more recent studies using electrophysiology demonstrate viroporins are not simply large pores but are *bona fide* viral ion channels, distinguishing them from viral fusions domains found on viral attachment and entry proteins (Gonzalez and Carrasco 2003; Nieva et al. 2012). Currently, the majority of viroporins are from mammalian viruses, with the exception of the K^+-selective channel (Kcv) expressed by chlorella viruses that infect unicellular algae (Kang et al. 2004). Additional viroporins come from avian reovirus (p10), and arboviruses that replicate in both insect and mammalian hosts (orbivirus NS3, ephemerovirus α-1), and it is likely that ion channels are also encoded by bacteriophage (e.g., holins/pinholins) as well. However, identification of ion channel activity through electrophysiological assays remains to be done for these proteins (Bodelon et al. 2002; Han and Harty 2004; Joubert et al. 2014; Young 2014). Although the Chlorella virus Kcv is clearly a viral ion channel, it has structural and functional features of bacterial and eukaryotic K^+ channels, suggesting a

Table 7.1 List of viruses and their viroporins

Viroporins		
Virus family	Virus	Viroporin
Orthomyxoviridae	Influenza A virus	M2
		PB1-F2
	Influenza B virus	BM2
	Influenza C virus	CM2
Flaviviridae	Hepatitis C virus	p7
	Dengue virus	preM/M
Retroviridae	HIV-1	Vpr
		Vpu
Togaviridae	Sindbis virus, Semliki forest virus, Ross River virus	6K
Coronaviridae	Severe acute respiratory syndrome virus (SARS), Human coronavirus 229E (CoV 229E), Mouse hepatitis virus (MHV), Infectious bronchitis virus (IBV)	Envelope (E)
	SARS	ORF3a
	CoV-229E	ORF4a
	SARS	ORF8a
Paramyxoviridae	Respiratory syncytial virus, Human metapneumovirus, Simian virus 5	SH
Reoviridae	Rotavirus (Serogroup A)	NSP4
Picornaviridae	Enterovirus 71	2B

shared evolutionary origin with conventional K^+ channels (Kang et al. 2004). This chapter will focus on the electrophysiology of mammalian viroporins (Table 7.1) that share only limited sequence homology with conventional host channels.

7.1.2 Structural Features and Classification of Viroporins

Viroporins come from a diverse set of virus families and there is little to no sequence homology among viroporins, making identification of these proteins challenging. However, all viroporins have a number of shared structural characteristics and it is by identification of these features that new viroporin candidates arise. Viroporins are relatively small (~60–275 amino acids) hydrophobic transmembrane proteins, which typically have highly basic regions of multiple lysine and arginine residues adjacent to the transmembrane domains. The basic regions are thought to aid in membrane association and invasion of the amphipathic α-helix, and the presence of the basic residues is typically well conserved (Gonzalez and Carrasco 2003). Viroporin activity is typically important for virus replication and/or pathogenesis because mutations or knockdown of viroporin expression result in reduced viral yield or attenuated disease (Nieva et al. 2012; Giorda and Hebert 2013).

In addition to primary sequence diversity, there is also considerable diversity in membrane topology among viroporins. They often have only short extra-membrane domains extending to the cytoplasm or into the intracellular membrane compartment or extracellular space, although there are examples of viroporins with long cytoplasmic domains, such as rotavirus NSP4 (Hyser et al. 2010). A classification system was recently developed based on the number of transmembrane segments (Class I or Class II) and the topology (Subclass A: N-terminus in the ER; Subclass B: N-terminus in the cytoplasm) (Nieva et al. 2012). More recent studies suggest the existence of Class III viroporins with three transmembrane segments, which includes the rotavirus NSP4 and SARS-CoV ORF3a proteins (Hyser et al. 2010; Lu et al. 2006). Thus far, no viroporin candidates with more than three transmembrane domains have been reported.

7.1.3 Non-electrophysiological Methods for Viroporin Characterization

The lack of sequence homology among viroporins makes genomic and metagenomic approaches for viroporin identification nearly impossible. A common functional hallmark of viroporins is the ability to induce plasma membrane permeability of host cells and *E. coli* upon recombinant expression or exogenous addition to liposomes [Reviewed in (Gonzalez and Carrasco 2003; Nieva et al. 2012; Giorda and Hebert 2013)]. Traditional assays to address viroporin-induced permeabilization have been developed using *E. coli* cell lysis, reduction in cell viability, or growth inhibition; permeabilization of cells or liposomes to β-galactosidase substrates (OMPG), translation-blocking compounds (hygromycin B) or fluorescent molecules (e.g., FITC dextrans); and fluorescence sensors to measure conductance of protons, Ca^{2+}, or redox state (Lama and Carrasco 1992; Taube et al. 2014; Agirre et al. 2002; Henkel et al. 2010; Hyser et al. 2012; Lin and Schroeder 2001). Many of these assays have been developed into screens for mutations that attenuate viroporin activity and chemical blockers of the channel (Agirre et al. 2002; Luscombe et al. 2010). These relatively simple assays serve as an important staging ground for understanding the viroporin's structure and identification of mutations that provide a basis for electrophysiology studies of channel activity.

7.1.4 Electrophysiology Methods and Challenges for Viroporins

Since viruses utilize host cell machinery for protein synthesis and replication, the study of viroporins utilizes the same electrophysiology methodologies as do conventional ion channels. However, there are a number of challenges associated with conducting and interpreting viroporin electrophysiology studies. First, because viroporins do not share sequence motifs that provide clues about ion selectivity (except for the HXXXW motif of proton channels, see below), it can be challenging to assign an observed conductance to the viroporin. For example, voltage clamp of viroporin-expressing *Xenopus leavis* oocytes have successfully demonstrated ion channel function of some viroporins, but the viroporin can also activate endogenous currents, which confound the data (Shimbo et al. 1995). Similarly, while planar lipid bilayer electrophysiology (PLB) is a popular method-of-choice for initial viroporins studies, the detergents and/or solvents used for viroporin purification can cause channel-like current artifacts in PLB experiments, again making rigorous demonstration that ion channel activity can be definitively assigned to the viroporin (Kelly et al. 2003). Thus, the most successful electrophysiology studies of viroporins have utilized multiple electrophysiology techniques, comparison of wild-type and viroporin-deficient mutant proteins, and a chemical inhibitor/blocker of viroporin function, if one is known.

7.1.4.1 Patch Clamp

Several patch clamp techniques have been used to study viroporins, with *X. leavis* oocytes being the most popular system to date and the system used in the seminal paper by Pinto et al. (1992) demonstrating IAV M2 is a viral ion channel. However, in a number of cases viroporin expression induced endogenous currents that could not be attributed to the viroporin itself (Shimbo et al. 1995). One study of HIV-1 Vpu in oocytes showed induction of a slowly activating, hyperpolarization-activated nonspecific cation (HANC) current [also called hyperpolarization-dependent inward cation current (I_{IN}) by others (Kuruma and Hartzell 1999)] that was interpreted to be a typical artifact of exogenous membrane proteins, and another study of the Sindbis virus 6K showed activation of endogenous calcium-activated chloride current (I_{Cl}) due to activation of store-operated calcium entry (SOCE) (Coady et al. 1998; Antoine et al. 2007). Similar currents were observed upon expression of HIV-1 Vpu (Schubert et al. 1996), influenza C virus M2 (CM2) (Hongo et al. 2004), and enterovirus 71 2B (Xie et al. 2011), but viroporin-defective mutants or blockers were used to demonstrate specificity. Such discrepancies are discussed individually below, but they highlight the need for cautious interpretation of data from oocytes when viroporins induce I_{IN} and/or I_{Cl}.

Fewer electrophysiology studies of mammalian cells expressing viroporins have been performed, likely because over-expression of viroporins causes rapid cytotoxicity and damages the cell (Madan 2008). As with oocytes, mammalian cell studies of the influenza A virus (IAV) M2 proton channel represent the seminal reports, first using CV-1 cells, an African green monkey kidney cell line (Wang et al. 1994) and subsequently using mouse erythroleukemia cells (MEL) stably expressing M2 (Chizhmakov et al. 1996). More recent studies on SARS CoV E protein, HIV-1 Vpu, and RSV SH protein utilized transient expression in the HEK293 or HEK293T cells (Pervushin et al. 2009; Bodelon et al. 2002; Gan et al. 2012). The low conductance of most viroporins studied thus far makes the study of macroscopic whole-cell currents an attractive alternative to PLB, provided the viroporin localizes to the plasma membrane and interference from cellular channels can be ruled out with viroporin-defective mutants and/or inhibitors.

One study has compared PLB to excised patches from unilamellar blisters of collapsed giant proteoliposomes to examine HCV p7 channel activity (Delcour et al. 1989; Montserret et al. 2010). Both techniques yielded similar results, including observation of multiple conductance states, which were attributed to either simultaneous insertion of multiple channels or channels composed of a different number of monomers (i.e., pentamers versus hexamers) (Montserret et al. 2010). Despite the more labor intensive preparation, liposome patch clamp has advantages over traditional PLB electrophysiology, such as direct control of the protein:lipid ratio, lack of continued channel insertion events, and the generally lower noise of patch clamp than PLB electrophysiology.

7.1.4.2 Planar Lipid Bilayer

The most common electrophysiology approach to study viroporins is planar lipid bilayer (PLB). The first studies of viroporins utilizing PLB were on the IAV M2 proton channel (Tosteson et al. 1994) and shortly after HIV-1 Vpr and Vpu, and the influenza B virus M2 (BM2) protein (Ewart et al. 1996; Piller et al. 1996; Sunstrom et al. 1996). While PLB studies have driven forward viroporin electrophysiology more than patch clamp studies, a disadvantage of using PLB for viroporins is that often the induced currents are highly variable without discernable conductance states, as illustrated for HCV p7 in Fig. 7.1a, but the same p7 peptide can also give rise to canonical unitary single channel states in separate experiments (Fig. 7.1b) (Premkumar et al. 2004). Further, the highly hydrophobic nature of most viroporins and the viroporin-domain peptides makes the choice of solvent important for generating high quality current recordings. Issues with solubility have been overcome by several different means, including reconstituting the viroporin into liposomes or addition of two flanking lysine resides (Torres et al. 2007). Nevertheless, clearly

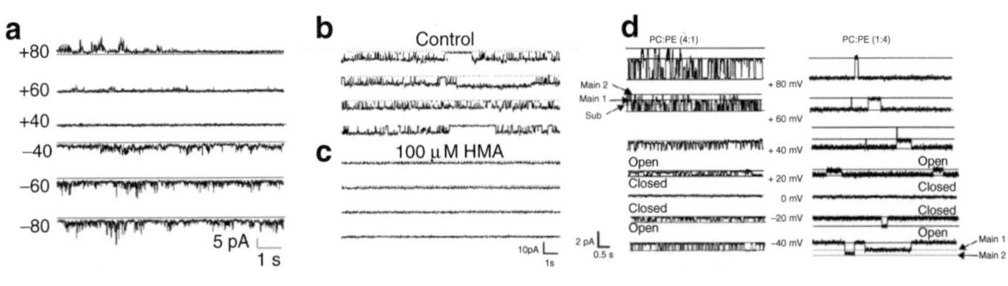

Fig. 7.1 Representative examples of HCV p7 viroporin channel activity in planar lipid bilayer. Synthetic p7 peptide was added to the *cis* bath resulting in (**a**) variable 'burst-like' currents or (**b**) unitary single channel currents. (**c**) The unitary single channels were completely blocked by 100 μM hexamethylene amiloride (HMA), but HMA only partially blocked the larger 'burst-like' currents (not shown) (Premkumar 2004). (**d**) Single channel activity of HCV p7 is altered by the lipid composition of the lipid bilayer. In PC-rich membranes (*left*) p7 has a higher open probability than in PE-rich membranes (*right*). PC-rich membrane also promoted slightly higher single channel conductance of the main conductance states and the subconductance state (*arrows*) (Whitfield 2011)

resolvable conductance states are observed for most viroporins, which may occur between 'burst events' (Henkel et al. 2010), enabling measurement of single channel conductance and ion permeability.

Interestingly, most viroporins display multiple distinct single channel conductance levels and subconductance states, which are observed within single traces and are reproducible between experiments (Chien et al. 2013; Ewart et al. 2002; Henkel et al. 2010; Montserret et al. 2010; Piller et al. 1999; Verdia-Baguena et al. 2013). This phenomenon is attributed to simultaneous opening of multiple channels, and/or activity of distinct channel oligomeric species, though few studies have been conducted. More recently, the lipid composition has been shown to significantly influence viroporin conductance. Changes in the phosphatidylcholine-to-phosphatidylethanolamine (PC:PE) ratio altered HCV p7 channel open probability and conductance states (Fig. 7.1d) (Whitfield et al. 2011), and similarly changes from neutral (PC) membrane to one containing negatively charged (PS) lipids reduced coronavirus E protein conductance by 50 % and increase cation selectivity (Verdia-Baguena et al. 2013).

7.2 Viroporin Electrophysiology

The functional characteristics of different viroporins vary substantially, with few similarities between virus families. Nevertheless, viroporins can be generally classified into three groups based on the ions they appear to conduct: (1) proton channels (e.g., influenza M2), (2) monovalent cation channels, some of which show ion selectivity (SARS CoV ORF3a), whereas others do not have strong selectivity (HIV-1 Vpu), and (3) divalent cation channels [(e.g., rotavirus NSP4 and picornavirus 2B, for which electrophysiology studies are not yet published (see below)]. This section contains a summary of the electrophysiology data for each viroporin in these viruses.

7.2.1 Viral Proton Channels

7.2.1.1 Influenza Virus M2

Influenza A virus (IAV) M2 was the first viral ion channel recorded using electrophysiology and is therefore the prototypical viral ion channel. The rationale for testing M2 for ion channel activity was based on amantadine and ramantadine inhibiting early steps in the low pH-dependent disassembly of viral ribonucleoprotein complexes during entry, as well as protecting the hemagglutinin protein (HA) from a premature conformational change during transit through the *trans* Golgi (Pinto et al. 1992). Amantadine resistant mutants mapped to the transmembrane domain of M2 protein, a small (~100 amino acid) tetrameric protein that is abundant in the Golgi and plasma membrane of host cells, and a minor component of the virion (23–60 copies) (Pinto et al. 1992), suggesting a direct role for M2 in modulating pH in infected cells. Whole-cell voltage clamp of *Xenopus* oocytes was performed using expressed wild-type M2 (A/Udorn/72) or an amantadine-resistant M2 mutant. A large inward current was detected in the WT M2-expressing cells. This current was greater than endogenous currents, was inhibited by 10 μM amantadine, and activated ~7-fold by decreasing the pH from 7.5 to 6.2; however, while similar currents were produced by the M2 mutants, they were not blocked by 100 μM amantadine (Pinto et al. 1992). Similar currents were observed when M2 from other IAV strains were expressed in oocytes (Wang et al. 1993), and when M2 was expressed in mammalian CV-1 cells expressing WT M2 (Wang et al. 1994). Recombinant M2 (Udorn) has ion channel activity when reconstituted into phosphatidylserine:phosphatidylethanolamine (PS:PE) bilayers (Tosteson et al. 1994) and these currents were also activated by low pH and inhibited by amantadine (Wang et al. 1994). In PLB, the M2 current was highly variable, with both small (50–90 pS) and large (40–600 pS) conductances in 150 mM NH_4Cl (Tosteson et al. 1994).

Initial reports suggested M2 is able to conduct protons, as well as group IA alkali cations, (Pinto et al. 1992; Shimbo et al. 1996; Tosteson et al. 1994), but more detailed investigation of ion selectivity by patch clamp of M2-expressing MEL cells (Chizhmakov et al. 1996) and oocytes and CV-1 cells (Mould et al. 2000) showed that M2 has high proton selectivity over Na^+ [$P_{H^+/Na^+} = 6 \times 10^7$ in MEL cells (Chizhmakov et al. 1996) and 2×10^6 in oocytes and CV-1 cells (Mould et al. 2000)]. Selective proton conductance was confirmed for bacterially expressed and purified M2 reconstituted into dimyristoyl phosphatidylcholine (DMPC) and dimyristoyl phosphatidylglycerol (DMPG) liposomes (1:5 w/w) and recorded in mixed lipid bilayers [phosphatidylethanolamine, phosphatidylcholine, phosphatidylserine, cholesterol (4:1:1:2 molar ratio)]. Single channel conductance in 1 mM HCl was 6 pS with a reversal potential (E_{rev}) close to E_H but E_{rev} did not vary upon changes in Cl^-, Na^+ or tetraethylammonium (TEA) concentrations (Vijayvergiya et al. 2004). In MEL cells, the estimated single channel conductance was <10 fA, though single channel currents could not be directly measured (Chizhmakov et al. 1996). In contrast to electrophysiological measurements of M2 conductance, liposomes-based studies using the fluorescent pH indicator pyranine measured the M2 single channel conductance to be 4.1×10^{-18} A (~4.1 aA) at pH 5.7; which is approximately 4 orders of magnitude lower than the lowest patch clamp measurement (Lin and Schroeder 2001). It is unclear why such a striking difference is observed between the electrophysiology and fluorescence based assays, but it the authors argue conductance below the noise level of patch clamp may lead to inaccurate estimates of M2 single channel conductance (Lin and Schroeder 2001)

M2 channel gating and proton selectivity are determined by the conserved HXXXW motif (aa 37–41) such that pH <6.5 at the ectodomain (i.e., extracellular or ER/Golgi/vesicle lumen) protonates His-37 thereby stabilizing an interaction with Trp-41, which acts as the primary gate (Okada et al. 2001; Tang et al. 2002). Substitution of His-37 with Gly, Ala, Ser, Thr, or Glu abolishes proton selectivity but ion channel activity is retained and proton selectivity is partially restored by including imidazole in the buffer. (Wang et al. 1995; Venkataraman et al. 2005).

M2 homologs from Influenza B and C viruses, BM2 and CM2, are also tetrameric one-pass membrane proteins (Class IA viroporin), and function similarly to M2 (Stewart and Pekosz 2012; Mould et al. 2003; Betakova and Hay 2007). When expressed in *Xenopus* oocytes BM2 showed external pH-dependent inward currents and strong selectivity for protons but was not inhibited by amantadine (Mould et al. 2003). M2 and BM2 share little sequence homology with the exception of the HXXXW motif, which validates their shared proton selectivity (Pinto and Lamb 2006). In contrast, CM2 expression in oocytes showed a slowly developing hyperpolarization activated inward current and was only weakly pH activated. This was a Cl^- current that was inhibited by the anion channel blocker DIDS but not amantadine, suggesting these were endogenous channels and not CM2 (Hongo et al. 2004; Shimbo et al. 1995). While CM2 does not contain the proton selective HXXXW motif, it was still able to protect HA from acid-induced conformational changes in the Golgi, albeit to a much lesser extent than M2 or BM2 (Betakova and Hay 2007). These biophysical studies have set the stage for an enormous scientific literature examining M2 structure, functions, and dynamics, as well as screens for new M2-blocker antiviral drugs (Loregian et al. 2014).

7.2.1.2 Hepatitis C Virus p7

Hepatitis C virus (HCV) is a member of the *Flaviviridae* family and a leading cause of cirrhosis and hepatocellular carcinoma. HCV structural and nonstructural proteins are synthesized as a single ER-associated polyprotein and cleaved by both viral and cellular proteases into the individual proteins. HCV p7 is a two-pass integral membrane protein located between the E2 envelope protein and nonstructural protein 2 (NS2) (Atoom et al. 2014). Subcellular distribution of p7 shows high levels in the ER but also localization to mitochondria or mitochondrial-adjacent membranes, and lipid droplets (Griffin et al. 2004; Vieyres et al. 2013).

Demonstration of p7 ion channel activity was performed with PLB experiments using bacterially expressed and purified GST- and His-tagged p7 (HCV J4 strain) inserted into PS:PC bilayers (Griffin et al. 2003). Addition of p7 either with or without a GST-tag induced poorly resolved bursts of current in symmetrical 0.1 M KCl and channel activity was increased by removal of the tag or measuring currents in 0.1 M $CaCl_2$. This current was completely inhibited by addition of 1 μM amantadine to both *cis* and *trans* chambers (Griffin et al. 2003). This initial study was confirmed by two groups who used synthetic peptides of full length p7 in PE:PC (4:1 w/w) (Pavlovic et al. 2003) or PE:PS:PC (5:3:2 w/w) (Premkumar et al. 2004) bilayers. Both groups observed large variable currents (Fig. 7.1a) with conductance ranging from ~100 pS to 2 nS (Pavlovic et al. 2003); however, some apparently single channel traces were recorded (Fig. 7.1b), enabling the identification of a 10–14 pS channel in either asymmetric KCl, NaCl, or $CaCl_2$ (Premkumar et al. 2004). Channel activity was selective for cations (Na^+ or K^+) over chloride; however, addition of

10 mM $CaCl_2$ increased Cl^- permeability (Premkumar et al. 2004). Long alkyl-chain iminosugar derivatives or hexamethylene amiloride (HMA), which also inhibits the HIV-1 Vpu channel, blocked p7 currents at high micromolar concentrations (Fig. 7.1c).

More consistent single channel activity was recorded using patch clamp of a full-length synthetic p7 peptide (HCV-J strain) reconstituted into asolectin liposomes. As before, variable currents were present, but two major conductance levels were discernable: 22 pS and 41 pS in 500 mM KCl (Montserret et al. 2010). Further, p7 is being used as a model ion channel to determine how lipid composition influences channel characteristics. Single channel traces in DOPC-rich [DOPC:POPE (4:1)] bilayers were compared to those in POPE-rich [DOPC:POPE (1:4)] bilayers. In the DOPC-rich bilayer, p7 exhibited a rapidly gating opening pattern with short opening times (~0.5 s), but had long (~8 s) openings with much lower frequency in the POPE-rich bilayer (Fig. 7.1d) (Whitfield et al. 2011). Despite these observed differences in channel kinetics, p7 had two main conductance states and one sub-conductance state; although the conductance levels were slightly greater in the PC-rich bilayer. However, lipid composition did not alter the voltage-independent character of p7 activity and p7 remained cation selective (Whitfield et al. 2011). Interestingly, coronavirus E protein also showed higher conductance in neutral lipid rich bilayers than in bilayers with charged lipids (Verdia-Baguena et al. 2012, 2013).

Despite p7 showing predominantly cation selective currents in PLB experiments, emerging data from cell culture studies suggest p7 to function as a proton channel, analogously to M2. First, the well characterized M2 blocker amantadine also inhibited p7 and p7 functionally complemented M2-deficient IAV to protect HA (Griffin et al. 2003; Wozniak et al. 2010). Further, p7 contains a highly conserved HXXXW/Y motif, similar to the M2 proton selectivity filter, the His-17 residue faces the pore and channel activity is inhibited by Cu^{2+} in a His-17 dependent manner (StGelais et al. 2007; Chew et al. 2009). However, substitutions of His-17 with Asn and Gln are present in many strains so this motif is not as strictly conserved in p7 (Meshkat et al. 2009). Thus far, proton conductance has not been demonstrated using electrophysiology but numerous drug screen studies have used PLB electrophysiology as validation for candidate p7 blockers, including the iminosugar derivatives and the drug BIT225, which are being tested for clinically relevant anti-HCV activity (Atoom et al. 2014).

7.2.2 Viral Monovalent Cation Channels

7.2.2.1 HIV-1

Human immunodeficiency virus-1 (HIV-1) encodes a series of accessory genes (*nef, vif, vpr* and *vpu*) in addition to the canonical retrovirus genes *gag, pol,* and *env*, and homologous or analogous genes are found in HIV-2 and simian immunodeficiency virus (SIV) (Strebel 2013). Two HIV-1 accessory proteins, called Viral protein regulatory (Vpr) and Viral protein unique (Vpu) have been demonstrated to have intrinsic ion channel activity, although to date the biological relevance of the observed channel function has not been determined (Strebel 2013, 2014). While channel activity of Vpr is controversial, Vpu has become a well-studied viral ion channel and used for many biophysical and structural studies (Strebel 2014). The electrophysiology studies of these viroporins are summarized below, as well as a discussion of the possible role ion channel activity may play during HIV-1 infection.

Vpr

HIV-1 Vpr is a 92 amino acid protein with a single predicted hydrophobic segment and multiple copies are incorporated into mature HIV-1 virions. Like many viral proteins, Vpr is multifunctional and assigned roles in HIV-1 replication and pathogenesis range from promotion of apoptosis, cell cycle arrest, regulation of HIV-1 reverse transcriptase, and modulation of host and viral gene transcription (Andrew and Strebel 2014).

The first evidence for intrinsic ion channel activity of Vpr utilized recombinant protein expressed in *E. coli* as a GST-tagged fusion protein, solubilized in 2 % CHAPS and purified by glutathione affinity chromatography, cleaved from the GST tag by thrombin and isolated by cation exchange chromatography as a detergent solubilized (0.5 % CHAPS) pure protein (Piller et al. 1996). This purified protein was added to bilayers composed of POPE:POPC (8:2 w/w) in *cis* chamber containing buffers of *cis*: 150 or 500 mM NaCl/*trans*: 50 mM NaCl. Purified Vpr (40–80 μL) induced poorly resolved, highly variable currents in 21 experiments, and appearance of currents required the *trans* chamber to be held at negative holding potentials, whereas positive holding potentials in the *trans* chamber rapidly inactivated currents. The reversal potential was between $^{+}30$ and −40 mV, leading to a relative permeability of P_{Na^+}/P_{Cl^-} of 5–12. Hyperimmune serum made against Vpr was able to neutralize the currents, which argued against current artifacts produced by CHAPS (Piller et al. 1996). A second study of Vpr truncations and mutations (also expressed in *E. coli*) localized the ion channel activity to the N-terminal 40 amino acids and identified Glu21 and Glu24 as potentially important for ion selectivity (Piller et al. 1999). However, other studies have focused on a downstream hydrophobic region (amino acids 52–83 or 52–96) that also has ion channel activity in PLB using either a conventional setup or the "dip-tip" method (Jacotot et al. 2001; Chen et al. 2010). This Vpr domain physically associates with the mitochondrial ANT transpoter and together these proteins cooperatively formed channels with larger conductance (190 pS) than Vpr alone (55 pS) (Jacotot et al. 2001). It is interesting to note that the channel activity of Vpr 52–96 has distinct conductance states than the N-terminal peptide or full-length protein, and it was postulated that the "burst" activity may be related to greater aggregation of monomers in aqueous environments or changes in oligomerization due to membrane curvature (Chen et al. 2010). However, the question of whether Vpr ion channel activity is biologically related to a known Vpr function remains to be determined (Andrew and Strebel 2014).

HIV-1 Vpu

HIV-1 Vpu was the third viral protein to be reported to have *bona fide* ion channel activity, after influenza virus M2 and HIV-1 Vpr; however, it has become one of the most intensely studied proteins, second only to IAV M2. Vpu is an ~81 amino acid, 16 kDa protein expressed from a *rev*-dependent bicistronic mRNA, which also encodes the *env* gene, and is co-translationally inserted into the ER membrane (Bour and Strebel 2003). Vpu is a type I integral membrane protein with a N-terminal hydrophobic membrane anchor comprised of residues 1–33. Like many HIV proteins, Vpu carries out multiple functions during infection, including down-regulation of CD4 expression through ERAD-mediated degradation, inhibition of NF-kB activation, sequestration and degradation of tetherin (a.k.a. BST-2) to increase HIV-1 particle release from infected cells, and ion channel activity of the N-terminal domain (Bour and Strebel 2003). While the molecular mechanisms of CD4 degradation are fairly well characterized, much work continues on how Vpu counteracts tetherin and the biological relevance of Vpu ion channel activity remains unsolved (Strebel 2014).

Vpu was initially postulated to be an ion channel due to topological similarities with the influenza A virus M2 proton channel and researchers speculated that Vpu may have an analogous role during HIV-1 replication. The first demonstration of Vpu ion channel activity utilized *E. coli* and PLB-based systems (Ewart et al. 1996). GST-tagged Vpu was expressed in *E. coli*, extracted from membranes using CHAPS, and affinity purified prior to thrombin-mediated removal of GST. Addition of purified Vpu to a bilayer composed of POPE:POPC (8:2) in asymmetric NaCl (500 mM NaCl *cis* and 50 mM NaCl *trans*) resulted in reproducible ion channel activity. Vpu showed a fivefold preference for monovalent cations (Na^+ or K^+) over anions, though conductivity of anions (Cl^- and $PO4^-$) were evident; however, as is often observed with viroporins, the conductivity varied drastically, from 14 to 280 pS, which was attributed to the amount of Vpu added to the chamber (Ewart et al. 1996).

These initial results were confirmed later that year using voltage clamp of *Xenopus* oocytes expressing wild-type Vpu or mutants with the conserved serine phosphorylation sites (aa 52 and 56) mutated to arginine (these are critical for down-regulation of host cell CD4). A mutant with a scrambled transmembrane domain that lacks channel activity was used as a control (Schubert et al. 1996). At 96 h post-injection of mRNA, the scrambled transmembrane mutant did not produce any current over background but WT Vpu produced a small (~130 nA), but consistent increase in inward current at −130 mV. In contrast, the Ser52Arg/Ser56Arg mutant resulted in a massively increased current (~600 nA) due to differences in expression between the wild-type and serine mutant proteins (Schubert et al. 1996). Since the serine-to-arginine mutant has the same transmembrane sequence as wild-type Vpu, the authors used this for characterization of Vpu channel activity in oocytes. The currents showed slow activation and no inactivation during the 2 s hyperpolarizing pulses. The current was primarily carried by Na^+ or K^+, with negligible contribution of divalent cations (Ca^{2+}) or anions (Cl^-, F^-, Br^-, or I^-) and was not blocked by the IAV M2 blocker amantadine or Cl^- channel blockers niflumic and flufenamic acids, providing evidence against activation of endogenous Cl^- channels by Vpu. Whole cell studies were confirmed using synthetic peptides to the Vpu transmembrane domain for either wild-type or the scrambled TM domain mutant. The peptides were applied to POPE:POPC bilayers and distinct single channel activity was observed (Schubert et al. 1996). As with oocytes, the permeability was selective for monovalent cations. A single channel conductance was determined as 12–48 pS in 0.5 M NaCl or 17–61 pS in 0.5 M KCl, much less variable than the previous bilayer study (Ewart et al. 1996). The scrambled transmembrane domain did not form discrete currents but caused erratic current fluctuations, demonstrating specificity of the observed channel function to the Vpu sequence (Schubert et al. 1996).

The simplistic single transmembrane topology of Vpu has also made this viroporin a model system for understanding the structure-function relationship of these minimalistic ion channels. Specific residue mutations within the Vpu transmembrane domain altered ion channel activity, particularly mutation of Ser24 to alanine or leucine caused a reduction in channel activity because this residue is needed to maintain a hydrated channel within the Vpu oligomeric bundle (Mehnert et al. 2008; Padhi et al. 2014). Further, mutation of the adjacent Trp23 to leucine significantly increased channel activity and open probability, but did not change ion selectivity, which remained selective for monovalent cations and followed the Eisenman series I (Mehnert et al. 2008). Interestingly, the same Trp23Leu mutation also disrupted Vpu-mediated downregulation of CD4 due to increase Vpu oligomerization, which is inactive for targeting CD4 for degradation (Magadan and Bonifacino 2012). The interplay between Vpu's role for targeting CD4 for degradation, sequestration of tetherin and the ion channel activity of the transmembrane domain are just beginning to be characterized (Strebel 2014).

Despite the initial studies, Vpu ion channel activity remained controversial for several years, particularly because another oocyte voltage-clamp study showed that wild-type Vpu induced the hyperpolarization-activated nonspecific cation current (HANC) often seen in oocytes, as well as it inhibited endogenous potassium currents rather than producing a *de novo* current (Coady et al. 1998). Further, Vpu blockers were not identified for several more years. Eventually, aliphatic modified amiloride derivatives such as hexamethylene amiloride (HMA) or dimethylamiloride (DMA)were found to block Vpu channel activity in PLB and to inhibit HIV-1 replication in cultured monocyte-derived macrophages (Ewart et al. 2004). HMA became the lead compound in a screen for novel drug-like molecules for blocking Vpu channel activity as candidate HIV-1 antiviral drugs, and from these screens the drug BIT225 was developed by Biotron Limited in Australia. BIT225 (40 μM) blocks Vpu channel activity in PLB and inhibited HIV-1 virus release for a range of strains in the low micromolar range (Khoury et al. 2010). Interestingly, BIT225 also inhibits the HCV p7 viroporin and has moved onto clinical trials against both HIV and HCV through Biotron Limited (Luscombe et al. 2010).

The biological relevance of Vpu ion channel activity remains unknown. Since the ability of Vpu to increase HIV-1 particle release mapped to the N-terminal hydrophobic domain, as does ion channel activity, it was the leading thought for many years that Vpu ion channel function was related to increasing particle release;

however, more recent evidence shows that it is the hydrophobic surface of the Vpu viroporin domain that interacts with tetherin, and the BIT225 Vpu blocker and mutations that block Vpu channel function do not disrupt tetherin down-regulation by Vpu (Bolduan et al. 2011; Kuhl et al. 2011). Current studies implicate Vpu in plasma membrane depolarization through both its intrinsic ion channel activity and through impaired host potassium channel targeting to the plasma membrane, which is caused through Vpu-mediated degradation of the TASK-1 potassium channel (Hsu et al. 2010, 2004). Nevertheless, little Vpu makes it to the plasma membrane and a direct role for Vpu in altering the membrane potential of T cells or monocyte derived macrophages has not been demonstrated. Further work is needed to determine the importance, if any, Vpu ion channel activity has for HIV-1 replication (Strebel 2014).

7.2.2.2 Alphavirus 6K

Alphaviruses are members of the *Togaviridae* family and include many emerging viruses of clinical importance for humans, particularly Chikungunya virus (Morrison 2014). The alphavirus structural proteins are produced from a subgenomic mRNA as a polyprotein and cleaved by both viral and cellular proteases into the individual proteins, PE2, 6K and E1. The 6K proteins are small (~58–61 amino acid) and hydrophobic, and are critical for membrane insertion and processing of the other structural proteins (Melton et al. 2002). The ability of the 6K protein to form an ion channel was speculated about even before the viroporin field emerged, and non-electrophysiological experiments using permeability of *E. coli* initially identified 6K as a viroporin (Ulug et al. 1984; Sanz et al. 1994). Electrophysiology of 6K was first examined using purified bacterially expressed protein and synthetic peptides of 6K from Ross River virus (RRV) and from Barmah Forest virus (BFV) (Melton et al. 2002). For the bacterially expressed and purified protein, the authors had to reconstitute purified GST-tagged protein solubilized in 400 mg/mL CHAPS into liposomes comprised of phosphatidylethanolamine, phosphatidylcholine, and phosphatidylserine (3:1:1 ratio) and CHAPS was removed by dialysis. Interestingly, while 6K had channel activity after the GST-tag was removed, the intact GST-fusion protein had no channel activity indicating the importance of how viroporins are expressed and purified to obtain a functional protein (Melton et al. 2002).

6K PLB experiments utilized membranes with an identical composition to the liposomes and asymmetric bath solutions (*cis*: 10 mM TES, 500 mM NaCl; *trans*: 10 mM TES, 60 mM NaCl). 6K-containing liposomes were added directly to the *cis* chamber and spontaneously incorporated into the bilayer, resulting in variable currents with conductance between 400 and 800 pS. Currents produced from both RRV and BFV 6K were qualitatively similar to each other, as well as for either the bacterially expressed protein or synthetic peptide (Melton et al. 2002). Channel activity was voltage-dependent, with greater activity at negative membrane voltages than positive. In asymmetric solutions the reversal potential was ~48 mV, revealing a 16-fold preference for Na^+ over Cl^-. The permeabilities of monovalent cations (Na^+ and K^+) were nearly equivalent, but a three to sixfold preference of Na^+ was found over Ca^{2+}. Addition of hyperimmune sera after channel activity was observed to abolish all currents and mutation of a conserved serine residue (Ser32) to proline reduced the monovalent/divalent selectivity, indicating Ser32 may be involved in the selectivity filter structure (Melton et al. 2002).

These initial PLB results were later confirmed using voltage-clamp of Sindbis virus 6K expressing *Xenopus* oocytes (Antoine et al. 2007). While expression of wild-type or *myc*-tagged 6K in oocytes did not induce currents attributable to 6K, it did induce endogenous currents similar to those seen for HIV-1 Vpu and influenza C M2 (Coady et al. 1998; Hongo et al. 2004). A deeper investigation of the currents demonstrated 6K increased both I_{IN} and I_{Cl} in oocytes (Antoine et al. 2007), including a 10.2-fold increase in I_{IN}, and a 2.9-fold increase in $I_{Cl1\text{-}S}$, 8.4-fold increase in I_{Cl2}, and a 12.2-fold increase in $I_{Cl1\text{-}T}$ (Kuruma and Hartzell 1999). Treating 6K-expressing oocytes with BAPTA-AM abolished the activation of both I_{IN} and $I_{Cl,}$ as did treating cells with low extracellular Cl^-, Ba^{2+}, Gd^{3+} or La^{3+} (Ca^{2+} channel blockers), niflumate (Cl^- channel blocker), caffeine (InsP3 inhibitor), and thapsigargin. Finally, the authors showed that 6K caused depletion of ER Ca^{2+} stores and a 13-fold induction of store-operated calcium entry (SOCE) (Antoine et al. 2007). These profound alterations in ion conductance led to shrinkage of the oocytes and cytotoxicity that were partially rescued by incubation in buffers that decreased ER Ca^{2+} release (e.g., presence of caffeine) or reduced SOCE (e.g., low Ca^{2+} medium) (Antoine et al. 2007). While the patch clamp studies in oocytes could not directly study 6K ion channel activity, they still provided useful information about how 6K may function during a natural infection; however, confirmation that 6K activates SOCE in mammalian cells has not been reported.

7.2.2.3 Coronavirus Viroporins

Coronavirus Envelope (E) Protein

The coronavirus (CoV) envelope (E) protein is a small (~9–12 kDa) integral membrane structural protein with a short N-terminal ectodomain and single transmembrane domain and plays a central role in coronavirus replication and particle assembly. While recombinant viruses with the E protein deleted are viable *in vitro*, they show deficient replication and a small plaque phenotype, suggesting that E protein is important, but not essential (Wilson et al. 2004). As with most viroporins, the small size and hydrophobicity of E protein led researchers to predict that it had ion channel activity. This was confirmed using synthetic peptides of full-length SARS-CoV E and a short peptide corresponding to the N-terminal domain (aa1 –40) that were dissolved in TFE and incorporated into bilayers composed of a lipid mix of POPE/POPS/POPC (3:1:1 ratio). Addition of either full-length or truncated peptide resulted in the induction of channel currents in >60 experiments, and initiation of channel activity was enhanced with a negative holding potential relative to the *cis* chamber. Channel activity tested in asymmetric buffers (*cis* 500 mM NaCl, trans:50 mM NaCl) had a reversal potential of ~48 mV, 52 pS conductance at −88 mV and 47 pS at −48 mV, and greater current at negative potentials. Both peptides were more selective for cations over anions and were more selective for Na^+ than K^+. To demonstrate specificity, stable channels were inhibited by 70 % upon addition of hyperimmune serum made to the N-terminal domain into the *cis* chamber (Wilson et al. 2004). E protein from other CoV strains also formed cation selective channels with relatively similar conductance levels, but the selectivity for Na^+ or K^+ varied (Wilson et al. 2006). Like HIV-1 Vpu and HCV p7, E protein channel activity was blocked by HMA (EC_{50} = 10.2 μM) and reduced MHV and human CoV-229E replication in cell culture with EC_{50} of 3.9 and 1.3 μM, respectively (Wilson et al. 2006).

Continued work on E protein channel activity have used synthetic peptides with flanking lysine residues added to improve solubility and this resulted in higher quality single channel activity for wild-type SARS-CoV E, but channel activity was abolished for point mutants, Asn15Ala and Val25Phe. The wild-type channel characteristics showed a Na^+ selective channel with 45 pS conductance in a 500 mM/5 mM asymmetric NaCl buffer system. Further, this study demonstrated channel activity was also inhibited by 100 μM amantadine (Torres et al. 2007). Finally, lipid composition influenced E protein channel activity, with negatively charged lipid bilayers (DPhPS) yielding channels with greater cation selectivity, weak selectivity of K^+ over Na^+, and higher conductance than those in neutral DPhPC lipid bilayers (Verdia-Baguena et al. 2012, 2013).

Despite considerable electrophysiology data on different CoV E proteins, the role ion channel activity has for CoV during infection remains unknown and while E protein can traffic to the plasma membrane, few electrophysiology studies in mammalian cells have been conducted. Whole-cell voltage clamp experiments using HEK293 cells transiently expressing SARS-CoV E found E protein induced moderate inward and large outward currents that were not observed in control cells. These currents were inhibited by HMA and did not show substantial selectivity between Na^+ and K^+, characteristics similar to PLB studies (Pervushin et al. 2009). However, others have found no E protein localization to the plasma membrane when expressed in HET293T cells and decreased, not increased, whole cell current (Nieto-Torres et al. 2011). Further, E protein disrupted host cell Na^+/K^+ ATPase α1 and ENaC channels; however, it is not known whether this is related to its ion channel function (Ji et al. 2009). Further work on the biological relevance of E protein channel activity is needed but E protein remains a potential target for antiviral drug development.

SARS-CoV ORF3a and ORF8

The SARS coronavirus (SARS-CoV) encodes two other viroporins, called ORF3a and ORF8. Like E protein, ion channel activity of these proteins has been identified, but their functional roles have not be determined. Their role is thought not to be critical for virus replication or virulence *in vivo*, because engineered virus or naturally occurring deletion mutants lacking ORF3a or ORF8, respectively, retain high replication and pathogenesis (DeDiego et al. 2014; Yount et al. 2005).

SARS ORF3a is the largest viroporin protein identified thus far, composed of 274 amino acids and three transmembrane domains, that is transported to the surface of virus-infected cells and is also incorporated as a minor component of mature virions. ORF3a folds into disulfide-stabilized homotetramers with a short N-terminal ectodomain and longer cytoplasmic tail. Based on topology and oligomerization, ORF3a was predicted to have ion channel activity, which was confirmed using voltage clamp in *Xenopus* oocytes expressing wild-type ORF3a and abrogation of channel activity for a cysteine-to-alanine mutant that failed to oligomerize (Lu et al. 2006). ORF3a increased K^+ currents in oocytes that are insensitive to typical K^+ channel blockers (tetraethylammonium and cesium) but was blocked by 10 mM Ba^{2+} in the bath buffer. Synthetic peptides of the three SARS ORF3a transmembrane domains have been tested in PLB experiments (POPE:DOPC, 1:4) to identify transmembrane segment 3 to be the main channel domain. While channel activity was influenced when the other transmembrane segments were mix in, they did not recapitulate the channel activity observed for full-length protein. Full-length ORF3a had single channel conductance levels of 3.0, 6.1 and 12.2 pS, and, in contrast with the oocyte voltage clamp data, was able to conduct Ca^{2+} with conductance of 10 and 58 pS in 500 mM $CaCl_2$ (Chien et al. 2013). However, whether Ca^{2+} conductance is a biologically relevant function of ORF3a remains unknown.

Similar K^+ channel activity was observed for ORF3 from PEDV and ORF4a from coronavirus 229E, and mutations in PEDV ORF3 found in attenuated strains correlated with decreased currents. In contrast to the K^+ selective conductance observed for SARS ORF3a, ORF4a conducted Na^+, K^+, Rb^+, Cs^+ to similar levels, while Li^+ partially reduced currents, indicating ORF4a is a non-selective monovalent cation channel (Zhang et al. 2014). This suggests that ORF3a and ORF4a may be functional homologs, and while they increase virion release, the mechanism by which this occurs is unknown (Wang et al. 2012; Zhang et al. 2014).

While SARS ORF3a is the largest viroporin identified, the SARS ORF8a viroporin is the smallest at only 39 amino acids (Chen et al. 2011). Channel recordings of an ORF8a synthetic peptide were performed in POPC bilayers at 38 °C (to simulate body temperature) in the presence of DTT because the protein contains four conserved cysteine residues that align along one face of the amphipathic alpha-helix. Currents were variable but with a primary conductance state of 8.8 pS with weak cation selectivity. Using computational modeling, ORF8a is expected to form a pentameric channel assembly; however, both the structure and ion channel function remain to be confirmed, as well as the possible importance of the channel function for virus replication (Chen et al. 2011).

7.2.2.4 Respiratory Syncytial Virus SH

Paramyxoviruses are nonsegmented negative stranded RNA viruses that include a number of key human pathogens, including Mumps virus (MuV) and human respiratory syncytial virus (hRSV) (Parks 2013, 24056173). Three membrane proteins are encoded by paramyxoviruses, the fusion protein (F), the glycoprotein (G) and small hydrophobic protein (SH) (McLellan et al. 2013). The SH protein is a 64–65 amino acid protein with a single transmembrane domain with the N-terminus remaining in the cytoplasm and the C-terminus forming the ectodomain. RSV SH homologs are found in parainfluenza virus 5, mumps virus and J paramyxovirus and human metapneumovirus; and while deletion of SH does not affect virus replication *in vitro*, it results in attenuation *in vivo* (Gan et al. 2008). Therefore the role of the SH protein remains unknown; however, SH is implicated in blocking TNF-alpha-mediated apoptosis and activation of the inflammasome (Triantafilou et al. 2013). Due to its short length, high hydrophobic nature and the small amounts found in virions, SH was predicted to be a viroporin and initially studied in the *E. coli*-based functional assays and shown to have viroporin-like characteristics (Perez et al. 1997).

Channel activity of a SH synthetic peptide was tested using PLB [POPE:POPS:POPC (5:3:2 v/v/v)] in asymmetric buffer (*cis*, 500 mM NaCl; *trans*, 50 mM NaCl) and found to have a cation-selective channel activity with a conductance of ~35 pS and reversal potential at −68.1 mV, near the sodium Nernst potential of 59 mV. Unlike the influenza A M2 proton channel, SH was not low-pH activated, but was in fact inhibited by decreasing the pH to 4.0 (Gan et al. 2008). Structural studies of *E. coli* expressed SH show formation of channel-like homopentameric rings and solved the NMR structure of SH in detergent micelles (Carter et al. 2010; Gan et al. 2012). Finally, using a liposome dye release-based drug screen, a potential RSV SH blocker called pyronin B was identified (Li et al. 2014). Pyronin B caused a 60 % inhibition of SH channel activity of purified full-length SH in PLBs, with a K_d of 6.8 μM. NMR and docking studies determined that pyronin B interacted with SH residues 39–44, and mutation of Ala39 to serine resulted in nearly complete resistance to pyronin B-mediated inhibition of channel activity (Li et al. 2014). Pyronin B was also able to block RSV replication in Vero cells, a monkey kidney cell line, making this a promising lead compound for a RSV antiviral drug.

In attempts to understand the biological role and molecular regulation of SH channel activity, whole cell patch clamp of SH-expressing HEK293T cells was performed, using artificial cerebrospinal fluid bath buffer and a potassium gluconate-based pipette buffer to mimic physiological conditions found *in vivo* (Gan et al. 2012). SH induced a fivefold higher current at 70 mV in low pH (5.5) than at neutral pH (7.4), and the reversal potential was close to 0 mV, indicating no selectivity between Na^+ and K^+. To examine the mechanism of the low-pH activation, mutants of two conserved histidine residues (His22 and His51) to alanine or phenylalanine were tested and had similar or increased (His22) currents at pH 5.5 relative to wild-type SH. Double mutations eliminated SH channel activity, suggesting the histidine residues are functionally important for SH but not whether they are involved in pH-dependent activation (Gan et al. 2012). However, the HEK293 patch clamp studies are not conclusive because the same group was unable to reproduce the low-pH activation of SH currents in a later study (Li et al. 2014). Thus, the biological role for SH channel activity during RSV infection, and whether channel activity is linked to dysregulation of cytokine signaling (e.g., TNF-alpha or inflammosomes) remains unknown (Triantafilou et al. 2013).

7.2.2.5 Influenza A Virus PB1-F2

While the influenza A virus M2 proton channel is the most well studied and well defined viral ion channel, another influenza A protein called PB1-F2 is gaining notoriety as a virulence factor in highly pathogenic influenza strains. PB1-F2 is encoded by the second genome segment by a $^{+}1$ alternate open reading frame in many influenza A strains, but is absent from influenza B virus. The functional role(s) of PB1-F2 are still being elucidated; however, at least one of its primary roles is the induction of cellular apoptosis by activation of the intrinsic apoptotic pathway (Chanturiya et al. 2004). Structural studies indicate PB1-F2 forms a positively charged amphipathic alpha-helix and has a high propensity for oligomerization. A mitochondrial localization sequence was first defined to be between amino acids 65–87 (Gibbs et al. 2003) and then refined to amino acids 46–75, with two conserved basic residues (Lys73 & Arg75) to be required for trafficking to mitochondria and association with the inner and outer mitochondrial membranes (Yamada et al. 2004). Localization of PB1-F2 to mitochondria resulted in a loss of mitochondrial membrane potential, release of cytochrome c and induction of apoptosis (Gibbs et al. 2003). The first study to examine the potential ion channel function of PB1-F2 used planar lipid bilayers of various compositions (Chanturiya et al. 2004). Addition of PB1-F2 to PLB chambers induced currents without typical unitary fluctuations of channels and the lack of transitions was not influenced by different lipid or buffer compositions; however, the observed currents were greatest with neutral lipids in monovalent cation-based solutions. PB1-F2 is selective for cations over anions, but addition of Ca^{2+} increased Cl^- permeability. Based on the membrane destabilization and lack of discrete channel-like behavior, PB1-F2 was thought to form lipidic pores rather than proteinaceous channels (Chanturiya et al. 2004). However, a recent study of a synthetic PB1-F2 peptide in phosphatidylcholine bilayers found that typical channel events were observed among the unresolved fast fluctuations. Traces containing unitary channel activity were occasionally interrupted by strong bursts of unresolved current and there appeared to be two distinct open levels (O_1 and O_2). In contrast with the previous study, PB1-F2 conducted anions better than cations, but did not discriminate between K^+ and Na^+ or Cl^- and gluconate, and was able to conduct Ca^{2+} (Henkel et al. 2010). More work is needed to determine whether the channel activity of PB1-F2 is directly responsible for induction of apoptosis through disruption of the mitochondria membrane potential, possibly by directly patching mitochondria from influenza A-infected cells or cells expressing PB1-F2.

7.2.2.6 Dengue Virus M

Dengue virus is a member of the Flaviviruses in the *Togaviridae* family. Dengue viruses cause up to 528 million infections per year, with approximately 20 % of these infections resulting in clinically significant disease, making it an emerging virus of significance to public health (Salazar et al. 2014). The Dengue virus membrane protein (M) is a transmembrane protein that is co-translationally inserted into the ER membrane and the mature 75-aa protein is produced from the preM precursor by furin cleavage (Hsieh et al. 2014). Dengue virus entry requires endosomal acidification, suggesting preM or M may have ion channel activity (Premkumar et al. 2005). PLB studies (PE:PS:PC, 5:3:2 ratio) of a 40 amino acid synthetic C-terminal peptide in asymmetric NaCl or KCl buffers (*cis*, 500 mM; *trans*, 50 mM) showed large and variable currents, with few single channel transitions (~6 pS) (Premkumar et al. 2005). The observed reversal potential of $^{+}$38 mV suggested K^+ selectivity over Cl^-. Both HMA and amantadine inhibited M channel activity (Premkumar et al. 2005). However, a recent study refuted the PLB data, as no significant conductance from preM/M was detected by voltage clamp of *Xenopus* oocytes (Wong et al. 2011). Given the few studies on preM/M, future studies are needed to determine whether preM/M forms a *bona fide* ion channel.

7.2.3 Viral Divalent Cation Conductance

A number of viruses and viroporins disrupt host cell Ca^{2+} homeostasis during infection; however, thus far, none of the viroporins that have been studied by electrophysiology has shown selectivity for divalent cations over monovalent cations (Zhou et al. 2009). Nevertheless, some degree of Ca^{2+} conductivity has been observed for some viroporins. PLB recordings of bacterially expressed p7 in 0.1 M $CaCl_2$ resulted in significantly larger currents than in KCl; however, whether p7 antagonists (e.g., HMA or amantadine) or channel-disrupting mutations block Ca^{2+} conductance was not studied (Griffin et al. 2003). A similar experimental set-up (*cis* 500 mM KCl, *trans* 500 mM $CaCl_2$) demonstrated that influenza A virus PB1-F2 was also able to conduct Ca^{2+} and a liposome-based assay using Fluo-3 loaded vesicles confirmed the ability of PB1-F2 to conduct Ca^{2+} into the liposome (Henkel et al. 2010).

In general, Ca^{2+} conductance by viroporins is demonstrated through indirect means, usually through fluorescent Ca^{2+} imaging to measure disruption in host cell Ca^{2+} homeostasis by examining the resulting activation of host cell channels due to viroporin-mediated Ca^{2+} currents. As detailed above, whole-cell voltage clamp experiments of Sindbis virus 6K in oocytes showed activation of SOCE and calcium-activated chloride channels, but current attributable to 6K was not detected (Antoine et al. 2007). Activation of SOCE by viroporin-mediated release of ER Ca^{2+} stores and activation of STIM1 may be a function shared by other ER-localized viroporins, as demonstrated for the rotavirus viroporin nonstructural protein 4 (NSP4) and predicted for the picornavirus 2B (van Kuppeveld et al. 1997; Hyser et al. 2013). The resulting Ca^{2+} fluxes affect multiple cell functions that facilitate virus replication and/or are involved in disease (Zhou et al. 2009).

For rotavirus, NSP4 viroporin activity is sufficient to disrupt Ca^{2+} homeostasis and therefore an excellent model to study the electrophysiology of Ca^{2+}-conducting viroporins (Hyser et al. 2013). Using both liposome patch clamp and PLB electrophysiology of a 44-amino acid synthetic NSP4 viroporin domain peptide (aa47-90), we have observed ion channel activity, including both the highly variable "burst-like" activity and unitary single channel conductance. In liposome patch clamp of NSP4 in asymmetric buffers (pipette 150 mM KCl; bath 75 mM CaC12), single channels were observed with both positive and negative pipette voltages, indicating NSP4 can conduct both K^+ and Ca^{2+}, although contribution of Cl^- cannot be ruled out (Hyser and Delcour, unpublished data, 2015). Of particular interest is how viroporins such as NSP4 and 2B, but not other ER-localized viroporins (e.g., HIV-1 Vpu), release ER Ca^{2+}. A highly conserved Ca^{2+} binding site in NSP4 may represent a regulatory motif, as it is found immediately adjacent to the NSP4 viroporin domain in the cytoplasmic coiled-coil domain (Chacko et al. 2011; Sastri et al. 2014). This motif is selective for Ca^{2+} and Ba^{2+}, but not Mg^{2+}, and binding is lost at pH <5.6, which corresponds with a conversion of the protein from a Ca^{2+}-bound tetramer to a Ca^{2+}-free pentamer with an open central channel (Sastri et al. 2014). Further electrophysiological studies are needed to establish whether these channel show any Ca^{2+} selectivity or whether they are non-selective channels that exploit the natural Ca^{2+} gradient across the ER membrane to provide the driving force for Ca^{2+} over other ions.

7.2.4 Viral Anion Channels

All currently identified viroporins preferentially conduct cations over anions; although high concentration of $CaCl_2$ or some transmembrane domain mutants alter the channel selectivity, but not to the extent that the viroporin becomes an anion channel (Griffin et al. 2003; Henkel et al. 2010). The primary exception to this have been recombinant viroporins from HIV-1 Vpu and alphavirus 6K expressed in *Xenopus* oocytes, where strong Cl^- currents were detected by whole cell voltage clamp (Coady et al. 1998; Melton et al. 2002; Antoine et al. 2007). Such was also the case for enterovirus 71 (EV71) 2B. Recombinant EV71 2B was expressed in *Xenopus* oocytes and large hyperpolarization-dependent currents were detected, but not for a transmembrane deletion mutant. Replacing Cl^- with gluconate reduced currents and the current was blocked by the well-characterized chloride channel blocker DIDs, while replacing Na^+ with NMDG had little effect, which suggested the invoked currents were carried by Cl^-. The authors postulated that the 2B anion channel activity may cause elevated Ca^{2+} indirectly by reducing ATP synthesis or by disrupting anion balance across the Golgi membrane, causing leakiness to Ca^{2+} (Xie et al. 2011). However, this study did not address the possible indirect activation of oocyte Ca^{2+}-activated chloride channels (CaCC), similar to what was determined for the Sindbis virus 6K protein expressed in oocytes (Antoine et al. 2007). These conflicting results underline the need for more studies of viroporins that disrupt host Ca^{2+} homeostasis. Further, if 2B is indirectly activating CaCC channels through induction of SOCE, then there are no candidate viroporin anion channels yet identified and this represents an interesting niche to be filled.

7.3 Summary and Future Directions

The first identification of viral ion channels occurred 23 years ago and the electrophysiological study of these proteins has accelerated rapidly ever since. However, it is safe to say that the viroporin field is still in its infancy, with many more questions remaining about viroporin biology to answer. Of primary importance to understanding viroporin electrophysiology is continued efforts to develop robust methods for expression, purification, and analysis of viroporins. This is particularly important because of the tendency of viroporins to induce highly variable "burst-like" currents rather than the canonical (and sought after) clear unitary conductance states (i.e., square-topped channel activity). While the burst-like activity is still likely true ion channel activity, it makes it more difficult to quantify the channel characteristics and therefore inpairs direct comparisons between different studies. This is further compounded by the fact that viroporins are able to adopt multiple oligomeric states, which is particularly true for peptides of the isolated transmembrane domains. Work to design and test peptides that adopt specific oligomeric states, such as with the Vpu tetramer versus pentamer, demonstrated these different oligomers may have different channel characteristics (Becker et al. 2004). Further, the evidence that the lipid environment and ionic concentration impact viroporin channel characteristics suggests these variables are important considerations for optimization of clear channel recordings (Whitfield et al. 2011; Verdia-Baguena et al. 2012, 2013). Establishing a systematic methodology for the reconstitution and testing of viroporins is needed to facilitate comparison between currently known viroporins, and to provide an experimentally robust framework for the testing of newly discovered viroporins, such as the papillomavirus E5 and ephemeral fever rhabdovirus α1 proteins, as well as many others (Wetherill et al. 2012; Joubert et al. 2014).

The majority of viroporins identified were tested due to similarities between a protein of interest and a known viroporin, typically based on transmembrane domain location and topology, and general sequence similarities in the viroporin signature motifs (polybasic domain and amphipathic α-helix). Using this methodology, the viroporins identified to date all have relatively non-specific cation channel activity, with the notable exception of IAV M2, which is a very specific proton channel (Chizhmakov et al. 1996). This raises the question of whether current methods to identify viroporins are too narrow and therefore only identifying those proteins that have similar functions, i.e., non-specific cation channels. While many classically defined viroporins have yet to be examined using electrophysiological techniques [reviewed in (Nieva et al. 2012; Giorda and Hebert 2013)], it is likely that we have only scratched the surface for identifying and characterizing viral ion channels. Given the fundamental importance ion channels play in cellular homeostasis, it is inconceivable to think viroporins are not found among viruses from all of the kingdoms of life, including among bacterial and archeal phages. A likely example in bacteriophage are pinholins, which induce cell lysis by depolarization of the host membrane but are too small to allow passage of endolysins like their larger cousins the holins (Young 2014). No candidate ion channels have been reported for archeal phages, and with the exception of the Kcv potassium channel in Phycodnaviridae, ion channels in protozoan and plate viruses remain unexplored. Since viroporins of mammalian viruses have proven to be a key aspect of pathogenesis, as well as exceptional models for minimalistic ion channel structure/function experiments, the search for viroporins from other groups of viruses represents a new challenge for collaboration between microbiologists and physiologists, particularly because electrophysiology equipment and expertise are not a typical part of the virologist's experimental armamentarium. Further, because there is little primary amino acid sequence similarity among viroporins, protein sequence alignments are unlikely to yield identification of promising candidates across kingdoms. Therefore, new ways to identify viroporin motifs, perhaps capitalizing on the biophysical requirements needed for ion channel formation (e.g., clustered basic residues and high amphipathicity), are needed to search through the huge amount of sequence data becoming available from next-generation genomic and metagenomic sequencing of environmental, animal and human samples.

In conclusion, the current electrophysiological knowledge of viroporins demonstrates that they are truly unconventional ion channels. This underlines the importance of studying viroporins in terms of basic biology and for biomedical research. Much can be learned about ion channel biophysics by studying viroporins, because while few viroporins are ion selective, they do appear to discriminate between cations and anions, which is remarkable given the simple structure of viroporins relative to conventional ion channels. Further, viroporins invariably play critical functional roles for virus replication and pathogenesis and in many cases, specific blockers make effective antiviral drugs. Therefore, continued work is needed to establish robust methods for viroporin identification, characterization of the ion channel activity, and screening for blockers that may be useful therapeutics to combat disease.

References

Agirre A, Barco A, Carrasco L, Nieva JL (2002) Viroporin-mediated membrane permeabilization. Pore formation by nonstructural poliovirus 2B protein. J Biol Chem 277:40434–40441

Andrew A, Strebel K (2014) HIV-1 accessory proteins: Vpu and Vif. Methods Mol Biol 1087:135–158

Antoine AF, Montpellier C, Cailliau K, Browaeys-Poly E, Vilain JP, Dubuisson J (2007) The alphavirus 6K protein activates endogenous ionic conductances when expressed in Xenopus oocytes. J Membr Biol 215:37–48

Atoom AM, Taylor NG, Russell RS (2014) The elusive function of the hepatitis C virus p7 protein. Virology 462–463:377–387

Becker CF, Oblatt-Montal M, Kochendoerfer GG, Montal M (2004) Chemical synthesis and single channel properties of tetrameric and pentameric TASPs (template-assembled synthetic proteins) derived from the transmembrane domain of HIV virus protein u (Vpu). J Biol Chem 279:17483–17489

Betakova T, Hay AJ (2007) Evidence that the CM2 protein of influenza C virus can modify the pH of the exocytic pathway of transfected cells. J Gen Virol 88:2291–2296

Bodelon G, Labrada L, Martinez-Costas J, Benavente J (2002) Modification of late membrane permeability in avian reovirus-infected cells: viroporin activity of the S1-encoded nonstructural p10 protein. J Biol Chem 277:17789–17796

Bolduan S, Votteler J, Lodermeyer V, Greiner T, Koppensteiner H, Schindler M, Thiel G, Schubert U (2011) Ion channel activity of HIV-1 Vpu is dispensable for counteraction of CD317. Virology 416:75–85
Bour S, Strebel K (2003) The HIV-1 Vpu protein: a multifunctional enhancer of viral particle release. Microbes Infect 5:1029–1039
Carter SD, Dent KC, Atkins E, Foster TL, Verow M, Gorny P, Harris M, Hiscox JA, Ranson NA, Griffin S, Barr JN (2010) Direct visualization of the small hydrophobic protein of human respiratory syncytial virus reveals the structural basis for membrane permeability. FEBS Lett 584:2786–2790
Chacko AR, Arifullah M, Sastri NP, Jeyakanthan J, Ueno G, Sekar K, Read RJ, Dodson EJ, Rao DC, Suguna K (2011) Novel pentameric structure of the diarrhea-inducing region of the rotavirus enterotoxigenic protein NSP4. J Virol 85:12721–12732
Chanturiya AN, Basanez G, Schubert U, Henklein P, Yewdell JW, Zimmerberg J (2004) p B1–F2, an influenza A virus-encoded proapoptotic mitochondrial protein, creates variably sized pores in planar lipid membranes. J Virol 78:6304–6312
Chen CP, Kremer C, Henklein P, Schubert U, Fink RH, Fischer WB (2010) Modulating the activity of the channel-forming segment of Vpr protein from HIV-1. Eur Biophys J 39:1089–1095
Chen CC, Kruger J, Sramala I, Hsu HJ, Henklein P, Chen YM, Fischer WB (2011) ORF8a of SARS-CoV forms an ion channel: experiments and molecular dynamics simulations. Biochim Biophys Acta 1808:572–579
Chew CF, Vijayan R, Chang J, Zitzmann N, Biggin PC (2009) Determination of pore-lining residues in the hepatitis C virus p7 protein. Biophys J 96:L10–L12
Chien TH, Chiang YL, Chen CP, Henklein P, Hanel K, Hwang IS, Willbold D, Fischer WB (2013) Assembling an ion channel: ORF 3a from SARS-CoV. Biopolymers 99:628–635
Chizhmakov IV, Geraghty FM, Ogden DC, Hayhurst A, Antoniou M, Hay AJ (1996) Selective proton permeability and pH regulation of the influenza virus M2 channel expressed in mouse erythroleukaemia cells. J Physiol 494(Pt 2):329–336
Coady MJ, Daniel NG, Tiganos E, Allain B, Friborg J, Lapointe JY, Cohen EA (1998) Effects of Vpu expression on Xenopus oocyte membrane conductance. Virology 244:39–49
DeDiego ML, Nieto-Torres JL, Jimenez-Guardeno JM, Regla-Nava JA, Castano-Rodriguez C, Fernandez-Delgado R, Usera F, Enjuanes L (2014) Coronavirus virulence genes with main focus on SARS-CoV envelope gene. Virus Res 194:124–137
Delcour AH, Martinac B, Adler J, Kung C (1989) Modified reconstitution method used in patch-clamp studies of Escherichia coli ion channels. Biophys J 56:631–636
Ewart GD, Sutherland T, Gage PW, Cox GB (1996) The Vpu protein of human immunodeficiency virus type 1 forms cation-selective ion channels. J Virol 70:7108–7115
Ewart GD, Mills K, Cox GB, Gage PW (2002) Amiloride derivatives block ion channel activity and enhancement of virus-like particle budding caused by HIV-1 protein Vpu. Eur Biophys J 31:26–35
Ewart GD, Nasr N, Naif H, Cox GB, Cunningham AL, Gage PW (2004) Potential new anti-human immunodeficiency virus type 1 compounds depress virus replication in cultured human macrophages. Antimicrob Agents Chemother 48:2325–2330
Gan SW, Ng L, Lin X, Gong X, Torres J (2008) Structure and ion channel activity of the human respiratory syncytial virus (hRSV) small hydrophobic protein transmembrane domain. Protein Sci 17:813–820
Gan SW, Tan E, Lin X, Yu D, Wang J, Tan GM, Vararattanavech A, Yeo CY, Soon CH, Soong TW, Pervushin K, Torres J (2012) The small hydrophobic protein of the human respiratory syncytial virus forms pentameric ion channels. J Biol Chem 287:24671–24689
Gibbs JS, Malide D, Hornung F, Bennink JR, Yewdell JW (2003) The influenza A virus PB1-F2 protein targets the inner mitochondrial membrane via a predicted basic amphipathic helix that disrupts mitochondrial function. J Virol 77:7214–7224
Giorda KM, Hebert DN (2013) Viroporins customize host cells for efficient viral propagation. DNA Cell Biol 32:557–564

Gonzalez ME, Carrasco L (2003) Viroporins. FEBS Lett 552:28–34
Griffin SD, Beales LP, Clarke DS, Worsfold O, Evans SD, Jaeger J, Harris MP, Rowlands DJ (2003) The p7 protein of hepatitis C virus forms an ion channel that is blocked by the antiviral drug, Amantadine. FEBS Lett 535:34–38
Griffin SD, Harvey R, Clarke DS, Barclay WS, Harris M, Rowlands DJ (2004) A conserved basic loop in hepatitis C virus p7 protein is required for amantadine-sensitive ion channel activity in mammalian cells but is dispensable for localization to mitochondria. J Gen Virol 85:451–461
Han Z, Harty RN (2004) The NS3 protein of bluetongue virus exhibits viroporin-like properties. J Biol Chem 279:43092–43097
Henkel M, Mitzner D, Henklein P, Meyer-Almes FJ, Moroni A, Difrancesco ML, Henkes LM, Kreim M, Kast SM, Schubert U, Thiel G (2010) The proapoptotic influenza A virus protein PB1-F2 forms a nonselective ion channel. PLoS One 5:e11112
Hongo S, Ishii K, Mori K, Takashita E, Muraki Y, Matsuzaki Y, Sugawara K (2004) Detection of ion channel activity in Xenopus laevis oocytes expressing Influenza C virus CM2 protein. Arch Virol 149:35–50
Hsieh SC, Wu YC, Zou G, Nerurkar VR, Shi PY, Wang WK (2014) Highly conserved residues in the helical domain of dengue virus type 1 precursor membrane protein are involved in assembly, precursor membrane (prM) protein cleavage, and entry. J Biol Chem 289:33149–33160
Hsu K, Seharaseyon J, Dong P, Bour S, Marban E (2004) Mutual functional destruction of HIV-1 Vpu and host TASK-1 channel. Mol Cell 14:259–267
Hsu K, Han J, Shinlapawittayatorn K, Deschenes I, Marban E (2010) Membrane potential depolarization as a triggering mechanism for Vpu-mediated HIV-1 release. Biophys J 99:1718–1725
Hyser JM, Collinson-Pautz MR, Utama B, Estes MK (2010) Rotavirus disrupts calcium homeostasis by NSP4 viroporin activity. MBio 1:e00265-10
Hyser JM, Utama B, Crawford SE, Estes MK (2012) Genetic divergence of rotavirus nonstructural protein 4 results in distinct serogroup-specific viroporin activity and intracellular punctate structure morphologies. J Virol 86:4921–4934
Hyser JM, Utama B, Crawford SE, Broughman JR, Estes MK (2013) Activation of the endoplasmic reticulum calcium sensor STIM1 and store-operated calcium entry by rotavirus requires NSP4 viroporin activity. J Virol 87:13579–13588
Jacotot E, Ferri KF, El HC, Brenner C, Druillennec S, Hoebeke J, Rustin P, Metivier D, Lenoir C, Geuskens M, Vieira HL, Loeffler M, Belzacq AS, Briand JP, Zamzami N, Edelman L, Xie ZH, Reed JC, Roques BP, Kroemer G (2001) Control of mitochondrial membrane permeabilization by adenine nucleotide translocator interacting with HIV-1 viral protein rR and Bcl-2. J Exp Med 193:509–519
Ji HL, Song W, Gao Z, Su XF, Nie HG, Jiang Y, Peng JB, He YX, Liao Y, Zhou YJ, Tousson A, Matalon S (2009) SARS-CoV proteins decrease levels and activity of human ENaC via activation of distinct PKC isoforms. Am J Physiol Lung Cell Mol Physiol 296:L372–L383
Joubert DA, Blasdell KR, Audsley MD, Trinidad L, Monaghan P, Dave KA, Lieu KG, Amos-Ritchie R, Jans DA, Moseley GW, Gorman JJ, Walker PJ (2014) Bovine ephemeral fever rhabdovirus alpha1 protein has viroporin-like properties and binds importin beta1 and importin 7. J Virol 88:1591–1603
Kang M, Moroni A, Gazzarrini S, DiFrancesco D, Thiel G, Severino M, Van Etten JL (2004) Small potassium ion channel proteins encoded by chlorella viruses. Proc Natl Acad Sci U S A 101:5318–5324
Kelly ML, Cook JA, Brown-Augsburger P, Heinz BA, Smith MC, Pinto LH (2003) Demonstrating the intrinsic ion channel activity of virally encoded proteins. FEBS Lett 552:61–67
Khoury G, Ewart G, Luscombe C, Miller M, Wilkinson J (2010) Antiviral efficacy of the novel compound BIT225 against HIV-1 release from human macrophages. Antimicrob Agents Chemother 54:835–845
Kuhl BD, Cheng V, Donahue DA, Sloan RD, Liang C, Wilkinson J, Wainberg MA (2011) The HIV-1 Vpu viroporin inhibitor BIT225 does not affect Vpu-mediated tetherin antagonism. PLoS One 6:e27660

Kuruma A, Hartzell HC (1999) Dynamics of calcium regulation of chloride currents in Xenopus oocytes. Am J Physiol 276:C161–C175
Lama J, Carrasco L (1992) Expression of poliovirus nonstructural proteins in Escherichia coli cells. Modification of membrane permeability induced by 2B and 3A. J Biol Chem 267:15932–15937
Lamb RA, Pinto LH (1997) Do Vpu and Vpr of human immunodeficiency virus type 1 and NB of influenza B virus have ion channel activities in the viral life cycles? Virology 229:1–11
Li Y, To J, Verdia-Baguena C, Dossena S, Surya W, Huang M, Paulmichl M, Liu DX, Aguilella VM, Torres J (2014) Inhibition of the human respiratory syncytial virus small hydrophobic protein and structural variations in a bicelle environment. J Virol 88:11899–11914
Lin TI, Schroeder C (2001) Definitive assignment of proton selectivity and attoampere unitary current to the M2 ion channel protein of influenza A virus. J Virol 75:3647–3656
Loregian A, Mercorelli B, Nannetti G, Compagnin C, Palu G (2014) Antiviral strategies against influenza virus: towards new therapeutic approaches. Cell Mol Life Sci 71:3659–3683
Lu W, Zheng BJ, Xu K, Schwarz W, Du L, Wong CK, Chen J, Duan S, Deubel V, Sun B (2006) Severe acute respiratory syndrome-associated coronavirus 3a protein forms an ion channel and modulates virus release. Proc Natl Acad Sci U S A 103:12540–12545
Luscombe CA, Huang Z, Murray MG, Miller M, Wilkinson J, Ewart GD (2010) A novel Hepatitis C virus p7 ion channel inhibitor, BIT225, inhibits bovine viral diarrhea virus in vitro and shows synergism with recombinant interferon-alpha-2b and nucleoside analogues. Antiviral Res 86:144–153
Madan, V., Castelló, A., & Carrasco, L. (2008). Viroporins from RNA viruses induce caspase-dependent apoptosis. Cellular Microbiology, 10(2), 437–451. doi:10.1111/j.1462-5822.2007.01057.x
Magadan JG, Bonifacino JS (2012) Transmembrane domain determinants of CD4 Downregulation by HIV-1 Vpu. J Virol 86:757–772
McLellan JS, Ray WC, Peeples ME (2013) Structure and function of respiratory syncytial virus surface glycoproteins. Curr Top Microbiol Immunol 372:83–104
Mehnert T, Routh A, Judge PJ, Lam YH, Fischer D, Watts A, Fischer WB (2008) Biophysical characterization of Vpu from HIV-1 suggests a channel-pore dualism. Proteins 70:1488–1497
Melton JV, Ewart GD, Weir RC, Board PG, Lee E, Gage PW (2002) Alphavirus 6K proteins form ion channels. J Biol Chem 277:46923–46931
Meshkat Z, Audsley M, Beyer C, Gowans EJ, Haqshenas G (2009) Reverse genetic analysis of a putative, influenza virus M2 HXXXW-like motif in the p7 protein of hepatitis C virus. J Viral Hepat 16:187–194
Montserret R, Saint N, Vanbelle C, Salvay AG, Simorre JP, Ebel C, Sapay N, Renisio JG, Bockmann A, Steinmann E, Pietschmann T, Dubuisson J, Chipot C, Penin F (2010) NMR structure and ion channel activity of the p7 protein from hepatitis C virus. J Biol Chem 285:31446–31461
Morrison TE (2014) Reemergence of chikungunya virus. J Virol 88:11644–11647
Mould JA, Drury JE, Frings SM, Kaupp UB, Pekosz A, Lamb RA, Pinto LH (2000) Permeation and activation of the M2 ion channel of influenza A virus. J Biol Chem 275:31038–31050
Mould JA, Paterson RG, Takeda M, Ohigashi Y, Venkataraman P, Lamb RA, Pinto LH (2003) Influenza B virus BM2 protein has ion channel activity that conducts protons across membranes. Dev Cell 5:175–184
Nieto-Torres JL, DeDiego ML, Alvarez E, Jimenez-Guardeno JM, Regla-Nava JA, Llorente M, Kremer L, Shuo S, Enjuanes L (2011) Subcellular location and topology of severe acute respiratory syndrome coronavirus envelope protein. Virology 415:69–82
Nieva JL, Madan V, Carrasco L (2012) Viroporins: structure and biological functions. Nat Rev Microbiol 10:563–574
Okada A, Miura T, Takeuchi H (2001) Protonation of histidine and histidine-tryptophan interaction in the activation of the M2 ion channel from influenza a virus. Biochemistry 40:6053–6060
Padhi S, Burri RR, Jameel S, Priyakumar UD (2014) Atomistic detailed mechanism and weak cation-conducting activity of HIV-1 Vpu revealed by free energy calculations. PLoS One 9:e112983

Pavlovic D, Neville DC, Argaud O, Blumberg B, Dwek RA, Fischer WB, Zitzmann N (2003) The hepatitis C virus p7 protein forms an ion channel that is inhibited by long-alkyl-chain iminosugar derivatives. Proc Natl Acad Sci U S A 100:6104–6108

Perez M, Garcia-Barreno B, Melero JA, Carrasco L, Guinea R (1997) Membrane permeability changes induced in Escherichia coli by the SH protein of human respiratory syncytial virus. Virology 235:342–351

Pervushin K, Tan E, Parthasarathy K, Lin X, Jiang FL, Yu D, Vararattanavech A, Soong TW, Liu DX, Torres J (2009) Structure and inhibition of the SARS coronavirus envelope protein ion channel. PLoS Pathog 5:e1000511

Piller SC, Ewart GD, Premkumar A, Cox GB, Gage PW (1996) Vpr protein of human immunodeficiency virus type 1 forms cation-selective channels in planar lipid bilayers. Proc Natl Acad Sci U S A 93:111–115

Piller SC, Ewart GD, Jans DA, Gage PW, Cox GB (1999) The amino-terminal region of Vpr from human immunodeficiency virus type 1 forms ion channels and kills neurons. J Virol 73:4230–4238

Pinto LH, Lamb RA (2006) The M2 proton channels of influenza A and B viruses. J Biol Chem 281:8997–9000

Pinto LH, Holsinger LJ, Lamb RA (1992) Influenza virus M2 protein has ion channel activity. Cell 69:517–528

Premkumar A, Wilson L, Ewart GD, Gage PW (2004) Cation-selective ion channels formed by p7 of hepatitis C virus are blocked by hexamethylene amiloride. FEBS Lett 557:99–103

Premkumar A, Horan CR, Gage PW (2005) Dengue virus M protein C-terminal peptide (DVM-C) forms ion channels. J Membr Biol 204:33–38

Salazar MI, del Angel RM, Lanz-Mendoza H, Ludert JE, Pando-Robles V (2014) The role of cell proteins in dengue virus infection. J Proteomics 111:6–15

Sanz MA, Perez L, Carrasco L (1994) Semliki Forest virus 6K protein modifies membrane permeability after inducible expression in Escherichia coli cells. J Biol Chem 269:12106–12110

Sastri NP, Viskovska M, Hyser JM, Tanner MR, Horton LB, Sankaran B, Prasad BV, Estes MK (2014) Structural plasticity of the coiled-coil domain of rotavirus NSP4. J Virol 88:13602–13612

Schubert U, Ferrer-Montiel AV, Oblatt-Montal M, Henklein P, Strebel K, Montal M (1996) Identification of an ion channel activity of the Vpu transmembrane domain and its involvement in the regulation of virus release from HIV-1-infected cells. FEBS Lett 398:12–18

Shimbo K, Brassard DL, Lamb RA, Pinto LH (1995) Viral and cellular small integral membrane proteins can modify ion channels endogenous to Xenopus oocytes. Biophys J 69:1819–1829

Shimbo K, Brassard DL, Lamb RA, Pinto LH (1996) Ion selectivity and activation of the M2 ion channel of influenza virus. Biophys J 70:1335–1346

Stewart SM, Pekosz A (2012) The influenza C virus CM2 protein can alter intracellular pH, and its transmembrane domain can substitute for that of the influenza A virus M2 protein and support infectious virus production. J Virol 86:1277–1281

StGelais C, Tuthill TJ, Clarke DS, Rowlands DJ, Harris M, Griffin S (2007) Inhibition of hepatitis C virus p7 membrane channels in a liposome-based assay system. Antiviral Res 76:48–58

Strebel K (2013a) HIV accessory proteins versus host restriction factors. Curr Opin Virol 3:692–699

Strebel K (2014) HIV-1 Vpu — an ion channel in search of a job, Biochimica et Biophysica Acta (BBA) - Biomembranes, 1838(4): 1074–1081, ISSN 0005-2736, http://dx.doi.org/10.1016/j.bbamem.2013.06.029.

Strebel K (2014) HIV-1 Vpu – an ion channel in search of a job. Biochim Biophys Acta 1838:1074–1081

Sunstrom NA, Premkumar LS, Premkumar A, Ewart G, Cox GB, Gage PW (1996) Ion channels formed by NB, an influenza B virus protein. J Membr Biol 150:127–132

Tang Y, Zaitseva F, Lamb RA, Pinto LH (2002) The gate of the influenza virus M2 proton channel is formed by a single tryptophan residue. J Biol Chem 277:39880–39886

Taube R, Alhadeff R, Assa D, Krugliak M, Arkin IT (2014) Bacteria-based analysis of HIV-1 Vpu channel activity. PLoS One 9:e105387

Torres J, Maheswari U, Parthasarathy K, Ng L, Liu DX, Gong X (2007) Conductance and amantadine binding of a pore formed by a lysine-flanked transmembrane domain of SARS coronavirus envelope protein. Protein Sci 16:2065–2071
Tosteson MT, Pinto LH, Holsinger LJ, Lamb RA (1994) Reconstitution of the influenza virus M2 ion channel in lipid bilayers. J Membr Biol 142:117–126
Triantafilou K, Kar S, Vakakis E, Kotecha S, Triantafilou M (2013) Human respiratory syncytial virus viroporin SH: a viral recognition pathway used by the host to signal inflammasome activation. Thorax 68:66–75
Ulug ET, Garry RF, Waite MR, Bose HR Jr (1984) Alterations in monovalent cation transport in Sindbis virus-infected chick cells. Virology 132:118–130
van Kuppeveld FJ, Melchers WJ, Kirkegaard K, Doedens JR (1997) Structure-function analysis of coxsackie B3 virus protein 2B. Virology 227:111–118
Venkataraman P, Lamb RA, Pinto LH (2005) Chemical rescue of histidine selectivity filter mutants of the M2 ion channel of influenza A virus. J Biol Chem 280:21463–21472
Verdia-Baguena C, Nieto-Torres JL, Alcaraz A, DeDiego ML, Torres J, Aguilella VM, Enjuanes L (2012) Coronavirus E protein forms ion channels with functionally and structurally-involved membrane lipids. Virology 432:485–494
Verdia-Baguena C, Nieto-Torres JL, Alcaraz A, DeDiego ML, Enjuanes L, Aguilella VM (2013) Analysis of SARS-CoV E protein ion channel activity by tuning the protein and lipid charge. Biochim Biophys Acta 1828:2026–2031
Vieyres G, Brohm C, Friesland M, Gentzsch J, Wolk B, Roingeard P, Steinmann E, Pietschmann T (2013) Subcellular localization and function of an epitope-tagged p7 viroporin in hepatitis C virus-producing cells. J Virol 87:1664–1678
Vijayvergiya V, Wilson R, Chorak A, Gao PF, Cross TA, Busath DD (2004) Proton conductance of influenza virus M2 protein in planar lipid bilayers. Biophys J 87:1697–1704
Wang C, Takeuchi K, Pinto LH, Lamb RA (1993) Ion channel activity of influenza A virus M2 protein: characterization of the amantadine block. J Virol 67:5585–5594
Wang C, Lamb RA, Pinto LH (1994) Direct measurement of the influenza A virus M2 protein ion channel activity in mammalian cells. Virology 205:133–140
Wang C, Lamb RA, Pinto LH (1995) Activation of the M2 ion channel of influenza virus: a role for the transmembrane domain histidine residue. Biophys J 69:1363–1371
Wang K, Lu W, Chen J, Xie S, Shi H, Hsu H, Yu W, Xu K, Bian C, Fischer WB, Schwarz W, Feng L, Sun B (2012) PEDV ORF3 encodes an ion channel protein and regulates virus production. FEBS Lett 586:384–391
Wetherill LF, Holmes KK, Verow M, Muller M, Howell G, Harris M, Fishwick C, Stonehouse N, Foster R, Blair GE, Griffin S, Macdonald A (2012) High-risk human papillomavirus E5 oncoprotein displays channel-forming activity sensitive to small-molecule inhibitors. J Virol 86:5341–5351
Whitfield T, Miles AJ, Scheinost JC, Offer J, Wentworth Jr,. P, Dwek RA, Wallace BA, Biggin PC, Zitzmann N (2011) The influence of different lipid environments on the structure and function of the hepatitis C virus p7 ion channel protein. Mol Membr Biol 28(5):254–264
Wilson L, McKinlay C, Gage P, Ewart G (2004) SARS coronavirus E protein forms cation-selective ion channels. Virology 330:322–331
Wilson L, Gage P, Ewart G (2006) Hexamethylene amiloride blocks E protein ion channels and inhibits coronavirus replication. Virology 353:294–306
Wong SS, Chebib M, Haqshenas G, Loveland B, Gowans EJ (2011) Dengue virus PrM/M proteins fail to show pH-dependent ion channel activity in Xenopus oocytes. Virology 412:83–90
Wozniak AL, Griffin S, Rowlands D, Harris M, Yi M, Lemon SM, Weinman SA (2010) Intracellular proton conductance of the hepatitis C virus p7 protein and its contribution to infectious virus production. PLoS Pathog 6:e1001087
Xie S, Wang K, Yu W, Lu W, Xu K, Wang J, Ye B, Schwarz W, Jin Q, Sun B (2011) DIDS blocks a chloride-dependent current that is mediated by the 2B protein of enterovirus 71. Cell Res 21:1271–1275

Yamada H, Chounan R, Higashi Y, Kurihara N, Kido H (2004) Mitochondrial targeting sequence of the influenza A virus PB1-F2 protein and its function in mitochondria. FEBS Lett 578:331–336
Young R (2014) Phage lysis: three steps, three choices, one outcome. J Microbiol 52:243–258
Yount B, Roberts RS, Sims AC, Deming D, Frieman MB, Sparks J, Denison MR, Davis N, Baric RS (2005) Severe acute respiratory syndrome coronavirus group-specific open reading frames encode nonessential functions for replication in cell cultures and mice. J Virol 79:14909–14922
Zhang R, Wang K, Lv W, Yu W, Xie S, Xu K, Schwarz W, Xiong S, Sun B (2014) The ORF4a protein of human coronavirus 229E functions as a viroporin that regulates viral production. Biochim Biophys Acta 1838:1088–1095
Zhou Y, Frey TK, Yang JJ (2009) Viral calciomics: interplays between Ca^{2+} and virus. Cell Calcium 46:1–17

Part III
Toxins and Antimicrobial Peptides

Chapter 8
Pore-Forming Colicins: Unusual Ion Channels – Unusually Regulated

Daria Stroukova and Jeremy H. Lakey

Abstract Colicins are toxins released by plasmid carrying *Escherichia coli* bacteria in order to kill bacteria not carrying the same plasmid. This selection is achieved since the plasmid bearing cells have an immunity protein that protects them from their own specific colicin. Colicins kill by peptidoglycan synthesis inhibition, nuclease digestion or pore formation. This review describes how the pore-forming colicins function, with reference to other types as needed. The function involves binding to target cells, translocation across the outer membrane and periplasm, insertion into the inner membrane and voltage gated pore formation. Due to this unusual pathway, the pores have an unusual and minimalist mode of pore formation that is still not fully defined. The immunity protein is an integral membrane protein of the inner membrane, which prevents formation of the open pore and thus protects the cell. Models for the interaction with the colicin pore are discussed which also give insights into protein-protein interactions in the lipid bilayer.

Keywords Pore-forming colicin • Outer membrane • Inner membrane • Gram-negative bacteria • Colicin immunity protein • Voltage gating

8.1 Introduction

Pore-formation in target membranes is a common mechanism used by toxins (Parker and Feil 2005). Whilst the majority of these toxins are likely to be bacterial, examples such as melittin in bee venom, actinoporins from sea anemones and perforins used by granulocytes to fight infection in eukaryotic organisms show that this strategy has been adopted widely in biology (Anderluh and Lakey 2010). Among bacterial toxins, most appear to be involved in pathogenic mechanisms and include molecules critical for progression of such diseases as diphtheria, anthrax and whooping cough. However, a separate group of molecules is involved in competition between bacteria and these are termed bacteriocins. These include both pore-forming and non-pore-forming toxins and range in size from the lantibiotics

D. Stroukova (✉) • J.H. Lakey
Institute for Cell and Molecular Biosciences, Newcastle University, Newcastle upon Tyne, UK
daria.stroukova@ncl.ac.uk

A.H. Delcour (ed.), *Electrophysiology of Unconventional Channels and Pores*, Springer Series in Biophysics 18, DOI 10.1007/978-3-319-20149-8_8

such as nisin (Guder et al. 2000), through microcins to large complex proteins such as the colicins, pesticins and klebicins (Kim et al. 2014). Whilst the lantibiotics are mostly aimed at gram-positive organisms, the latter examples are all targeted against gram-negative bacteria, which, by virtue of their additional outer membrane, pose an extra challenge to the toxins trying to kill them. The microcins as their name implies are small (<100 amino acids) and are either intrinsically disordered or characterised by disulphide bridges and covalent post-translational modifications, such as the adoption of the structure of an iron siderophore, which permit them to be taken up by the specialised iron uptake systems of the outer membrane (Mathavan and Beis 2012). The larger bacteriocins are highly species specific, being involved in intra specific competition, and are targeted at closely related cells (Lakey et al. 1994b). Because of this, these toxins are named after the species they target thus colicins kill *E. coli*, pesticins kill *Yersinia pestis* and Klebicins kill *Klebsiella spp*.

Here we will concentrate on the best known family, the colicins. In addition to those that kill by pore formation, there are also colicins that kill by nuclease activity and peptidoglycan synthesis inhibition (Cascales et al. 2007). Importantly, the genes that code for colicins are mostly carried on plasmids and are responsible for maintaining the plasmid in the population. To do this they impart a selective advantage upon their hosts. The mechanism for this is twofold. Firstly, the plasmids code for the toxin, which is normally produced by a minority of cells in response to stresses such as nutrient shortage or DNA damage. The producer cells lyse (Cavard 1991) and thus die in the process but, if successful, cause the demise of many non-plasmid carrying competitor cells because the remaining plasmid bearing cells are resistant to their own colicin. This resistance is due to the presence of immunity proteins, which are produced constitutively within the cell and are thus ready at all times to neutralise incoming toxins. The immunity proteins for nucleases and pore-forming colicins are very different in structure and mode of action, but the result is the same. After release of colicins into the environment, it is the plasmid bearing cells, which have not gone down the colicin-producing path and then withstood the colicins released by their peers, which are best placed to expand their numbers.

In this article we will describe the unusual channel properties of pore-forming colicins and also show how, by requiring an immunity protein that inhibits the pore-forming toxicity, these molecules have also developed a unique method to modulate ion channel function.

8.1.1 Similarities and Differences Between Pore–Forming and Catalytic Colicins

All colicins are three domain proteins, with each domain responsible for one of three separate functions (Cascales et al. 2007). The toxic function, whether pore-forming or enzymatic, is always found in the C-terminal domain, the recognition of the outer membrane receptor is found within the central receptor binding region and the N-terminal domain contains the translocation (T-)domain. This last region has binding sites for both the outer membrane translocator proteins and periplasmic receptors. It is characterised by being either partly or fully disordered to the extent that it is often incompletely resolved in X-ray structures (Jakes and Cramer 2012) (Fig. 8.1). Binding of such a flexible structure to its target receptor usually involves a disorder to order transition of a linear epitope within this domain (Hecht et al. 2008, 2012). It has been shown that when different domains are swapped between colicins they retain their overall toxicity and thus these domains can be considered largely independent in function (Benedetti et al. 1991). Moreover, as we shall see later, the pore-forming (Lakey et al. 1991) and catalytic domains (Moore et al. 1995) can be isolated *in vitro* and their toxic functions preserved and studied. The structures of the catalytic and pore-forming domains are very different to each other with no obvious homology. On the other hand within the latter group there is strong structural homology (Fig. 8.1), hinting at a single common ancestor domain that was recruited to colicins at some early stage in their evolution (Lakey and Slatin 2001).

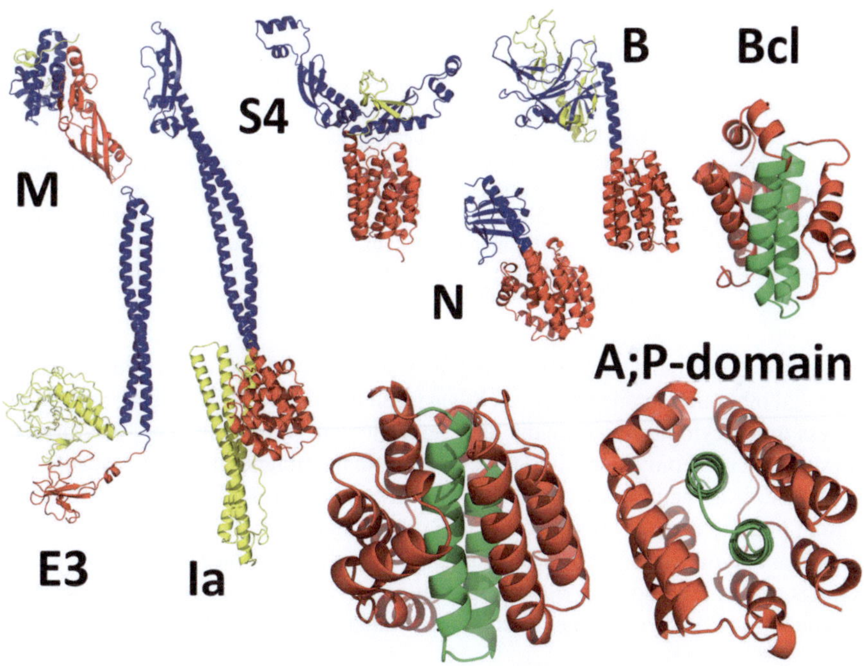

Fig. 8.1 Pore-forming colicins and related structures. (PDB references in parenthesis). For the pore-forming colicins Ia (1CII) (Wiener et al. 1997), N (1A87) (Vetter et al. 1998), B (1RH1) (Hilsenbeck et al. 2004) & S4 (3FEW) (Arnold et al. 2009), the helical pore-forming domain is shown in *red*, the receptor binding domain in *blue* and the translocation domain (which may be incomplete due to disorder e.g. colicin N) in *yellow*. The colicins E3 (3EIP) (Soelaiman et al. 2001) and M (3DA4) (Zeth et al. 2008) which target RNA and peptidoglycan respectively are shown with the same colour scheme revealing the dissimilar toxic domains. The first pore-forming domain to be defined, colicin A (1COL) (Parker et al. 1989), is shown enlarged in two orientations with the hydrophobic helical hairpin in *green*. Finally, the structurally homologous Bcl-XL (1LXL) (Muchmore et al. 1996) apoptosis regulator protein is shown with the same colour scheme as colicin A

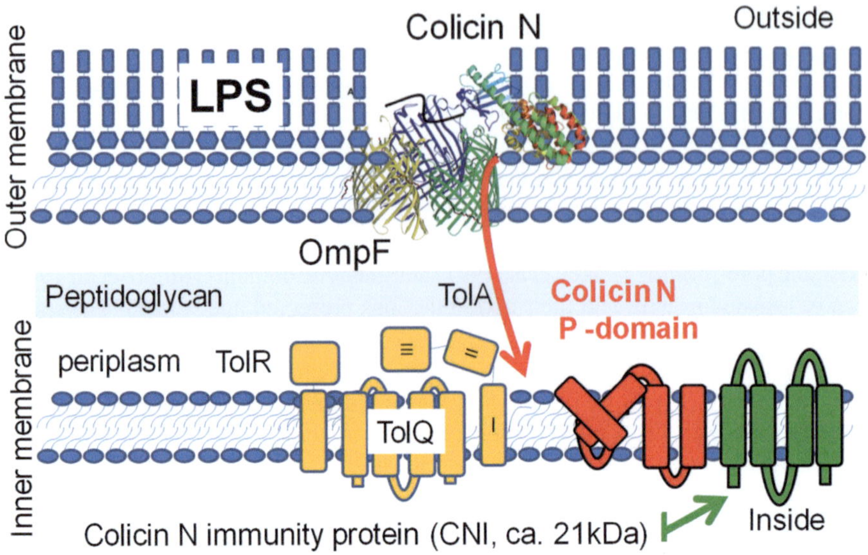

Fig. 8.2 The colicin N translocation pathway. Using colicin N as an example, the X-ray derived models of OmpF and colicin N are shown in a schematic outer membrane with LPS in the outer leaflet and phospholipids in the inner. Only the periplasmic proteins used by colicin N are shown and other colicins will employ a wider range. Direct binding to TolA soluble domains has been demonstrated but genetics shows that TolQ and TolR are also needed. Also in the inner membrane is the immunity protein (*green*) and this is shown next to a cartoon of the inserted P-domain (*red*). This is shown separate from the remaining domains for clarity because we do not know where they are when the P-domain is in the inner membrane

For any colicin, the challenge is to deliver the toxic domain to the gram-negative bacterial inner membrane (Fig. 8.2). We assume that the nuclease toxicity must gain access to the cytoplasm where the target nucleic acids are, whilst the pore-formers presumably need only insert into the cytoplasmic membrane. In either case, the initial barrier is the same, the outer membrane (Kim et al. 2014). This is a unique structure in biology and we know of only one other vaguely similar structure, the outer membrane of mitochondria to which the gram-negative structure is distantly related. This is due to the endosymbiotic origin of eukaryotic cells during which bacteria invaded their hosts to become mitochondrial ancestors (Mueller et al. 2011).

8.1.2 Outer Membrane Structure

Since the pore-forming domains, which have the ability to switch between soluble and membrane bound states, have to traverse two membranes, it is useful to consider the first barrier in detail. The gram-negative outer membrane is a highly asymmetric structure containing, in its outer leaflet, a characteristic lipid, lipopolysaccharide (LPS), and in its inner leaflet a mixture of more generic lipids such as phosphatidylethanolamine, phosphatidylglycerol and phosphatidylserine (Fig. 8.2) (Clifton et al. 2013). LPS consists of a core molecule, lipid A which comprises a phosphorylated glucosamine dimer covalently linked to six or seven acyl chains, which vary between 12 and 16 carbons in length, which anchor the LPS molecule to the hydrophobic interior of the outer membrane. Attached to the glucosamine head-group of Lipid A, and facing the outer surface, is the core oligosaccharide region, which can be further broken down into the inner and outer core. The inner core is composed of the sugars L-glycero-D-manno-heptose (Hep) and 3-deoxy-D-manno-octulsonic acid (Kdo) and the outer core region is composed of glucose and glucosamines. Attached to the core is the O-antigen region, the largest part of LPS and composed of a repeating chain of oligosaccharides (Le Brun et al. 2013). The charge on the gram-negative bacterial outer membrane surface is strongly negative due to the high concentration of phosphate and carboxylate groups. However, this leads to the binding of divalent cations, such as calcium and magnesium, which are essential in the stability of the LPS layer.

Embedded in this asymmetric bilayer are the outer membrane proteins (Omp) which have roles in nutrient uptake, cell adhesion and motility as well as being essential for colicin function (Fig. 8.2). The outer membrane proteins are all β-barrels and they play two distinct roles in colicin uptake although one protein may undertake both roles. The first role is that of receptor and the best known examples, by virtue of having high resolution X-ray structures, are BtuB binding to E colicins (enzymatic) (Kurisu et al. 2003) and Cir binding to colicin Ia (ColIA) (pore former) (Buchanan et al. 2007). In each case, the receptor-binding domain is a coiled coil structure that inserts into the Omp such that the colicin projects from the surface at an oblique angle. Both cytotoxic and translocation domains are thus exposed and this may allow the translocation domain to interact with the next required Omp, the translocator. For the enzymatic E colicins (2–9) and pore-forming colicins A, K, N, S4, U and 28b this is the trimeric porin OmpF (Kim et al. 2014) and the disordered N-terminus has been elegantly demonstrated to bind via two channels in adjacent monomers in such a way as to present a third disordered binding epitope to its internal receptor TolB (Housden et al. 2013). This supports previous evidence using electrophysiological methods which showed that the N-termini of several colicins can block the channels of OmpF proteins reconstituted in artificial planar bilayers (Zakharov et al. 2004). For Colicin Ia it has been shown by careful and insightful experiments that a second copy of the receptor, Cir, is recruited to enable translocation (Jakes and Finkelstein 2010).

This stage of the translocation is the most mysterious; how does a large, and presumably unwanted protein structure, penetrate a barrier (Fig. 8.2) which is often able to resist small antibiotics? The requirement for translocator proteins has inspired several hypotheses concerning their role. Proteins like Cir have large internal plug domains that fill an otherwise large β-barrel channel and it has been suggested that colicin binding may remove the plug to allow for colicin passage across the outer membrane (Udho et al. 2009). OmpF on the other hand has a small pore and it could be envisaged that insertion of the unfolded N-terminus is the first stage of a complete translocation by this route, which would however require complete unfolding. Both enzymatic (E 2–9) and pore formers (A, K, N, U) share the OmpF route so their methods of translocation would appear to be similar (Kim et al. 2014).

8.2 Pore-Forming Colicins

All pore-forming colicins contain a similar pore-forming domain (P-domain), which consists of about 200 amino acid residues in ten α-helices (Fig. 8.1). Two of these α-helices (8&9) are hydrophobic, form a helical hairpin and showed up on hydropathy plots before the first structure was solved by Parker et al. (1989). The remaining helices are largely amphipathic and surround the hydrophobic hairpin; an arrangement that suggested how a water-soluble protein could convert into a membrane bound form. This transformation is dealt with in a subsequent section. By possessing one or more buried α-helices, colicin pore-forming domains were seen to be analogous to other toxins that make a similar water-soluble to membrane-bound transition, such as diphtheria toxin and *Bacillus thuringiensis* δ-endotoxin (Parker and Feil 2005). Later it became apparent that the structures of apoptosis regulator proteins of the Bcl and Bax families shared a very similar three-dimensional structure to the colicin pore-forming domains (Fig. 8.1). This was not evident from their primary structures and only became evident after the determination of the Bcl-XL structure (Muchmore et al. 1996). Sequence analysis shows that the colicin domains are highly homologous and divide into two main families PI and PII. If one takes into account subtle differences in the structure of the latter group, the known pore-forming domains can be divided into three divisions PI (S4, A, B, N, U, Y), PIIa (E1, 5, K, 10) and PIIb (Alveicin B, Pyocin S5, Ia and Ib) (Arnold et al. 2009). The main difference is that the PII group are smaller and this is largely due to a shorter helical hairpin. This separation will be seen to be important when we later discuss the role of immunity proteins in pore-forming colicin regulation.

Most known pore-forming domains are basic with pI above neutral. This is common to many membrane-binding peptides since most membranes bear a net negative charge that arises from the mixture of anionic and zwitterionic lipids that they contain. However, two colicins (A and B) have acidic pore-forming domains with pI close to 4.5. This is even more surprising when one sees that their most similar homologue is the pore-forming domain of colicin N (ColN), which has a pI of ≈10. This apparent anomaly becomes interesting when the behaviour of these pore-forming domains *in vitro* and *in vivo* is examined. Many pore-forming domains insert into membranes more efficiently at low pH but for colicin A this was always more profound and reproducible. It was then shown that insertion into membranes of colicin A (ColA) (and later ColB) was directly correlated with the formation of an acidic molten globule state which colicin N does not undergo (van der Goot et al. 1991; Evans et al. 1996). Finally, it was revealed that colicin A required acidic lipids in the inner membrane of target bacteria whereas colicin N did not (van der Goot et al. 1993). Thus, colicins A and B appear to have diverged from the other pore formers at some stage to make a subtle change in their translocation and pore-formation pathway. In their pore-forming domains this change involves the introduction of acidic aspartate residues at the N-termini of several α- helices which provide stabilising N-cap side-chains. Upon protonation, which occurs at pH values below 4.5, this stabilisation is lost and the destabilised α-helices cause the formation of a molten globule state which accelerates membrane insertion (Huang et al. 2012). In colicin N the aspartate residues are largely replaced by asparagines which stabilise the helices at all pH values (Fridd and Lakey 2002). Since the periplasm does not provide a sufficiently low pH for the destabilisation of colicins A and B they must occupy a localised acidic lipid environment to destabilise their stable native fold and insert into membranes. The discovery that colicin A is less active in strains that do not produce the acidic lipid phosphatidylglycerol (van der Goot et al. 1993) concurs with the previous measurement of the 1.6 unit reduction of local pH by acidic membrane lipids (van der Goot et al. 1991) to indicate that colicin A may be unfolded within the surface layer of the membrane. With our current understanding of lateral localisation of lipids in membranes (Renner and Weibel 2011) it may be that there are negatively charged islands that provide the required low pH. We do not know what provokes a similar destabilisation in the acid indifferent majority of pore-formers but it may simply be the result of unfolding during translocation across the outer membrane. It is known that translocation can be inhibited by the insertion of artificial disulphide cross-links into the pore-forming domain and that unfolding of colicins in urea accelerates their killing kinetics (Benedetti et al. 1992; Duche et al. 1994a). Thus, all colicin pore-forming domains are likely to arrive in the periplasm in an unfolded state, in which case the role of acidic unfolding in colicins A and B pore-forming domains is less clear.

A clue to the translocated state of the pore-forming domain has come from studies of the pore-forming colicin N, which has several unique features (Jakes 2014) which make it a useful research tool and which suggest some alternative solutions to the outer membrane translocation problem. Firstly, it is the smallest pore-forming colicin, comprising only 387 amino acids, and lacking the long α-helices found in the receptor binding domains of other colicins. This was instrumental to allow colicin N to become the second complete colicin to be solved by X-ray crystallography (Vetter et al. 1998) (after Colicin Ia) (Wiener et al. 1997) although the disorder in the T-domain ensured that only residues from 90 to 387 were resolved. Secondly, it only requires outer membrane protein F (OmpF) and core proteins (AQR) of the Tol complex to function and, thirdly, its translocation domain consists solely of intrinsically unfolded residues enabling it to be well mapped by mutagenesis and biophysical techniques including NMR (Anderluh et al. 2003).

Studies in the presence of both SDS detergent and LPS showed that the pore-forming domain binds to its translocator protein, OmpF (Dover et al. 2000), and a later examination of two-dimensional crystals by electron microscopy revealed colicin N on the outside of OmpF displacing the tightly bound LPS which can co-purify with this membrane protein (Baboolal et al. 2008). Finally, neutron scattering resolved the 3D structure and showed clearly that the pore-forming domain was bound in the cleft between two monomers of the OmpF trimer (Clifton et al. 2012).

This *in vitro* structure has yet to be verified *in vivo* but, in the case of pore-forming colicins, it does provide a means to transport a hydrophobic domain across the outer membrane, without the need to thread it through the strongly charged and hydrophilic core of the OmpF channel. More recently, it was shown that colicin N is the only colicin to use LPS as its outer membrane receptor (Johnson et al. 2014). Since it binds to the core region of LPS, described above, this interaction places it in an ideal position to traverse the OM via the LPS-OmpF interface. By doing so, it would emerge into the periplasm in an unfolded state and this may enable the pore-forming domains to undergo the conformational changes required to form a channel in the inner membrane. Thus two possible routes exist for colicin translocation across the outer membrane through a membrane channel or via the protein lipid interface. It is certain that colicin translocation (T-)domains can traverse the OmpF channel and it remains to be seen if the remaining 300–500 amino acid residues fully unfold to follow them through the channel pore. How the proteins traverse the periplasm is yet to be understood but it is known that only the seven C-terminal helices are required to form a channel (Baty et al. 1990) and that mutations which alter the electrostatic field around this domain have significant effects upon the channel formation kinetics (Lakey et al. 1994a).

8.2.1 Models of Pore Formation

The ion channels formed by colicins are still somewhat mysterious (Lakey and Slatin 2001). They are easy to study in so called black lipid membranes (BLM) since they can be added in their soluble form and will form channels spontaneously. Channel opening occurs only at positive potentials on the side of colicin addition and can be reversed by changing polarity. Colicin binding to the membrane occurs independently of membrane potential but its less clear if the transition to the inserted closed form is potential independent. As a result experiments show an increase in channel numbers as time progresses. The initial opening of channels is slow as it requires *de novo* protein recruitment whilst the reopening of closed channels is fast showing that reversing the polarity does not extricate the colicin from the membrane. Opening and closing kinetics are faster at higher potentials. This behaviour is summarised in Fig. 8.3. Whilst it is easy to investigate the closed membrane bound form of the protein, the open channel form requires the application of an electric field across the membrane and this is difficult to achieve in the lipid micelles and vesicles used in spectroscopy. Based upon the concentration dependence of the killing kinetics and ion conductance it appears that colicins act as monomers. However, they have been seen to form convincing trimeric pores in electron crystallography analysis of two-dimensional crystals (Greig et al. 2009).

Due to the lack of structural information, several models have been proposed for the open pore of colicins. As described earlier, all pore-forming domains consist of ten helices, with eight amphipathic helices surrounding two hydrophobic ones. The first model, the umbrella model was developed after the hydrophobic helical hairpin

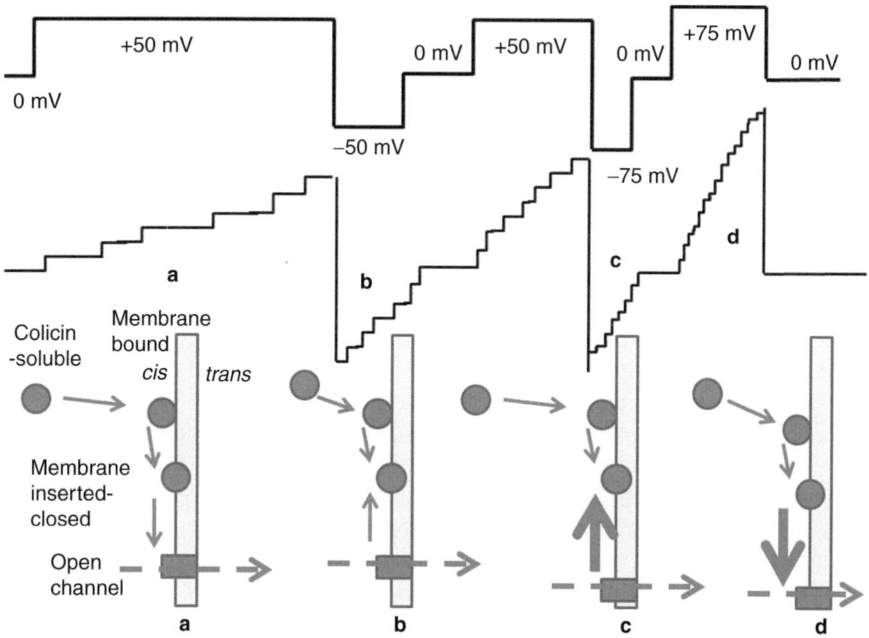

Fig. 8.3 Colicin channel activity *in vitro*. (**a**) When added to the buffer on the *cis* side of an artificial planar lipid bilayer (black lipid membrane (BLM)) the soluble colicins (or even just their pore–forming domains) bind to and insert into the lipid bilayer and adopt a membrane inserted closed channel state. To open the channels a *cis* positive potential is applied and the channels open relatively slowly. These slow kinetics may be due to *de novo* insertion of the pore forming domain into the membrane. If the potential is reversed (**b**) then the channels close by reverting to a membrane inserted closed state. Proteins are not removed from the membrane as a further reversal in potential polarity shows that channels open much more rapidly than in (**a**) and the total number of channels continues to increase. If the potential is reversed and increased in magnitude (**c**) then channel closure is more rapid and this is also seen with subsequent channel opening when the potential is again reversed (**d**)

was discovered in colicin A and suggests that the hydrophobic helices 8 and 9 insert into the inner membrane while the remaining helices spread out on top of the inner membrane, like an umbrella (Parker et al. 1989). The open channel was then proposed to be formed by the voltage-induced insertion of the surface bound helices. This was supported by FRET measurements, which showed that helices 1&2 moved away from the rest of the protein (Lakey et al. 1991). Further experiments however failed to show complete separation of the helical hairpin from the remaining helices (Lakey et al. 1993). A different model, the penknife, was then suggested in which only helices 1 and 2 move away from the hydrophobic core, while the rest of the molecule with its exposed hydrophobic face sinks into the membrane, including helices 8 and 9 (Lakey et al. 1993; Duche et al. 1994b). Since the helical hairpin does not have many polar or charged residues at its tip, there is nothing to maintain it in a perpendicular transmembrane orientation so it is likely that a dynamic structure exists with various orientations. This would account for the exposure of the tip in the closed channel state in the biotin experiments, described later, and the failure to determine a FRET average distance in agreement with the transmembrane format (Lakey et al. 1993). FRET and quenching experiments with colicin E1 (Zakharov et al. 1998a, b, c; Lindeberg et al. 2000; Padmavathi and Steinhoff 2008) have proposed a modification of the umbrella model. Supposedly, the molecule rearranges and unfolds before inserting helices 8 and 9 into the inner membrane, while the amphipathic helices remain on the membrane surface in a spiral-like orientation. The models agree that the structure must be flexible to allow insertion of helices 8 and 9 and channel formation (Zakharov and Cramer 2002). As described above colicins A and B seem to interact with the membrane surface in a molten-globule state, losing the tertiary structure but maintaining the secondary structure (Vandergoot et al. 1991; Evans et al. 1996). However, this form of interaction could not be confirmed for colicin E1 (ColE1) (Schendel and Cramer 1994), and ColN (Evans et al. 1996). Both of these colicins interact poorly with membranes even at low pH and their "translocated state" could best be imitated *in vitro* by pre-incubation with a non-ionic detergent which causes rapid channel formation presumably by opening up the pore-forming domain ready for insertion (Dover et al. 2000). Several studies using domain swapping or deletion have attempted to determine the P-domain regions essential for channel formation. Helices 1–3 of ColE1 and ColA can be removed without losing activity and some parts of helices 4 and 5 might not be crucial (Collarini et al. 1987; Baty et al. 1990; Lindeberg et al. 2000; Nardi et al. 2001b). However, some truncated P-domains form channels with different properties. For ColIa and ColA, helices 2–5 were determined as unnecessary (Qiu et al. 1996; Slatin et al. 2004). The finding that only 132 residues are needed for a wild type channel activity and that the channel can conduct ions of 8 Å or larger (Lakey and Slatin 2001) means that constructing a realistic model has been difficult.

The most informative approach was developed by the group of Alan Finkelstein who, using planar lipid bilayers, were able to detect the parts of the colicin pore-forming domain that traversed the lipid bilayer upon application of a transmembrane electrical potential difference (Jakes et al. 1990; Slatin et al. 1994; Qiu et al. 1996; Kienker et al. 1997). The method involved the creation of colicin pore-forming domain mutants that had a single cysteine amino acid residue inserted at various strategic sites within the domain. Each cysteine residue was then biotinylated to yield a unique site on each protein that could interact quasi-irreversibly with the large tetrameric protein streptavidin. When streptavidin was placed on the opposite (trans) side of the bilayer to that which the colicin had been added (cis-side) it acted as a trap for any biotin moieties which were transported across during channel formation. Thus, when a membrane potential of +50 mV (relative to the side of colicin addition) is applied, it is normal to see a slow opening of multiple colicin channels (Fig. 8.3). In 1 M NaCl these have conductances of 20–50 pS and will remain open until the membrane potential is reduced or reversed, whereupon they close with kinetics proportional to the change in the membrane potential (Fig. 8.3). If a channel contains a voltage-translocated biotin group bound to a trans streptavidin protein it cannot close upon membrane potential reversal and this could easily be detected as a channel which conducts ions when the membrane potential is reversed. In this way a series of mutants were tested and the parts of the colicin, which traversed the membrane during channel opening, were mapped. The result of a series of studies showed that the colicin channel is highly dynamic and plastic with a broad range of possible conformations. Furthermore, even previously unexpected highly charged regions seem able to traverse the bilayer easily and foreign epitopes could also be translocated (Lakey and Slatin 2001). Importantly, the method showed that the inter-helical loop at the tip of helix 8 and 9 of ColIa is exposed to the cytosol in the channel's open state and in, at least some, closed states.

Finally, to overcome the problem of having too little protein to make an active channel, it has been suggested that the colicin deforms the lipid bilayer to make a so-called lipid toroidal pore (Sobko et al. 2004, 2006). Despite discrepancies between the various models, the interactions of the hydrophobic hairpin with the lipid tails of the inner membrane remain undisputed and are supported by *in vivo* experiments on ColE1 (Song et al. 1991) and *in vitro* experiments on ColIa (Kienker et al. 1997).

8.3 Unusual Ion Channel Regulation: Colicin Immunity Proteins

8.3.1 The Three Types of Colicin Immunity Proteins

Colicin producers need an effective and efficient protection mechanism against their own toxin. For all colicins, this protection is provided by the highly specific immunity proteins that interact with the cytotoxic domain; however, the nature of the interaction is defined by the colicin type. Immunity proteins of nuclease colicins attach to the toxic domain and prevent its function as soon as the colicins are produced (Cascales et al. 2007), to avoid nucleic acid degradation in the producer cell. The affinity constant for this reaction is extremely large and the colicin is thus released with a bound immunity protein, which remains attached even when the colicin inserts into the target cell. Clearly, the immunity protein must be released in order for toxicity to occur and, considering the rapid kinetics of the toxicity, the very slow dissociation rate posed a serious problem (Kleanthous et al. 1998). However, it has recently been shown that a relatively gentle directional force can disrupt this complex and provides a clue for how the directional effects of translocation may be felt by the proteins involved (Farrance et al. 2013). The nuclease domain of colicin E7 has been shown to insert into and form channels in planar lipid bilayers (Mosbahi et al. 2002). Whilst there is no evidence that this permeability increase contributes to cell death, it does support the possibility that the nuclease domains translocate through the lipid bilayer in a flexible state. This may only be possible when the immunity protein, with its firm grip on the structure, is removed (Mosbahi et al. 2006). Since the immunity protein also has to protect cells from nuclease colicins carrying the same plasmid as the producer cells, it is likely that freely diffusing versions of the immunity protein are present in the cytoplasm and rapidly inactivate any incoming toxic domains.

The immunity protein of Colicin M (ColM) is anchored to the cytoplasmic membrane in the periplasm, which is the site of action of this, the only known colicin to act as a peptidoglycan synthesis inhibitor (Olschlager and Braun 1987). In a similar way, the mode of action of immunity proteins to pore-forming colicins must be defined by their target. Thus, these voltage dependent ion channels are controlled by interactions with another membrane protein. This has also been shown to occur in some ion channel proteins in eukaryotes, most notably the interaction of minK/I_{sK} with voltage gated ion channels (Barhanin et al. 1999). Thus the study of colicin pore-forming immunity proteins can shed light upon a type of molecular interaction that may have relevance to other fields of biology. The ability to study it in bacteria and the possibility of expressing the components with good yield make this an attractive model system for in-membrane molecular recognition studies. This area was first explored through the studies of the glycophorin dimer (Lemmon et al. 1994) but it remains under developed compared to our knowledge of how soluble protein-protein interactions are driven.

8.3.2 Expression and Location of Immunity Proteins Against Pore Forming Colicins

Immunity proteins to pore-forming colicins are thus located in the inner membrane (Goldman et al. 1985; Geli et al. 1988; Schramm et al. 1988) (Fig. 8.2) and allow cells to live at a colicin concentration of 10^4–10^7 times greater than that needed to kill sensitive cells (Cascales et al. 2007). However, this immunity can be overcome if the amount of colicin is further increased (Levisohn et al. 1968; Smajs et al. 2006). They are regulated by their own promoter and are expressed constitutively as shown for ColA, ColB and ColK immunity proteins (Lloubes et al. 1986; Geli et al. 1988; Schramm et al. 1988; Mulec et al. 2003) (Fig. 8.4). This is in stark contrast to the tightly regulated but strong over expression of the colicin toxins themselves, which are regulated by the SOS response. A relatively homogenous expression of the immunity protein in most of the cell population was shown (Mulec et al. 2003) by fusing GFP to the immunity protein to colicin K (CKI-GFP), however ColK-GFP (which reported on the presence of the toxin) was only expressed in 3 % of all cells. Expression levels of immunity proteins to pore-forming toxins are relatively low so a limited number of immunity proteins seem to protect cells against large numbers of exogenous colicins (Song and Cramer 1991).

8.3.3 How Do Immunity Proteins Work?

In order to answer the fundamental question of how the immunity proteins efficiently exhibit such a high level of specific protection, the mode of interaction between a colicin and its immunity protein must be determined and so research has focused on determining the site of interaction. This work is easily done in whole

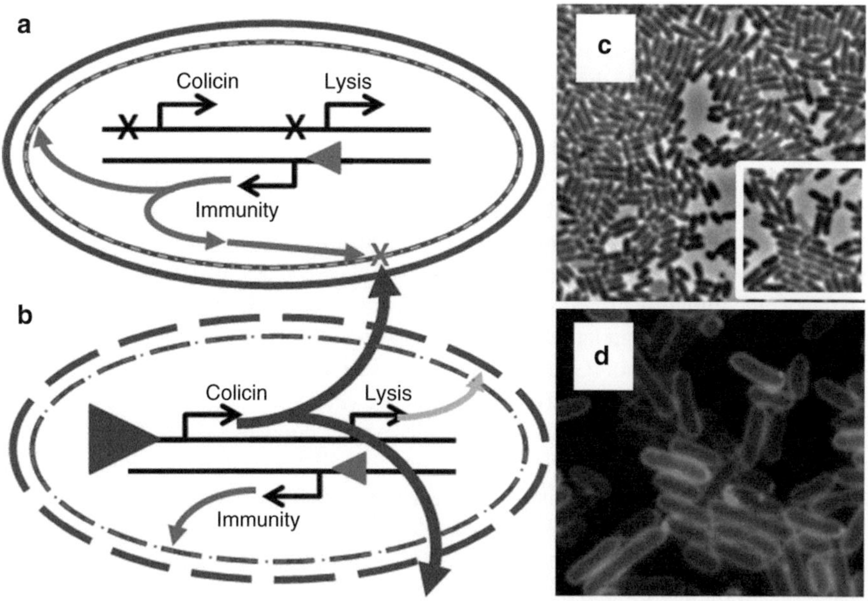

Fig. 8.4 The colicin operon in action. In normal conditions (**a**) most pore-forming colicin operons only express the immunity protein from a weak constitutive promoter. The resulting protein inserts into the inner membrane. Under stress conditions, the strong promoter of the colicin is activated (**b**) and the resulting RNA message contains, in about 10 % of cases, the region coding for the lysis protein. This lipoprotein destabilises the bacterial membranes causing a quasi-lysis that allows the release of colicin into the medium. The immunity protein cannot protect these producer cells from this damage (and death) but in neighbouring plasmid bearing cells the immunity protein can prevent killing by the released toxin. (**c**) Phase contrast image of *E. coli* bacteria expressing a GFP fusion with colicin N immunity protein (*E. coli* C41 with 1 mM IPTG after 24 h) Section shown in D is marked. (**d**) Enlarged fluorescence image of CNI-GFP showing membrane localisation of the immunity protein fusion

cells since the colicin toxicity is clearly monitored and the majority of data results from novel protein constructs coupled with tests for cell growth. Because interactions of other colicin domains (Translocation- and Receptor-binding domains) with the immunity protein have largely been excluded, it is likely that the interaction of the immunity protein with its cognate colicin is directly influenced by how the pore is formed.

Domain swapping experiments in colicins Ia and Ib (Mankovich et al. 1986; Benedetti et al. 1991), as well as A and E1 (Bishop et al. 1985), have shown that the immunity proteins interact exclusively with the P-domain and neutralise its lethal action through an, as yet, unknown mechanism. It was shown that the P-domain of colicin A expressed in the cytoplasm was non-toxic presumably because the polarity of the membrane potential difference is opposite to that required to open the channel (Espesset et al. 1994). When a periplasmic targeting signal sequence was added to the N-terminus, the P-domain became active. The toxicity was prevented when the ColA immunity protein (CAI) was expressed in the same cells. Interestingly, the toxicity was independent of the Tol proteins normally required. This system was later used to study P-domain function (Duche 2002). The P-domain of colicin N in the cytoplasm is toxic even without a signal sequence, which may indicate that it somehow gains access to the periplasm across the inner membrane by another means. This toxicity is specific as it can be inhibited by the colicin N immunity protein (CNI) but not CAI (Fridd et al. 2002). In the presence of CNI large amounts of full length ColN or its P-domain, e.g. 100 mg/l of culture, can be produced and thus the amount of P-domain in the cytoplasm vastly exceeds the amount of immunity protein in the membrane. Thus, the numbers of pore-forming domains in a toxic form must be very low, or the number of sites at which they can act are sufficiently limited, so that a few immunity proteins can guard these effectively (Fridd et al. 2002). Others have suggested that rapid lateral diffusion is enough to ensure rapid neutralisation of the pore-forming domain (Geli and Lazdunski 1992; Zhang and Cramer 1993).

Although the P-domains have high sequence (Parker et al. 1992) and structural similarity, the immunity proteins are specific to their cognate colicin (Fig. 8.5). The reasons for this specificity are at first sight unclear since it seems likely that a broad immunity to several colicins might offer a selective advantage. However, Riley (1993) proposed it as the mechanism that drives colicin diversity, which has gone a long way to explain the reasons for strict specificity between colicins and their immunity proteins. The immunity protein interacting regions are located in areas of high sequence divergence in P-domains, which would support a two-step evolutionary model (Riley et al. 2000). This is supported by a gradual shift in immunity specificity from one colicin to the other, if specificity-determining residues are exchanged one at a time. This molecular evolution process has thus created clear specificities that have been used both in sequence swapping (Geli and Lazdunski 1992; Pilsl and Braun 1995; Lindeberg and Cramer 2001) and mutagenesis studies (Zhang and Cramer 1993) to determine the P-domain regions which interact with the immunity protein.

Nevertheless some cross-specificity does occur as in the case of colicin 10 (Col10) immunity protein which provides some resistance against ColE1, but interestingly the ColE1 immunity protein provides no measurable resistance against Col10 (Lindeberg and Cramer 2001). Finally, in spite of the clear assays available, the data interpretation has been made difficult by our lack of knowledge of the structure of the inserted and open channel states.

8.3.4 Anti-pore Forming Immunity Proteins Comprise Two Families

Based on sequence similarity and resulting hydropathy profiles (Fig. 8.5), immunity proteins to pore-forming colicins can be divided into A-type (A, B, N and U, Y and S4) and E1-type (E1, 5, K, 10, Ia and Ib) which map in turn onto the PI and PII type pore-forming domains mentioned previously (Arnold et al. 2009). The A–type is predicted to have 4 transmembrane helices with both ends in the cytoplasm (Schramm et al. 1988; Geli et al. 1989a; Pilsl et al. 1998) and the E1-type to have 3 transmembrane domains (Song and Cramer 1991) with the N–terminus located in the cytoplasm and the C-terminus in the periplasm (Song and Cramer 1991; Pilsl and Braun 1995). The topology of CNI was previously unknown largely because the

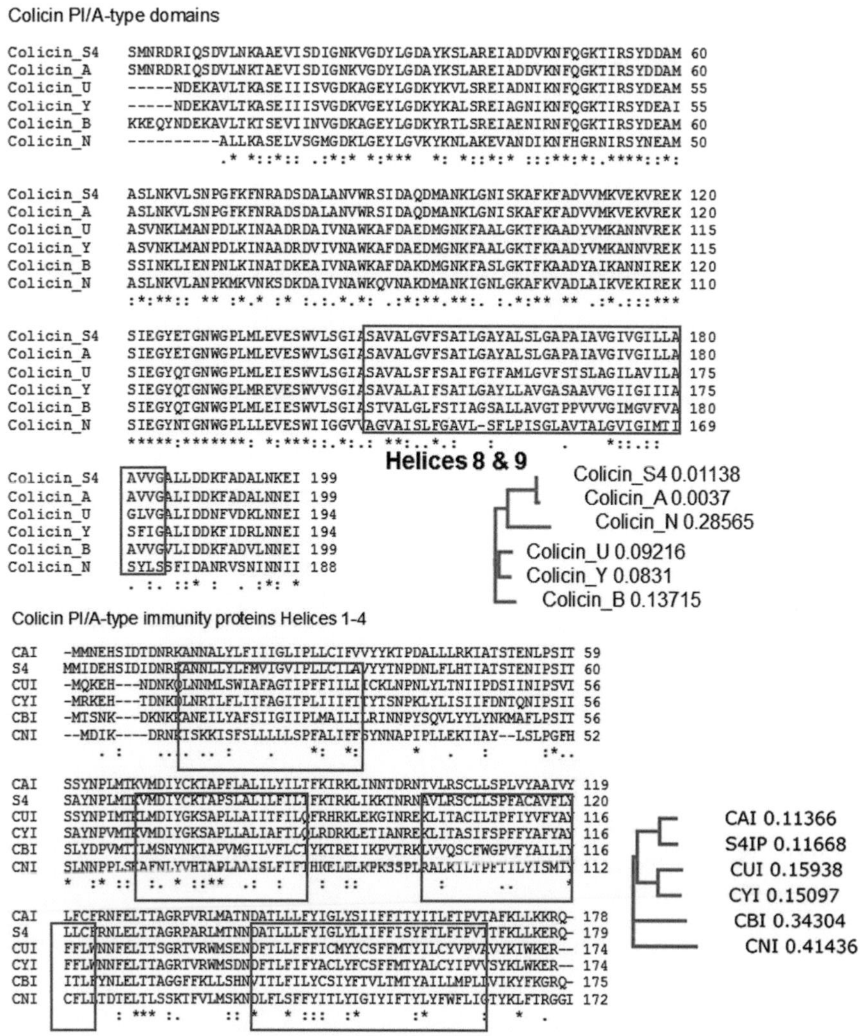

Fig. 8.5 Alignments and phylogenetic trees of type A/PI colicins and their immunity proteins. The colicin P-domains are highly homologous with high levels of sequence similarity. They do however only respond to their cognate immunity proteins. This is the result of an evolutionary process that drives differentiation and the alignments here show, along with the phylogenetic trees, how this group of colicins and their immunity proteins have co-diverged. The alignments were derived using the program Clustal (Larkin et al. 2007). The boxes define the known hydrophobic helical hairpin (Helices 8&9) in the colicins and the predicted transmembrane helices in the immunity proteins, for which high resolution structures are not available. The figures on the trees show the number of substitutions as a proportion of the length of the alignment

published sequence (Pugsley 1988) has an error which introduces a premature stop codon. We have determined the full-length sequence (Stroukova et al. unpublished data) which predicts a 4 transmembrane helix structure expected of A-type immunity proteins (Fig. 8.6). To confirm this, a fully functional fusion with GFP to the CNI

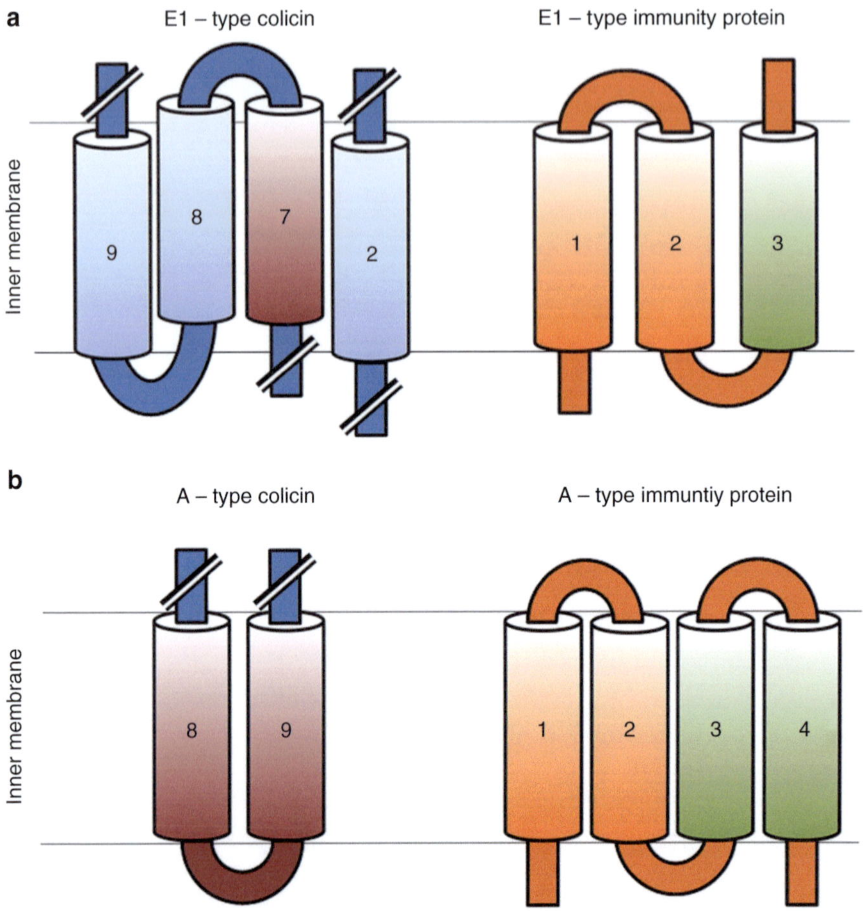

Fig. 8.6 Models representing transmembrane helices of E1 – and A – type colicins and their immunity proteins. Based on Zhang and Cramer (1993); Bohme et al. (2009); Sobko et al. (2009); Zhang et al. (2010); Honigmann et al. (2012). Cytosolic and periplasmic domains have been omitted for the colicins. (**a**) Helix 7 (*red*) of E1 – type colicins probably interacts with helix 3 (*green*) of the E1 – type immunity protein, when the channel is in an open state. (**b**) Helices 8 and 9 of A-type colicins, and in some cases also the connecting loop, probably interact with helix 3 and 4 (*green*) of their respective A-type immunity proteins (Smajs et al. 2006, 2008)

C-terminus showed that the C-terminus is located in the cytoplasm (Fig. 8.4) because GFP does not fluoresce in the periplasm (Feilmeier et al. 2000; Drew et al. 2002).

Colicin relationships and evolution have been analysed by comparing DNA and protein sequences, identifying DNA sequence polymorphisms in natural isolates as well as by experimental evolution and mathematical modelling (Cascales et al. 2007). It is proposed that to diversify, pore-forming colicins have mainly relied on recombination and domain shuffling, while nuclease colicins have mainly evolved through single base mutations (Fig. 8.5). Within each group, colicins generally share a very high sequence similarity, e.g. 50–97 % DNA sequence similarity in nuclease colicins (Cascales et al. 2007), with the greatest differences lying in the region interacting with the immunity protein.

This suggests a two-step evolution model. Firstly, mutations in the immunity proteins occur, broadening immunity and providing a survival advantage. Then mutations in the colicin gene change the killing mechanism, producing a new kind of colicin. By its ability to overcome other colicins' immunities, this new type can spread through the population until a new equilibrium is attained. Further mutations create a group of closely related colicins and immunity proteins (Riley 1993). The creation of new pore-forming colicins by recombination must depend upon their abundance within the population because it requires a certain variety and concentration of templates to occur simultaneously in the same environment. A later study using colicin Y has shown that pore-forming domains may also show diversification by single mutation changes (Riley et al. 2000).

8.3.5 Possible Interactions with the Colicin Translocation Machinery

To understand the structure of the P-domain immunity protein interaction it is useful to know if any other proteins are involved. As this interaction is within the inner membrane, the only likely additional proteins are those of the translocation machinery (Song and Cramer 1991). For group A colicins such as colicin N, these may be TolA, Q and R (Fig. 8.2) whilst for group B colicins they may be TonB, ExbB and ExbD. However, in studies where Translocation and Receptor-binding domains have been swapped between colicins in order to send the P-domain through a different translocation system, the immunity protein cognate to the P-domain still neutralised killing (Zhang and Cramer 1993). This indicates that the translocation system does not play a significant role in the function of the immunity protein.

8.3.6 How the Immunity Proteins Protect Against Pore Forming Colicins?

The immunity proteins to ColA (PI division) and ColE1 (PII) are the best studied examples and can be considered case studies for their respective (4 or 3helix) types of immunity protein. Zhang and Cramer (Zhang and Cramer 1993) mapped the main areas of ColE1 that control the interaction with the ColE1 immunity protein. These "hot-spots" comprised residues V441-W460 and A471-A488, with residues Ala474, Ser477, His440, Phe443 and Gly444 being key. ColE1 proteins mutated in these areas showed reduced lethality against non-immune cells but were able to kill cells immune to the wild-type colicin E1. Sobko et al. (2009) identified His440 also as an important residue for colicin channel formation. The five residues critical for immunity protein recognition are located in helices 6 and 7 which are believed to be located on the periplasmic side of the inner membrane when the channel is in a closed state but their conformation may change once the channel is open (Vandergoot et al. 1991; Evans et al. 1996; Sobko et al. 2009). The membrane inserted colicin E1 immunity protein is therefore, likely to interact with ColE1 in its membrane inserted open state

(Zhang and Cramer 1993). Furthermore, the authors disprove a previous assumption (Song and Cramer 1991) that charged domains of the immunity protein, i.e. the termini and cytosolic and periplasmic loops, are involved in interactions with ColE1.

A later study focused on finding specificity determinants which make ColE1 immunity protein specific to its colicin rather than residues crucial for its activity. This tested which mutations in ColE1 caused it to be inhibited by closely related Col 10 immunity protein (Lindeberg and Cramer 2001). It confirmed that residues between 419 and 501 on ColE1 are important for the immunity interaction and suggested that 448, 470, 472 and 474 are the specificity determinants. Unsurprisingly, they lie in the same region as the residues which determine ColE1 immunity recognition (Zhang and Cramer 1993), but interestingly, they are different.

As an example of P I-type colicins, the hydrophobic helical hairpin of the ColA P-domain has been identified as the region of interaction with the immunity protein (Espesset et al. 1996; Pilsl et al. 1998; Nardi et al. 2001a). For interaction to occur, only this region needs to be inserted into the membrane (Espesset et al. 1996; Nardi et al. 2001a) and, in contrast to the ColE1/ PII type colicins, the open state is not required.

The most recent description of Colicin A immunity protein (CAI) structure and interactions with ColA (Zhang et al. 2010) shows that CAI exists as a dimer *in vivo,* which is formed through disulphide bonds between helix 3 of each molecule. CAI has 4 cysteine residues. It is proposed that the other 3 cysteine residues form intramolecular bonds between helix 1 and 2 or helix 2 and 3 to stabilise the molecule. They suggest that all these bonds are broken when ColA is added. This occurs either through the external application of full-length ColA or expression of the signal peptide P-domain fusion mentioned above. Furthermore, Zhang et al. (2010) confirmed that channel activity is not required for interaction, because an inactive P-domain, which can insert the hydrophobic helical hairpin into the inner membrane but cannot form a channel, also breaks the inter- and intramolecular disulphide bonds. This indicates that the immunity protein binds to ColA while the channel is in a closed state and might prevent it from.

Smajs et al. (2006) investigated the colicin Y immunity protein (CYI) interaction with Colicin Y using helix swapping, followed by more precise site-directed conversion of Colicin U immunity protein residues into the respective CYI residues. Most useful here is the discovery that Colicin U is inactivated by CUI and CYI, but Colicin Y (ColY) is only inactivated by CYI, not CUI. They obtained mutants of CUI that could inactivate both ColU and ColY, constituting gain of function mutations. The amino acids responsible for the immunity protein interaction with ColY are S104, S107, F110, A112 on helix 3 and A159 on helix 4 and are located close to the cytoplasmic membrane where they could interact with the hydrophobic hairpin of the pore forming domain. Interestingly, at least three mutations, (C104S, T107S & I159A) are needed to confer immunity to CYI and single mutations cause no change in specificity. The fact that a CUI with four mutations (S104, S107, F110, A112, A159) is able to inactivate both ColY and ColU suggests that conserved residues are responsible for binding while variable residues define specificity. This is similar to the observations made by for ColE1 (Zhang and Cramer 1993; Lindeberg and Cramer 2001).

Subsequently, Smajs et al. (2008) confirmed that that residues located on helix 8 in the hydrophobic hairpin of ColY interact with residues on helix 3 and 4 of

CYI. These pairings and an energy minimised model led to the conclusion that helix-helix interactions play a crucial role in the inactivation of ColY by CYI. This is considered to be the general mechanism for PI type immunity.

The studies of colicin A and E1 immunity proteins together with the investigation of channel formation (Sobko et al. 2009, 2010) have thus led to the proposal of two different models of immunity protein action (Fig. 8.6). PII/E1-type immunity proteins interact with colicin molecules when the channel is in an open state conformation because interaction between helices while it is in a closed state does not match the assumed structural arrangement of the molecules in the membrane. A-type immunity proteins interact with their colicins while the channel is in a closed state. Both models of interaction are heavily based on the assumed structure of colicin molecules in the membrane and may only be validated when experiments showing the membrane bound P-domain structure alone or in complex with the immunity protein are successful.

All successful studies of immunity proteins to pore-forming colicins have so far been *in vivo. In vitro* data is very sparse due to the hydrophobicity of immunity proteins and the fact that they are integral membrane proteins. Even where overexpression and purification have been successful (Mankovich et al. 1986; Geli et al. 1989b; Shirabe et al. 1993; Taylor et al. 2000), a demonstration of activity or even interaction with purified colicins has been unsuccessful (Geli et al. 1989b).

8.4 Conclusion

Colicin pore-forming domains, which create voltage-gated channels in lipid bilayer membranes were once thought of as possible models for the voltage gated channels in excitable cells. In the years since then they have shown that they interact with membranes in a unique fashion and surprisingly their closest functional homologs appear to be proteins involved in apoptosis. Studies of colicins have revealed a great deal about the translocation of proteins across membranes and studies of their immunity proteins have begun to show how ion channels may be regulated by protein-protein interactions within the membrane. Due to their ease of study and the effective *in vivo* assays pore-forming colicins are likely to continue to teach us new aspects of membrane protein function.

References

Anderluh G, Lakey JH (eds) (2010) Proteins; membrane binding and pore formation, 1st edn. Springer, Heidelberg

Anderluh G, Hong Q, Boetzel R, MacDonald C, Moore GR, Virden R, Lakey JH (2003) Concerted folding and binding of a flexible colicin domain to its periplasmic receptor TolA. J Biol Chem 278:21860–21868

Arnold T, Zeth K, Linke D (2009) Structure and function of colicin S4, a colicin with a duplicated receptor-binding domain. J Biol Chem 284(10):6403–6413

Baboolal TG, Conroy MJ, Gill K, Ridley H, Visudtiphole V, Bullough PA, Lakey JH (2008) Colicin N binds to the periphery of its receptor and translocator, outer membrane protein F. Structure 16(3):371–379

Barhanin J, Romey G, Lazdunski M (1999) IsK: a novel type of potassium channel regulatory subunit. Potass Ion Chan 46:67–84

Baty D, Lakey J, Pattus F, Lazdunski C (1990) A 136-amino-acid-residue COOH-terminal fragment of colicin-a is endowed with ionophoric activity. Eur J Biochem 189(2):409–413

Benedetti H, Frenette M, Baty D, Knibiehler M, Pattus F, Lazdunski C (1991) Individual domains of colicins confer specificity in colicin uptake, in pore-properties and in immunity requirement. J Mol Biol 217(3):429–439

Benedetti H, Lloubes R, Lazdunski C, Letellier L (1992) Colicin A unfolds during its translocation in Escherichia coli cells and spans the whole cell envelope when its pore has formed. EMBO J 11(2):441–447

Bishop LJ, Bjes ES, Davidson VL, Cramer WA (1985) Localization of the immunity protein-reactive domain in unmodified and chemically modified COOH-terminal peptides of colicin E1. J Bacteriol 164(1):237–244

Bohme S, Padmavathi PVL, Holterhues J, Ouchni F, Klare JP, Steinhoff HJ (2009) Topology of the amphipathic helices of the colicin A pore-forming domain in E. coli lipid membranes studied by pulse EPR. Phys Chem Chem Phys 11(31):6770–6777

Buchanan SK, Lukacik P, Grizot S, Ghirlando R, Ali MM, Barnard TJ, Jakes KS, Kienker PK, Esser L (2007) Structure of colicin I receptor bound to the R-domain of colicin Ia: implications for protein import. EMBO J 26:2594–2604

Cascales E, Buchanan SK, Duche D, Kleanthous C, Lloubes R, Postle K, Riley M, Slatin S, Cavard D (2007) Colicin biology. Microbiol Mol Biol Rev 71(1):158–229

Cavard D (1991) Synthesis and functioning of the colicin E1 lysis protein: comparison with the colicin A lysis protein. J Bacteriol 173(1):191–196

Clifton LA, Johnson CL, Solovyova AS, Callow P, Weiss KL, Ridley H, Le Brun AP, Kinane CJ, Webster JRP, Holt SA, Lakey JH (2012) Low resolution structure and dynamics of a colicin-receptor complex determined by neutron scattering. J Biol Chem 287(1):337–346

Clifton LA, Skoda MWA, Daulton EL, Hughes AV, Le Brun AP, Lakey JH, Holt SA (2013) Asymmetric phospholipid: lipopolysaccharide bilayers; a Gram-negative bacterial outer membrane mimic. J R Soc Interface 10(89):20130810

Collarini M, Amblard G, Lazdunski C, Pattus F (1987) Gating processes of channels induced by colicin-a, its C-terminal fragment and colicin-E1 in planar lipid bilayers. Eur Biophys J 14(3):147–153

Dover LG, Evans LJ, Fridd SL, Bainbridge G, Raggett EM, Lakey JH (2000) Colicin pore-forming domains bind to Escherichia coli trimeric porins. Biochemistry 39(29):8632–8637

Drew D, Sjostrand D, Nilsson J, Urbig T, Chin CN, de Gier JW, von Heijne G (2002) Rapid topology mapping of Escherichia coli inner-membrane proteins by prediction and PhoA/GFP fusion analysis. Proc Natl Acad Sci U S A 99(5):2690–2695

Duche D (2002) The pore-forming domain of colicin A fused to a signal peptide: a tool for studying pore-formation and inhibition. Biochimie 84(5–6):455–464

Duche D, Baty D, Chartier M, Letellier L (1994a) Unfolding of colicin A during its translocation through the *Escherichia coli* envelope as demonstrated by disulfide bond engineering. J Biol Chem 269(40):24820–24825

Duche D, Parker MW, Gonzalezmanas JM, Pattus F, Baty D (1994b) Uncoupled steps of the colicin-a pore formation demonstrated by disulfide bond engineering. J Biol Chem 269(9):6332–6339

Espesset D, Corda Y, Cunningham K, Benedetti H, Lloubes R, Lazdunski C, Geli V (1994) The colicin A pore-forming domain fused to mitochondrial intermembrane space sorting signals can be functionally inserted into the Escherichia coli plasma membrane by a mechanism that bypasses the Tol proteins. Mol Microbiol 13(6):1121–1131

Espesset D, Duche D, Baty D, Geli V (1996) The channel domain of colicin A is inhibited by its immunity protein through direct interaction in the Escherichia coli inner membrane. EMBO J 15(10):2356–2364

Evans LJA, Goble ML, Hales KA, Lakey JH (1996) Different sensitivities to acid denaturation within a family of proteins; implications for acid unfolding and membrane translocation. Biochemistry 35(40):13180–13185

Farrance OE, Hann E, Kaminska R, Housden NG, Derrington SR, Kleanthous C, Radford SE, Brockwell DJ (2013) A force-activated trip switch triggers rapid dissociation of a colicin from its immunity protein. PLoS Biol 11(2):e1001489

Feilmeier BJ, Iseminger G, Schroeder D, Webber H, Phillips GJ (2000) Green fluorescent protein functions as a reporter for protein localization in Escherichia coli. J Bacteriol 182(14):4068–4076

Fridd SL, Lakey JH (2002) Surface aspartate residues are essential for the stability of colicin A P-domain: a mechanism for the formation of an acidic molten-globule. Biochemistry 41(5):1579–1586

Fridd SL, Gokce I, Lakey JH (2002) High level expression of His-tagged colicin pore-forming domains and reflections on the sites for pore formation in the inner membrane. Biochimie 84(5–6):477–483

Geli V, Lazdunski C (1992) An alpha-helical hydrophobic hairpin as a specific determinant in protein-protein interaction occurring in Escherichia coli colicin A and B immunity systems. J Bacteriol 174(20):6432–6437

Geli V, Baty D, Lazdunski C (1988) Use of a foreign epitope as a tag for the localization of minor proteins within a cell – the case of the immunity protein to colicin-A. Proc Natl Acad Sci U S A 85(3):689–693

Geli V, Baty D, Pattus F, Lazdunski C (1989a) Topology and function of the integral membrane protein conferring immunity to colicin A. Mol Microbiol 3(5):679–687

Geli V, Knibiehler M, Bernadac A, Lazdunski C (1989b) Purification and reconstitution into liposomes of an integral membrane-protein conferring immunity to colicin-A. FEMS Microbiol Lett 60(2):239–244

Goldman K, Suit JL, Kayalar C (1985) Identification of the plasmid-encoded immunity protein for colicin E1 in the inner membrane of Escherichia coli. FEBS Lett 190(2):319–323

Greig SL, Radjainia M, Mitra AK (2009) Oligomeric structure of colicin Ia channel in lipid bilayer membranes. J Biol Chem 284(24):16126–16134

Guder A, Wiedemann I, Sahl HG (2000) Posttranslationally modified bacteriocins – the lantibiotics. Biopolymers 55(1):62–73

Hecht O, Ridley H, Boetzel R, Lewin A, Cull N, Chalton DA, Lakey JH, Moore GR (2008) Self-recognition by an intrinsically disordered protein. FEBS Lett 582(17):2673–2677

Hecht O, Macdonald C, Moore GR (2012) Intrinsically disordered proteins: lessons from colicins. Biochem Soc Trans 40:1534–1538

Hilsenbeck JL, Park H, Chen G, Youn B, Postle K, Kang C (2004) Crystal structure of the cytotoxic bacterial protein colicin B at 2.5Å resolution. Mol Microbiol 51(3):711–720

Honigmann A, Pulagam LP, Sippach M, Bartsch P, Steinhoff HJ, Wagner R (2012) A high resolution electro-optical approach for investigating transition of soluble proteins to integral membrane proteins probed by colicin A. Biochem Biophys Res Commun 427(2):385–391

Housden NG, Hopper JTS, Lukoyanova N, Rodriguez-Larrea D, Wojdyla JA, Klein A, Kaminska R, Bayley H, Saibil HR, Robinson CV, Kleanthous C (2013) Intrinsically disordered protein threads through the bacterial outer-membrane porin OmpF. Science 340(6140):1570–1574

Huang Y, Le Brun A, Soliakov A, MacDonald C, Moore G, Lakey JH (2012) Helix N-cap Asp are the pH trigger for colicin a membrane insertion. Biophys J 102(3):656A

Jakes KS (2014) Daring to be different: colicin N finds another way. Mol Microbiol 92(3):435–439

Jakes KS, Cramer WA (2012) Border crossings: colicins and transporters. Annu Rev Genet 46(46):209–231

Jakes KS, Finkelstein A (2010) The colicin Ia receptor, Cir, is also the translocator for colicin Ia. Mol Microbiol 75(3):567–578

Jakes KS, Abrams CK, Finkelstein A, Slatin SL (1990) Alteration of the pH-dependent ion selectivity of the colicin E1 channel by site-directed mutagenesis. J Biol Chem 265(12):6984–6991

Johnson CL, Ridley H, Marchetti R, Silipo A, Griffin DC, Crawford L, Bonev B, Molinaro A, Lakey JH (2014) The antibacterial toxin colicin N binds to the inner core of lipopolysaccharide and close to its translocator protein. Mol Microbiol 92(3):440–452
Kienker PK, Qiu XQ, Slatin SL, Finkelstein A, Jakes KS (1997) Transmembrane insertion of the colicin Ia hydrophobic hairpin. J Membr Biol 157(1):27–37
Kim YC, Tarr AW, Penfold CN (2014) Colicin import into E. coli cells: a model system for insights into the import mechanisms of bacteriocins. Biochim Biophys Acta 1843(8):1717–1731
Kleanthous C, Hemmings AM, Moore GR, James R (1998) Immunity proteins and their specificity for endonuclease colicins: telling right from wrong in protein-protein recognition. Mol Microbiol 28(2):227–233
Kurisu G, Zakharov SD, Zhalnina MV, Bano S, Eroukova VY, Rokitskaya TI, Antonenko YN, Wiener MC, Cramer WA (2003) The structure of BtuB with bound colicin E3 R-domain implies a translocon. Nat Struct Biol 10(11):948–954
Lakey JH, Slatin SL (2001) Pore-forming colicins and their relatives. In: Van Der Goot FG (ed) Pore-forming toxins. Springer, Heidelberg, pp 131–161
Lakey JH, Massotte D, Heitz F, Dasseux JL, Faucon JF, Parker MW, Pattus F (1991) Membrane insertion of the pore-forming domain of colicin A. A spectroscopic study. Eur J Biochem 196(3):599–607
Lakey JH, Duche D, Gonzalezmanas JM, Baty D, Pattus F (1993) Fluorescence energy-transfer distance measurements – the hydrophobic helical hairpin of colicin-a in the membrane-bound state. J Mol Biol 230(3):1055–1067
Lakey JH, Parker MW, Gonzalez Manas JM, Duche D, Vriend G, Baty D, Pattus F (1994a) The role of electrostatic charge in the membrane insertion of colicin A: calculation and mutation. Eur J Biochem 220:155–163
Lakey JH, van der Goot FG, Pattus F (1994b) All in the family: the toxic activity of colicins. Toxicology 87:85–108
Larkin MA, Blackshields G, Brown NP, Chenna R, McGettigan PA, McWilliam H, Valentin F, Wallace IM, Wilm A, Lopez R, Thompson JD, Gibson TJ, Higgins DG (2007) Clustal W and clustal X version 2.0. Bioinformatics 23(21):2947–2948
Le Brun AP, Clifton LA, Halbert CE, Lin B, Meron M, Holden PJ, Lakey JH, Holt SA (2013) Structural characterization of a model gram-negative bacterial surface using lipopolysaccharides from rough strains of Escherichia coli. Biomacromolecules 14(6):2014–2022
Lemmon MA, Treutlein HR, Adams PD, Brunger AT, Engelman DM (1994) A dimerisation motif for transmembrane alpha helices. Nat Struct Biol 1(3):157–163
Levisohn R, Konisky J, Nomura M (1968) Interaction of colicins with bacterial cells. IV. Immunity breakdown studied with colicins Ia and Ib. J Bacteriol 96(3):811–821
Lindeberg M, Cramer WA (2001) Identification of specific residues in colicin E1 involved in immunity protein recognition. J Bacteriol 183(6):2132–2136
Lindeberg M, Zakharov SD, Cramer WA (2000) Unfolding pathway of the colicin E1 channel protein on a membrane surface. Biophys J 78(1):176a
Lloubes R, Baty D, Lazdunski C (1986) The promoters of the genes for colicin production, release and immunity in the ColA plasmid: effects of convergent transcription and Lex A protein. Nucleic Acids Res 14(6):2621–2636
Mankovich JA, Hsu CH, Konisky J (1986) DNA and amino acid sequence analysis of structural and immunity genes of colicins Ia and Ib. J Bacteriol 168(1):228–236
Mathavan I, Beis K (2012) The role of bacterial membrane proteins in the internalization of microcin MccJ25 and MccB17. Biochem Soc Trans 40:1539–1543
Moore GR, Osborne MJ, Whittaker S, Videler H, Lian LY, Pommer A, wallis R, James R, Kleanthous C (1995) Mechanism of action of the cytotoxic domain of colicin e9 and its inhibition by its cognate immunity protein. J Cell Biochem:31. Supplement 21B
Mosbahi K, Lemaitre C, Keeble AH, Mobasheri H, Morel B, James R, Moore GR, Lea E, Kleanthous C (2002) The cytotoxic domain of colicin E9 is a channel-forming endonuclease. Nat Struct Biol 9(6):476–484

Mosbahi K, Walker D, James R, Moore GR, Kleanthous C (2006) Global structural rearrangement of the cell penetrating ribonuclease colicin E3 on interaction with phospholipid membranes. Protein Sci 15(3):620–627
Muchmore SW, Sattler M, Liang H, Meadows RP, Harlan JE, Yoon HS, Nettesheim D, Chang BS, Thompson CB, Wong SL, Ng SC, Fesik SW (1996) X-ray and nmr structure of human bcl-x(l), an inhibitor of programmed cell-death. Nature 381(6580):335–341
Mueller JEN, Papic D, Ulrich T, Grin I, Schuetz M, Oberhettinger P, Tommassen J, Linke D, Dimmer KS, Autenrieth IB, Rapaport D (2011) Mitochondria can recognize and assemble fragments of a beta-barrel structure. Mol Biol Cell 22(10):1638–1647
Mulec J, Podlesek Z, Mrak P, Kopitar A, Ihan A, Zgur-Bertok D (2003) A cka-gfp transcriptional fusion reveals that the colicin K activity gene is induced in only 3 percent of the population. J Bacteriol 185(2):654–659
Nardi A, Corda Y, Baty D, Duche D (2001a) Colicin A immunity protein interacts with the hydrophobic helical hairpin of the colicin A channel domain in the Escherichia coli inner membrane. J Bacteriol 183(22):6721–6725
Nardi A, Slatin SL, Baty D, Duche D (2001b) The C-terminal half of the colicin A pore-forming domain is active in vivo and in vitro. J Mol Biol 307(5):1293–1303
Olschlager T, Braun V (1987) Sequence, expression, and localization of the immunity protein for colicin-M. J Bacteriol 169(10):4765–4769
Padmavathi PVL, Steinhoff HJ (2008) Conformation of the closed channel state of colicin a in proteoliposomes: an umbrella model. J Mol Biol 378(1):204–214
Parker MW, Feil SC (2005) Pore-forming protein toxins: from structure to function. Prog Biophys Mol Biol 88(1):91–142
Parker MW, Pattus F, Tucker AD, Tsernoglou D (1989) Structure of the membrane-pore-forming fragment of colicin A. Nature 337(6202):93–96
Parker MW, Postma JPM, Pattus F, Tucker AD, Tsernoglou D (1992) Refined structure of the pore-forming domain of colicin-a at 2-bullet-4 angstrom resolution. J Mol Biol 224(3):639–657
Pilsl H, Braun V (1995) Evidence that the immunity protein inactivates colicin-5 immediately prior to the formation of the transmembrane channel. J Bacteriol 177(23):6966–6972
Pilsl H, Smajs D, Braun V (1998) The tip of the hydrophobic hairpin of colicin U is dispensable for colicin U activity but is important for interaction with the immunity protein. J Bacteriol 180(16):4111–4115
Pugsley AP (1988) The immunity and lysis genes of ColN plasmid pCHAP4. Mol Gen Genet 211(2):335–341
Qiu XQ, Jakes KS, Kienker PK, Finkelstein A, Slatin SL (1996) Major transmembrane movement associated with colicin Ia channel gating. J Gen Physiol 107(3):313–328
Renner LD, Weibel DB (2011) Cardiolipin microdomains localize to negatively curved regions of Escherichia coli membranes. Proc Natl Acad Sci U S A 108(15):6264–6269
Riley MA (1993) Positive selection for colicin diversity in bacteria. Mol Biol Evol 10(5):1048–1059
Riley MA, Cadavid L, Collett MS, Neely MN, Adams MD, Phillips CM, Neel JV, Friedman D (2000) The newly characterized colicin Y provides evidence of positive selection in pore-former colicin diversification. Microbiology 146:1671–1677
Schendel SL, Cramer WA (1994) On the nature of the unfolded intermediate in the in-vitro transition of the colicin E1 channel domain from the aqueous to the membrane phase. Protein Sci 3(12):2272–2279
Schramm E, Olschlager T, Troger W, Braun V (1988) Sequence, expression and localization of the immunity protein for colicin B. Mol Gen Genet 211(1):176–182
Shirabe K, Yamada M, Merrill AR, Cramer WA, Nakazawa A (1993) Overproduction and purification of the colicin-E1 immunity protein. Plasmid 29(3):236–240
Slatin SL, Qiu XQ, Jakes KS, Finkelstein A (1994) Identification of a translocated protein segment in a voltage-dependent channel. Nature 371:158–161
Slatin SL, Duche D, Kienker PK, Baty D (2004) Gating movements of colicin A and colicin Ia are different. J Membr Biol 202(2):73–83

Smajs D, Matejkova P, Weinstock GM (2006) Recognition of pore-forming colicin Y by its cognate immunity protein. FEMS Microbiol Lett 258(1):108–113
Smajs D, Dolezalova M, Macek P, Zidek L (2008) Inactivation of colicin Y by intramembrane helix-helix interaction with its immunity protein. FEBS J 275(21):5325–5331
Sobko AA, Kotova EA, Antonenko YN, Zakharov SD, Cramer WA (2004) Effect of lipids with different spontaneous curvature on the channel activity of colicin E1: evidence in favor of a toroidal pore. FEBS Lett 576(1–2):205–210
Sobko AA, Kotova EA, Antonenko YN, Zakharov SD, Cramer WA (2006) Lipid dependence of the channel properties of a colicin E1-lipid toroidal pore. J Biol Chem 281(20):14408–14416
Sobko AA, Rokitskaya TI, Kotova EA (2009) Histidine 440 controls the opening of colicin E1 channels in a lipid-dependent manner. Biochim Biophys Acta 1788(9):1962–1966
Sobko AA, Kovalchuk SI, Kotova EA, Antonenko YN (2010) Induction of lipid flip-flop by colicin E1-a hallmark of proteolipidic pore formation in liposome membranes. Biochemistry (Mosc) 75(6):728–733
Soelaiman S, Jakes K, Wu N, Li CM, Shoham M (2001) Crystal structure of colicin E3: implications for cell entry and ribosome inactivation. Mol Cell 8(5):1053–1062
Song HY, Cramer WA (1991) Membrane topography of Cole1 gene-products – the immunity protein. J Bacteriol 173(9):2935–2943
Song HY, Cohen FS, Cramer WA (1991) Membrane topography of Cole1 gene-products – the hydrophobic anchor of the colicin-E1 channel is a helical hairpin. J Bacteriol 173(9):2927–2934
Taylor RM, Zakharov SD, Heymann JB, Girvin ME, Cramer WA (2000) Folded state of the integral membrane colicin E1 immunity protein in solvents of mixed polarity. Biochemistry 39(40):12131–12139
Udho E, Jakes KS, Buchanan SK, James KJ, Jiang XX, Klebba PE, Finkelstein A (2009) Reconstitution of bacterial outer membrane TonB-dependent transporters in planar lipid bilayer membranes. Proc Natl Acad Sci U S A 106(51):21990–21995
van der Goot FG, González-Mañas JM, Lakey JH, Pattus F (1991) A 'molten-globule' membrane-insertion intermediate of the pore-forming domain of colicin A. Nature 354(6352):408–410
van der Goot FG, Didat N, Pattus F, Dowhan W, Letellier L (1993) Role of acidic lipids in the translocation and channel activity of colicins A and N in *Escherichia coli* cells. Eur J Biochem 213:217–221
Vandergoot FG, Gonzalezmanas JM, Lakey JH, Pattus F (1991) A molten-globule membrane-insertion intermediate of the pore-forming domain of colicin-A. Nature 354(6352):408–410
Vetter IR, Parker MW, Tucker AD, Lakey JH, Pattus F, Tsernoglou D (1998) Crystal structure of a colicin N fragment suggests a model for toxicity. Structure 6:863–874
Wiener M, Freymann D, Ghosh P, Stroud RM (1997) Crystal structure of colicin Ia. Nature 385(6615):461–464
Zakharov SD, Cramer WA (2002) Insertion intermediates of pore-forming colicins in membrane two-dimensional space. Biochimie 84(5–6):465–475
Zakharov SD, Lindeberg M, Cramer WA (1998a) Kinetic phases in membrane binding-insertion of the colicin E1 channel domain. Biophys J 74(2):A228
Zakharov SD, Lindeberg M, Griko Y, Salamon Z, Tollin G, Cramer WA (1998b) Conformational changes of the colicin E1 channel domain at the membrane surface. Biophys J 74(2):A228
Zakharov SD, Lindeberg M, Griko Y, Salamon Z, Tollin G, Prendergast FG, Cramer WA (1998c) Membrane-bound state of the colicin E1 channel domain as an extended two-dimensional helical array. Proc Natl Acad Sci U S A 95(8):4282–4287
Zakharov SD, Eroukova VY, Rokitskaya TI, Zhalnina MV, Sharma O, Loll PJ, Zgurskaya HI, Antonenko YN, Cramer WA (2004) Colicin occlusion of OmpF and TolC channels: outer membrane translocons for colicin import. Biophys J 87(6):3901–3911
Zeth K, Roemer C, Patzer SI, Braun V (2008) Crystal structure of colicin M, a novel phosphatase specifically imported by Escherichia coli. J Biol Chem 283(37):25324–25331
Zhang YL, Cramer WA (1993) Intramembrane helix-helix interactions as the basis of inhibition of the colicin E1 ion channel by its immunity protein. J Biol Chem 268(14):10176–10184
Zhang XYZ, Lloubes R, Duche D (2010) Channel domain of colicin A modifies the dimeric organization of its immunity protein. J Biol Chem 285(49):38053–38061

Chapter 9
Anthrax Toxin Protective Antigen Forms an Unusual Channel That Unfolds and Translocates Proteins Across Membranes

Bryan A. Krantz

Abstract Anthrax toxin is one of two major virulence factors secreted by pathogenic *Bacillus anthracis*, the etiologic agent of anthrax. Because inhalational anthrax is highly fatal, the agent has been weaponized for biowarfare and bioterrorism. Anthrax toxin is comprised of three individually nontoxic proteins, protective antigen (PA), lethal factor (LF), and edema factor (EF). But, to physiologically function, these individual subunits assemble into potent cytotoxins containing PA plus LF and/or EF. The PA component oligomerizes into a heptamer or octamer, which can insert into a host cell membrane to form a protein translocase channel. Under a transmembrane proton gradient driving force, LF and EF translocate through the narrow PA channel into the cytosol of the host cell. The narrowness of the channel necessitates that LF and EF unfold during translocation. This channel is unusual in this respect, because it contains its own unfoldase and translocase machinery. Highly nonspecific and dynamic clamp sites in the PA channel catalyze these activities. Anthrax toxin has been used extensively as a biophysical model to interrogate the molecular basis of translocation-coupled unfolding and translocation. It is being actively targeted by therapeutics to inhibit its function. New biotechnological adaptations use the toxin as a cancer therapy and generalized protein delivery vehicle.

Keywords Anthrax toxin • Protein translocation • Unfolding • Proton gradient

9.1 Anthrax

B. anthracis is the only pathogen in its genus known to cause epidemic disease in mammals, including humans. Attributed to its global reach are its names: Greek, ἀνθραξ ("anthrax" meaning "coal"); German, *Milzbrand* ("spleen fire")

B.A. Krantz
Department of Microbial Pathogenesis, School of Dentistry,
University of Maryland – Baltimore, Baltimore, MD 21201, USA
e-mail: bkrantz@umaryland.edu

A.H. Delcour (ed.), *Electrophysiology of Unconventional Channels and Pores*,
Springer Series in Biophysics 18, DOI 10.1007/978-3-319-20149-8_9

and *Hadernkrankheit* ("rag disease"); European, "Black Bane"; French, *charbon* ("charcoal"); Siberian Plague; Lodiana fever; and Indian, Pali Plague (Turnbull 1996). Anthrax was the first infectious disease to be linked to a specific bacterium, definitively establishing the Henle-Koch Postulates (Evans 1976). The disease is zoonotic and transmitted from animals to humans through three canonical infectious routes: cutaneous (~95 % of cases), inhalational (~5 %), and gastrointestinal (<1 %) (Mock and Fouet 2001; Koehler 2009).

9.1.1 Pathogenesis

Inhalation anthrax is the most pathogenic form of the disease. Inhaled *B. anthracis* spores germinate following phagocytosis, disseminating to nearby regional lymph nodes and the red pulp of the spleen. Lymphatic dissemination precedes hematogenous spread and terminal haemorrhagic septicemia. The high bacterial load of terminal anthrax (~10^9 bacilli ml^{-1}) is favorable to its life cycle, especially given that it increases the probability that extremely stable infectious spores will be deposited into the soils surrounding an animal carcass for the next victim to contact, inhale, or ingest (Beyer and Turnbull 2009). Endowing *B. anthracis* with its extreme pathogenicity are its virulence plasmids, pXO1 and pXO2, which produce two major virulence factors, anthrax toxin and a poly-γ-D-glutamate (γDPGA) capsule, respectively.

9.1.2 Vaccine

To thwart anthrax's march at the start of the twentieth century timely vaccination programs were instituted. Modernity and industrial-scale agricultural processing and global and railway transportation of animals, hides, and furs led to the heightened severity of worldwide epidemics (Turnbull 1991). Pasteur's prototypic bacterial vaccine worked as the encapsulation virulence plasmid, pXO2, was heat-cured, attenuating the bacillus' virulence (Mikesell et al. 1983). Thus the anthrax toxin protective antigen (PA) was likely present in the earliest toxigenic cellular vaccine. More readily deployable and practical livestock vaccinations were required, and Max Sterne later isolated the toxigenic but "uncapsulated" 34 F2 strain (Sterne 1939), which today is the live-spore vaccine used for livestock (Turnbull 1991).

9.1.3 Biowarfare and Bioterrorism

Bioweaponization has again made anthrax an existential threat to civilization. Humankind's tragic attraction to chemical and biological weapons is rooted in the Greek, τόξον (toxon, meaning bow or arrow) (Mayor 2008). Anthrax weaponization is not hard to imagine given its infamous history as a ravager of civilizations, its ability to manifest itself asymptomatically in the early-stages of infection, its rapid progression to acute disease, and its ability to be widely disseminated in the war theater as a highly infectious airborne aerosol of resilient spores. Correspondingly, the eminent threat of biowarfare intensified anthrax vaccinology research. Live-spore vaccines used in the former Soviet Union and China for military were not without risk. In Britain, an acellular culture filtrate of *B. anthracis* 34 F2 containing anthrax toxin PA, was prepared and combined with the adjuvant alum to make Anthrax Vaccine Precipitated (Belton and Strange 1954; Strange and Belton 1954; Wright et al. 1954). The American acellular vaccine, Anthrax Vaccine Adsorbed, now BioThrax (Pittman et al. 2002; Wasserman et al. 2003), is safe and is the best pre-exposure defense against inhalational anthrax (Friedlander et al. 1999).

Anthrax biowarfare and bioterrorism incidents throughout the twentieth century included clandestine German and highly organized Japanese (Williams and Wallace 1989) efforts in World War I and II, respectively. These incidents and subsequent bioweapons buildups during the Cold War led to the signing of the 1972 Biological Weapons Convention. Nonetheless, rogue nations, terrorist groups, and disturbed individuals unfortunately either maintained underground bioweapons efforts or carried out bioterrorist attacks. During the Amerithrax bioterrorism attack of 2001, spore-laden letters were sent in two separate mailings via the US Postal system to two US senators in Washington D.C., television and print media centers in New York City, NY, and a publisher in Boca Raton, FL, leading all told to 11 inhalation and 11 cutaneous infections, which ultimately caused 5 fatalities (Jernigan et al. 2002). Cleanup costs were $1 billion due to the tenacity of the spores in the environment. Cutting-edge genomic evidence (Rasko et al. 2011) identified the strain and flask from which the spores were derived. After the prime suspect committed suicide, the FBI closed its case.

9.2 Anthrax Virulence Factors

B. anthracis would be a benign soil bacillus were it not for its two large virulence plasmids, pXO1 (Mikesell et al. 1983) and pXO2 (Green et al. 1985; Uchida et al. 1985). These plasmids express its two principle virulence factors: the tripartite toxin, anthrax toxin, encoded on pXO1; and a unique peptidoglycan-anchored γDPGA capsule encoded on pXO2. The toxin dampens the innate and adaptive immune response in myriad of cell types, induces hypotension, assists in dissemination of the bacillus, and causes edema and a shock-like death (Baldari et al. 2006; Frankel et al. 2009; Moayeri and Leppla 2009). The capsule also functions to evade immune cell activities by thwarting phagocytosis.

9.2.1 Anthrax Toxin

Anthrax toxin, originally discovered by Harry Smith in the 1950s (Smith and Keppie 1954), is a binary toxin (Gill 1978; Barth et al. 2004) comprised of three ~80–90 kDa proteins, called lethal factor (LF), edema factor (EF), and protective antigen (PA)

(Frankel et al. 2009; Collier 2009). LF and EF are the two enzymatically active A moieties; PA is the cell-binding and channel-forming B moiety, making anthrax toxin an A_2B binary toxin. LF is a 90-kDa Zn^{2+} protease that cleaves numerous mitogen-activated protein kinase kinases (Duesbery et al. 1998; Vitale et al. 2000) and the innate immunity inflammasome sensor, NLRP1 (Levinsohn et al. 2012; Chavarría-Smith and Vance 2013), thus interfering with many host cell signaling pathways. EF is a 89-kDa Ca^{2+} and calmodulin-dependent adenylyl cyclase (Leppla 1982), which increases the concentration of cyclic-AMP, causing edema and facilitating dissemination (Dumetz et al. 2011). Crystal structures of LF and EF (Pannifer et al. 2001; Drum et al. 2002) and PA (Petosa et al. 1997) are known. PA is an 83-kDa Ca^{2+}-binding proprotein that contains four domains, D1, D2, D3 and D4

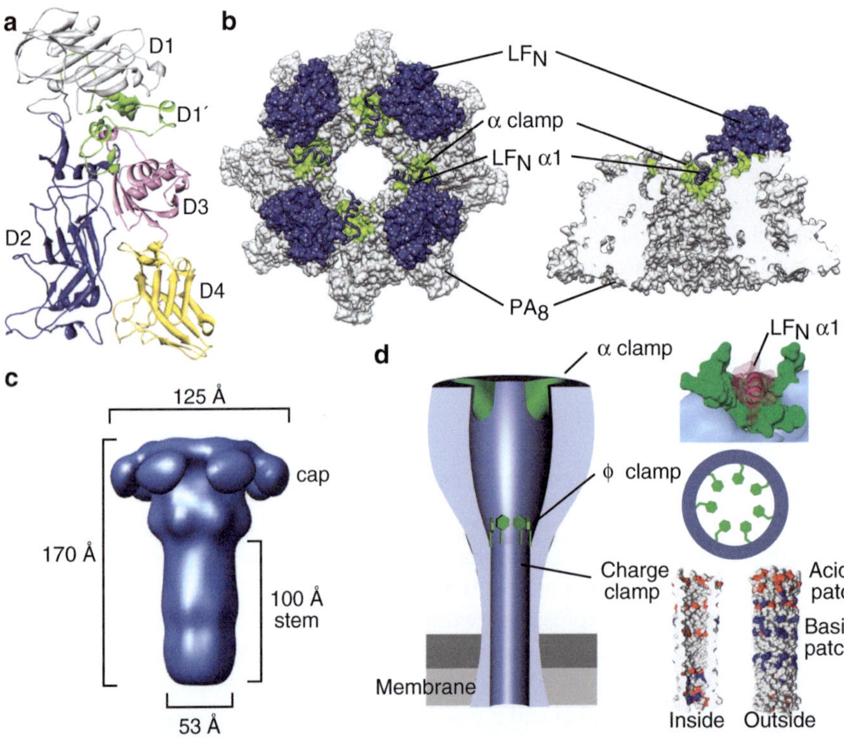

Fig. 9.1 Monomeric and oligomeric PA structures. (**a**) Ribbons depiction of the 83-kDa monomeric PA proprotein (Feld et al. 2012b) colored by domain: the prodomain (D1, *gray*); LF/EF binding domain (D1′, *green*); channel-forming domain (D2, *blue*); oligomerization domain (D3, *magenta*); and receptor-binding domain (D4, *gold*). (**b**) *Left*, axial view of the PA_8 prechannel (*gray space-fill*) co-complex with four LF_N (*blue*) (Feld et al. 2010). LF_N α1 (*blue ribbon*) is docked into the α clamp (*green space-fill*). *Right*, sagittal section of the PA_8 prechannel with one LF_N shown for clarity. (**c**) PA_7 channel EM structure (*blue surface*) with indicated dimensions (Katayama et al. 2008). (**d**) *Left*, PA channel cartoon (*gray*) illustrating α- (*green*), ϕ- (*green*), and charge-clamp sites (Feld et al. 2012a). *Right*, clamp-site details and putative modes of action based on the structure/function of the α clamp-LF_N α1 interaction (Feld et al. 2010), functional and spectroscopic measurements of the ϕ clamp (Krantz et al. 2005), or functional analysis of an electrostatic model of the β-barrel charge clamp (Wynia-Smith et al. 2012)

(Fig. 9.1a). The D1′ domain is what remains of the D1 prodomain upon proteolytic cleavage. Cleaved PA oligomerizes, assembles with LF and EF, and ultimately transforms into a translocase channel that delivers LF and EF into the cytosol of host cells. Hence, while PA, LF, and EF are individually nontoxic, the binary combinations, PA + LF and PA + EF, are toxic and known as anthrax lethal toxin (LeTx) and edema toxin (EdTx), respectively.

Holotoxin Assembly To function, LeTx or EdTx holotoxin complexes must first self-assemble. Two proposed assembly pathways are known: on cell-surfaces (Collier and Young 2003; Collier 2009) or in serum (Ezzell and Abshire 1992; Panchal et al. 2005; Moayeri et al. 2007; Ezzell et al. 2009; Kintzer et al. 2009, 2010a, b). In the former, PA binds to a host-cell-surface receptors, called the anthrax toxin receptors (ANTXR1 and ANTXR2) (Bradley et al. 2001; Scobie et al. 2003). PA is a proprotein that must be cleaved between D1 and D1' (Fig. 9.1a) by a furin or furin-like protease (Klimpel et al. 1992; Molloy et al. 1992) to produce the 20-kDa (PA_{20}) and 63-kDa (PA_{63}) fragments. Similar fragments are also observed with trypsin nicking (Blaustein et al. 1989; Christensen et al. 2005). PA_{63} remains bound to the cell-surface receptor, allowing it to assemble at the cell surface to make ring-shaped heptameric (PA_7) (Milne et al. 1994; Petosa et al. 1997; Lacy et al. 2004a) and octameric (PA_8) (Kintzer et al. 2009, 2010a; Feld et al. 2010) prechannel oligomers (Fig. 9.1b). Prechannel oligomers then bind LF and EF to form holotoxin complexes.

In the serum of rabbits and guinea pigs, a protease resides there that can nick PA to form PA_{63} (Ezzell and Abshire 1992; Panchal et al. 2005). In bovine serum, this cleavage can allow PA_{63} to self-assemble with LF or EF to make assembled PA_7- and PA_8-prechannel LeTx or EdTx complexes (Kintzer et al. 2009, 2010a). The putative mammalian serum protease activity varies according to species (rats versus mice) (Moayeri et al. 2007), however, suggesting this mechanism may not be generalizable (Bann 2012). Nevertheless, several other highly active proteases are part of the *B. anthracis* secretome, and they can activate PA (Pflughoeft et al. 2014).

Host-Cell Receptor Binding Following assembly, LeTx or EdTx complexes bind to cells via an extremely tight, sub-picomolar interaction with the anthrax toxin receptor (ANTXR) (Wigelsworth et al. 2004), as recently reviewed (van der Goot and Young 2009). Two ANTXR have been described, tumor endothelial marker 8 (TEM8) (Bradley et al. 2001) and capillary morphogenesis protein 2 (CMG2) (Scobie et al. 2003), but these have been renamed to ANTXR1 and ANTXR2, respectively (Young and Collier 2007). Another potential receptor, heterodimeric integrin β1, was also recently described (Martchenko et al. 2010). An ANTXR1/2 co-receptor was also described (Wei et al. 2006), albeit knockouts of the receptor show that it is not required in a mouse infection model (Young et al. 2007; Ryan and Young 2008). The widely studied receptors, ANTXR1 and ANTXR2, typically function as binding sites for the extracellular matrix, i.e., collagen type IV (Nanda et al. 2004) and basal lamina (Bell et al. 2001), respectively. Mutations in ANTXR2, for example, can lead to the rare genetic disorders juvenile hyaline fibromatosis and infantile systemic hyalinosis (Hanks et al. 2003). In vivo ANTXR2 is the major receptor mediating not only the pathology of toxemia but also the susceptibility to spore infection (Liu et al. 2009). The affinity of PA for either of these receptors is much greater than the physiological substrates, allowing PA to compete with natural ligands. The structure of ANTXR2 is known both in isolation (Lacy et al. 2004b) and in complex with PA monomer (Santelli et al. 2004) and the PA_7 prechannel (Lacy et al. 2004a). ANTXR interactions are key to assembling (Kintzer et al. 2010b) and stabilizing holotoxin complexes (Miller et al. 1999; Lacy et al. 2004a; Kintzer et al. 2010b, 2012). A crystal structure of ANTXR1 (Fu et al. 2010) has also been reported.

Channel Formation Upon endocytosis, the toxin-bearing vesicles traffic to the lysosome. Subsequent acidification of the endosomal compartment is essential to toxin action (Friedlander 1986). Lower pH conditions drive a conformational change in the PA oligomer, transforming it from the prechannel to the channel state (Blaustein et al. 1989; Miller et al. 1999; Kintzer et al. 2012). Overall, the putative channel is mushroom-shaped in its architecture (Fig. 9.1c), sharing much in common structurally with the known prechannel structures, notably the pore-facing loops localize similarly (Krantz et al. 2005). However, there are two major differences: first, the phenylalanine-clamp (ϕ-clamp) pore-facing loop, most critical to facilitating protein translocation, converges from a wider 25–30 Å diameter opening in the prechannel to a much narrower <10 Å diameter opening in the channel (Fig. 9.1d) (Krantz et al. 2005); and second, additional loops in a Greek-key motif unfurl and refold into a large ~100-Å-long, 14- or 16-stranded β barrel (Petosa et al. 1997; Nassi et al. 2002). Models of the β barrel (Nguyen 2004) are consistent with it being 12–15 Å in diameter, or about as wide as an α helix (Krantz et al. 2004), as seen in an autotransporter structure (Oomen et al. 2004).

LF/EF Unfolding and Translocation Due to the aforementioned steric constraints, translocation of folded full-length LF and EF necessitates unfolding (Krantz et al. 2004, 2005). Endosomal acidification creates a destabilizing environment for LF and EF's amino-terminal domains (LF_N and EF_N), facilitating their unfolding (Krantz et al. 2004). Furthermore, endosomal acidification establishes a proton motive force (PMF), which is comprised of the proton gradient chemical potential (ΔpH) and an electrical potential (Δψ) (Thoren et al. 2009; Brown et al. 2011). The ΔpH can drive the translocation of LF and EF (Krantz et al. 2006). Largely, in the endosomal context, where the latter Δψ is estimated to be minimal, it is the former ΔpH chemical gradient that drives translocation, and it was shown that the ΔpH on its own is sufficient (Brown et al. 2011). The ΔpH is critical to unfolding LF and EF as well as translocating their unfolded chains (Thoren et al. 2009). The mechanism of force transduction is largely consistent with a charge-state Brownian ratchet (Krantz et al. 2006; Basilio et al. 2009; Thoren et al. 2009; Pentelute et al. 2010; Pentelute et al. 2011; Brown et al. 2011; Wynia-Smith et al. 2012), albeit there are features of that model that do not explain all aspects of translocation, notably the high proton dependence of substrate unfolding (Thoren et al. 2009; Brown et al. 2011; Feld et al. 2012a). Following translocation into the cytosol, LF and EF refold and carry out their enzymatic functions, which serve to disrupt the host cell's normal physiology.

9.2.2 *γDPGA Capsule*

Pathogenic strains of *B. anthracis* also secrete a γDPGA virulence factor (Candela and Fouet 2006), which consists of long, linear polypeptides (50–200 kDa) of D-Glu linked via amide linkages between the γ-carboxylate and α-amino group of adjacent monomers (Bruckner et al. 1953). Several gene products, Cap A-E, on the pXO2 virulence plasmid produce the thick capsule layer of γDPGA that surrounds the bacillus. While γDPGA was first isolated from *B. anthracis*, the genes responsible for its synthesis were determined by homology to *B. subtilis* (Ashiuchi et al. 2001). Cap A, Cap B, Cap C, and Cap E form the membrane-embedded molecular machinery that produces and secretes γDPGA (Candela and Fouet 2005; Candela et al. 2005). Cap D, anchors γDPGA to the peptidoglycan cell wall, and it also cleaves γDPGA into dissociable fragments; these function to greatly augment virulence and allow the bacillus to evade host innate immune defenses (Uchida et al. 1993; Candela and Fouet 2005). During infection, serum levels of γDPGA polymers reach up to 1–2 mg ml^{-1} in vivo (Boyer et al. 2009). The individual anthrax toxin components are secreted through the hydrogel layer of γDPGA capsule surrounding the bacillus. The free capsule polymers can interact with PA in vivo (Ezzell et al. 2009), and PA, LF and EF in vitro, and these interactions can facilitate holotoxin assembly and stability (Kintzer et al. 2012). γDPGA polymers can be either toxin-activating (Jang et al. 2011) or toxin-deactivating (Kintzer et al. 2012) depending on the molecular weight of the γDPGA and its preparation procedures.

9.3 PA Oligomerization

Assembled anthrax toxin complexes have been studied for many years with some controversy surrounding their stoichiometry and heterogeneity. Nevertheless, recent research has determined two possible PA oligomer stoichiometries and demonstrated new biochemical methods to isolate and detect particular complexes for physiological and biophysical studies.

9.3.1 *PA Heptamer*

The earliest work showed that the 83-kDa PA component can be cleaved by trypsin to make nicked PA (nPA); and nicking yields 63-kDa and 20-kDa fragments, called PA_{63} and PA_{20}, respectively (Blaustein et al. 1989). PA_{63} was isolated from PA_{20} via mono Q anion-exchange chromatography; and PA_{63}, but not PA_{20}, formed cation-conducting channels in planar lipid bilayers (Blaustein et al. 1989). Anion-exchange purified PA_{63} also yields a stable oligomer, which was homoheptameric (PA_7) by electron microscopy (Milne et al. 1994). Furthermore, PA oligomers were also observed in cell extracts, although the stoichiometry of the cell-surface-assembled complexes was not characterized (Milne et al. 1994). The oligomer was subsequently crystallized initially to low resolution (Petosa et al. 1997) and then to 3.6-Å resolution (Lacy et al. 2004a) and shown in either case to be a homoheptamer. The crystal structure reveals a large surface comprised of the D1′ domain and the twin Ca^{2+}-ion-binding sites, where LF and EF were known to bind. There is extensively buried surface between D1′, D2, and parts of D3 between PA monomers. The inside of the oligomeric ring is defined by D2 and key pore loops shown to be critical for protein translocation via the PA channel. The D4 domains are located on the outside of the oligomer, and the D4 metal ion-dependent adhesion site, which is key for binding to the anthrax toxin receptor, is located on the bottom face opposite of the D1′ LF/EF-binding face (Lacy et al. 2004a).

To determine the PA:LF holotoxin co-complex stoichiometry, PA_7 oligomers were extensively characterized by radioisotope-labeling, light-scattering, and ultracentrifugation studies (Mogridge et al. 2002a; Wigelsworth et al. 2004; Phillips et al. 2013). To simplify matters, the PA-binding, amino-terminal domain of LF (LF_N) was used. When PA_7 oligomers form co-complexes with LF_N or LF the stoichiometry is consistent with 7:3 and the PA_7LF_3 complex (Mogridge et al. 2002a). Mutagenesis studies on PA revealed that binding interactions between PA and LF_N occurred at PA_2 dimer interfaces (Cunningham et al. 2002; Mogridge et al. 2002b), hence explaining the observed 7:3 stoichiometry. A prior study using native polyacrylamide gel electrophoresis, however, suggested that nPA and LF form co-complexes with a 1:1 stoichiometry (Singh et al. 1999).

9.3.2 PA Oligomer Heterogeneity

Electrospray ionization mass spectrometry (ESI-MS) determined that when nPA co-assembles with LF_N it forms two distinct oligomers, the well-known heptamer (PA_7) and a unique octamer (PA_8), at 7:3 and 8:4 PA:LF_N stoichiometries, respectively (Kintzer et al. 2009). Two even-numbered intermediates, PA_2LF_N and $PA_4(LF_N)_2$, were also observed by ESI-MS. The unique PA_8 species was not observed in prior radioisotope-labeling, light-scattering, crystallographic, and EM studies, because Q anion-exchange purified PA oligomers (the dominant form studied) are >95 % PA_7 and contain only a small percentage (~5 %) of PA_8 oligomers (Kintzer et al. 2009).

The PA_8 oligomer assembly mechanism was characterized further by EM and electrophysiology. EM was used to estimate the ratio of PA_7:PA_8 oligomers, which was ~2:1 when nPA was co-assembled with LF_N, EF_N, or dimeric anthrax toxin receptor (Kintzer et al. 2009). Moreover, when carboxy-terminal His_6-tagged PA was assembled upon Chinese hamster ovary (CHO) cells and subsequently purified by His_6-affinity chromatography, it was also found that it formed a similar 2:1 ratio of PA_7:PA_8. Planar lipid bilayer electrophysiology studies showed that PA oligomer samples enriched in the PA_8 species revealed a 20–30 % subpopulation of channels that had ~10 % greater conductance (Kintzer et al. 2009). These larger channels translocate LF_N, EF_N, LF and EF at similar rates to the PA_7 oligomer, revealing PA_8 oligomers are physiologically active translocases.

9.3.3 PA Octamer

The PA_8 crystal structure was solved to 3.2 Å using a membrane-insertion loop deleted construct ($PA^{\Delta MIL}$) (Kintzer et al. 2009). $PA^{\Delta MIL}$ unlike wild-type PA enabled purification of a high population (28 %) of free PA_8 oligomers by Q anion-exchange. A unique property of the PA_8 complex is that is it more pH and thermostable than the PA_7 complex, and hence $(PA^{\Delta MIL})_7$ complexes were removed from $(PA^{\Delta MIL})_8$ by precipitating them away from a heterogenous mixture by treatment with ~10 % ethanol and mildly acidic conditions (pH 5.7). This procedure yielded homogeneous $(PA^{\Delta MIL})_8$ complexes that crystallized in a square-planar architecture. $(PA^{\Delta MIL})_8$ contains more extensively buried interface between adjacent PA subunits than the PA_7 structure, when accounting for the difference in the MIL between the two constructs. Based on the known PA_7 oligomer structure, it was suspected that the interaction of the MIL with neighboring PA subunits at the D4-D2 inter-PA interface could alter the oligomeric stoichiometry. This hypothesis was tested by chemically cross-linking the D4-D2 intra-PA interface. Depending upon the cross-link length, different oligomeric ratios could be produced, which favored the PA_8 architecture (Feld et al. 2012b). An innovative synthetic octamerization strategy was devised by mutating the known D2-D3 oligomerization interface (Mogridge et al. 2001) in two separate PA monomer constructs, which could form a mutant, but functional, PA-PA interface (Phillips et al. 2013). When these two mutant PA are combined then only the even-numbered PA_8 oligomers can form. This method works well and supports the physiological activity previously reported for WT PA_8 complexes (Kintzer et al. 2010a).

9.3.4 Unique Serum Stability of PA_8

Currently, WT PA_8 oligomers can only be isolated as co-complexes with LF_N, EF_N, LF and EF. For example, a heterogenous 2:1 mixture of $PA_7(LF_N)_3$ and $PA_8(LF_N)_4$ is first produced by co-assembling nPA with LF_N at pH 8, 0 °C. Then the mixture is pH- and temperature-challenged at pH 7, 37 °C, causing the $PA_7(LF_N)_3$ complexes to precipitate as prematurely formed, insoluble channels (Kintzer et al. 2010a). EF_N, LF and EF co-complexes with PA_8 can be produced by ligand exchange using purified $PA_8(LF_N)_4$ and an excess of the former (Kintzer et al. 2010a, 2012). Due to this stability difference, PA_8LF_4 predominated over PA_7LF_3 complexes in defibrinated bovine serum at physiological pH and temperature (37 °C) (Kintzer et al. 2010a). Moreover, the macrophage cytolysis activities of the PA_7LF_3 and PA_8LF_4 complexes were compared after incubations in defibrinated bovine serum finding that the PA_8LF_4 complex exhibited a 30-min half life, whereas the PA_7LF_3 complex was significantly less stable with a <5 min half life (Kintzer et al. 2010a). Hence, PA_8 toxin complexes are more thermostable than the PA_7 toxin complexes under conditions encountered in mammalian serum.

On cell surfaces, by contrast, PA oligomer complexes contact the soluble integrin-like domain of the anthrax toxin receptor (Lacy et al. 2004b). This interaction occurs via ANTXR2 and D2 and D4 of PA (Lacy et al. 2004a; Santelli et al. 2004). Since D2 and D4 are effectively stapled together by the ANTXR2 interaction, PA is unable to form the channel state readily unless the pH is dropped to about 5.5 (Lacy et al. 2004a). Receptor stabilization is roughly equivalent for PA_7 and PA_8 complexes; therefore, the stabilization advantage of the PA_8 architecture over the PA_7 architecture resides only within the milieu of the serum (Kintzer et al. 2010b). Since all of the observed PA in the serum of animal infection models late in infection is proteolytically activated as PA_{63} (Ezzell and Abshire 1992; Panchal et al. 2005; Ezzell et al. 2009), then significant serum-based assembly of anthrax toxin is possible, and octamer formation in the serum is another potential route of anthrax toxin assembly (Kintzer et al. 2009, 2010b). Detailed examination of serum-based assembly and the oligomeric architecture of toxin complexes in anthrax infection models are future areas of research.

9.4 Holotoxin Assembly

LF and EF bind to PA through a homologous ~250-residue PA-binding domain (referred to as LF_N and EF_N, respectively), and these domains share 35 % sequence identity and 55 % similarity (Quinn et al. 1991). The stoichiometry of assembled co-complexes with LF, EF, LF_N, or EF_N depends on the identity of the PA oligomer, where PA_7 binds three LF and/or EF (Mogridge et al. 2002a; Wigelsworth et al. 2004; Kintzer et al. 2009, 2010a) and PA_8 binds four LF and/or EF (Fig. 9.1b) (Kintzer et al. 2009, 2010a; Feld et al. 2010). Heterologous fusions to the C-terminal end of the PA-binding domain, LF_N, was sufficient to deliver proteins into host cells in a PA-dependent manner, suggesting that the domain is necessary and sufficient for translocation (Arora and Leppla 1993; Milne et al. 1995). However, later studies, determined that amino-terminal His_6, poly-lysine, and poly-arginine tags were also sufficient to deliver heterologous proteins into cells in a PA-dependent manner, albeit higher concentrations of heterologous substrate were required than with the LF_N fusions (Milne et al. 1995; Blanke et al. 1996). While not understood at the time, these seemingly opposing sets of results revealed that PA oligomers contain two distinct binding subsites for LF and EF. One subsite was the tight-affinity site specific for the homologous folded domains of EF_N and LF_N, and the other subsite was a lower-affinity site that was *far less specific* and could bind unstructured peptides (like His_6 tags).

LeTx Prechannel Core Complex Structure Many subsequent mutagenesis studies confirmed that the binding sites comprised of specific contacts between PA and LF_N and EF_N (Cunningham et al. 2002; Lacy et al. 2002; Mogridge et al. 2002b;

Lacy et al. 2005; Melnyk et al. 2006; Feld et al. 2010). Hydrogen exchange mass spectrometry (Melnyk et al. 2006) and rigid-body computational docking algorithms (Schueler-Furman et al. 2005; Lacy et al. 2005) were used to deduce the structure of a co-complex between PA oligomers and LF_N. These models did not resolve why poly-cationic sequences (by themselves) were sufficient to localize heterologous proteins to PA oligomers. Also the computationally-docked model (Lacy et al. 2005) did not reconcile why PA point mutations (notably R178A) disrupted LF_N binding (Cunningham et al. 2002).

The PA_8 oligomer was exploited as a structural platform to answer these questions due to its enhanced thermostability and C4 symmetry. The structure (Fig. 9.1b) solved to 3.1 Å was representative of the $PA_8(LF_N)_4$ co-complex (Feld et al. 2010). Unlike the isolated structure of LF_N (Pannifer et al. 2001), the LF_N subunits in the $PA_8(LF_N)_4$ co-complex are in a partially unfolded conformation. This co-complex is one of the few examples, where the unfolding process has been visualized. The first α helix (α1) and β strand (β1) of LF_N are in a partly unfolded conformation that is dissociated about 40 Å from the remainder of the folded structure (Feld et al. 2010). Hence, there are two distinct binding subsites for each LF_N on the surface of the PA_8 oligomer. In one subsite, the $\alpha 1/\beta 1$ portion of LF_N docks into a deep cleft between PA subunits, called the "α clamp"; and in the other subsite, the carboxy-terminal portion of of LF_N binds in a separate subsite contained within a single PA subunit (Feld et al. 2010). While the latter subsite was well defined in mutagenesis and computer modeling studies (Lacy et al. 2005), the former α-clamp site was not. This discrepancy was due to the fact the modeling approach employed (Schueler-Furman et al. 2005) did not allow for conformational changes, such as partial unfolding.

The walls of the α-clamp cleft are comprised of the twin Ca^{2+} binding sites, which are fully-conserved within all known PA homologs (in other *Clostridium* and *Bacillus* binary toxins). The twin Ca^{2+} binding sites are such that their side chains are pointed inward toward the Ca^{2+} ions, leaving mostly backbone surface area contacts with the α1 helix. By virtue of the larger than average backbone exposure, the interactions with the α clamp are very nonspecific (Feld et al. 2010). The proposed mechanism of nonspecific helix recognition is consistent with the broad substrate specificity of calmodulin, a well known adapter, which uses a pair of twin Ca^{2+} sites to recognize α helices (Meador et al. 1992, 1993; Shen et al. 2005). Hence nonspecific and nonhomologous poly-cationic sequences also bind to the α-clamp cleft remarkably well (Feld et al. 2010), explaining prior observations that poly-cationic tracks are necessary and sufficient for heterologous protein delivery (Blanke et al. 1996).

9.5 PA Channel

Recent biophysical efforts have elucidated the basic structural features of the PA channel (Fig. 9.1c, d). The suggested resemblance of the PA channel to staphylococcal α hemolysin (Song et al. 1996; Benson et al. 1998) was based on morphological similarities not homology. The PA channel also possesses a ring-shaped oligomeric architecture, is SDS-resistent, and utilizes a hollow β-barrel domain to penetrate the membrane (Benson et al. 1998; Nassi et al. 2002). From modeling studies (Nguyen 2004), the β-barrel stem is extended, and it likely can accommodate structures as wide as an α helix (10–15 Å-wide) (Krantz et al. 2004; Oomen et al. 2004). The inward-facing pore loops within the soluble prechannel oligomer are also inward-facing within the cap of the channel (Krantz et al. 2005). By EM, the PA_7 channel is indeed mushroom-shaped and approximately 170 Å tall × 125 Å wide at its maximum dimensions (Fig. 9.1c) (Katayama et al. 2008). The wider, cap is about 70 Å long, and like the prechannel, the topmost surface contains the LF/EF binding sites. The 100 Å long stem beneath the cap inserts into the membrane bilayer (Fig. 9.1d) (Katayama et al. 2010).

9.5.1 Peptide Clamps as Unfoldase and Translocase Active Sites

Functional electrophysiology studies laid the early groundwork in mapping the major active sites in the PA translocase, referred to collectively as "peptide clamps" (Fig. 9.1d) (Thoren and Krantz 2011; Feld et al. 2012a). These peptide-clamp sites have been empirically shown to catalyze the translocation of LF and EF into cells (Krantz et al. 2005; Feld et al. 2010; Wynia-Smith et al. 2012). Based on the locations of these clamps, the PA translocase channel can be divided into three sections: (i) the substrate docking surface on the top of the cap, which contains the "α clamp" (Feld et al. 2010); (ii) a critical hydrophobic constriction point with at opening near the base of the cap containing a ring of F427 residues, called the "ϕ clamp" (Krantz et al. 2005); and (iii) the highly anionic charged region in the solvophilic β-barrel stem portion, called the "charge clamp" (Wynia-Smith et al. 2012). Nonspecific substrate recognition is a key feature of a peptide translocase, because as the protein translocates, the segment of polypeptide sequence within the transporter will change continuously during translocation (Krantz et al. 2005; Thoren and Krantz 2011; Brown et al. 2011; Feld et al. 2012a; Wynia-Smith et al. 2012). In general, these clamps allow the PA channel to interact with a translocating polypeptide substrate via broad chemical or morphological features, imparting a high degree of nonspecificity (Thoren and Krantz 2011; Feld et al. 2012a). It is no surprise, therefore, that functional clamp sites are observed throughout the length of the channel.

Peptide-Clamp Paradox The observed peptide clamp sites (Fig. 9.1d) may on their surface appear paradoxical: on one hand tight binding to clamp sites may impede protein translocation; on the other, when clamp sites are functionally disrupted, protein unfolding and translocation are kinetically inaccessible at physiological driving forces (Krantz et al. 2005; Feld et al. 2010; Thoren and Krantz 2011; Brown et al. 2011; Wynia-Smith et al. 2012). Because functioning clamp sites lower the activation energy of protein translocation, then these sites likely bind and release the translocating chain in a dynamic manner, much like an enzyme binds its substrate and releases product during catalysis. A protonation and deprotonation dynamic of the substrate, for example, could modulate interaction between the channel and a charge-clamp site. Under these conditions, a clamp site, therefore, is not a static binding site found in a more simplistic antibody-epitope interaction. Rather, a clamp site more closely resembles a dynamic binding site for polypeptide, where, for example, a clamp site can be modulated from a higher-affinity binding mode to a lower-affinity binding mode. Considering the structural plasticity of polypeptides and molecular machines, these putative dynamics are reasonable.

9.5.2 Transition to the Channel State

As the details of the structure of the PA channel have become apparent, much interest has been given to the acidic pH-induced conformational change from the prechannel to the channel (Bann 2012). Mutagenesis studies established a series of dominant-negative inhibitor (DNI) point mutations in PA that could allow for PA assembly into prechannel oligomers but would prevent formation of the channel state (Sellman et al. 2001b; Mourez et al. 2003). These possess therapeutic potential as a single copy of a DNI PA in an otherwise wild-type PA oligomer rendered the toxin harmless in vitro (Janowiak et al. 2009) and in a toxemia animal model (Sellman et al. 2001a). Hence DNI mutations provided the insight that certain molecular interactions occur in the channel that are less critical to prechannel oligomerization. 2-fluorohistidine substitution, however, did not effect the pH dependence of PA channel formation (Wimalasena et al. 2007), contrary to prior expectations that His residues sensed endosomal acidification to trigger the channel transition (Blaustein et al. 1989; Petosa et al. 1997).

Among the DNIs, the more notable D425K, F427A, and K397D mutations localize within the known PA_7 and PA_8 prechannel structures in adjacent pore-facing loops: the ϕ-clamp loop (2β10-2β11 loop) and the 2β7-2β8 loop. Biophysical characterizations of these sites demonstrate considerable inward narrowing and convergence of these loops as the channel forms. Electron paramagnetic resonance spectroscopy showed Cys-spin-labeled ϕ-clamp sites collapsed to a narrow pore of dimensions <10 Å (Krantz et al. 2005). Furthermore, bilayer-electrophysiology (Krantz et al. 2005), circular-dichroism (CD) spectroscopy (Kintzer et al. 2012), and liposome-pore-formation (Sun et al. 2008) assays revealed many mutations (such as F427A) slowed, but did not prevent, channel formation, whereas others (F427G, F427D, and F427R) blocked pore formation (Sun et al. 2008). Phylogenetic comparisons to the *Clostridium* binary toxins revealed covariance at two positions (D426Q and K397Q). PA K397Q channels were less functional translocases, but they could be complemented with the D426Q mutation, suggesting that the ϕ-clamp site and the 2β7-2β8 loop engaged each other through inter-PA contacts (Melnyk and Collier 2006). In summary, a major step in channel formation is the collapse and assembly of the ϕ clamp and adjacent 2β7-2β8 loop.

ϕ-clamp assembly, however, does not appear to be the only rate-limiting aspect of the mechanism of channel formation. The interfacial D2-D4 contacts on the periphery also must be disrupted not only within each individual PA subunit (intra-PA) but also between the membrane-insertion loop and a D4 in an adjacent PA subunit (inter-PA) (Kintzer et al. 2012). The intra-PA D2-D4 contacts occur in a water-coordinated interface between the two domains, where mutations in the interface can increase the pH threshold for channel formation in bilayers and by CD (Kintzer et al. 2012). Consistent with this view ANTXR2 (Lacy et al. 2004a; Sun et al. 2007; Kintzer et al. 2010b, 2012) or γ-DPGA (Kintzer et al. 2012) binding to PA oligomers stabilizes the intra-PA D2-D4 interface, and either of these ligands can significantly limit the rate of channel formation. Not surprisingly the loss of structure in intra- and inter-D2-D4 contacts is needed, because the membrane-insertion loops (2β2–2β3) situated between D2 and D4 must detach, unfold, and refold into the membrane-spanning β-barrel. The final step of PA channel formation involves the penetration of the β-barrel across the hydrophobic lipid bilayer as deduced from the D315A point mutation, which does not block channel formation in solution by CD, but does block insertion into planar lipid bilayers (Kintzer et al. 2012).

Thus while other pathways have been suggested (Bann 2012), a possible mechanism of prechannel-to-channel conversion (Kintzer et al. 2012) first requires the detachment of the various intra- and inter-D2-D4 interfaces in the oligomer; this dissociation or unfolding is followed by the collapse and proper organization of the ϕ-clamp site; an organized ϕ-clamp site then initiates the folding of the β-barrel; lastly, the barrel must penetrate the bilayer to form a functional translocase channel. The logic of the proposed scheme is not only based on functional mutagenesis studies but also on the idea that the prechannel must partially unfold before it can refold into the channel.

9.6 Translocation

Anthrax toxin is an excellent model system to study transmembrane protein translocation in part because the three subunits of the toxin can be expressed recombinantly and worked on in isolation. PA channels can be studied using whole-cell (Wolfe et al. 2005) and planar lipid bilayer electrophysiology (Finkelstein 2009) using a painted-membrane technique (Mueller et al. 1963). Single PA channels, when inserted into planar bilayers, form an open and ungated cation-conducting channel at modest potentials (Blaustein et al. 1989, 1990; Krantz et al. 2005). But when small molecule blocking ions (Blaustein and Finkelstein 1990; Blaustein et al. 1990; Orlik et al. 2005; Krantz et al. 2005) or protein substrates, such as LF_N, are added, the conductance is blocked (Zhang et al. 2004a, b; Krantz et al. 2006; Thoren et al. 2009; Basilio et al. 2011b). Upon raising the driving force, substrate translocation may be monitored as the time-dependent increase in conductance. In an LF_N translocation experiment under symmetrical pH 5.6, PA_7 oligomer is added to the cis side to form channels; upon stabilization of the current, the cis compartment is perfused; LF_N is added to block conductance and the cis compartment is prefused a second time, removing excess LF_N; and finally the voltage is raised, driving LF_N unfolding and translocation (Fig. 9.2a). Translocation is observed as PA channel conductance is restored and can be replotted to show a kinetic time course (Fig. 9.2b). With this method, the applied driving force can be externally controlled and continuously adjusted to identify key barriers in the translocation pathway (Zhang et al. 2004b; Krantz et al. 2006; Thoren et al. 2009; Thoren and Krantz 2011; Brown et al. 2011; Wynia-Smith et al. 2012).

A second and perhaps more powerful technique has emerged to study protein translocation at the single-channel level (Kintzer et al. 2009; Thoren et al. 2009; Basilio et al. 2011b). With this method, a single PA channel can be inserted into a planar lipid bilayer, and peptide or protein substrate translocations can be inferred from the step-like openings and closings of the channel (Fig. 9.2c). Dwell times (τ) can be determined for when the channel is closed (τ_C) or open (τ_O). The τ_C values report on unblocking processes, including translocation and dissociation of the protein from the channel, whereas τ_O values report on the binding kinetics of the substrate to the channel. While individual measured τ values are stochastic, they can be analyzed statistically using the cumulative distribution function (CDF) (Fig. 9.2d). When combined with a driving-force analysis, translocation can be studied independently of dissociation. Single-channel assays have been used to determine pore size of the PA channel to deduce relative populations of PA_7 and PA_8 oligomers and determine their ability to translocate proteins (Kintzer et al. 2009); they have been used to observe transient intermediates during docking (Brown et al. 2011) and translocation (Thoren et al. 2009); and they have been employed to dissect the unusual S-shaped kinetics of translocation (Basilio et al. 2011b). The only disadvantage of single-channel analyses of translocation is the time investment of the investigator in the assays. Otherwise, there are tremendous advantages: single-channel assays, for example, can elucidate more complex chemical kinetic mechanisms, observe intermediate conductance states, and determine the order of transient kinetic intermediates.

Topologically, LF and EF initiate translocation starting from their unstructured 20–30 residue-long amino-terminal ends (Zhang et al. 2004a). Albeit when polycationic tracks are added to the carboxy-terminal end of LF_N or heterologous fusions (Milne et al. 1995), the protein can also enter the PA channel and translocate into cells at 10- to 100-fold reduced efficiency (Sharma and Collier 2014). Nevertheless, because the amino-terminal ends are unstructured, it is unlikely LF and EF translocate to any significant extent from their carboxy-terminal ends, since the latter would require their unfolding to produce significant unstructured carboxy-terminal peptide to initiate translocation. Initiation (or docking) is charge and sequence (Pentelute et al. 2010, 2011; Brown et al. 2011; Wynia-Smith et al. 2012); length (Zhang et al. 2004a); pH (Brown et al. 2011); and PA-channel dependent at both the ϕ- (Krantz et al. 2005) and α-clamp sites (Feld et al. 2010). Empirically, LF, EF, LF_N and EF_N translocation kinetics are highly complex and cannot be fit to

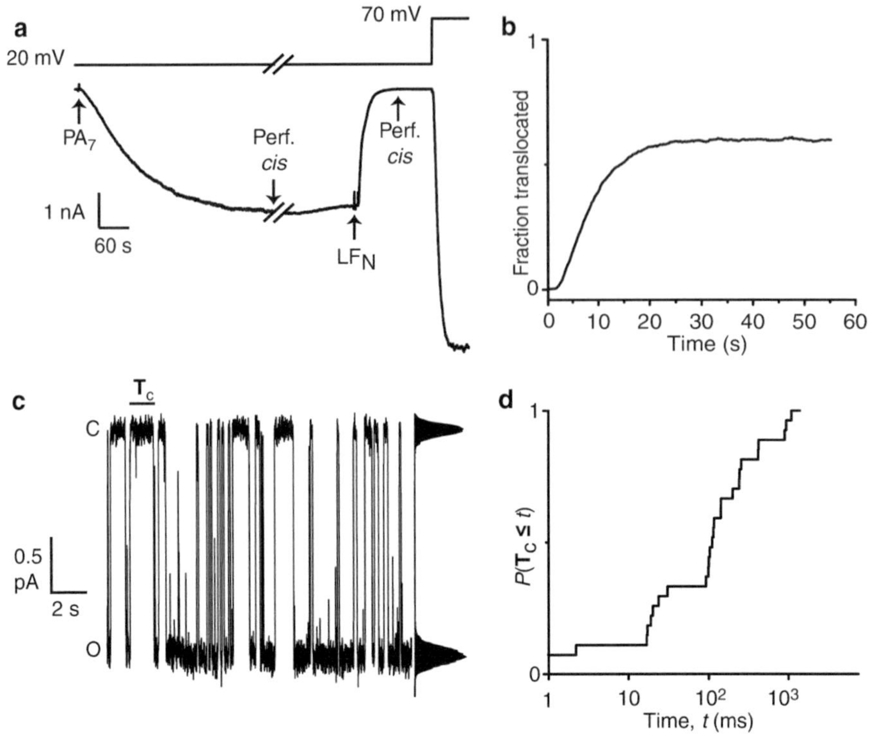

Fig. 9.2 Electrophysiological studies of translocation. (**a**) Ensemble planar lipid bilayer translocation experiment is shown for a bilayer formed between 1 mL chamber volumes and bathed in symmetric universal bilayer buffer (UBB) at pH 5.6 (Thoren et al. 2009). PA_7 channel formation was recorded as the increase in conductance at a $\Delta\psi$ of 20 mV (cis positive). Upon stabilization, the cis side was perfused (perf.) with 10 mL of fresh UBB. Then LF_N was added to the cis side, causing an exponential decrease in current as LF_N docks inside the PA channels. Following a second perfusion of the cis side, the voltage was raised, driving the translocation of LF_N. (**b**) To better observe the S-shaped translocation kinetics, the translocation phase from panel (**a**) is replotted. Time zero is set to the time when $\Delta\psi$ was raised to 70 mV. Fraction translocated is given as the level of current observed divided by the expected current, assuming a linear current-voltage relationship. (**c**) A single-channel translocation experiment under a 2-unit ΔpH and $\Delta\psi$ of 0 mV in asymmetrical KCl and pH conditions: cis buffer was 100 mM KCl, pH 5.6; trans buffer was pH 7.6 with no added KCl. A synthetic peptide (LF residues 1–50) was added to a single PA channel, and two-state opened (*O*) and closed (*C*) channel conductances were observed where a closed dwell time (τ_C) is highlighted. (**d**) τ_C from the recording in panel (**c**) were statistically analyzed as the CDF, $P(\tau_C \leq t)$, or probability a τ_C is less than or equal to time, t

simple single-exponential functions; rather these kinetics are S-shaped (Fig. 9.2b) and indicative of many consecutive steps (Zhang et al. 2004b; Krantz et al. 2005, 2006; Thoren et al. 2009; Brown et al. 2011; Wynia-Smith et al. 2012). The relative lifetime of these complex kinetic records has been quantified by measuring the time for half of the substrate to translocate ($t_{½}$) (Krantz et al. 2006).

9.6.1 Driving Forces and Barriers

For anthrax toxin, protein unfolding and translocation are driven by the PMF comprised of a $\Delta\psi$ and a ΔpH (Thoren et al. 2009). The electrical potential, $\Delta\psi$, relates to the activation energy of translocation, $\Delta G^{\ddagger}(\Delta\psi)$

$$\Delta G^{\ddagger}(\Delta\psi) = \Delta G^{\ddagger\circ} + zF\Delta\psi \tag{9.1}$$

$\Delta G^{\ddagger\circ}$ is the $\Delta G^{\ddagger}$ when $\Delta\psi$ is 0 mV; F is Faraday's constant; and z is the charge required to cross the limiting barrier. Similarly, a chemical potential-modulated energy barrier, $\Delta G^{\ddagger}(\Delta pH)$, relates to ΔpH

$$\Delta G^{\ddagger}(\Delta pH) = \Delta G^{\ddagger\circ} + 2.3nRT\Delta pH \tag{9.2}$$

$\Delta G^{\ddagger\circ}$ is thc $\Delta G^{\ddagger}$ when ΔpH is 0 units; n is the number of protons used to cross the limiting barrier; R is the gas constant; and T is the temperature. $\Delta G^{\ddagger}(\Delta\psi, \Delta pH)$ values can be estimated by computing $t_{1/2}$ values (in seconds) extracted from primary kinetic records under different applied potentials, $\Delta\psi$ or ΔpH, where c is a one-second constant

$$\Delta G^{\ddagger}(\Delta\psi, \Delta pH) = RT \ln t_{1/2}/c \tag{9.3}$$

In principle and in practice, either of these driving forces can affect the activation energy of the two major barrier-crossing steps in the mechanism of translocation (Thoren et al. 2009). A pure $\Delta\psi$ can drive the translocation of the small substrate, LF_N (Zhang et al. 2004b; Krantz et al. 2005, 2006; Thoren and Krantz 2011). However, EF_N, LF, and EF translocate poorly under a pure $\Delta\psi$ (Krantz et al. 2006; Wynia-Smith et al. 2012), and efficient translocation of these substrates also requires a ΔpH (Krantz et al. 2006; Kintzer et al. 2009; Feld et al. 2010). A diode-like rectifying blockade of PA channels with LF was reported under a pure $\Delta\psi$, even at high driving potentials, demonstrating inefficient translocation under that condition (Halverson et al. 2005). The ΔpH by itself, on the other hand, is sufficient for translocation, and mutagenesis of LF_N suggests protonation and deprotonation of Asp and Glu residues play a key role in harnessing the ΔpH (Brown et al. 2011). Hence the PA channel is a proton/protein symporter (Krantz et al. 2006; Brown et al. 2011; Wynia-Smith et al. 2012).

Overall, in the translocation mechanism, there are two major limiting barriers (Thoren et al. 2009). In $\Delta G^{\ddagger}(\Delta\psi, \Delta pH)$ versus driving force ($\Delta\psi$ or ΔpH) plots, the observed curves are not linear as would be expected for a simple single-barrier crossing kinetic mechanism. (Of course, a single-barrier model cannot be expected to fully describe translocation records that are S-shaped.) Instead, $\Delta G^{\ddagger}$ versus $\Delta\psi$ or ΔpH plots are boomerang-shaped with two asymptotic linear extremes, indicating that there are two major limiting barriers in the translocation mechanism (Thoren et al. 2009). One barrier has a high driving-force dependence, i.e., a high z value or n value for $\Delta\psi$-dependent and ΔpH-dependent translocation, respectively. The second has a ~10-fold lower diving-force dependence. A two-major-limiting-barriers model was fit using mean-transit-time theory for two sequential first-order reactions (Fersht 1998; Thoren et al. 2009). From a series of destabilizing point mutations in LF_N, the high driving-force-dependent barrier was shown to be limited by protein unfolding, but the lower-force-dependent barrier was not dependent on unfolding. This two-barrier model pertains only to the steps following substrate docking and initiation as deduced from single-channel experiments (Thoren et al. 2009). The oversimplification of the two-major-limiting-barriers model is that the folding and translocation processes could have multiple similar height barriers in succession, which is likely the case for the latter lower-force unfolding-independent translocation barrier. (See arguments below).

Barrier-less Drift-diffusion Model of Translocation An alternative barrier-less model of translocation, the drift-diffusion model, has been applied to LF_N translocation (Basilio et al. 2011b). Based on electrodiffusion, the model represents LF_N as a rigid and positively charged rod dominated by Brownian motion and an applied $\Delta\psi$. Over a narrow $\Delta\psi$ range (45–60 mV) the model fits the S-shaped single-channel translocation records remarkably well. Yet, fits to the semi-natural-log $t_{1/2}$ versus $\Delta\psi$ plot are crudely linear. By comparison, when a larger dynamic range of $\Delta\psi$ records (from 30 to 150 mV) are analyzed, however, the empirical data are not linear on a semi-log plot ($RT \ln t_{1/2}$ versus $\Delta\psi$), but rather the empirical relation is boomerang shaped with two linear extrema (Krantz et al. 2006; Thoren et al. 2009; Brown et al. 2011; Wynia-Smith et al. 2012). The experimental results of Basilio (Basilio et al. 2011b) are important in that they imply the post-unfolding step(s) contain many small barriers. But, the barrier-less drift-diffusion model, while a nice addition, is overly simplified, since it ignores unfolding, a process shown to limit translocation (Wesche et al. 1998; Zhang et al. 2004b; Thoren et al. 2009).

Toward a Unifying Barrier Model A unifying interpretation is that translocation is universally barrier limited. This hypothesis is cogent with myriad of biochemical processes: protein folding and unfolding, peptide binding and dissociation, and enzyme catalysis (Fersht 1998). At low potentials <50 mV, the major limiting barrier(s) is (are) dominated by unfolding (Thoren et al. 2009; Brown et al. 2011). This notion is evident in that destabilizing point mutations shift the observed $\Delta G^{\ddagger}$ versus $\Delta\psi$ curves at voltages <50 mV but not at higher potentials (Thoren et al. 2009). At higher potentials, translocation is not limited by unfolding, but rather it is logically hypothesized to be limited by the movement of the unfolded chain through the channel (Thoren et al. 2009). To maintain S-shaped kinetics, i.e., when the latter translocation barrier is limiting at higher potentials, a unifying picture may be that translocation requires the crossing of many equivalent height barriers, which must be traversed in serial/consecutive order. Kinetic schemes comprised of a series of irreversible steps with similar lifetimes may be analyzed using a gamma-function (Feller 2008), and these models yield lag-phase S-shaped kinetics (Floyd et al. 2008). A multi-barrier-limited interpretation also explains the use of an extremely slow diffusion constant (D) of

10^{-11} cm s^{-1} in the drift-diffusion model (Basilio et al. 2011b) (much slower than the D of 10^{-8} cm s^{-1} expected for free diffusion through a narrow pore (Simon et al. 1992; Chauwin et al. 1998)). The slower D estimated by the drift-diffusion model for PA suggests the channel binds and releases the peptide chain (likely at the peptide-clamp sites) in a repetitious manner along the length of the protein substrate (Krantz et al. 2005; Thoren and Krantz 2011; Feld et al. 2012a). Consistently, other slow D constants of 10^{-15} cm s^{-1} have been estimated for mitochondrial protein import and were rationalized as being due to peptide-pore interactions (Chauwin et al. 1998). In summary, binding and release steps are barrier-limited events immediately analogous to peptide-protein binding and dissociation reactions.

A unifying model of unfolding and translocation will be broadly relevant to understanding transmembrane protein translocation in the cell and mitochondrion (Simon et al. 1992; Chauwin et al. 1998; Huang et al. 1999) and the unfolding and translocation by ring-shaped oligomeric ATP-dependent disaggregase, unfoldase, and degradase molecular machines (Burton et al. 2001; Kenniston et al. 2003). At the very highest potentials, the observed rate of PA translocation begins to plateau in a way consistent with the maximal rates observed in many other ATP-dependent molecular machines that unfold and translocate proteins in the cell (Thoren et al. 2009). Of particular note, single-molecule translocation studies, which measure displacements and forces, show step-wise displacement, consistent with nm-length movements during translocation (Aubin-Tam et al. 2011; Maillard et al. 2011). Step-wise displacements in the anthrax toxin translocation mechanism would manifest in a series of consecutive barrier crossings and hence the empirically observed S-shaped kinetics.

9.6.2 Translocation-Coupled Unfolding

The earliest indication that protein unfolding limited protein translocation came from cellular studies of anthrax toxin translocation using heterologous LF_N fusions (Wesche et al. 1998). Addition of fold-stabilizing ligands and disulfide bonds further supported this view that unfolding was required prior to translocation (Wesche et al. 1998; Zhang et al. 2004b; Juris et al. 2007). Protons, on the other hand, are likely a key physiological ligand that serves to assist in unfolding LF_N and EF_N through the destabilization of their native states (Krantz et al. 2004). Correspondingly, $\Delta G^{\ddagger}$ values for translocation at low driving forces (~50 mV) correlate with equilibrium changes in stability (Krantz et al. 2006). Because the unfolding-limited barrier coincides with the force-dependent barrier, the unfolding transition state (TS) was mapped onto the structure of LF_N using 21 different destabilizing point mutations (Thoren et al. 2009). Upon applying protein folding ϕ analysis (Fersht 1998), it was found that the unfolding TS is asymmetrically localized in the β-sheet subdomain of LF_N; hence, the β-sheet subdomain is the "mechanical breakpoint" (Thoren et al. 2009) (according to nomenclature used in mechanical unfolding studies (Crampton and Brockwell 2010)). Once the rate-limiting β-sheet-subdomain structure is ruptured the rest of the LF_N unfolds (Thoren et al. 2009).

The aforementioned translocation-coupled unfolding mechanism deduced from the study of the destabilized LF_N mutants (Thoren et al. 2009) is also supported well by the crystal structure of the $PA_8(LF_N)_4$ prechannel core complex (Feld et al. 2010). The $\alpha1$ helix and $\beta1$ strand of LF_N are unfolded in that prechannel structure, and this partially unfolded starting state was verified in the channel-LF_N co-complex using planar bilayer electrophysiology (Feld et al. 2010). One interesting destabilizing mutation in LF_N (M40A), the only outlier in that study, exhibited an unusually negative translocation-coupled unfolding ϕ value (Thoren et al. 2009). The negative ϕ value was indicative of slow translocation despite the fact that LF_N M40A is destabilized. This residue is in the center of the $\alpha1$ helix, which is situated in the α-clamp site on the surface of the $PA_8(LF_N)_4$ structure. Mutations in this site affect the stability of the co-complex, where M40A caused LF_N to bind more tightly to the PA channel (Feld et al. 2010). Tight binding correlated with slower translocation for multiple substitutions at M40. Furthermore, B-factor analysis and fluorescence anisotropy studies corroborated that the partial unfolding of the $\alpha1/\beta1$ structure leads to increased disorder in the β-sheet subdomain (Feld et al. 2010). Therefore, prior to translocation, these interactions with the α clamp initiate some destabilization to the key rate-limiting structure in the LF_N substrate.

9.6.3 Brownian-Ratchet Mechanism

A Brownian-ratchet (Astumian 1997), or charge-state-ratchet (Krantz et al. 2006), model of translocation is one possible explanation for how the ΔpH may be harnessed by the PA channel to produce physical force on the substrate protein (Fig. 9.3a). In this mechanism, acidic residues in the substrate protonate on the lower pH side upon entering the PA channel. Protonation is required because the channel is cation-selective (Blaustein et al. 1989) and hence anion-repulsive to deprotonated Asp/Glu residues (Krantz et al. 2006). Brownian motion then dominates the chain within the channel allowing it to move past the major cation-selective site. As protonated acidic groups emerge beyond the major cation-selective (i.e., anion-repulsive) site in the channel via Brownian motion, they then deprotonate down gradient to the higher pH side, thus reforming a net anion-anion repulsion between the chain and the channel. Therefore, the asymmetric ΔpH condition effectively enforces the direction of protonation and deprotonation, providing directionality to translocation.

Empirical evidence for the ΔpH-driven charge-state ratchet has come from 2-sulfonato-ethyl-methanethiosulfonate (MTSES) modification of Cys residues or semisynthetic chemistry, either of which added an SO_3^- modification to LF_N (Basilio et al. 2009; Pentelute et al. 2010). These modifications are effectively non-protonatable anionic sites at physiological pH. Many, but not all, of these SO_3^--modified LF_N showed significant decreases in the rate of translocation and suggested anionic sites were blocked by the strong anion-repulsive features of the channel. The positional effects of these SO_3^- modifications were complex, appearing to combine sterics and local electrostatics (Basilio et al. 2009). Additional semisynthetic peptide ligation studies showed positive and negative charges were required to utilize the ΔpH (Pentelute et al. 2011).

Detailed mutagenesis of LF_N showed replacement of all anionic residues on the amino terminus up to residue 85 slowed the relative rate of ΔpH-driven translocation 100–1000 fold (Brown et al. 2011). An LF_N with an uncharged amino terminus (i.e., no D, E, H, K, or R residues) was similarly deficient, but when only Asp/Glu residues were reintroduced at various positions, especially between residues 20 and 30 (the "20s cassette"), then translocation could be restored. Reintroduction of positively charged residues on their own did not stimulate ΔpH-driven translocation. The position of the 20 cassette is critical to driving the unfolding of LF_N. When an uncharged spacer was inserted between the 20s cassette and the folded portion of LF_N, then the rate of translocation was greatly impeded. However, this uncharged-spacer defect could be reversed with the LF_N L145A (Thoren et al. 2009) destabilizing point mutation (Brown et al. 2011). A chimera study was conducted, which recombined the amino-terminal ends of LF_N and EF_N, with the goal to understand why EF_N translocated 100-fold slower than LF_N (Wynia-Smith et al. 2012). The major difference in the ΔpH-driven translocation of EF_N is a general lack of negatively charged residues in EF_N's effective "20s cassette" (using LF_N residue numbering). When too many Asp/Glu residues were added to EF_N's effective "20s cassette", however, translocation slowed; therefore, an appropriate balance of charges is needed in the force-generation cassette (Wynia-Smith et al. 2012). Thus the highly charged 20 cassette is the ΔpH force-transduction cassette, and its position and Asp/Glu charge composition are required for ΔpH-driven translocation.

A charge-clamp site was electrostatically modeled to the upper portion of the β barrel; and when the strongly anionic clamp was mutated to neutral residues, cation-selectivity was reduced, and Δψ- and ΔpH-driven translocation was impeded (Wynia-Smith et al. 2012). While the major electrostatic sites both in the substrate and the channel have been identified, there are some curious deficiencies with the Brownian-ratchet mechanism when applied to anthrax toxin translocation. First, the location of the 20s cassette relative to the folded domain greatly affected ΔpH-driven translocation, suggesting mechanical force generation has significant positional requirements (Brown et al. 2011). A positional requirement is not anticipated in a Brownian-ratchet model, because this ratchet does not generate large enough force to unfold proteins, and the channel would need to wait for the protein to unfold prior to translocation (Glick 1995). However, if the channel were to wait for unfolding to equilibrate prior to translocating (as suggested (Basilio et al. 2011b)), then the location of the 20s cassette would not greatly influence the translocation kinetics (Brown et al. 2011). The next section considers these perceived deficiencies and provides an update to the overall mechanism.

9.6.4 *Molecular Basis of Force Transduction*

A key aim in the investigation of ΔpH-driven protein translocation is to understand the molecular basis of force transduction. The charge-state Brownian ratchet (Krantz et al. 2006) is compelling, but the energies garnered from a Brownian ratchet are near *RT*. These energies are lower than what may be needed to unfold LF and EF on a reasonable timescale. The protein unfolding step is more highly proton dependent than the subsequent translocation step; moreover, unfolding possesses a strong positional requirement for anionic, proton-binding residues in the 20s cassette of LF_N. A complementary view is that there are two phases to this force-transduction mechanism (Brown et al. 2011; Feld et al. 2012b; Wynia-Smith et al. 2012). The first phase is a proton-binding step that manifests in a force-generating power stroke required for unfolding, and the latter phase contains the salient features of the charge-state Brownian ratchet.

The mechanics of force transduction requires some consideration of the conformation of the translocating polypeptide chain. There are two different views. One view, the "extended-chain model" (Basilio et al. 2011a), is that the translocating chain is acted upon in the extended-chain configuration, where ϕ/ψ angles would be 180°/180°. In that model for significant force to be applied to the load (the folded protein) the chain would need to be fully extended. Another view, the "helix-compaction model" or "proton-helix engine model" (Feld et al. 2012a) considers that the translocating chain could bind protons and convert from a more extended conformation to a more helical one (Fig. 9.3b). This interconversion would substantially contract the end-to-end length of a polypeptide about 1.5-2 Å per residue, allowing for force to be applied to the load. Alternatively a formed helix may be electrostatically pulled into the channel via interactions with the positive end of its helical dipole and other residue charges induced upon protonation. Either mechanism would yield a similar net favorable displacement on the load. While the former extended-chain model considers the chain to remain in a single conformation, the latter model considers that the translocating chain cycles between contracted α-helical and extended conformations in myriad serial steps along the length of the substrate chain.

The helix-compaction model (Fig. 9.3b) has emerged from recent data showing helical structure in the α clamp of the prechannel co-complex with LF_N (Feld et al. 2010). Also surprisingly slow kinetics (orders of magnitude slower than LF_N translocation) are observed in streptavidin-trapping experiments of a highly extended 33-residue portion of LF_N in the PA channel (Basilio et al. 2011a). In the current picture, the various peptide-clamp sites in the PA channel coordinate their activities to promote the cyclic turnover of translocating chain from the extended-chain state to the collapsed and compacted helical state. Proton binding would either lead to helical compaction of a small portion of the translocating chain into the PA channel or the net attraction of a preformed helix deeper into the channel. For either possibility, proton-driven compaction or attraction may then be the major force generation step or "power stroke" (Glick 1995) in this mechanism, explaining the positional effects of the 20s cassette observed in the unfolding of LF_N (Brown et al. 2011). Subsequently, the now helical segment would then have to convert to a more extended state. This transition could be driven by a putative conversion of the ϕ-clamp site to a more constricted state (only accommodating to extended chain). This conversion is suggested by work done on the ϕ-clamp-loop network in the PA channel (Melnyk and Collier 2006), and also by the fact that the ϕ-clamp loop has been solved in two different structures in the two different PA_8 complexes (3HVD

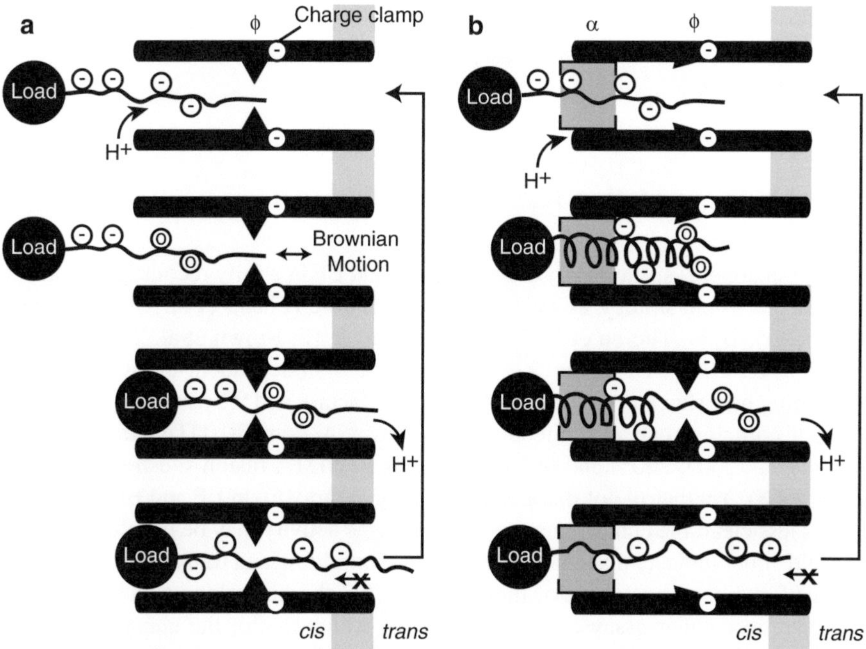

Fig. 9.3 ΔpH force-transduction models. (**a**) Charge-state Brownian ratchet (Krantz et al. 2006) as described in the text. The ϕ clamp constriction, charge clamp, proton binding (H^+), and Brownian motion (↔) are indicated. Peptide chain is solely in the extended-chain configuration (Basilio et al. 2011a). (**b**) Helix-compaction model as described in the text (Feld et al. 2012a). The α clamp is also indicated. Peptide chain is either α-helical (*coiled loops*) or extended chain. The displacement of the "load" (*black sphere*) in either model is illustrative and meant only to indicate application of a productive force that could unfold or translocate a protein substrate

(Kintzer et al. 2009) and 3KWV (Feld et al. 2010)) (Feld et al. 2012a). Once in the extended-chain state, the charge-state Brownian-ratchet phase of the mechanism ensues. Interaction at the the α-clamp and ϕ-clamp sites may serve as backstops to prevent retrotranslocation during chain extension past the charge clamp, thus favoring that the polypeptide release its protons down gradient instead (Fig. 9.3b). The now anionic peptide will be trapped as the charge-clamp will repel the peptide, preventing its retrotranslocation (Krantz et al. 2006; Brown et al. 2011; Wynia-Smith et al. 2012). The transport cycle can then repeat on the next segment of polypeptide.

Allosteric associations between proton binding and the α, ϕ, and charge clamps may be likely pivotal to the latter mechanism (Feld et al. 2012a). For example, the ϕ clamp may be triggered to change state when the α clamp is either occupied or vacated; and/or the degree of protonation of the substrate or clamp site may trigger a change in state. If proton binding directly drives clamp transitions, then uncharged sequences could be moved through the channel, albeit without the benefit of the charge clamp acting to prevent backsliding. Overall, this conceptualization should serve as a framework for continued experimentation. Future work should address the allostery and mechanics of the ϕ-, α- and charge-clamp sites both structurally through EM, NMR and crystallography and functionally through single-channel electrophysiology.

9.7 Co-translocation Factors from the Host

The PA channel is sufficient to translocate LF and EF in a fully reconstituted planar bilayer electrophysiology assay at room temperature (Zhang et al. 2004b; Krantz et al. 2005, 2006; Thoren et al. 2009; Feld et al. 2010; Brown et al. 2011; Wynia-Smith et al. 2012). However, ATP-dependent and -independent cytosolic factors (including heat shock protein 90, thioredoxin reductase and the β subunit of the coat protein complex) can expedite diphtheria toxin A domain (DTA) translocation (Ratts et al. 2003, 2005) and the heterologous LF_N-DTA fusion substrate (Tamayo et al. 2008). Yet the role of these chaperones in anthrax toxin LF and EF translocation is less clear; the enhancement of DTA enzyme activity may be peculiar to DTA; it could either indicate that chaperones assisted unfolding and translocation or chaperones functioned post-translocationally to refold the heterologous substrate. For instance, in cellular assays, drug compounds, which inhibit the chaperones, heat shock protein 90, and cyclophilin, block LF_N-DTA but not LF translocation (Dmochewitz et al. 2011). If host-cell chaperones significantly assist LF and EF translocation, it is unlikely that cellular chaperones could operate in pure isolation because a portion of LF or EF must translocate via the PA channel in order to then allow a host chaperone to engage and complete translocation.

9.8 Therapeutics and Biotechnologies

There are many technologies poised to either inhibit or exploit the PA channel as a new protein-delivery platform. Since the studies of tetraalkylammonium ion blockade (Blaustein and Finkelstein 1990), a variety of other tetra-substituted ammonium and phosphoniums and other cationic small molecules have been tested as potential inhibitors of the channel (Orlik et al. 2005; Krantz et al. 2005). Derivatives of β-cyclodextrin have been successfully utilized as potent inhibitors of the staphylococcal α-toxin pore-forming channel (Gu et al. 1999). Numerous similar β-cyclodextrin derivatives (Nestorovich and Bezrukov 2014) were synthesized with multiple cationic sites to complement the anionic charge of the PA channel; and channel blockade appears to be a primary mode of action (Karginov et al. 2005). The β-cyclodextrin approach is applicable to other homologous *Clostridium* binary toxin channels (Bezrukov et al. 2012). The PA channel is also being exploited: to treat cancer through targeted cytotoxicity (Wein et al. 2013; Phillips et al. 2013; McCluskey and Collier 2013), to deliver antigens and active proteins into T cells (Shaw and Starnbach 2008), and to dissect innate immune sensing pathways (Kofoed and Vance 2011; Zhao et al. 2011; von Moltke et al. 2012). New protein engineering approaches with native peptide ligation will allow this protein-delivery science to scale to highly flexible screening platforms (Liao et al. 2014). These are just some of the many new and exciting PA channel therapeutics and biotechnologies.

References

Arora N, Leppla SH (1993) Residues 1-254 of anthrax toxin lethal factor are sufficient to cause cellular uptake of fused polypeptides. J Biol Chem 268:3334–3341

Ashiuchi M, Nawa C, Kamei T et al (2001) Physiological and biochemical characteristics of poly gamma-glutamate synthetase complex of Bacillus subtilis. Eur J Biochem FEBS 268:5321–5328

Astumian RD (1997) Thermodynamics and kinetics of a Brownian motor. Science 276:917–922

Aubin-Tam ME, Olivares AO, Sauer RT et al (2011) Single-molecule protein unfolding and translocation by an ATP-fueled proteolytic machine. Cell 145:257–267. doi:10.1016/j.cell.2011.03.036, S0092-8674(11)00313-8 [pii]

Baldari CT, Tonello F, Paccani SR, Montecucco C (2006) Anthrax toxins: a paradigm of bacterial immune suppression. Trends Immunol 27:434–440. doi:10.1016/j.it.2006.07.002

Bann JG (2012) Anthrax toxin protective antigen-Insights into molecular switching from prepore to pore. Protein Sci 21:1–12. doi:10.1002/pro.752

Barth H, Aktories K, Popoff MR, Stiles BG (2004) Binary bacterial toxins: biochemistry, biology, and applications of common Clostridium and Bacillus proteins. Microbiol Mol Biol Rev 68:373–402

Basilio D, Jennings-Antipov LD, Jakes KS, Finkelstein A (2011a) Trapping a translocating protein within the anthrax toxin channel: implications for the secondary structure of permeating proteins. J Gen Physiol 137:343–356. doi:10.1085/jgp.201010578, jgp.201010578 [pii]

Basilio D, Juris SJ, Collier RJ, Finkelstein A (2009) Evidence for a proton-protein symport mechanism in the anthrax toxin channel. J Gen Physiol 133:307–314

Basilio D, Kienker PK, Briggs SW, Finkelstein A (2011b) A kinetic analysis of protein transport through the anthrax toxin channel. J Gen Physiol 137:521–531. doi:10.1085/jgp.201110627, jgp.201110627 [pii]

Bell SE, Mavila A, Salazar R et al (2001) Differential gene expression during capillary morphogenesis in 3D collagen matrices: regulated expression of genes involved in basement membrane matrix assembly, cell cycle progression, cellular differentiation and G-protein signaling. J Cell Sci 114:2755–2773

Belton FC, Strange RE (1954) Studies on a protective antigen produced in vitro from Bacillus anthracis: medium and methods of production. Br J Exp Pathol 35:144–152

Benson EL, Huynh PD, Finkelstein A, Collier RJ (1998) Identification of residues lining the anthrax protective antigen channel. Biochemistry 37:3941–3948

Beyer W, Turnbull PCB (2009) Anthrax in animals. Mol Aspects Med 30:481–489. doi:10.1016/j.mam.2009.08.004

Bezrukov SM, Liu X, Karginov VA et al (2012) Interactions of high-affinity cationic blockers with the translocation pores of B. anthracis, C. botulinum, and C. perfringens binary toxins. Biophys J 103:1208–1217. doi:10.1016/j.bpj.2012.07.050

Blanke SR, Milne JC, Benson EL, Collier RJ (1996) Fused polycationic peptide mediates delivery of diphtheria toxin A chain to the cytosol in the presence of anthrax protective antigen. Proc Natl Acad Sci U S A 93:8437–8442

Blaustein RO, Finkelstein A (1990) Voltage-dependent block of anthrax toxin channels in planar phospholipid bilayer membranes by symmetric tetraalkylammonium ions. Effects on macroscopic conductance. J Gen Physiol 96:905–919

Blaustein RO, Koehler TM, Collier RJ, Finkelstein A (1989) Anthrax toxin: channel-forming activity of protective antigen in planar phospholipid bilayers. Proc Natl Acad Sci U S A 86:2209–2213

Blaustein RO, Lea EJ, Finkelstein A (1990) Voltage-dependent block of anthrax toxin channels in planar phospholipid bilayer membranes by symmetric tetraalkylammonium ions. Single-channel analysis. J Gen Physiol 96:921–942

Boyer AE, Quinn CP, Hoffmaster AR et al (2009) Kinetics of lethal factor and poly-D-glutamic acid antigenemia during inhalation anthrax in rhesus macaques. Infect Immun 77:3432–3441. doi:10.1128/IAI.00346-09, IAI.00346-09 [pii]

Bradley KA, Mogridge J, Mourez M et al (2001) Identification of the cellular receptor for anthrax toxin. Nature 414:225–229

Brown MJ, Thoren KL, Krantz BA (2011) Charge requirements for proton gradient-driven translocation of anthrax toxin. J Biol Chem 286:23189–23199. doi:10.1074/jbc.M111.231167, M111.231167 [pii]

Bruckner V, Kovacs J, Denes G (1953) Structure of poly-D-glutamic acid isolated from capsulated strains of *B. anthracis*. Nature 172:508

Burton RE, Siddiqui SM, Kim YI et al (2001) Effects of protein stability and structure on substrate processing by the ClpXP unfolding and degradation machine. EMBO J 20:3092–3100

Candela T, Fouet A (2005) *Bacillus anthracis* CapD, belonging to the γ-glutamyltranspeptidase family, is required for the covalent anchoring of capsule to peptidoglycan. Mol Microbiol 57:717–726

Candela T, Fouet A (2006) Poly-gamma-glutamate in bacteria. Mol Microbiol 60:1091–1098. doi:10.1111/j.1365-2958.2006.05179.x

Candela T, Mock M, Fouet A (2005) CapE, a 47-amino-acid peptide, is necessary for Bacillus anthracis polyglutamate capsule synthesis. J Bacteriol 187:7765–7772. doi:10.1128/JB.187.22.7765-7772.2005

Chauwin JF, Oster G, Glick BS (1998) Strong precursor-pore interactions constrain models for mitochondrial protein import. Biophys J 74:1732–1743

Chavarría-Smith J, Vance RE (2013) Direct proteolytic cleavage of NLRP1B is necessary and sufficient for inflammasome activation by anthrax lethal factor. PLoS Pathog 9, e1003452. doi:10.1371/journal.ppat.1003452

Christensen KA, Krantz BA, Melnyk RA, Collier RJ (2005) Interaction of the 20 kDa and 63 kDa fragments of anthrax protective antigen: kinetics and thermodynamics. Biochemistry 44:1047–1053. doi:10.1021/bi047791s

Collier RJ (2009) Membrane translocation by anthrax toxin. Mol Aspects Med 30:413–422. doi:10.1016/j.mam.2009.06.003, S0098-2997(09)00035-1 [pii]

Collier RJ, Young JA (2003) Anthrax toxin. Annu Rev Cell Dev Biol 19:45–70

Crampton N, Brockwell DJ (2010) Unravelling the design principles for single protein mechanical strength. Curr Opin Struct Biol 20:508–517. doi:10.1016/j.sbi.2010.05.005, S0959-440X(10)00075-8 [pii]

Cunningham K, Lacy DB, Mogridge J, Collier RJ (2002) Mapping the lethal factor and edema factor binding sites on oligomeric anthrax protective antigen. Proc Natl Acad Sci U S A 99:7049–7053

Gill DM (1978) Seven toxic peptides that cross cell membranes. In: Jeljaszewicz J, Wadstrom T (eds) Bacterial Toxins and Cell Membranes. Academic, New York, pp 291–332

Dmochewitz L, Lillich M, Kaiser E et al (2011) Role of CypA and Hsp90 in membrane translocation mediated by anthrax protective antigen. Cell Microbiol 13:359–373. doi:10.1111/j.1462-5822.2010.01539.x

Drum CL, Yan SZ, Bard J et al (2002) Structural basis for the activation of anthrax adenylyl cyclase exotoxin by calmodulin. Nature 415:396–402

Duesbery NS, Webb CP, Leppla SH et al (1998) Proteolytic inactivation of MAP-kinase-kinase by anthrax lethal factor. Science 280:734–737

Dumetz F, Jouvion G, Khun H et al (2011) Noninvasive imaging technologies reveal edema toxin as a key virulence factor in anthrax. Am J Pathol 178:2523–2535. doi:10.1016/j.ajpath.2011.02.027

Evans AS (1976) Causation and disease: the Henle-Koch postulates revisited. Yale J Biol Med 49:175–195
Ezzell JW, Abshire TG (1992) Serum protease cleavage of *Bacillus anthracis* protective antigen. J Gen Microbiol 138:543–549
Ezzell JW, Abshire TG, Panchal R et al (2009) Association of *Bacillus anthracis* capsule with lethal toxin during experimental infection. Infect Immun 77:749–755. doi:10.1128/IAI.00764-08, IAI.00764-08 [pii]
Feld GK, Brown MJ, Krantz BA (2012a) Ratcheting up protein translocation with anthrax toxin. Protein Sci 21:606–624. doi:10.1002/pro.2052
Feld GK, Kintzer AF, Tang II et al (2012b) Domain flexibility modulates the heterogeneous assembly mechanism of anthrax toxin protective antigen. J Mol Biol 415:159–174. doi:10.1016/j.jmb.2011.10.035, S0022-2836(11)01172-7 [pii]
Feld GK, Thoren KL, Kintzer AF et al (2010) Structural basis for the unfolding of anthrax lethal factor by protective antigen oligomers. Nat Struct Mol Biol 17:1383–1390. doi:10.1038/nsmb.1923, nsmb.1923 [pii]
Feller W (2008) An introduction to probability theory and its applications. Wiley, Hoboken
Fersht AR (1998) Structure and mechanism in protein science: a guide to enzyme catalysis and protein folding. Freeman, New York
Finkelstein A (2009) Proton-coupled protein transport through the anthrax toxin channel. Philos Trans R Soc Lond B Biol Sci 364:209–215. doi:10.1098/rstb.2008.0126, 74702356P024100L [pii]
Floyd DL, Ragains JR, Skehel JJ et al (2008) Single-particle kinetics of influenza virus membrane fusion. Proc Natl Acad Sci U S A 105:15382–15387. doi:10.1073/pnas.0807771105
Frankel AE, Kuo S-R, Dostal D et al (2009) Pathophysiology of anthrax. Front Biosci Landmark Ed 14:4516–4524
Friedlander AM (1986) Macrophages are sensitive to anthrax lethal toxin through an acid-dependent process. J Biol Chem 261:7123–7126
Friedlander AM, Pittman PR, Parker GW (1999) Anthrax vaccine: evidence for safety and efficacy against inhalational anthrax. JAMA 282:2104–2106
Fu S, Tong X, Cai C et al (2010) The structure of tumor endothelial marker 8 (TEM8) extracellular domain and implications for its receptor function for recognizing anthrax toxin. PLoS One 5, e11203. doi:10.1371/journal.pone.0011203
Glick BS (1995) Can Hsp70 proteins act as force-generating motors? Cell 80:11–14
Green BD, Battisti L, Koehler TM et al (1985) Demonstration of a capsule plasmid in *Bacillus anthracis*. Infect Immun 49:291–297
Gu LQ, Braha O, Conlan S et al (1999) Stochastic sensing of organic analytes by a pore-forming protein containing a molecular adapter. Nature 398:686–690
Halverson KM, Panchal RG, Nguyen TL et al (2005) Anthrax biosensor: protective antigen ion channel asymmetric blockade. J Biol Chem 280:34056–34062
Hanks S, Adams S, Douglas J et al (2003) Mutations in the gene encoding capillary morphogenesis protein 2 cause juvenile hyaline fibromatosis and infantile systemic hyalinosis. Am J Hum Genet 73:791–800
Huang S, Ratliff KS, Schwartz MP et al (1999) Mitochondria unfold precursor proteins by unraveling them from their N-termini. Nat Struct Biol 6:1132–1138
Jang J, Cho M, Chun JH et al (2011) The poly-γ-D-glutamic acid capsule of *Bacillus anthracis* enhances lethal toxin activity. Infect Immun 79:3846–3854. doi:10.1128/IAI.01145-10, IAI.01145-10 [pii]
Janowiak BE, Finkelstein A, Collier RJ (2009) An approach to characterizing single-subunit mutations in multimeric prepores and pores of anthrax protective antigen. Protein Sci 18:348–358. doi:10.1002/pro.35
Jernigan DB, Raghunathan PL, Bell BP et al (2002) Investigation of bioterrorism-related anthrax, United States, 2001: epidemiologic findings. Emerg Infect Dis 8:1019–1028. doi:10.3201/eid0810.020353
Juris SJ, Melnyk RA, Bolcome RE et al (2007) Cross-linked forms of the isolated N-terminal domain of the lethal factor are potent inhibitors of anthrax toxin. Infect Immun 75:5052–5058. doi:10.1128/IAI.00490-07, IAI.00490-07 [pii]

Karginov VA, Nestorovich EM, Moayeri M et al (2005) Blocking anthrax lethal toxin at the protective antigen channel by using structure-inspired drug design. Proc Natl Acad Sci U S A 102:15075–15080. doi:10.1073/pnas.0507488102

Katayama H, Janowiak BE, Brzozowski M et al (2008) GroEL as a molecular scaffold for structural analysis of the anthrax toxin pore. Nat Struct Mol Biol 15:754–760. doi:10.1038/nsmb.1442, nsmb.1442 [pii]

Katayama H, Wang J, Tama F et al (2010) Three-dimensional structure of the anthrax toxin pore inserted into lipid nanodiscs and lipid vesicles. Proc Natl Acad Sci U S A 107:3453–3457

Kenniston JA, Baker TA, Fernandez JM, Sauer RT (2003) Linkage between ATP consumption and mechanical unfolding during the protein processing reactions of an AAA+ degradation machine. Cell 114:511–520

Kintzer AF, Sterling HJ, Tang II et al (2010a) Anthrax toxin receptor drives protective antigen oligomerization and stabilizes the heptameric and octameric oligomer by a similar mechanism. PLoS One 5, e13888

Kintzer AF, Sterling HJ, Tang II et al (2010b) Role of the protective antigen octamer in the molecular mechanism of anthrax lethal toxin stabilization in plasma. J Mol Biol 399:741–758. doi:10.1016/j.jmb.2010.04.041, S0022-2836(10)00429-8 [pii]

Kintzer AF, Tang II, Schawel AK et al (2012) Anthrax toxin protective antigen integrates poly-γ-d-glutamate and pH signals to sense the optimal environment for channel formation. Proc Natl Acad Sci U S A 109:18378–18383. doi:10.1073/pnas.1208280109, 1208280109 [pii]

Kintzer AF, Thoren KL, Sterling HJ et al (2009) The protective antigen component of anthrax toxin forms functional octameric complexes. J Mol Biol 392:614–629. doi:10.1016/j.jmb.2009.07.037, S0022-2836(09)00876-6 [pii]

Klimpel KR, Molloy SS, Thomas G, Leppla SH (1992) Anthrax toxin protective antigen is activated by a cell surface protease with the sequence specificity and catalytic properties of furin. Proc Natl Acad Sci U S A 89:10277–10281

Koehler TM (2009) Bacillus anthracis physiology and genetics. Mol Aspects Med 30:386–396. doi:10.1016/j.mam.2009.07.004, S0098-2997(09)00052-1 [pii]

Kofoed EM, Vance RE (2011) Innate immune recognition of bacterial ligands by NAIPs determines inflammasome specificity. Nature 477:592–595. doi:10.1038/nature10394, nature10394 [pii]

Krantz BA, Finkelstein A, Collier RJ (2006) Protein translocation through the anthrax toxin transmembrane pore is driven by a proton gradient. J Mol Biol 355:968–979

Krantz BA, Melnyk RA, Zhang S et al (2005) A phenylalanine clamp catalyzes protein translocation through the anthrax toxin pore. Science 309:777–781

Krantz BA, Trivedi AD, Cunningham K et al (2004) Acid-induced unfolding of the amino-terminal domains of the lethal and edema factors of anthrax toxin. J Mol Biol 344:739–756

Lacy DB, Lin HC, Melnyk RA et al (2005) A model of anthrax toxin lethal factor bound to protective antigen. Proc Natl Acad Sci U S A 102:16409–16414

Lacy DB, Mourez M, Fouassier A, Collier RJ (2002) Mapping the anthrax protective antigen binding site on the lethal and edema factors. J Biol Chem 277:3006–3010

Lacy DB, Wigelsworth DJ, Melnyk RA et al (2004a) Structure of heptameric protective antigen bound to an anthrax toxin receptor: a role for receptor in pH-dependent pore formation. Proc Natl Acad Sci U S A 101:13147–13151

Lacy DB, Wigelsworth DJ, Scobie HM et al (2004b) Crystal structure of the von Willebrand factor A domain of human capillary morphogenesis protein 2: an anthrax toxin receptor. Proc Natl Acad Sci U S A 101:6367–6372. doi:10.1073/pnas.0401506101 [doi] 0401506101 [pii]

Leppla SH (1982) Anthrax toxin edema factor: a bacterial adenylate cyclase that increases cyclic AMP concentrations of eukaryotic cells. Proc Natl Acad Sci U S A 79:3162–3166

Levinsohn JL, Newman ZL, Hellmich KA et al (2012) Anthrax lethal factor cleavage of Nlrp1 is required for activation of the inflammasome. PLoS Pathog 8:e1002638. doi:10.1371/journal.ppat.1002638, PPATHOGENS-D-12-00046 [pii]

Liao X, Rabideau AE, Pentelute BL (2014) Delivery of antibody mimics into mammalian cells via anthrax toxin protective antigen. Chembiochem 15:2458–2466. doi:10.1002/cbic.201402290

Liu S, Crown D, Miller-Randolph S et al (2009) Capillary morphogenesis protein-2 is the major receptor mediating lethality of anthrax toxin in vivo. Proc Natl Acad Sci U S A 106:12424–12429. doi:10.1073/pnas.0905409106

Maillard RA, Chistol G, Sen M et al (2011) ClpX(P) generates mechanical force to unfold and translocate its protein substrates. Cell 145:459–469. doi:10.1016/j.cell.2011.04.010, S0092-8674(11)00429-6 [pii]

Martchenko M, Jeong S-Y, Cohen SN (2010) Heterodimeric integrin complexes containing beta1-integrin promote internalization and lethality of anthrax toxin. Proc Natl Acad Sci U S A 107:15583–15588. doi:10.1073/pnas.1010145107

Mayor A (2008) Greek fire, poison arrows, and scorpion bombs: biological & chemical warfare in the ancient world. Overlook Press, New York

McCluskey AJ, Collier RJ (2013) Receptor-directed chimeric toxins created by sortase-mediated protein fusion. Mol Cancer Ther 12:2273–2281. doi:10.1158/1535-7163.MCT-13-0358

Meador WE, Means AR, Quiocho FA (1992) Target enzyme recognition by calmodulin: 2.4 a structure of a calmodulin-peptide complex. Science 257:1251–1255

Meador WE, Means AR, Quiocho FA (1993) Modulation of calmodulin plasticity in molecular recognition on the basis of x-ray structures. Science 262:1718–1721

Melnyk RA, Collier RJ (2006) A loop network within the anthrax toxin pore positions the phenylalanine clamp in an active conformation. Proc Natl Acad Sci U S A 103:9802–9807. doi:10.1073/pnas.0604000103, 0604000103 [pii]

Melnyk RA, Hewitt KM, Lacy DB et al (2006) Structural determinants for the binding of anthrax lethal factor to oligomeric protective antigen. J Biol Chem 281:1630–1635

Mikesell P, Ivins BE, Ristroph JD, Dreier TM (1983) Evidence for plasmid-mediated toxin production in Bacillus anthracis. Infect Immun 39:371–376

Miller CJ, Elliott JL, Collier RJ (1999) Anthrax protective antigen: prepore-to-pore conversion. Biochemistry 38:10432–10441

Milne JC, Blanke SR, Hanna PC, Collier RJ (1995) Protective antigen-binding domain of anthrax lethal factor mediates translocation of a heterologous protein fused to its amino- or carboxy-terminus. Mol Microbiol 15:661–666

Milne JC, Furlong D, Hanna PC et al (1994) Anthrax protective antigen forms oligomers during intoxication of mammalian cells. J Biol Chem 269:20607–20612

Moayeri M, Leppla SH (2009) Cellular and systemic effects of anthrax lethal toxin and edema toxin. Mol Aspects Med 30:439–455. doi:10.1016/j.mam.2009.07.003, S0098-2997(09)00051-X [pii]

Moayeri M, Wiggins JF, Leppla SH (2007) Anthrax protective antigen cleavage and clearance from the blood of mice and rats. Infect Immun 75:5175–5184. doi:10.1128/IAI.00719-07, IAI.00719-07 [pii]

Mock M, Fouet A (2001) Anthrax. Annu Rev Microbiol 55:647–671

Mogridge J, Cunningham K, Collier RJ (2002a) Stoichiometry of anthrax toxin complexes. Biochemistry 41:1079–1082

Mogridge J, Cunningham K, Lacy DB et al (2002b) The lethal and edema factors of anthrax toxin bind only to oligomeric forms of the protective antigen. Proc Natl Acad Sci U S A 99:7045–7048

Mogridge J, Mourez M, Collier RJ (2001) Involvement of domain 3 in oligomerization by the protective antigen moiety of anthrax toxin. J Bacteriol 183:2111–2116

Molloy SS, Bresnahan PA, Leppla SH et al (1992) Human furin is a calcium-dependent serine endoprotease that recognizes the sequence Arg-X-X-Arg and efficiently cleaves anthrax toxin protective antigen. J Biol Chem 267:16396–16402

Mourez M, Yan M, Lacy DB et al (2003) Mapping dominant-negative mutations of anthrax protective antigen by scanning mutagenesis. Proc Natl Acad Sci U S A 100:13803–13808

Mueller P, Rudin DO, Tien HT, Westcott WC (1963) Methods for the formation of single bimolecular lipid membranes in aqueous solution. J Phys Chem 67:534–535

Nanda A, Carson-Walter EB, Seaman S et al (2004) TEM8 interacts with the cleaved C5 domain of collagen alpha 3(VI). Cancer Res 64:817–820

Nassi S, Collier RJ, Finkelstein A (2002) PA_{63} channel of anthrax toxin: an extended β-barrel. Biochemistry 41:1445–1450
Nestorovich EM, Bezrukov SM (2014) Designing inhibitors of anthrax toxin. Expert Opin Drug Discov 9:299–318. doi:10.1517/17460441.2014.877884
Nguyen TL (2004) Three-dimensional model of the pore form of anthrax protective antigen. Structure and biological implications. J Biomol Struct Dyn 22:253–265
Oomen CJ, Van Ulsen P, Van Gelder P et al (2004) Structure of the translocator domain of a bacterial autotransporter. EMBO J 23:1257–1266
Orlik F, Schiffler B, Benz R (2005) Anthrax toxin protective antigen: inhibition of channel function by chloroquine and related compounds and study of binding kinetics using the current noise analysis. Biophys J 88:1715–1724
Panchal RG, Halverson KM, Ribot W et al (2005) Purified Bacillus anthracis lethal toxin complex formed in vitro and during infection exhibits functional and biological activity. J Biol Chem 280:10834–10839. doi:10.1074/jbc.M412210200
Pannifer AD, Wong TY, Schwarzenbacher R et al (2001) Crystal structure of the anthrax lethal factor. Nature 414:229–233
Pentelute BL, Barker AP, Janowiak BE et al (2010) A semisynthesis platform for investigating structure-function relationships in the N-terminal domain of the anthrax Lethal Factor. ACS Chem Biol 5:359–364. doi:10.1021/cb100003r
Pentelute BL, Sharma O, Collier RJ (2011) Chemical dissection of protein translocation through the anthrax toxin pore. Angew Chem Int Ed Engl 50:2294–2296. doi:10.1002/anie.201006460
Petosa C, Collier RJ, Klimpel KR et al (1997) Crystal structure of the anthrax toxin protective antigen. Nature 385:833–838
Pflughoeft KJ, Swick MC, Engler DA et al (2014) Modulation of the Bacillus anthracis secretome by the immune inhibitor A1 protease. J Bacteriol 196:424–435. doi:10.1128/JB.00690-13
Phillips DD, Fattah RJ, Crown D et al (2013) Engineering anthrax toxin variants that exclusively form octamers and their application to targeting tumors. J Biol Chem 288:9058–9065. doi:10.1074/jbc.M113.452110
Pittman PR, Kim-Ahn G, Pifat DY et al (2002) Anthrax vaccine: immunogenicity and safety of a dose-reduction, route-change comparison study in humans. Vaccine 20:1412–1420
Quinn CP, Singh Y, Klimpel KR, Leppla SH (1991) Functional mapping of anthrax toxin lethal factor by in-frame insertion mutagenesis. J Biol Chem 266:20124–20130
Rasko DA, Worsham PL, Abshire TG et al (2011) Bacillus anthracis comparative genome analysis in support of the Amerithrax investigation. Proc Natl Acad Sci U S A 108:5027–5032. doi:10.1073/pnas.1016657108, 1016657108 [pii]
Ratts R, Trujillo C, Bharti A et al (2005) A conserved motif in transmembrane helix 1 of diphtheria toxin mediates catalytic domain delivery to the cytosol. Proc Natl Acad Sci U S A 102:15635–15640. doi:10.1073/pnas.0504937102, 0504937102 [pii]
Ratts R, Zeng H, Berg EA et al (2003) The cytosolic entry of diphtheria toxin catalytic domain requires a host cell cytosolic translocation factor complex. J Cell Biol 160:1139–1150. doi:10.1083/jcb.200210028, jcb.200210028 [pii]
Ryan PL, Young JA (2008) Evidence against a human cell-specific role for LRP6 in anthrax toxin entry. PLoS One 3:e1817. doi:10.1371/journal.pone.0001817
Santelli E, Bankston LA, Leppla SH, Liddington RC (2004) Crystal structure of a complex between anthrax toxin and its host cell receptor. Nature 430:905–908
Schueler-Furman O, Wang C, Baker D (2005) Progress in protein-protein docking: atomic resolution predictions in the CAPRI experiment using RosettaDock with an improved treatment of side-chain flexibility. Proteins 60:187–194. doi:10.1002/prot.20556
Scobie HM, Rainey GJ, Bradley KA, Young JA (2003) Human capillary morphogenesis protein 2 functions as an anthrax toxin receptor. Proc Natl Acad Sci U S A 100:5170–5174. doi:10.1073/pnas.0431098100, 0431098100 [pii]
Sellman BR, Mourez M, Collier RJ (2001a) Dominant-negative mutants of a toxin subunit: an approach to therapy of anthrax. Science 292:695–697

Sellman BR, Nassi S, Collier RJ (2001b) Point mutations in anthrax protective antigen that block translocation. J Biol Chem 276:8371–8376

Sharma O, Collier RJ (2014) Polylysine-mediated translocation of the diphtheria toxin catalytic domain through the anthrax protective antigen pore. Biochemistry 53:6934–6940. doi:10.1021/bi500985v

Shaw CA, Starnbach MN (2008) Antigen delivered by anthrax lethal toxin induces the development of memory CD8+ T cells that can be rapidly boosted and display effector functions. Infect Immun 76:1214–1222. doi:10.1128/IAI.01208-07

Shen Y, Zhukovskaya NL, Guo Q et al (2005) Calcium-independent calmodulin binding and two-metal-ion catalytic mechanism of anthrax edema factor. EMBO J 24:929–941

Simon SM, Peskin CS, Oster GF (1992) What drives the translocation of proteins? Proc Natl Acad Sci U S A 89:3770–3774

Singh Y, Klimpel KR, Goel S et al (1999) Oligomerization of anthrax toxin protective antigen and binding of lethal factor during endocytic uptake into mammalian cells. Infect Immun 67:1853–1859

Smith H, Keppie J (1954) Observations on experimental anthrax: demonstration of a specific lethal factor produced in vivo by *Bacillus anthracis*. Nature 173:689

Song L, Hobaugh MR, Shustak C et al (1996) Structure of staphylococcal α-hemolysin, a heptameric transmembrane pore. Science 274:1859–1866

Sterne M (1939) The use of anthrax vaccines prepared from avirulent (uncapsulated) variants of *Bacillus anthracis*. Ondersteport J Vet Sci Anim Indust 13:307–312

Strange RE, Belton FC (1954) Studies on a protective antigen produced in vitro from Bacillus anthracis: purification and chemistry of the antigen. Br J Exp Pathol 35:153–165

Sun J, Lang AE, Aktories K, Collier RJ (2008) Phenylalanine-427 of anthrax protective antigen functions in both pore formation and protein translocation. Proc Natl Acad Sci U S A 105:4346–4351. doi:10.1073/pnas.0800701105, 0800701105 [pii]

Sun J, Vernier G, Wigelsworth DJ, Collier RJ (2007) Insertion of anthrax protective antigen into liposomal membranes: effects of a receptor. J Biol Chem 282:1059–1065

Tamayo AG, Bharti A, Trujillo C et al (2008) COPI coatomer complex proteins facilitate the translocation of anthrax lethal factor across vesicular membranes in vitro. Proc Natl Acad Sci U S A 105:5254–5259. doi:10.1073/pnas.0710100105, 0710100105 [pii]

Thoren KL, Krantz BA (2011) The unfolding story of anthrax toxin translocation. Mol Microbiol 80:588–595. doi:10.1111/j.1365-2958.2011.07614.x

Thoren KL, Worden EJ, Yassif JM, Krantz BA (2009) Lethal factor unfolding is the most force-dependent step of anthrax toxin translocation. Proc Natl Acad Sci U S A 106:21555–21560. doi:10.1073/pnas.0905880106, 0905880106 [pii]

Turnbull PC (1991) Anthrax vaccines: past, present and future. Vaccine 9:533–539

Turnbull PCB (1996) Bacillus Medical Microbiology. University of Texas Medical Branch at Galveston, Galveston

Uchida I, Makino S, Sasakawa C et al (1993) Identification of a novel gene, dep, associated with depolymerization of the capsular polymer in *Bacillus anthracis*. Mol Microbiol 9:487–496

Uchida I, Sekizaki T, Hashimoto K, Terakado N (1985) Association of the encapsulation of Bacillus anthracis with a 60 megadalton plasmid. J Gen Microbiol 131:363–367

Van der Goot G, Young JA (2009) Receptors of anthrax toxin and cell entry. Mol Aspects Med 30:406–412. doi:10.1016/j.mam.2009.08.007, S0098-2997(09)00059-4 [pii]

Vitale G, Bernardi L, Napolitani G et al (2000) Susceptibility of mitogen-activated protein kinase kinase family members to proteolysis by anthrax lethal factor. Biochem J 352:739–745

Von Moltke J, Trinidad NJ, Moayeri M et al (2012) Rapid induction of inflammatory lipid mediators by the inflammasome in vivo. Nature 490:107–111. doi:10.1038/nature11351

Wasserman GM, Grabenstein JD, Pittman PR et al (2003) Analysis of adverse events after anthrax immunization in US Army medical personnel. J Occup Environ Med Am Coll Occup Environ Med 45:222–233

Wein AN, Liu S, Zhang Y et al (2013) Tumor therapy with a urokinase plasminogen activator-activated anthrax lethal toxin alone and in combination with paclitaxel. Invest New Drugs 31:206–212. doi:10.1007/s10637-012-9847-1

Wei W, Lu Q, Chaudry GJ et al (2006) The LDL receptor-related protein LRP6 mediates internalization and lethality of anthrax toxin. Cell 124:1141–1154. doi:10.1016/j.cell.2005.12.045, S0092-8674(06)00199-1 [pii]

Wesche J, Elliott JL, Falnes PO et al (1998) Characterization of membrane translocation by anthrax protective antigen. Biochemistry 37:15737–15746

Wigelsworth DJ, Krantz BA, Christensen KA et al (2004) Binding stoichiometry and kinetics of the interaction of a human anthrax toxin receptor, CMG2, with protective antigen. J Biol Chem 279:23349–23356

Williams P, Wallace D (1989) Unit 731: Japan's secret biological warfare in World War II. Free Press, New York

Wimalasena DS, Cramer JC, Janowiak BE et al (2007) Effect of 2-fluorohistidine labeling of the anthrax protective antigen on stability, pore formation, and translocation. Biochemistry 46:14928–14936. doi:10.1021/bi701763z

Wolfe JT, Krantz BA, Rainey GJ et al (2005) Whole-cell voltage clamp measurements of anthrax toxin pore current. J Biol Chem 280:39417–39422

Wright GG, Hedberg MA, Slein JB (1954) Studies on immunity in anthrax. III. Elaboration of protective antigen in a chemically defined, non-protein medium. J Immunol 72:263–269

Wynia-Smith SL, Brown MJ, Chirichella G et al (2012) Electrostatic ratchet in the protective antigen channel promotes anthrax toxin translocation. J Biol Chem 287:43753–43764. doi:10.1074/jbc.M112.419598, M112.419598 [pii]

Young JA, Collier RJ (2007) Anthrax toxin: receptor binding, internalization, pore formation, and translocation. Annu Rev Biochem 76:243–265. doi:10.1146/annurev.biochem.75.103004.142728

Young JJ, Bromberg-White JL, Zylstra C et al (2007) LRP5 and LRP6 are not required for protective antigen-mediated internalization or lethality of anthrax lethal toxin. PLoS Pathog 3, e27. doi:10.1371/journal.ppat.0030027, 07-PLPA-RA-0017 [pii]

Zhang S, Finkelstein A, Collier RJ (2004a) Evidence that translocation of anthrax toxin's lethal factor is initiated by entry of its N terminus into the protective antigen channel. Proc Natl Acad Sci U S A 101:16756–16761

Zhang S, Udho E, Wu Z et al (2004b) Protein translocation through anthrax toxin channels formed in planar lipid bilayers. Biophys J 87:3842–3849

Zhao Y, Yang J, Shi J et al (2011) The NLRC4 inflammasome receptors for bacterial flagellin and type III secretion apparatus. Nature 477:596–600. doi:10.1038/nature10510

Chapter 10
Staphylococcal β-barrel Pore-Forming Toxins: Mushrooms That Breach the Greasy Barrier

Jack Fredrick Gugel and Liviu Movileanu

Abstract *Staphylococcus aureus* exhibits a myriad of virulence elements, including β-barrel pore-forming toxins (β-PFTs). The primary mission of these protein toxins is to destroy the physical and chemical gradients across the membrane of the targeted cell by generating well-defined transmembrane pores, ultimately causing the cell death. Such a form of biomolecular attack is a ubiquitous membrane-perforation mechanism in numerous organisms, including bacterial systems and eukaryotes. One unusual commonality of the β-PFTs is their amphipathic nature, enabling sophisticated conformational alterations that are required for their transit from the secreting to attacked cell. Intriguingly enough, proteinaceous toxins are secreted as a hydrophilic form. Then, they must navigate within the aqueous phase between the two cells and ultimately breach the hydrophobic barrier posed by the susceptible cell membrane. The archetype of these non-enzymatic staphylococcal β-PFTs is the homoheptameric α-hemolysin (αHL) protein. Moreover, *S. aureus* has the ability to secrete up to four heteromeric, bi-component β-PFTs. Although the homomeric and heteromeric β-PFTs are related in sequence, homology, and structure, they demonstrate distinct biophysical features.

Keywords Unitary conductance • Polymer partitioning • Cysteine scanning mutagenesis • Bi-component toxin protein • Channel gating • Rational protein design • Subunit stoichiometry • Chemical modification

---

J.F. Gugel
Department of Physics, Syracuse University,
201 Physics Building, Syracuse, NY 13244-1130, USA

L. Movileanu (✉)
Department of Physics, Syracuse University,
201 Physics Building, Syracuse, NY 13244-1130, USA

Structural Biology, Biochemistry, and Biophysics Program, Syracuse University,
111 College Place, Syracuse, NY 13244-4100, USA

The Syracuse Biomaterials Institute, Syracuse University,
121 Link Hall, Syracuse, NY 13244, USA
e-mail: lmovilea@syr.edu

A.H. Delcour (ed.), *Electrophysiology of Unconventional Channels and Pores*,
Springer Series in Biophysics 18, DOI 10.1007/978-3-319-20149-8_10

10.1 What Do the β-PFTs Have and Not Have in Common?

Transmembrane protein pores represent cytotoxic factors expressed by various bacterial organisms, including *Escherichia coli*, *S. aureus*, *Bacillus cereus*, *Vibrio cholerae*, *Streptococcus pneumoniae*, and *Aeromonas hydrophila* (Parker and Feil 2005; Gonzalez et al. 2008; Iacovache et al. 2008; Bischofberger et al. 2009; Iacovache et al. 2010; Los et al. 2013). The focus of this book chapter is to provide a comprehensive overview of bacterial β-barrel pore-forming exotoxins (β-PFTs) secreted by *S. aureus* (Heuck et al. 2001; Menestrina et al. 2001, 2003; Comai et al. 2002; DuMont and Torres 2014; Iacovache et al. 2010; Otto 2014). This organism is a human pathogen that produces a number of β-PFTs. An interesting trait of these proteins is their ability to undergo complex structural transformations in such a way that they adapt to both the aqueous phase and hydrophobic environments. Therefore, these protein toxins cannot be placed into either the category of hydrophilic polypeptides or hydrophobic transmembrane proteins. In addition, the existence of such proteins questions the biophysical concept that a unique protein sequence determines a unique three-dimensional structure. This demonstrates that the nature of the bathing solvent is crucial for establishing the protein structure. The major commonality of these β-PFTs is that they are first synthesized as hydrophilic monomers, which then travel until they reach the membrane of the susceptible cell. The membrane represents a convenient environment for monomer attachment to a surface as well as oligomerization and pre-pore formation. The final step of these transformations is the pore formation within the hydrophobic membrane.

The structure of the pore is a β-barrel that spans the lipid membrane and is made by an even number of anti-parallel β-strands. The hydrophobic residues of the β-strands are oriented towards the external side (e.g., that of the lipid bilayer), whereas the hydrophilic side chains are oriented towards the β-barrel interior, conferring a hydrophilic environment for the transport of diverse water-soluble nutrients. Moreover, because the anti-parallel β-strands are connected through a network of numerous hydrogen bonds, the β-barrel structure shows an unusual mechanical, electrical and thermodynamic stability (White and Wimley 1999; Wimley 2003), an attribute that has been extensively used in membrane protein design (Bayley and Cremer 2001; Bayley and Jayasinghe 2004; Bayley et al. 2004; Kang et al. 2005; Jung et al. 2006; Howorka and Siwy 2008; Movileanu 2008, 2009; Howorka and Siwy 2009; Majd et al. 2010; Siwy and Howorka 2010; Mayer and Yang 2013).

We now know that β-PFTs are multimeric protein complexes. However, one feature that they do not share is the identity of the participating complex subunits: some toxins are homomeric, whereas others are heteromeric. An immediate example of the homomeric β-PFTs is the αHL protein pore (Song et al. 1996). The heteromeric β-PFTs contain two distinct polypeptides. That is why they are sometimes called bi-component β-PFTs (Ferreras et al. 1998; Pedelacq et al. 1999; Werner et al. 2002; Sugawara-Tomita et al. 2002; Menestrina et al. 2003; Potrich et al. 2009; Yamashita et al. 2011). Therefore, there must be different kinetics and dynamics of the complex formation of these proteins. This is schematically illustrated below.

$$\alpha_1^{\,u} \rightarrow \alpha_1^{\,b} \rightarrow \alpha_1^{\,b*} \rightarrow \alpha_7^{\,pp} \rightarrow \alpha_7^{\,p} \quad (10.1)$$

$$\alpha_1^{\,u} + \beta_1^{\,u} \rightarrow \alpha_1^{\,b} + \beta_1^{\,u} \rightarrow (\alpha_1\beta_1)^{b} \rightarrow (\alpha_1\beta_1)^{b*} \rightarrow (\alpha_n\beta_n)^{pp} \rightarrow (\alpha_n\beta_n)^{p} \quad (10.2)$$

On the top row, the steps of oligomerization and pore formation of the αHL protein are shown (Scheme 10.1). α_1^{u} indicates a membrane-unbound, water-soluble αHL monomer. α_1^{b} is a membrane-bound αHL monomer, whereas α_1^{b*} is a membrane-bound αHL monomer in an activated conformation, a state that precedes the non-lytic formation of the homomeric αHL pre-pore (α_7^{pp}). The final state is the insertion of the stem domain of the homomer into the lipid bilayer of the susceptible cell, generating a transmembrane β-barrel protein pore (α_7^{p}). In contrast, the kinetics of pore formation of a bi-component β-PFT undergoes a more complex scheme (the bottom row, Scheme 10.2). Both types of bi-component toxin monomers are secreted as a hydrophilic form (α_1^{u} and β_1^{u}). Then, they navigate to the membrane of the susceptible cell. One class of monomers shows specific interactions with the membrane (α_1^{b}) before the heterodimerization at the membrane surface ($(\alpha_1\beta_1)^{b}$). This is followed by the reorientation of the heterodimer in an activated conformation ($(\alpha_1\beta_1)^{b*}$) that enables the pre-pore formation ($(\alpha_n\beta_n)^{pp}$). Finally, an insertion of the stem into the membrane generates a transmembrane β-barrel pore of a bi-component β-PFT ($(\alpha_n\beta_n)^{p}$). The steps that lead from the unbound monomer to the insertion of a transmembrane protein pore are likely closely similar for all β-PFTs, as shown above. Such similarity suggests that they form a unique protein superfamily, including β-PFTs produced by other bacterial systems. Examples are aerolysin of *A. hydrophila*, cytolysin of *V. cholera*, *B. cereus* hemolysin II, and others. However, neither homomeric nor heteromeric β-PFTs of *S. aureus* include an enzymatic domain, such as in the case of the protective antigen of the anthrax toxin.

10.2 α-Hemolysin (αHL) is an Archetype of the Homomeric β-PFTs

It is likely that the most known β-PFT is staphylococcal αHL protein (Song et al. 1996; Gouaux 1998; Menestrina et al. 2001, 2003; Montoya and Gouaux 2003). This is a well-studied toxin because it exhibits virulence activity on a broad variety of mammalian cells, such as granulocytes and erythrocytes (Los et al. 2013). Such a protein complex is secreted as a water-soluble, 293-residue monomer, but it does assemble as a heptamer on model and cell membranes (Gouaux et al. 1994; Song et al. 1996; Cheley et al. 1997; Fang et al. 1997; Krasilnikov et al. 2000) (Table 10.1). The heptamer is a mushroom-shaped protein complex, in which each subunit contributes two anti-parallel β strands to the formation of the 14-stranded β barrel (Fig. 10.1a). The structure of the αHL protein is divided into three distinct domains: stem, rim, and cap. The cap domain, whose external diameter is ~100 Å, is located within the aqueous phase. The overall length of the complex from one opening to the other is ~100 Å, out of which the β-barrel domain (the stem domain) measures ~52 Å.

Table 10.1 Structural features of staphylococcal β-PFTs

Protein	Monomer mass (kDa)	Oligomer size	Available 3D structure	Barrel diameter (Å)	Barrel length (Å)	Lumen length (Å)
α-Hemolysin (αHL)	33.2	Homoheptamer (14 β strands)[a] Homohexamer (12 β strands)[b]	Detergent-solubilized heptamer[c]	26[c]	52[c]	100[c]
Leukocidin (Luk) LukF LukS	34 33	Hetero-octamer (16 β strands)[d]	NA[e]	NA[e]	NA[e]	NA[e]
γ-Hemolysin (γHL) LukF Hlg2	34 32	Hetero-heptamer (14 β strands)[f] Hetero-octamer (16 β strands)[g]	Detergent-solubilized octamer[g]	25.5[g]	47[g]	93[g]

[a]Gouaux et al. (1994); Song et al. (1996); Cheley et al. (1997); Fang et al. (1997); Krasilnikov et al. (2000)
[b]Under some experimental contexts, the αHL protein forms a hexameric structure (Czajkowsky et al. 1998)
[c]Song et al. (1996)
[d]Miles et al. (2002b); Jayasinghe and Bayley (2005)
[e]Not available
[f]Sugawara-Tomita et al. (2002)
[g]Yamashita et al. (2011)

The internal diameter of the barrel, excluding the side chains of the internal residues, is ~26 Å (Fig. 10.1b). However, if we include the side chains of the residues within the pore lumen, αHL exhibits a strongly varying internal diameter from ~16 Å, within the constricted region of the β barrel, to ~46 Å, within the cap domain. The average internal diameter of the β-barrel is ~20 Å.

It is quite interesting that almost the entire structure of the staphylococcal αHL protein is made of anti-parallel β strands, conferring an unusual mechanical, electrical and thermal stability. We will discuss below the critical importance of this trait, which has been heavily used for the transformation of the area of single-molecule biophysics. The heptameric structural organization is not challenged, as this has been determined either directly or indirectly by several independent approaches under various experimental contexts. The high-resolution, X-ray structure of the detergent-solubilized αHL protein revealed its heptameric structure (Song et al. 1996) (Fig. 10.1).

Collaborative efforts between Hagan Bayley's and Jie Yang's groups have produced an unquestionable demonstration that the αHL protein pore assembles as a heptamer (Cheley et al. 1997; Fang et al. 1997). Using atomic force microscopy (AFM) imaging of a β barrel-truncation αHL mutant, they showed that the stoichiometry of αHL remains heptameric even under the experimental contexts of the AFM approach (Fig. 10.2). Later, a high-impact methodology was developed by coupling chemical modification, protein engineering, SDS-PAGE gel analysis, and single-channel electrical recordings. In this case, targeted, engineered, or chemically-modified

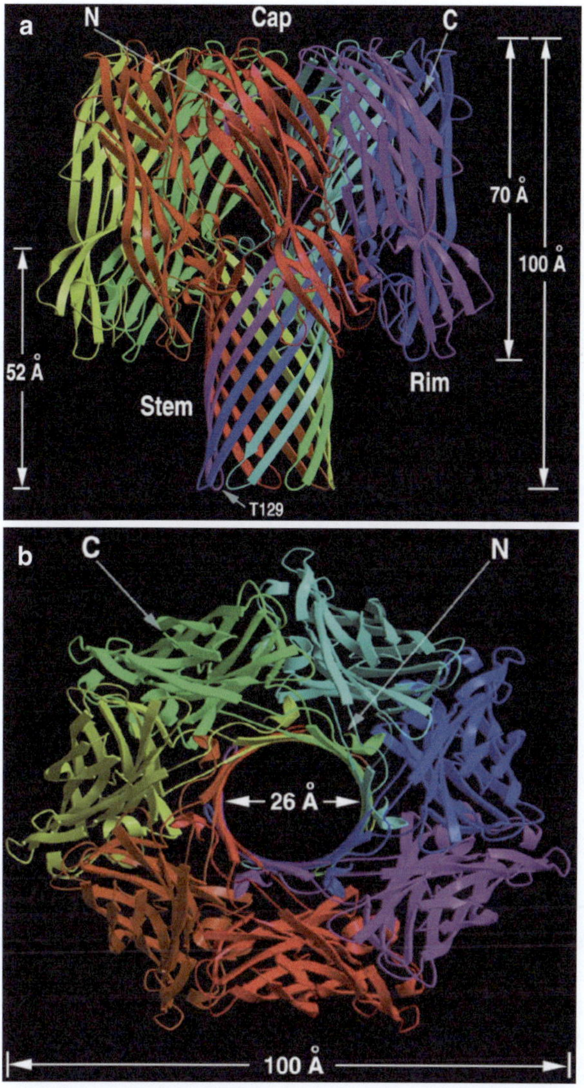

Fig. 10.1 αHL is presented using a ribbon representation. Here, individual protomers are illustrated by different colors. (**a**) This is a side view that is parallel to the putative lipid membrane; (**b**) This is a top view. It reveals the seven-fold symmetry of the protein. The αHL protein model was generated using 7ahl.pdb (Reproduced, with permission, from reference (Song et al. 1996))

αHL heteromers were separated and visualized as resulting protein band products of combinations and permutations of the unmodified and native subunits of a heptameric protein (Braha et al. 1997; Howorka et al. 2000; Movileanu et al. 2000). Single-channel electrical recordings and analysis permitted the obtaining of informative data pertinent to the desired heteromers. These spectacular experiments were only possible owing to a great feature of this protein: its stability in SDS up to 65 °C, a stunning

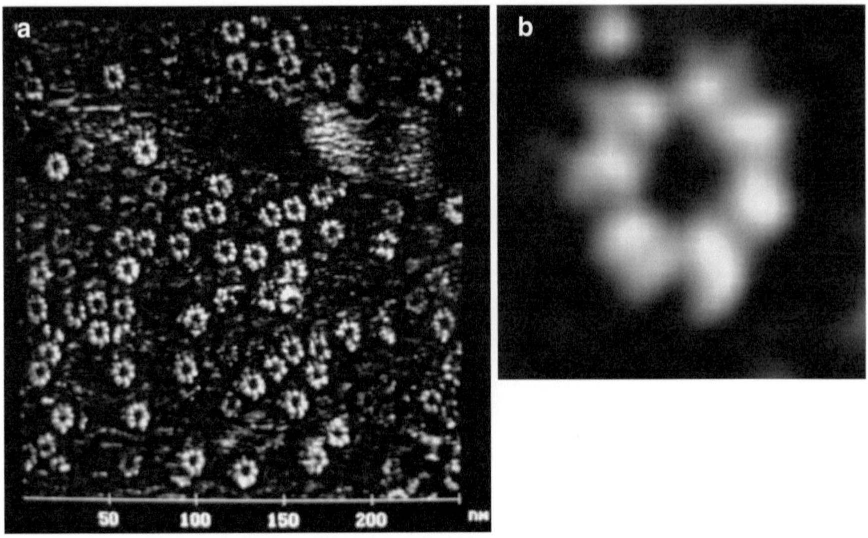

Fig. 10.2 The use of atomic force microscopy (AFM) imaging to show that the αHL protein pore is a homoheptameric β-PFT. (**a**) An αHL mutant, made by truncation of the native β-barrel, was attached on an AFM mica grid as a heptamer. The horizontal scale bar is marked in nanometers; (**b**) An AFM image of an individual β-barrel truncation αHL mutant (Reproduced, with permission, from reference (Cheley et al. 1997))

discovery made at an earlier time (Walker and Bayley 1995). Later in this chapter, we show that this exceptional feature has also been employed to determine the subunit composition of a bi-component β-PFT (Miles et al. 2002b).

Moreover, Krasilnikov and coworkers (Krasilnikov et al. 2000) demonstrated that the αHL protein is a heptamer under experimental contexts of reconstitution into a planar lipid bilayer. They used targeted cysteine mutagenesis, sulfhydryl-directed reagents, and single-channel electrical recordings (Sackmann and Neher 1995). Each chemical reaction between an engineered cysteine sulfhydryl and a sulfhydryl-directed reagent molecule produced a distinctive current blockade. A total of seven reagent-induced current blockades were noted during the same single-channel experiment, confirming the heptameric structure of αHL. We will show below that such a methodology worked well for the determination of the subunit composition of a staphylococcal bi-component β-PFT (Miles et al. 2002b).

10.3 Why Did αHL Become Such a Popular Transmembrane Protein Pore?

Speaking about staphylococcal αHL protein pore in a book chapter raises our obligation to provide a clear message to a non-expert reader that protein biochemistry and electrophysiological studies involving this protein have transformed the area of single-molecule biophysics. In addition, extensive studies using αHL impacted our understanding of membrane protein folding, stability, and design. Moreover, αHL was heavily utilized in studies of bionanotechnology (Majd et al. 2010), particularly with implications in high-resolution DNA (Astier et al. 2005; Bayley 2006; Branton et al. 2008) and protein (Movileanu 2009; Mayer and Yang 2013) detection and analysis. Therefore, it is impossible to mention all of the outstanding articles involving this protein. However, it needs to be said that such major αHL-related transformations in these rapidly changing areas were primarily driven by many of the advantageous characteristics of this protein. The single-channel conductance of αHL is ~260 pS and 960 pS in 300 mM KCl (Howorka et al. 2000) and 1000 M KCl (Movileanu et al. 2003), respectively (Table 10.2). The αHL protein pore is weakly anion selective ($P_{Cl}/P_K = 1.5$, pH 7.0) (Menestrina 1986), exhibiting gating fluctuations at low, physiological salt concentrations (Mohammad and Movileanu 2010) and acidic pH (Kasianowicz and Bezrukov 1995; Mohammad et al. 2012). This protein self-assembles on planar lipid bilayers into a single orientation (Menestrina 1986; Bezrukov and Kasianowicz 1993).

The αHL pore maintains its quiet, large-conductance open state for long periods of time even under extreme conditions of pH (from 4 to 11) (Gu and Bayley 2000; Misakian and Kasianowicz 2003), transmembrane potential (Menestrina 1986; Korchev et al. 1995), temperature (up to 95 °C) (Kang et al. 2005; Jung et al. 2006), salt concentration (several molar) (Bezrukov et al. 2004; Rodrigues et al. 2008) or osmotic pressure (a few tens of % (w/v) PEGs) (Bezrukov et al. 1996; Krasilnikov and Bezrukov 2004; Gu et al. 2003). The transport processes of molecules across this protein pore can be readily designed and implemented, by imposing different thermodynamic driving forces for the translocating molecules, such as transmembrane potentials, transmembrane difference in pH, ionic strength, disulfide-thiol

Table 10.2 Electrophysiological traits of staphylococcal β-PFTs

Protein	Conductance (pS)	Ion Selectivity	Permeability ratio (P_K/P_{Cl})
α-Hemolysin (αHL)	260[a] 960[b] 92[c]	Weakly anion-selective	0.6[c] 0.8[d]
Leukocidin (LukF/LukS)	2540[d]	Weakly cation-selective	1.7[e]
γ-Hemolysin (γHL) HlgA/HlgB HlgC/HlgB	 115[f] 192[f]	 Weakly cation-selective	 1.3[g] 3.6[g]

[a]This value was determined in 300 mM KCl, 5 mM Tris.HCl, pH 7.0 (Howorka et al. 2000)
[b]This value was determined in 1000 mM KCl, 10 mM Tris.HCl, pH 7.5 (Movileanu et al. 2003)
[c]This value was determined in 100 mM NaCl, pH 7.0 (Menestrina 1986)
[d]This value was determined in 1000 mM KCl, 5 mM HEPES, pH 7.4 (Miles et al. 2001)
[e]These values were determined when the *cis* and *trans* chambers contained 1000 mM and 200 mM KCl, respectively. The buffer solution contained 5 mM HEPES, pH 7.4. The *cis* chamber was grounded (Miles et al. 2001)
[f]This value was determined in 100 mM NaCl, 20 mM HEPES, pH 7.0 (Comai et al. 2002)
[g]These values were determined when the chambers contained 20 (*cis*) and 200 (*trans*) mM KCl, respectively. The buffer solution contained 20 mM HEPES, pH 7.0. The *cis* chamber was grounded (Comai et al. 2002)

exchange or other physical and chemical gradients, as both sides of the membranes are fully accessible in a planar lipid membrane format. It was found that αHL may serve as a permeation pathway for a broad range of molecules, including highly flexible and neutral PEGs of up to ~3 kDa (Bezrukov et al. 1996; Movileanu et al. 2001; Movileanu and Bayley 2001; Krasilnikov and Bezrukov 2004; Krasilnikov et al. 2006), positively charged peptides (Movileanu et al. 2005; Sutherland et al. 2005), negatively charged single stranded RNAs and DNAs of greater molecular mass (~ tens of kDa) (Kasianowicz et al. 1996; Akeson et al. 1999; Meller et al. 2000), and large proteins in unfolded conformation (Rodriguez-Larrea and Bayley 2013; Nivala et al. 2013). In addition, the αHL pore interacts with cyclodextrins (Gu et al. 1999, 2003), cyclic peptides (Sanchez-Quesada et al. 2000), and crown ethers (Bezrukov et al. 2004), some of which were used as molecular adapters for the design and creation of various molecular biosensors. Later, αHL was used to obtain a mechanistic understating of the binding interactions between enzymes and DNA (Hornblower et al. 2007; Benner et al. 2007), with great expectations for DNA sequencing (Bayley 2006; Branton et al. 2008; Cherf et al. 2012). The use of αHL in polymer translocation studies stimulated a new realm called *nanopore biophysics*. For example, the current blockades induced by polymer partitioning into a single αHL protein pore were indicative of the polymer properties, which included the polymer length, persistence length, repetitive unit composition, and secondary structure. Moreover, these single-channel parameters strongly depended on other physical traits of the lipid bilayer-pore system, such as temperature, transmembrane potential, ionic strength, pH, and osmotic pressure. Therefore, it is not surprising that such fundamental polymer-pore studies were rapidly extended to applicative areas of bionanotechnology (Bayley et al. 2004; Bayley and Jayasinghe 2004; Maglia et al. 2008, 2009; Hall et al. 2010; Wang et al. 2011, 2013a, b; Gurnev and Nestorovich 2014; Kusters et al. 2014).

In Fig. 10.3a, we illustrate the sites at which native residues were substituted by cysteines. Systematic examinations of the chemical reactions between sulfhydryl-directed PEGs and individual cysteine residues have provided information on both the single polymer dynamics under confined nanopore environment (Howorka et al. 2000; Movileanu et al. 2000; Movileanu and Bayley 2001) as well as the internal geometry of the pore interior (Bezrukov 2000; Movileanu et al. 2001). A wealth of theoretical and computational studies was already available at that time (Muthukumar 1999; De Gennes 1999a, 1999b, 1999c; Brochard-Wyart et al. 2000). Later, more extensive theoretical examinations were stimulated by experimental studies using single αHL protein nanopores (Slonkina and Kolomeisky 2003; Tian and Andricioaei 2005; Kong and Muthukumar 2005; Kolomeisky 2007; Goodrich et al. 2007; Muthukumar 2007; Kolomeisky 2008; Maffeo et al. 2012). These experimental studies have provided a new methodology for the systematic exploration of the lumen of the transmembrane protein pores. In addition, such experiments have enabled the direct determination of the number of subunits of a desired cysteine αHL mutant that interact with a specific polymer of a given size (Movileanu et al. 2003).

For example, a single-channel trace acquired with $T117C_7$, a homoheptamer cysteine αHL mutant (Fig. 10.3a), is shown in Fig. 10.3b, c. When a sample of sulfhydryl-directed PEG-1 kDa was applied to the chamber, brief current blockades were observed, representing transient movements of the polymer into the nanopore interior. This pattern was followed by a permanent current blockade (Fig. 10.3b, state 1), which indicated that one sulfhydryl-directed PEG-1 kDa covalently reacted with a cysteine sulfhydryl. The amplitude of the transient current blockades was comparable with the amplitude of the permanent current blockade, suggesting that indeed a single PEG-1 kDa produced a current blockade, whose size corresponds to 2/3 that of the unitary conductance of the αHL protein pore. Again, the open state 1

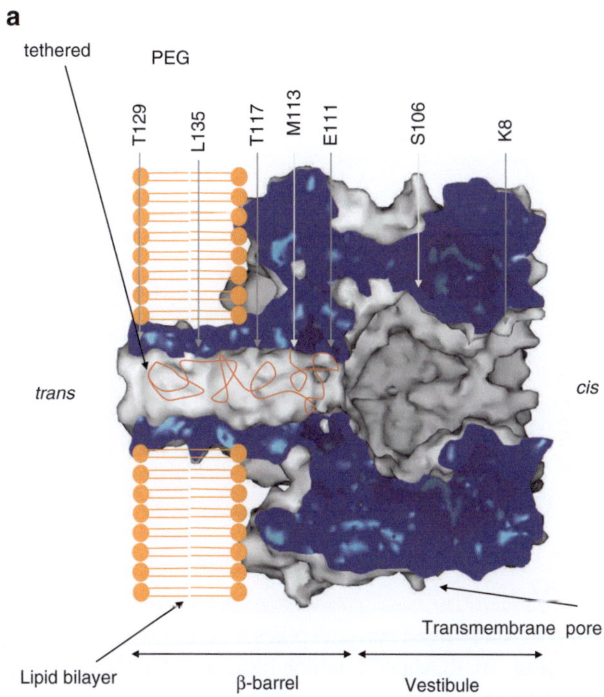

Fig. 10.3 Examples of single-channel electrical recordings monitoring single-molecule chemical reactions of individual sulfhydryl-directed PEGs-1 kDa reagent with a cysteine αHL mutant. (**a**) Cross-sectional view of the αHL protein pore illustrating sites where single cysteines were engineered for the exploration of polymer partitioning. The partitioning of poly(ethylene glycol)s (PEGs) of different sizes was examined by covalent modification of the pore lumen. The chemical reaction occurred between the reactive sulfhydryl-containing group of the PEG and the individual cysteine sulfhydryl of each reconstituted αHL mutant. Seven cysteine αHL mutants were examined at the locations labeled in the figure; (**b**) The cysteine was engineered within the β-barrel region of the pore, in position T117. This single-channel current resulted from voltage-induced ion translocation through the pore. The trace indicates permanent current blockades produced by individual, reacted polymers with cysteine sulfhydryls. Two distinguished permanent current blockades showed that two independent polymers reacted with distinct cysteine sulfhydryls within the pore lumen; (**c**) Application of the reducing agent to the chamber facilitated the cleavage of both PEGs from the pore lumen, producing a full recovery of the unitary conductance of αHL. In (**B**) and (**C**), the buffer solution contained 1 M KCl, 10 mM Tris.HCl, and 100 μM EDTA, pH 8.5. The applied transmembrane potential was +100 mV. The single-channel electrical traces were low-pass Bessel filtered at 10 kHz (Reproduced, with permission, from references (Movileanu et al. 2003) and (Movileanu and Bayley 2001). Copyright © (2001) National Academy of Sciences, U.S.A)

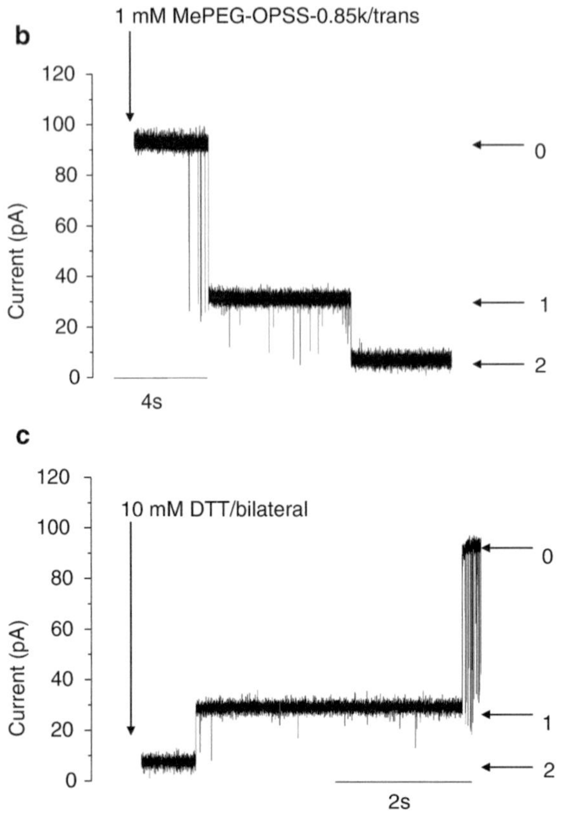

Fig. 10.3 (continued)

was accompanied by short-lived current spikes and followed by a second permanent current blockade (Fig. 10.3b, state 2), which indicated that a second polymer reacted with a cysteine sulfhydryl. Therefore, only two PEGs-1 kDa were accommodated at position T117C. These single covalent reactions were reversed upon addition of the reducing agent to the chamber (Fig. 10.3c). This example revealed the informative power of single-channel electrical recordings of the modified αHL protein pore, providing a clear indication on what is the total molecular size of the sulfhydryl-directed PEGs reacted at a desired location. Why would the two types of current blockade events have different amplitudes? The first reacted polymer reduced the ionic flux about 2/3 of that corresponding to the unitary conductance, whereas the second reacted polymer altered the ionic flux about 1/3 of that corresponding to the unitary conductance. This distinction might be determined by a difference in the local conformation of the reacted PEGs-1 kDa. Another possibility was that only a fraction of the second reactive polymer partitioned into the pore lumen, leaving the remaining fraction outside the pore.

PEGs produced a longer current blockade through the αHL protein pore lumen at increased salt concentrations, likely owing to electro-osmotic effect (Krasilnikov et al. 2006; Rodrigues et al. 2008). Taking into account this finding, John Kasianowicz and

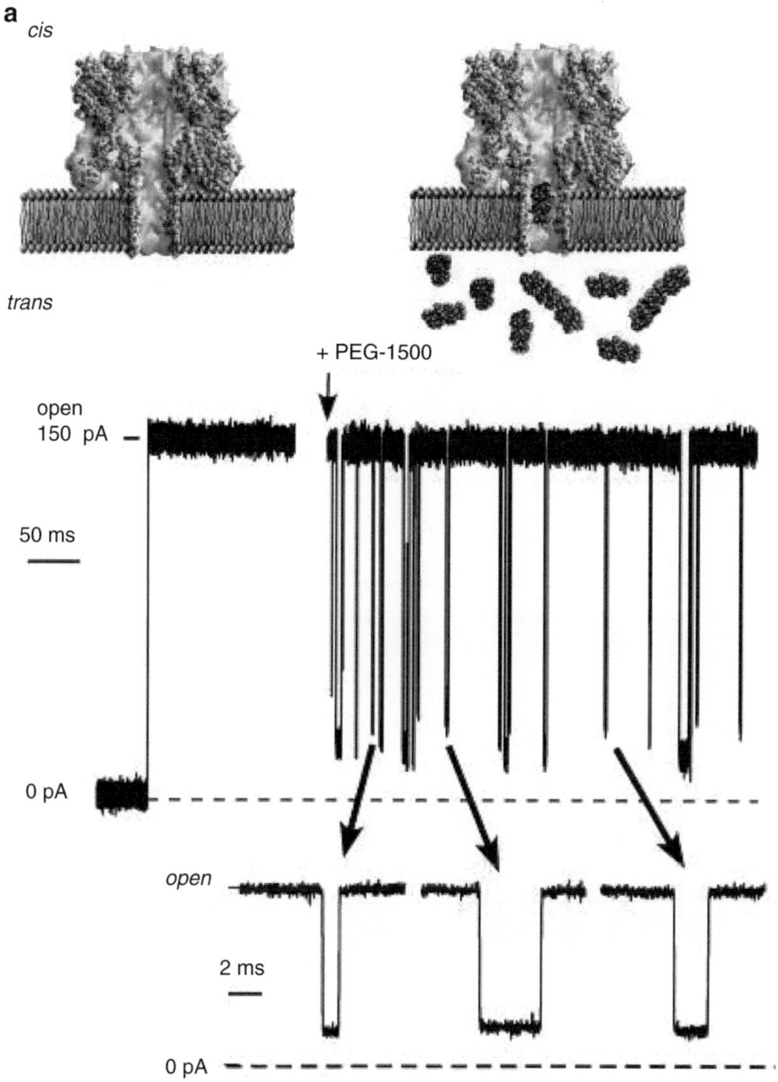

Fig. 10.4 Development of single-molecule mass spectrometry using a single protein nanopore. (**a**) The αHL protein pore was inserted into a planar lipid bilayer, enabling a quiet single-channel current. Application of a sample of polydisperse PEG to the chamber produced well-defined current blockages, the nature of which depended on the length of the PEG molecules as well as their concentration. The horizontal dashed lines show the level of the zero current; (**b**) A single αHL protein nanopore can be used to resolve the monodisperse (mPEG; with a molecular weight of 1.3 kDa; lower panels) and polydisperse (pPEG; with a molecular weight of 1.5 kDa; upper panels) PEG samples. The standard histograms of the current amplitudes indicated either a single, broad distribution (the top panel) or two well-separated and narrow distributions (the bottom panel). The buffer solution contained 4 M KCl, 5 mM Tris, pH 7.5. The applied transmembrane potential was +40 mV. In the bottom trace of (**b**), the peak given by the long-lived current blockades was interpreted in terms of impurities present in the mPEG sample (Reproduced, with permission, from reference (Robertson et al. 2007). Copyright © (2007) National Academy of Sciences, U.S.A)

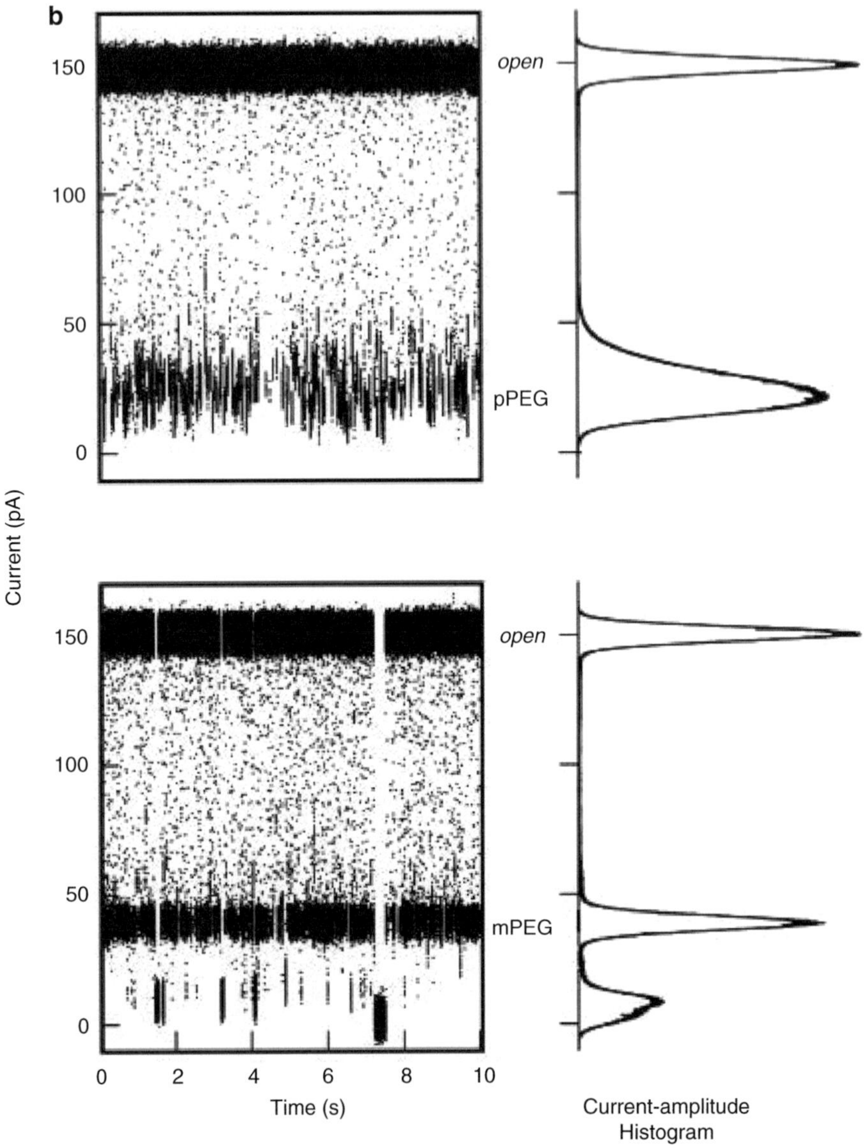

Fig. 10.4 (continued)

co-workers have developed an elegant nanopore-based, single-molecule mass spectrometry for sizing flexible polymers in solution (Robertson et al. 2007, 2010; Reiner et al. 2010; Baaken et al. 2011). The stunning discovery was that, under experimental conditions of high salt concentration, PEG molecules produced distinctive mass-dependent transient current blockades when they partitioned into the αHL protein pore.

In Fig. 10.4a, a quiet single-channel current is shown in the absence of PEG, but frequent and reversible current transitions were observed in the presence of PEGs added to the chamber. A very high salt concentration in the chamber was advantageously used for both increasing signal-to-noise ratio due to increase in the unitary conductance and prolonging the residence times due to local osmotic pressure on trapped PEGs into the pore lumen. In Fig. 10.4b, the top trace shows the open-state current and current blockades of varying amplitude made by a polydisperse sample of PEG (pPEG), whereas the bottom trace indicate uniform current blockades produced by a monodisperse sample of PEG (mPEG). In contrast to Fig. 10.4a, the time scale is compressed, so that the current blockades cannot be observed in full in Fig. 10.4b. In addition, all inter-event intervals, otherwise distinguishable in Fig. 10.4a, were removed in Fig. 10.4b. Such a single-channel electrical trace representation enabled an easier determination that the pPEG histogram distribution is noisier than that acquired with mPEG. pPEG sample contained a mixture of polymer molecules with a number of repetitive units varying from 25 to 50. However, average molecular weight of pPEG was similar to that of mPEG. There is a clear distinction between the results obtained with mPEG and pPEG. The current-amplitude histograms demonstrated a single and broad distribution with pPEG. In contrast, the current-amplitude histogram obtained with mPEG revealed two narrower distributions. The second level of the current blockades, given by the long-lived events, in the experiment using mPEG, was interpreted in terms of impurities in the PEG sample. Using the mean current value of each blockade, histograms with multiple current-blockade distributions correlated with multiple molecular-mass PEGs, demonstrating the informative power of these measurements for discriminating various PEGs within the same sample.

Remarkably, this method has proven the ability to resolve the repetitive unit of PEG, which was beyond the previous attempts (Bezrukov et al. 1996; Movileanu et al. 2003). The residence time of PEG within the pore lumen was increasing with the increase in the molecular mass of PEG in a monotonic fashion. It was also demonstrated that sizing and discriminating analytes in real time can be accomplished in the case of neutral polymeric species. We do not anticipate challenges in applying this methodology for other neutral, water-soluble and flexible polymers. Modelling the transient current blockades as signal responses given by an equivalent electrical circuit has permitted an additional level of analysis of analyte-produced events, increasing the resolution of the residence times produced by flexible polymers and open sub-states at sub-picoampere level (Balijepalli et al. 2014). Finally, it is worth mentioning that all traits of the αHL protein pore ignited persistent interest of computational biophysicists in the modeling of ion transport across channels and pores, a primary goal of modern electrophysiology (Noskov et al. 2004; Aksimentiev and Schulten 2005; Wells et al. 2007).

10.4 Structural Properties of Leukotoxin Monomers

S. aureus secretes not only homomeric, but also heteromeric β-PFTs. For example, a mandatory interaction between two distinct monomers, one that is part of the class F polypeptides and one that belongs to the class S polypeptides, results in the formation of heteromeric or bi-component β-PFTs. They include leukocidins (Luk), γ-hemolysins (γHL), and Pantom-Valentine leukocidins (PVL) (Ferreras et al. 1998;

Olson et al. 1999; Pedelacq et al. 1999; Prevost et al. 2001; Sugawara-Tomita et al. 2002; Guillet et al. 2004; Yoong and Torres 2013; Alonzo and Torres 2014). No experimental evidence is available on direct interactions between two similar monomers of class F or two similar monomers of class S. Therefore, homomeric β-PFTs containing either class F or class S polypeptides have never been detected. Because each class of leukotoxin polypeptides includes several members, four bi-component β-PFTs have been determined so far (Yoong and Torres 2013; Alonzo and Torres 2014). PVL complexes are formed by LukS-PV and LukF-PV subunits. The γHL protein pores are made by either HlgA and HlgB subunits or HlgC and HlgB subunits (Comai et al. 2002). LukED leukotoxins consist of LukE and LukD polypeptides. LukAB/LukGH toxins are made by either LukA and LukB or LukH and LukG.

The diversity of bi-component β-PFTs is also consistent with their distinctive lytic activities with respect to various cell membrane models (Yoong and Torres 2013; DuMont and Torres 2014; Alonzo and Torres 2014). All four types of leukocidins produce the death of human neutrophils, whereas only LukED and γHL induce lysis against erythrocytes (Yoong and Torres 2013). The mechanisms by which the individual monomers bind to the cell membrane surface, then oligomerize and assemble into a transmembrane β-barrel protein pore are not yet completely understood. These bi-component β-PFTs efficiently assemble on lipid membranes either in the absence (Miles et al. 2001, 2002b; Jayasinghe and Bayley 2005; Holden et al. 2006) or presence of cellular receptors located on the surface of the susceptible cells (Yoong and Torres 2013; DuMont and Torres 2014; Alonzo and Torres 2014).

In Fig. 10.5, we illustrate the relationships among the polypeptides of the homomeric and heteromeric β-PFTs. The sequence identity of the members of either class with that of αHL is always less than 30 %. There is an increased sequence identity among the members within the same class: 71–79 % and 59–79 % for the class F and S polypeptides, respectively (Gouaux et al. 1997). The X-ray crystal structure of LukF reveals that its fold looks closely similar to that of the αHL protomer in the heptameric complex, except for two domains that are intimately involved in the protomer-protomer contacts and in the binding of the monomer to the membrane surface (Olson et al. 1999). These are the amino latch and pre-stem domains and they are involved in the protein-protein and protein-membrane interfaces, respectively. Parallel studies by Pédelacq and colleagues revealed an almost identical fold of the water-soluble LukF-PV polypeptide (Pedelacq et al. 1999). Later, the same group was able to determine the structure of the water-soluble LukS-PV polypeptide at 2.0 Å resolution (Guillet et al. 2004). Interestingly, it was found that the overall structure of LukS-PV is identical to that of the water-soluble LukF monomer and the αHL protomer, except for the domain that interfaces its interaction with the lipid membrane (the rim domain, Fig. 10.1a). This finding suggested a different propensity of LukS-PV to interact in a distinct fashion with the membrane surface of the susceptible cells.

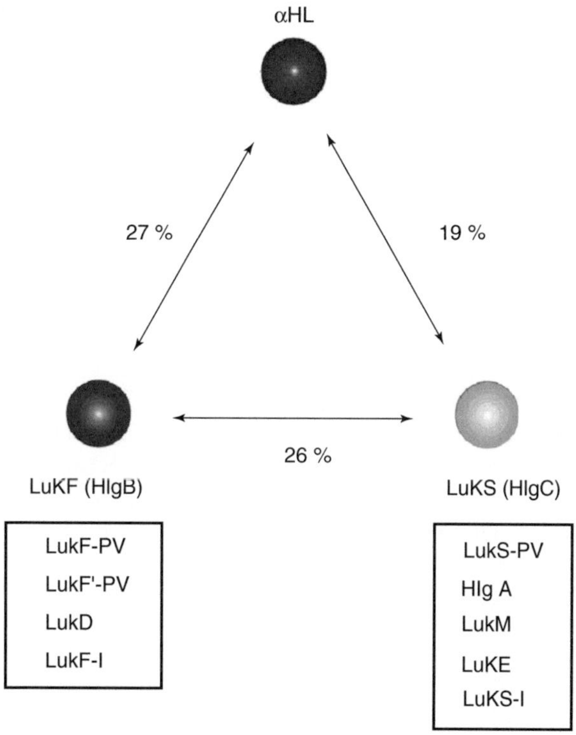

Fig. 10.5 Residue sequence relationships among αHL, LukF (HlgB), and LukS (HlgC). The boxes indicate other components of the F and S classes of the staphylococcal bi-component β-PFTs. HlgA, HlgB, and HlgC are polypeptides corresponding to γHL (Reproduced, with permission, from Reference (Miles et al. 2002b))

10.5 Electrophysiological Traits of the Staphylococcal Bi-component β-PFTs

Based on similarities in protein folding between each of the leukotoxin monomers with the αHL protomers (Olson et al. 1999; Pedelacq et al. 1999; Guillet et al. 2004), the staphylococcal bi-component β-PFTs were expected to form large-conductance pores in all model and cell membranes. In accord with this anticipation, Miles and colleagues found that the bi-component Luk protein complex (LukF/LukS) forms transmembrane pores on planar lipid membranes with a single-channel conductance of ~2540 pS in 1000 mM KCl, 5 mM HEPES, pH 7.4 (Table 10.2) (Miles et al. 2001). The single-channel conductance of the LukF/LukS protein pore is ~3-fold greater than that of the αHL protein pore. The value of the unitary conductance of the heptameric hemolysin II, a β-PFT of *B. cereus*, measured under similar experimental contexts, is 630 pS (Miles et al. 2002a). A large difference between the single-channel conductance of the LukF/LukS and αHL protein pores or the LukF/LukS and *B. cereus* hemolysin II protein pores suggested that the number of participating subunits to the LukF/LukS protein pore complex is greater than that corresponding to the homomeric β-PFTs.

Moreover, systematic measurements of the reversal potentials under various ionic gradients have revealed that the LukF/LukS pores represent weakly cation selective pathways across the lipid bilayers (Table 10.2). This weak cation selectivity is surprising, given the lack of negative charges within the narrowest region of the lumen, the β-barrel part of the pore, except for a single residue, Asp-122 of LukF. Another mechanism of this selectivity might be the cluster of acidic residues within the cap domain, facing the opening of the pore and in the proximity of the N-termini of both LukF and LukS polypeptides (Comai et al. 2002). Such negatively charged residues do not exist within the same location of the αHL polypeptide. The LukF/LukS protein pores exhibit voltage-induced current gating at positive transmembrane potentials greater than 60 mV, and in 1000 mM KCl, 5 mM HEPES, pH 7.4, but a remarkable stability of the open-state current at large negative applied transmembrane potentials (e.g., 120 mV or greater than this value). Again, this is in contrast to αHL protein pore, whose open-state current remains stable for long periods at positive applied potentials up to +160 mV, but undergoes gating fluctuations at negative applied transmembrane potentials greater than 100 mV (Movileanu et al. 2005). It might be possible that the gating mechanisms of the two β-PFTs are different. In the case of the αHL protein, brief current flickering at positive potentials and at low, physiological salt concentrations are likely caused by the localized fluctuations of the structurally flexible glycine-rich turn of the β-barrel domain of the pore (Mohammad and Movileanu 2010).

Extensive studies on other leukotoxins have confirmed increased unitary conductance and weak cation selectivity. For instance, Comai and colleagues determined that γHLs, consisting of either HlgA/HlgB or HlgC/HlgB, exhibit a unitary conductance of either 115 or 192 pS, respectively, in 100 mM NaCl (Table 10.2) (Comai et al. 2002). These values are greater than that measured with αHL under similar experimental conditions (~92 pS) (Menestrina 1986). Using PEGs in osmotic-protection experiments, the effective pore diameter of γHLs was 20–24 Å (Ferreras et al. 1998). The above-mentioned hypothesis that the weak cation selectivity is produced by a cluster of acidic residues located within the cap domain near the pore opening was confirmed by single- and multiple-site mutagenesis (Comai et al. 2002). Indeed, by engineering charge-reversal mutations of negative → positive charge, namely converting all acidic side chains of the cluster into basic residues, the unitary conductance and ion selectivity of αHL were obtained. Therefore, this cluster of negative charges in γHLs represents a selectivity filter.

10.6 Subunit Composition of Staphylococcal Bi-component β-PFTs

Given extensive biochemical and biophysical studies employing staphylococcal bi-component β-PFTs, along with detailed structural information about the leukotoxin oligomers, one would immediately ask what is the subunit stoichiometry of the protein pore complex? Because the single-channel conductance of the LukF/LukS protein complex is about three times greater than that of αHL (Miles et al. 2002b), it is likely that the number of participating subunits of the LukF/LukS protein complex is greater than seven. Heteromeric composition of the LukF/LukS protein complex has been addressed using two independent methodologies: (i) gel-shift electrophoresis assay, and (ii) single-channel electrical recordings, along with chemical modification at strategic locations of the pore interior (Miles et al. 2002b).

In the first approach, individual LukF and LukS polypeptides were fused to the C-terminal, 94-residue extension of the *B. cereus* hemolysin II (TL). Other β-PFTs lack this C-terminal extension. The primary motivation for this protein design was the TL-induced alteration of the gel-shift mobility of the LukF and LukS polypeptides could be noted on an SDS-PAGE gel. This strategy was coupled with the structural stability of the oligomers in SDS-PAGE gels. Such an approach enabled the *in vitro* transcription and translation (IVTT) of the protein complex assembled on rabbit red blood cell membranes (rrbcm), and then SDS-PAGE gel separation of the engineered heteromer complexes containing subunits with or without TL. In this way, various heteromers containing either LukF-TL and LukS subunits or LukF and LukS-TL subunits were resolved on the SDS-PAGE gels.

For example, in the first experiment heteromers containing various ratios of wild-type LukF (WT LukF) and LukF-TL with wild-type LukS (WT LukS) were produced by cotranslation in the presence rrbcm. Bands with increments in electrophoretic mobility were observed. Four new bands were observed in addition to the band corresponding to WT LukF/WT LukS. The fastest band was attributed to WT LukF/WT LukS. The TL-induced, slow-mobility bands were produced by heteromers with either one, two, three, or four LukF-TL subunits. This finding indicated that the LukF/LukS protein complex contains four LukF subunits. In the second experiment, a similar approach was employed to examine the contribution of LukS to the heteromeric complex. It was determined that LukF/LukS contains four LukS subunits as well. In brief, such gel-shift electrophoresis assay on SDS-PAGE gels, along with protein design, indicated that the LukS/LukF protein complex is an octamer.

In the second approach, targeted chemical modification of engineered cysteines within the pore lumen, along with single-channel electrical recordings were employed (Fig. 10.6). To some extent, this second approach was quite related to the previous test. This is becuase the gel-shift assay used for the engineered heteromers was replaced by the unitary conductance change produced by individual chemical reactions of a sulfhydryl-specific reagent sodium (2-sulfonatoethyl)methanethiosulfonate (MTSES) molecule with an engineered cysteine sulfhydryl of a Luk polypeptide within the complex (either LukF or LukS). These studies proved that both LukF-S124C/WT LukS and WT LukF/LukS-A122C underwent four distinct current blockades when MTSES was applied to the bilayer chamber. Recovery of the unitary conductance was achieved by the addition of the DTT reducing reagent. This conductance recovery was observed in the form of four step-wise current upshifts, suggesting that the LukS/LukF protein complex is an octamer.

Independently of and in parallel with these studies, Sugawara-Tomita and colleagues, using high-resolution electron microscopy and chemical cross-linking experiments, discovered that γHL assembles in a stochastic fashion as a heptamer of either 4 LukF subunits and 3 Hlg2 subunits or 3 LukF subunits and 4 Hlg2 sub-

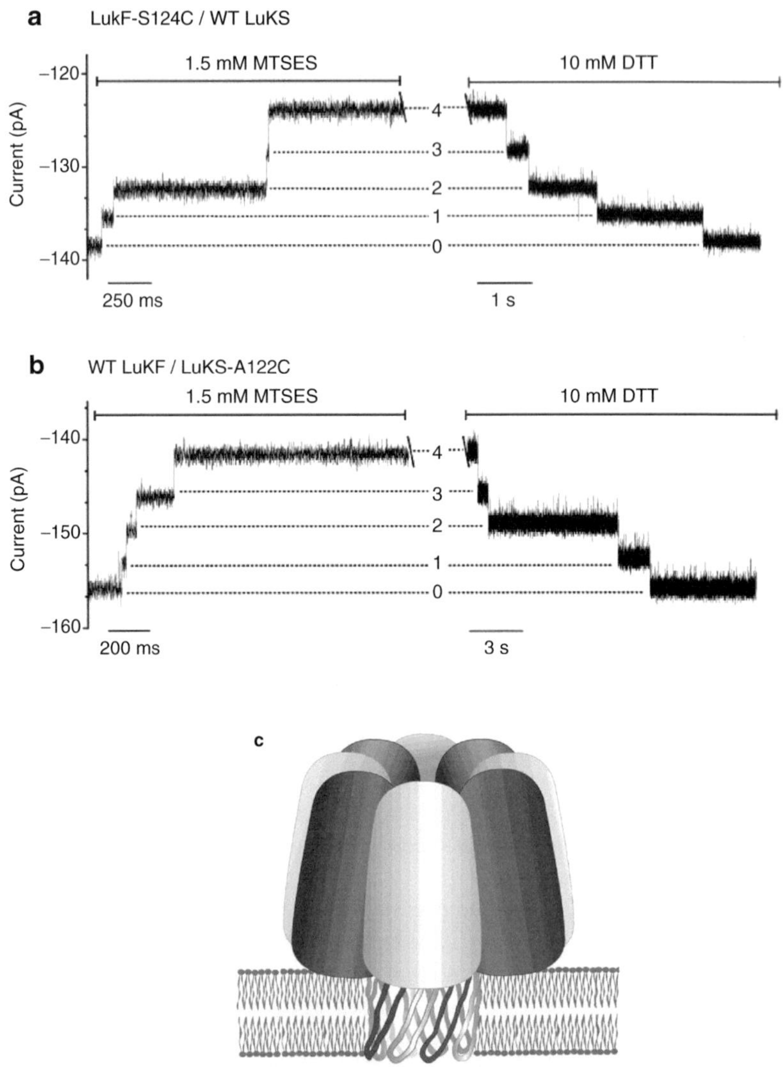

Fig. 10.6 Subunit stoichiometry of a bi-component leukocidin (LukF/LukS) protein complex can be determined using single-channel electrical recordings, membrane protein design, and chemical modification. (**a**) Single-channel electrical recordings probing cysteine Luk mutant LukF-S124C/ WT LukS reacted with MTSES; (**b**) Single-channel electrical recordings probing cysteine Luk mutant WT LukF/LukS A122C reacted with MTSES. Here, WT stands for wild type. Permanent cleavage of MTSES from the protein surface was achieved after direct application of the DTT reducing agent to the chamber. The number of reacted individual cysteine sulfhydryls is displayed on the central part of each panel, in between the traces; (**c**) A simplified cartoon of the model of the staphylococcal LukF/LukS pore complex. Each type of polypeptide contributes four subunits to the entire octameric protein complex. LukF is marked in dark grey, whereas LukS is marked in light grey. This model suggests that the leukocidin channel is a 16-stranded β-barrel with an alternating distribution of LukF and LukS, featuring a four-fold symmetry (Table 10.1). In (**a**, **b**), the buffer solution contained 1 M KCl, 50 mM Tris.HCl, 200 μM DTT, 100 μM EDTA, pH 8. The applied transmembrane potential was −60 mV. The single-channel electrical traces were low-pass Bessel filtered at 0.5 kHz (Reproduced, with permission, from Reference (Miles et al. 2002b))

units (Sugawara-Tomita et al. 2002). Such parallel studies from two different groups, offering two distinct outcomes to the same question, have ignited controversies over the true subunit composition of the staphylococcal bi-component β-PFTs. This was especially true, given the lack of the availability of the high-resolution X-ray crystal structure of any leukotoxin complex. Further work by Jayasinghe and Bayley, using chemical cross-linking and clever protein engineering, has confirmed that the LukF/LukS complex is indeed an octamer formed by four LukF and four LukS subunits (Jayasinghe and Bayley 2005). Moreover, using the fusion of LukF with LukS, they showed that the Luk oligomer features a four-fold symmetry with alternating LukF and LukS subunits.

More recently, Tanaka and Kamio's groups have been able to crystalize γHL protein complex at 2.5 Å resolution (Yamashita et al. 2011). This was indeed a breakthrough in the field of staphylococcal β-PFTs, because it was the first crystal structure of a bi-component β-PFT, and it was the first structure of a heteromeric transmembrane β-barrel protein complex. Remarkably, the structure of the complex revealed an octamer with a four-fold symmetry axis about which LukF and Hlg2 alternate. This discovery is consistent with the previous biophysical and biochemical examinations in Bayley's group (Miles et al. 2002b; Jayasinghe and Bayley 2005) and contrasts with early high-resolution electron microscopy and chemical cross-linking experiments using γHL (Sugawara-Tomita et al. 2002). The structure of γHL features a mushroom shape, closely similar with αHL, with the largest diameter of the cap measuring 114 Å, whereas the length of the pore lumen is 93 Å. Interestingly, the internal diameter of the cylindrical β-barrel domain, excluding the side chains of the internal residues is ~25.5 Å, again similar to that of αHL (~26 Å), whereas its total length is ~47 Å. It is surprising that despite its additional subunit, as compared to αHL, the internal dimensions as well as the overall internal shape of the lumen resemble those of αHL. This finding is in accord with a much smaller difference between the unitary conductance of γHL (~115 pS, HlgA/HlgB) and αHL (~92 pS) recorded in 100 mM NaCl (Table 10.2) (Menestrina 1986; Comai et al. 2002). Moreover, a smaller conductance of αHL might be explained by the presence of very bulky residues (e.g., Met-113, Lys-147) forming the central constriction of the pore lumen (Song et al. 1996). Each γHL protomer is made by three domains: cap, rim and stem.

Given this structural information on the staphylococcal bi-component γHL octamer, Yamashita and colleagues have proposed a schematic model for its momomer binding, dimerization, pre-pore formation, and pore assembly (Yamashita et al. 2011) (Fig. 10.7). The monomer binding to the surface is mediated by LukF via a cluster of hydrophobic residues. This hydrophobic cluster includes Tyr72, Trp257, Phe260, and Tyr261 (marked in red). Other residues that are marked in green, including Arg-198, form hydrophilic interface contacts with the headgroups of the bilayer lipids, just in the proximity of the membrane surface. Interface 1, which is exposed upon LukF binding to the membrane, mediates the contact interactions for the dimerization process of LukF-Hlg2. The dimerization undergoes a reorientation process, which activates the dimer for its participation in the pre-pore formation. This process is followed by the pre-stem unfolding from the cap domain,

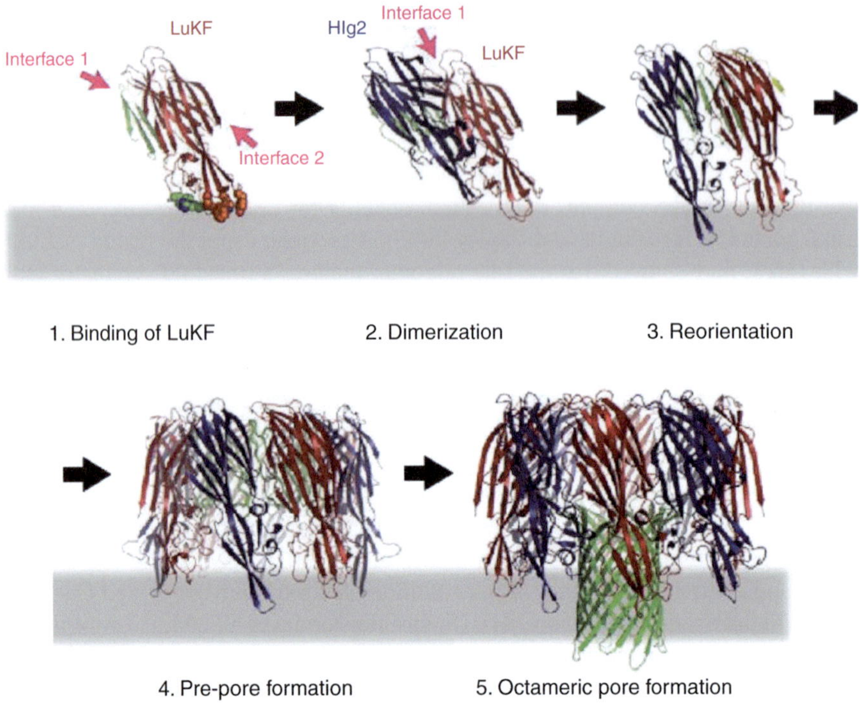

Fig. 10.7 A step-wise model of the oligomerization and assembly of γHL. This model reveals that the initiation process is governed by the spontaneous binding (1) of LukF to the surface of the lipid bilayer, which is illustrated by a gray horizontal bar. The binding of LukF to the membrane surface occurs via a hydrophobic cluster of residues, which includes Tyr72, Trp257, Phe260, and Tyr261 (indicated in *red*). This process is also mediated by Trp177 and Arg198 (marked by *green spheres*). This initial step is followed by a dimerization process (2) of the LukF-Hlg2 complex. Then, the heterodimer undergoes a re-orientation transformation (3) to bring about the dimer from an inactivated to an activated state. Such a transition enables a rapid heteromer growth into tetramers, hexamers, and ultimately octamers, generating the pre-pore formation (4). This process will lead to the complete assembly of the octameric γHL transmembrane pore (5), featuring a four-fold symmetry (Reproduced, with permission, from Reference (Yamashita et al. 2011). Copyright © (2011) National Academy of Sciences, U.S.A.)

triggering its full partitioning into the hydrophobic side of the bilayer as a bi-component transmembrane β-barrel pore.

10.7 Conclusions

This chapter represents a very brief overview of the extensive developments on the structure and function of staphylococcal β-PFTs. The αHL protein is indeed the archetype of the β-PFTs. There are many commonalities of these transmembrane β-barrel proteins. They form mushroom-shaped pore structures on lipid bilayers of model and cell membranes. The cap domain serves at least two tasks: (i) a large hydrophilic part of the polypeptide makes it water soluble during its navigation from the secreting to attacked cell, and (ii) a large geometrical obstruction would control the partitioning of the β-barrel part into the bilayer and would anchor the protein into the lipid membrane via its rim domain. This particular structure is attained after complex structural transformations of the monomer, in the case of the homomeric toxin, or the dimer, in the case of the bi-component β-PFTs. The transition of water-soluble monomeric form to the transmembrane pore shows the richness and versatility of the protein scaffold for unusual adaptations from the aqueous phase to hydrophobic milieu of the lipid bilayer. The diversity of staphylococcal bi-component β-PFTs has evolutionary been accomplished by pivotal mutations at the protomer-protomer and monomer-membrane interfaces. Staphylococcal β-PFTs are weakly selective with a preference for anions and cations for homomeric and bi-component toxins, respectively. The latter is mainly determined by the presence of an electrostatic selectivity filter, which consists of a cluster of negatively charged residues, on the pore opening near the cap domain. Recent functional and crystallographic studies represent a solid platform for further protein engineering developments of these β-PFTs with great prospects in numerous areas of fundamental and clinical biomedical nanotechnology.

Acknowledgments We are grateful to members of the Movileanu laboratory for their constructive comments. We realized the challenging nature of writing a chapter about a β-barrel toxin that has transformed the area of nanopore biophysics. Therefore, we regret that due to space limitations we were unable to introduce all exciting publications pertinent to this field. This work was supported by the National Institutes of Health, Grant GM088403 (to L.M.).

References

Akeson M, Branton D, Kasianowicz JJ, Brandin E, Deamer DW (1999) Microsecond time-scale discrimination among polycytidylic acid, polyadenylic acid, and polyuridylic acid as homopolymers or as segments within single RNA molecules. Biophys J 77:3227–3233

Aksimentiev A, Schulten K (2005) Imaging {alpha}-Hemolysin with Molecular Dynamics: Ionic Conductance, Osmotic Permeability, and the Electrostatic Potential Map. Biophys J 88:3745–3761

Alonzo F III, Torres VJ (2014) The bicomponent pore-forming leucocidins of Staphylococcus aureus. Microbiol Mol Biol Rev 78:199–230

Astier Y, Bayley H, Howorka S (2005) Protein components for nanodevices. Curr Opin Chem Biol 9:576–584

Baaken G, Ankri N, Schuler AK, Ruhe J, Behrends JC (2011) Nanopore-based single-molecule mass spectrometry on a lipid membrane microarray. ACS Nano 5:8080–8088

Balijepalli A, Ettedgui J, Cornio AT, Robertson JW, Cheung KP, Kasianowicz JJ, Vaz C (2014) Quantifying short-lived events in multistate ionic current measurements. ACS Nano 8:1547–1553

Bayley H (2006) Sequencing single molecules of DNA. Curr Opin Chem Biol 10:628–637

Bayley H, Braha O, Cheley S, Gu LQ (2004) Engineered nanopores. In: Mirkin CA, Niemeyer CM (eds) NanoBiotechnology. Wiley-VCH Verlag GmbH & Co. KGaA, Weinheim, pp 93–112

Bayley H, Cremer PS (2001) Stochastic sensors inspired by biology. Nature 413:226–230

Bayley H, Jayasinghe L (2004) Functional engineered channels and pores (review). Mol Membr Biol 21:209–220

Benner S, Chen RJ, Wilson NA, Abu-Shumays R, Hurt N, Lieberman KR, Deamer DW, Dunbar WB, Akeson M (2007) Sequence-specific detection of individual DNA polymerase complexes in real time using a nanopore. Nat Nanotechnol 2:718–724

Bezrukov SM (2000) Ion channels as molecular Coulter counters to probe metabolite transport. J Membr Biol 174:1–13

Bezrukov SM, Kasianowicz JJ (1993) Current noise reveals protonation kinetics and number of ionizable sites in an open protein ion channel. Phys Rev Lett 70:2352–2355

Bezrukov SM, Krasilnikov OV, Yuldasheva LN, Berezhkovskii AM, Rodrigues CG (2004) Field-Dependent Effect of Crown Ether (18-Crown-6) on Ionic Conductance of {alpha}-Hemolysin Channels. Biophys J 87:3162–3171

Bezrukov SM, Vodyanoy I, Brutyan RA, Kasianowicz JJ (1996) Dynamics and free energy of polymers partitioning into a nanoscale pore. Macromolecules 29:8517–8522

Bischofberger M, Gonzalez MR, van der Goot FG (2009) Membrane injury by pore-forming proteins. Curr Opin Cell Biol 21:589–595

Braha O, Walker B, Cheley S, Kasianowicz JJ, Song LZ, Gouaux JE, Bayley H (1997) Designed protein pores as components for biosensors. Chem Biol 4:497–505

Branton D, Deamer DW, Marziali A, Bayley H, Benner SA, Butler T, Di Ventra M, Garaj S, Hibbs A, Huang X, Jovanovich SB, Krstic PS, Lindsay S, Ling XS, Mastrangelo CH, Meller A, Oliver JS, Pershin YV, Ramsey JM, Riehn R, Soni GV, Tabard-Cossa V, Wanunu M, Wiggin M, Schloss JA (2008) The potential and challenges of nanopore sequencing. Nat Biotechnol 26:1146–1153

Brochard-Wyart F, De Gennes P-G, Sandre O (2000) Transient pores in stretched vesicles: role of leak-out. Physica A 278:32–51

Cheley S, Malghani MS, Song LZ, Hobaugh M, Gouaux JE, Yang J, Bayley H (1997) Spontaneous oligomerization of a staphylococcal alpha-hemolysin conformationally constrained by removal of residues that form the transmembrane beta-barrel. Protein Eng 10:1433–1443

Cherf GM, Lieberman KR, Rashid H, Lam CE, Karplus K, Akeson M (2012) Automated forward and reverse ratcheting of DNA in a nanopore at 5-A precision. Nat Biotechnol 30:344–348

Comai M, Serra MD, Coraiola M, Werner S, Colin DA, Monteil H, Prevost G, Menestrina G (2002) Protein engineering modulates the transport properties and ion selectivity of the pores formed by staphylococcal gamma- haemolysins in lipid membranes. Molecular Microbiology 44:1251–1267

Czajkowsky DM, Sheng ST, Shao ZF (1998) Staphylococcal alpha-hemolysin can form hexamers in phospholipid bilayers. J Mol Biol 276:325–330

De Gennes P-G (1999a) Flexible polymers in nanopores. Adv Polym Sci 138:91–105

De Gennes P-G (1999b) Passive entry of a DNA molecule into a small pore. Proc Natl Acad Sci U S A 96:7262–7264

De Gennes P-G (1999c) Problems of DNA entry into a cell. Physica A 274:1–7

DuMont AL, Torres VJ (2014) Cell targeting by the Staphylococcus aureus pore-forming toxins: it's not just about lipids. Trends Microbiol 22:21–27

Fang Y, Cheley S, Bayley H, Yang J (1997) The heptameric prepore of a Staphylococcal alpha-hemolysin mutant in lipid bilayers imaged by atomic force microscopy. Biochemistry 36:9518–9522

Ferreras M, Hoper F, Dalla Serra M, Colin DA, Prevost G, Menestrina G (1998) The interaction of Staphylococcus aureus bi-component gamma- hemolysins and leucocidins with cells and lipid membranes. Biochim Biophys Acta -Biomembr 1414:108–126

Gonzalez MR, Bischofberger M, Pernot L, van der Goot FG, Freche B (2008) Bacterial pore-forming toxins: the (w)hole story? Cell Mol Life Sci 65:493–507

Goodrich CP, Kirmizialtin S, Huyghues-Despointes BM, Zhu AP, Scholtz JM, Makarov DE, Movileanu L (2007) Single-molecule electrophoresis of beta-hairpin peptides by electrical recordings and Langevin dynamics simulations. J Phys Chem B 111:3332–3335

Gouaux E (1998) Alpha-hemolysin from staphylococcus aureus: an archetype of beta-barrel, channel-forming toxins. J Struct Biol 121:110–122

Gouaux E, Hobaugh M, Song LZ (1997) Alpha-hemolysin, gamma-hemolysin, and leukocidin from staphylococcus aureus: distant in sequence but similar in structure. Protein Sci 6:2631–2635

Gouaux JE, Braha O, Hobaugh MR, Song LZ, Cheley S, Shustak C, Bayley H (1994) Subunit stoichiometry of staphylococcal Alfa-hemolysin in crystals and on membranes - a heptameric transmembrane pore. Proc Natl Acad Sci U S A 91:12828–12831

Gu LQ, Bayley H (2000) Interaction of the noncovalent molecular adapter, beta- cyclodextrin, with the staphylococcal alpha-hemolysin pore. Biophys J 79:1967–1975

Gu LQ, Braha O, Conlan S, Cheley S, Bayley H (1999) Stochastic sensing of organic analytes by a pore-forming protein containing a molecular adapter. Nature 398:686–690

Gu LQ, Cheley S, Bayley H (2003) Electroosmotic enhancement of the binding of a neutral molecule to a transmembrane pore. Proc Natl Acad Sci U S A 100:15498–15503

Guillet V, Roblin P, Werner S, Coraiola M, Menestrina G, Monteil H, Prevost G, Mourey L (2004) Crystal structure of leucotoxin S component: new insight into the Staphylococcal beta-barrel pore-forming toxins. J Biol Chem 279:41028–41037

Gurnev PA, Nestorovich EM (2014) Channel-forming bacterial toxins in biosensing and macromolecule delivery. Toxins (Basel) 6:2483–2540

Hall AR, Scott A, Rotem D, Mehta KK, Bayley H, Dekker C (2010) Hybrid pore formation by directed insertion of alpha-haemolysin into solid-state nanopores. Nat Nanotechnol 5:874–877

Heuck AP, Tweten RK, Johnson AE (2001) Beta-Barrel pore-forming toxins: intriguing dimorphic proteins. Biochemistry 40:9065–9073

Holden MA, Jayasinghe L, Daltrop O, Mason A, Bayley H (2006) Direct transfer of membrane proteins from bacteria to planar bilayers for rapid screening by single-channel recording. Nat Chem Biol 2:314–318

Hornblower B, Coombs A, Whitaker RD, Kolomeisky A, Picone SJ, Meller A, Akeson M (2007) Single-molecule analysis of DNA-protein complexes using nanopores. Nat Methods 4:315–317

Howorka S, Movileanu L, Lu XF, Magnon M, Cheley S, Braha O, Bayley H (2000) A protein pore with a single polymer chain tethered within the lumen. J Am Chem Soc 122:2411–2416

Howorka S, Siwy Z (2008) Nanopores: generation, engineering and single-molecule applications. In: Hinterdorfer P (ed) Handbook of single-molecule biophysics. Springer, New York

Howorka S, Siwy Z (2009) Nanopore analytics: sensing of single molecules. Chem Soc Rev 38:2360–2384

Iacovache I, Bischofberger M, van der Goot FG (2010) Structure and assembly of pore-forming proteins. Curr Opin Struct Biol 20:241–246

Iacovache I, van der Goot FG, Pernot L (2008) Pore formation: an ancient yet complex form of attack. Biochim Biophys Acta 1778:1611–1623

Jayasinghe L, Bayley H (2005) The leukocidin pore: evidence for an octamer with four LukF subunits and four LukS subunits alternating around a central axis. Protein Sci 14:2550–2561

Jung Y, Bayley H, Movileanu L (2006) Temperature-responsive protein pores. J Am Chem Soc 128:15332–15340

Kang XF, Gu LQ, Cheley S, Bayley H (2005) Single protein pores containing molecular adapters at high temperatures. Angew Chem Int Ed Engl 44:1495–1499

Kasianowicz JJ, Bezrukov SM (1995) Protonation dynamics of the alpha-toxin ion-channel from spectral-analysis of ph-dependent current fluctuations. Biophys J 69:94–105

Kasianowicz JJ, Brandin E, Branton D, Deamer DW (1996) Characterization of individual polynucleotide molecules using a membrane channel. Proc Natl Acad Sci U S A 93:13770–13773

Kolomeisky AB (2007) Channel-facilitated molecular transport across membranes: attraction, repulsion, and asymmetry. Phys Rev Lett 98:048105

Kolomeisky AB (2008) How polymers translocate through pores: memory is important. Biophys J 94:1547–1548

Kong CY, Muthukumar M (2005) Simulations of stochastic sensing of proteins. J Am Chem Soc 127:18252–18261
Korchev YE, Alder GM, Bakhramov A, Bashford CL, Joomun BS, Sviderskaya EV, Usherwood PN, Pasternak CA (1995) Staphylococcus aureus alpha-toxin-induced pores: channel-like behavior in lipid bilayers and patch clamped cells. J Membr Biol 143:143–151
Krasilnikov OV, Bezrukov SM (2004) Polymer partitioning from nonideal solutions into protein voids. Macromolecules 37:2650–2657
Krasilnikov OV, Merzlyak PG, Yuldasheva LN, Rodrigues CG, Bhakdi S, Valeva A (2000) Electrophysiological evidence for heptameric stoichiometry of ion channels formed by Staphylococcus aureus alpha-toxin in planar lipid bilayers. Molecular Microbiology 37:1372–1378
Krasilnikov OV, Rodrigues CG, Bezrukov SM (2006) Single polymer molecules in a protein nanopore in the limit of a strong polymer-pore attraction. Phys Rev Lett 97:018301
Kusters I, van Oijen AM, Driessen AJ (2014) Membrane-on-a-chip: microstructured silicon/silicon-dioxide chips for high-throughput screening of membrane transport and viral membrane fusion. ACS Nano 8:3380–3392
Los FC, Randis TM, Aroian RV, Ratner AJ (2013) Role of pore-forming toxins in bacterial infectious diseases. Microbiol Mol Biol Rev 77:173–207
Maffeo C, Bhattacharya S, Yoo J, Wells D, Aksimentiev A (2012) Modeling and simulation of ion channels. Chem Rev 112:6250–6284
Maglia G, Heron AJ, Hwang WL, Holden MA, Mikhailova E, Li Q, Cheley S, Bayley H (2009) Droplet networks with incorporated protein diodes show collective properties. Nat Nanotechnol 4:437–440
Maglia G, Restrepo MR, Mikhailova E, Bayley H (2008) Enhanced translocation of single DNA molecules through {alpha}-hemolysin nanopores by manipulation of internal charge. Proc Natl Acad Sci U S A 105:19720–19725
Majd S, Yusko EC, Billeh YN, Macrae MX, Yang J, Mayer M (2010) Applications of biological pores in nanomedicine, sensing, and nanoelectronics. Curr Opin Biotechnol 21:439–476
Mayer M, Yang J (2013) Engineered ion channels as emerging tools for chemical biology. Acc Chem Res 46:2998–3008
Meller A, Nivon L, Brandin E, Golovchenko J, Branton D (2000) Rapid nanopore discrimination between single polynucleotide molecules. Proc Natl Acad Sci U S A 97:1079–1084
Menestrina G (1986) Ionic channels formed by staphylococcus-aureus alpha-toxin – voltage-dependent inhibition by divalent and trivalent cations. J Membr Biol 90:177–190
Menestrina G, Dalla Serra M, Prevost G (2001) Mode of action of beta-barrel pore-forming toxins of the staphylococcal alpha-hemolysin family. Toxicon 39:1661–1672
Menestrina G, Dalla SM, Comai M, Coraiola M, Viero G, Werner S, Colin DA, Monteil H, Prevost G (2003) Ion channels and bacterial infection: the case of beta-barrel pore-forming protein toxins of Staphylococcus aureus. FEBS Lett 552:54–60
Miles G, Bayley H, Cheley S (2002a) Properties of Bacillus cereus hemolysin II: a heptameric transmembrane pore. Protein Sci 11:1813–1824
Miles G, Cheley S, Braha O, Bayley H (2001) The staphylococcal leukocidin bicomponent toxin forms large ionic channels. Biochemistry 40:8514–8522
Miles G, Movileanu L, Bayley H (2002b) Subunit composition of a bicomponent toxin: staphylococcal leukocidin forms an octameric transmembrane pore. Protein Sci 11:894–902
Misakian M, Kasianowicz JJ (2003) Electrostatic influence on ion transport through the alphaHL channel. J Membr Biol 195:137–146
Mohammad MM, Iyer R, Howard KR, McPike MP, Borer PN, Movileanu L (2012) Engineering a rigid protein tunnel for biomolecular detection. J Am Chem Soc 134:9521–9531
Mohammad MM, Movileanu L (2010) Impact of distant charge reversals within a robust beta-barrel protein pore. J Phys Chem B 114:8750–8759
Montoya M, Gouaux E (2003) Beta-barrel membrane protein folding and structure viewed through the lens of alpha-hemolysin. Biochim Biophys Acta 1609:19–27

Movileanu L (2008) Squeezing a single polypeptide through a nanopore. Soft Matter 4:925–931

Movileanu L (2009) Interrogating single proteins through nanopores: challenges and opportunities. Trends Biotechnol 27:333–341

Movileanu L, Bayley H (2001) Partitioning of a polymer into a nanoscopic protein pore obeys a simple scaling law. Proc Natl Acad Sci U S A 98:10137–10141

Movileanu L, Cheley S, Bayley H (2003) Partitioning of individual flexible polymers into a nanoscopic protein pore. Biophys J 85:897–910

Movileanu L, Cheley S, Howorka S, Braha O, Bayley H (2001) Location of a constriction in the lumen of a transmembrane pore by targeted covalent attachment of polymer molecules. J Gen Physiol 117:239–251

Movileanu L, Howorka S, Braha O, Bayley H (2000) Detecting protein analytes that modulate transmembrane movement of a polymer chain within a single protein pore. Nat Biotechnol 18:1091–1095

Movileanu L, Schmittschmitt JP, Scholtz JM, Bayley H (2005) Interactions of the peptides with a protein pore. Biophys J 89:1030–1045

Muthukumar M (1999) Polymer translocation through a hole. J Chem Phys 111:10371–10374

Muthukumar M (2007) Mechanism of DNA transport through pores. Annu Rev Biophys Biomol Struct 36:435–450

Nivala J, Marks DB, Akeson M (2013) Unfoldase-mediated protein translocation through an alpha-hemolysin nanopore. Nat Biotechnol 31:247–250

Noskov SY, Im W, Roux B (2004) Ion permeation through the alpha-hemolysin channel: theoretical studies based on Brownian dynamics and Poisson-Nernst-Plank electrodiffusion theory. Biophys J 87:2299–2309

Olson R, Nariya H, Yokota K, Kamio Y, Gouaux E (1999) Crystal structure of Staphylococcal LukF delineates conformational changes accompanying formation of a transmembrane channel. Nat Struct Biol 6:134–140

Otto M (2014) Staphylococcus aureus toxins. Curr Opin Microbiol 17:32–37

Parker MW, Feil SC (2005) Pore-forming protein toxins: from structure to function. Prog Biophys Mol Biol 88:91–142

Pedelacq JD, Maveyraud L, Prevost G, Baba-Moussa L, Gonzalez A, Courcelle E, Shepard W, Monteil H, Samama JP, Mourey L (1999) The structure of a Staphylococcus aureus leucocidin component (LukF-PV) reveals the fold of the water-soluble species of a family of transmembrane pore-forming toxins. Structure 7:277–287

Potrich C, Bastiani H, Colin DA, Huck S, Prevost G, Dalla SM (2009) The influence of membrane lipids in Staphylococcus aureus gamma-hemolysins pore formation. J Membr Biol 227:13–24

Prevost G, Mourey L, Colin DA, Menestrina G (2001) Staphylococcal pore-forming toxins. Pore-Forming Toxins 257:53–83

Reiner JE, Kasianowicz JJ, Nablo BJ, Robertson JW (2010) Theory for polymer analysis using nanopore-based single-molecule mass spectrometry. Proc Natl Acad Sci U S A 107:12080–12085

Robertson JW, Kasianowicz JJ, Reiner JE (2010) Changes in ion channel geometry resolved to sub-angstrom precision via single molecule mass spectrometry. J Phys Condens Matter 22:454108

Robertson JW, Rodrigues CG, Stanford VM, Rubinson KA, Krasilnikov OV, Kasianowicz JJ (2007) Single-molecule mass spectrometry in solution using a solitary nanopore. Proc Natl Acad Sci U S A 104:8207–8211

Rodrigues CG, Machado DC, Chevtchenko SF, Krasilnikov OV (2008) Mechanism of KCl enhancement in detection of nonionic polymers by nanopore sensors. Biophys J 95:5186–5192

Rodriguez-Larrea D, Bayley H (2013) Multistep protein unfolding during nanopore translocation. Nat Nanotechnol 8:288–295

Sackmann B, Neher E (1995) Single-channel recording. Kluwer Academic/Plenum Publishers, New York

Sanchez-Quesada J, Ghadiri MR, Bayley H, Braha O (2000) Cyclic peptides as molecular adapters for a pore-forming protein. J Am Chem Soc 122:11757–11766

Siwy ZS, Howorka S (2010) Engineered voltage-responsive nanopores. Chem Soc Rev 39:1115–1132

Slonkina E, Kolomeisky AB (2003) Polymer translocation through a long nanopore. J Chem Phys 118:7112–7118

Song LZ, Hobaugh MR, Shustak C, Cheley S, Bayley H, Gouaux JE (1996) Structure of staphylococcal alpha-hemolysin, a heptameric transmembrane pore. Science 274:1859–1866

Sugawara-Tomita N, Tomita T, Kamio Y (2002) Stochastic assembly of two-component staphylococcal gamma- hemolysin into heteroheptameric transmembrane pores with alternate subunit arrangements in ratios of 3: 4 and 4: 3. J Bacteriol 184:4747–4756

Sutherland TC, Long YT, Stefureac RI, Bediako-Amoa I, Kraatz HB, Lee JS (2005) Structure of peptides investigated by nanopore analysis. Nano Lett 4:1273–1277

Tian P, Andricioaei I (2005) Repetitive pulling catalyzes co-translocational unfolding of barnase during import through a mitochondrial pore. J Mol Biol 350:1017–1034

Walker B, Bayley H (1995) Restoration of pore-forming activity in staphylococcal alfa-hemolysin by targeted covalent modification. Protein Eng 8:491–495

Wang HY, Gu Z, Cao C, Wang J, Long YT (2013a) Analysis of a single alpha-synuclein fibrillation by the Interaction with a Protein Nanopore. Anal Chem 85:8254–8261

Wang S, Haque F, Rychahou PG, Evers BM, Guo P (2013b) Engineered nanopore of Phi29 DNA-packaging motor for real-time detection of single colon cancer specific antibody in serum. ACS Nano 7:9814–9822

Wang Y, Zheng D, Tan Q, Wang MX, Gu LQ (2011) Nanopore-based detection of circulating microRNAs in lung cancer patients. Nat Nanotechnol 6:668–674

Wells DB, Abramkina V, Aksimentiev A (2007) Exploring transmembrane transport through alpha-hemolysin with grid-steered molecular dynamics. J Chem Phys 127:125101

Werner S, Colin DA, Coraiola M, Menestrina G, Monteil H, Prevost G (2002) Retrieving biological activity from LukF-PV mutants combined with different S components implies compatibility between the stem domains of these staphylococcal bicomponent leucotoxins. Infect Immun 70:1310–1318

White SH, Wimley WC (1999) Membrane protein folding and stability: physical principles. Annu Rev Biophys Biomol Struct 28:319–365

Wimley WC (2003) The versatile beta-barrel membrane protein. Curr Opin Struct Biol 13:404–411

Yamashita K, Kawai Y, Tanaka Y, Hirano N, Kaneko J, Tomita N, Ohta M, Kamio Y, Yao M, Tanaka I (2011) Crystal structure of the octameric pore of staphylococcal gamma-hemolysin reveals the beta-barrel pore formation mechanism by two components. Proc Natl Acad Sci U S A 108:17314–17319

Yoong P, Torres VJ (2013) The effects of Staphylococcus aureus leukotoxins on the host: cell lysis and beyond. Curr Opin Microbiol 16:63–69

Chapter 11
Properties of Pores Formed by Cholesterol-Dependent Cytolysins and Actinoporins

Nejc Rojko, Manuela Zanetti, Gregor Anderluh, and Mauro Dalla Serra

Abstract Planar lipid membrane (PLM) measurements allow direct observation of pores in model lipid membranes. This biophysical approach was very important for our understanding of how transmembrane pores are formed by cholesterol-dependent cytolysins (CDCs), a toxin family from pathogenic bacteria, and actinoporins, cytolysins from sea anemones. In this review we discuss current knowledge of pore formation by these two protein families and how the PLM approach revealed mechanisms by which these two unrelated protein families porate membranes. Interestingly, for both toxins, the protein portion constituting the pore walls has an alpha helical configuration in the secreted water soluble form. This structure is maintained for actinoporins in the membrane inserted configuration, while the pore of CDCs necessitates a drastic change in secondary structure, which transforms to beta hairpins in the membrane. Both proteins are able to form toroidal proteo-lipid pores.

Keywords Cholesterol-dependent cytolysins • Actinoporins • Pore • Planar lipid membrane

---

N. Rojko
Laboratory for Molecular Biology and Nanobiotechnology,
National Institute of Chemistry, Hajdrihova 19, Ljubljana 1000, Slovenia
e-mail: nejc.rojko@ki.si

M. Zanetti • M. Dalla Serra
Laboratory of Biomolecular Sequence and Structure Analysis for Health, Istituto di Biofisica, Consiglio Nazionale delle Ricerche & Fondazione Bruno Kessler,
via alla Cascata 56/C, Trento 38123, Italy
e-mail: mzanetti@fbk.eu; mauro.dallaserra@cnr.it

G. Anderluh (✉)
Laboratory for Molecular Biology and Nanobiotechnology,
National Institute of Chemistry, Hajdrihova 19, Ljubljana 1000, Slovenia

Department of Biology, Biotechnical Faculty, University of Ljubljana,
Jamnikarjeva 101, Ljubljana 1000, Slovenia
e-mail: gregor.anderluh@ki.si

A.H. Delcour (ed.), *Electrophysiology of Unconventional Channels and Pores*,
Springer Series in Biophysics 18, DOI 10.1007/978-3-319-20149-8_11

11.1 Introduction

Pore-forming toxins (PFT) are the largest and best characterized class of protein toxins. For example, they account for roughly 30 % of all known bacterial protein toxins (Iacovache et al. 2008). They are potent virulence factors evolved during ancient wars among competing organisms for defense or preying purposes. PFTs are well represented in all life kingdoms, being produced also by higher organisms. Interestingly, similar structures and functions are also adopted by components of the immune system, like perforin and complement C9 (Gilbert et al. 2013). The cell plasma membrane is a complex system which isolates the cell from the surrounding environment, conferring the cell with its own identity and permitting its vital functions. So, not surprisingly the plasma membrane of the target cell is often the primary site of action of many PFTs. Close to the site of infection, or where PFT concentration is high enough, their main effect is to punch poorly selective nanometer-sized holes into the target membrane, causing osmotic imbalance, possibly leading to cell death. The formation of a PFT lesion can be normally schematized in the following four steps: (i) the toxin is secreted as water-soluble monomer, which (ii) binds to the target membrane where (iii) it starts to oligomerize; when the oligomers reach a critical size (iv) they undergo a conformational change which permits them to drill holes into the membrane (Anderluh and Lakey 2008; Menestrina et al. 2001). PFTs could be classified according to the producing organism, the dimension of their lesion, the secondary structure of the membrane inserted configuration adopted by the protein element constituting the pore, or the structure of the pore that is formed (Anderluh and Lakey 2008; Dalla Serra and Tejuca 2011).

Regarding pore structure, two conformations are possible: a purely proteinic channel or a protein-lipid mixed arrangement (Gilbert et al. 2014). A barrel-stave pore is constituted by proteins that could adopt either β-hairpin or α-helix structures, and do not include lipids as constitutive elements of pore walls (Fig. 11.1). Proteo-lipid pores are characterized by the co-presence of lipidic and proteinic elements in the pore walls, where lipids could either intercalate between protein monomers stabilizing their interaction, or being confined to the free edge of arc shaped pores. In both cases the lipid lamellar structure is destroyed and lipids should bend, possibly assuming a toroidal shape. The lipid participation in toroidal pore structures has been reported for many antimicrobial peptides (Matsuzaki 1998; Yang et al. 2001), as well as larger proteins (Metkar et al. 2015; Valcarcel et al. 2001; Anderluh et al. 2003; Xu et al. 2013; Garcia-Saez et al. 2006). More details on formation, properties and significance of proteo-lipid pores may be found in a recent review paper (Gilbert et al. 2014).

A very powerful approach to gain insight into the biophysical characteristics of the conducting unit at the single molecule level is the electrophysiological method. Since these exogenous pore formers are able to self-aggregate onto a lipid membrane in the absence of a specific receptor, the use of purely lipid planar model membranes (PLM) is very useful. With this model system it is possible to

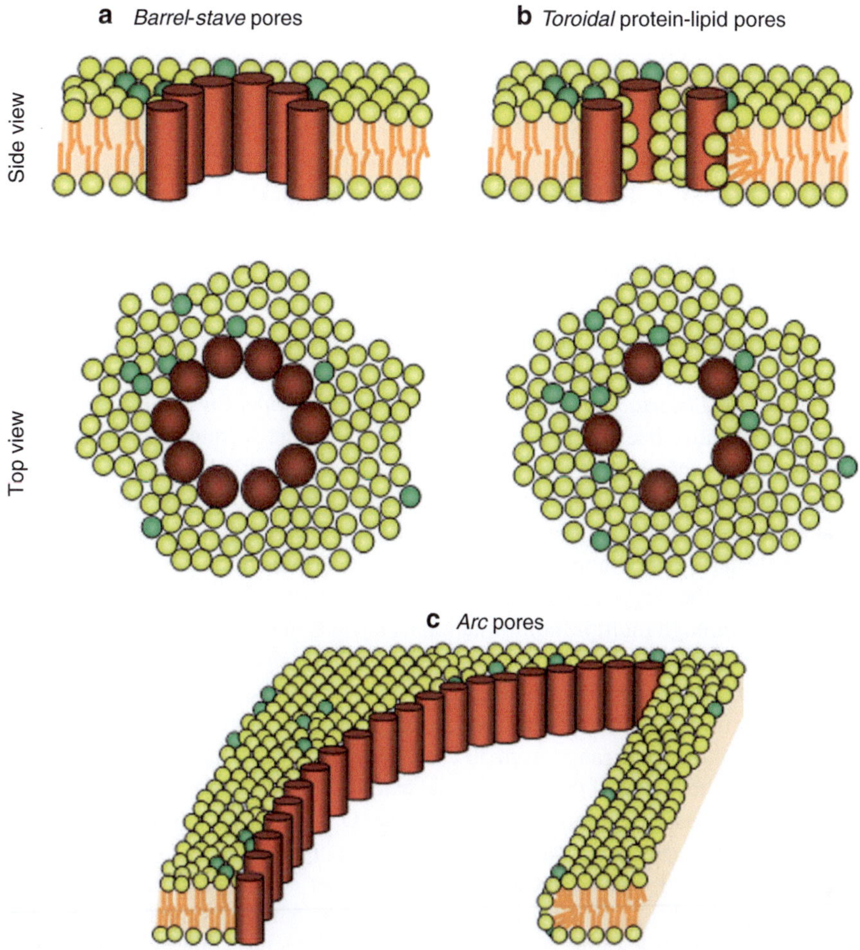

Fig. 11.1 Pore formation in biological membranes. There are different possibilities of how pores may be formed in biological membranes. Barrel-stave pores (**a**) are formed by either α-helices clusters or β-barrels. Here a protein component lines the walls of the pore. In toroidal (**b**) and arc (**c**) pores, at least part of the pore lumen is lined by headgroups of the membrane lipids (Reproduced from (Dalla Serra and Tejuca 2011) with permission)

have free access to both compartments, mimicking the cytoplasmic and the external cell environments; different symmetric or asymmetric lipid compositions are allowed; many experimental conditions can be easily changed, like temperature, salt and buffer composition, as well as introduction of analytes able to interact with the conductive pore. In general the electrophysiological characteristics of the pore can be determined, like single pore conductance, current-voltage relationship and ion selectivity (Dalla Serra and Menestrina 2000). Here we discuss how the PLM approach helped to describe the ability of the bacterial PFT family of cholesterol-dependent cytolysins (CDCs) and actinoporins from sea anemones to generate proteolipidic nanopores into lipid membranes, with a focus on electrophysiological properties of such pores.

11.2 Cholesterol-Dependent-Cytolysins (CDCs)

11.2.1 CDC Properties

CDCs are a large family of β-barrel pore-forming toxins (Hotze and Tweten 2012) that are secreted by Gram-positive (these include species from *Clostridium*, *Streptococcus*, *Listeria*, *Arcanobacterium*, *Gardnerella*, *Bacillus* and *Lactobacillus*) and Gram-negative bacteria (Hotze et al. 2013). CDCs are secreted as water-soluble monomers that once bound to the target cell, will form a circular homo-oligomeric complex containing up to 40 monomers (Hotze and Tweten 2012). The pore-forming mechanism of the CDCs is a multistep process that involves recognition and binding to the cholesterol-containing membrane by domain 4 of the toxin, oligomerisation of monomers on the target cell membrane to form a large oligomeric complex, and penetration, through multiple conformational changes, of the structure into the membrane to form a transmembrane β-barrel pore with a diameter of around 25 nm (Gilbert 2005; Hotze and Tweten 2012). The members of the CDC family share high similarity in primary structure, with 40 to 70 % identity. The molecular weights of mature CDC monomers range from 50 to 72 kDa. This broad range is due mostly to differences in the length of their N-terminal extensions, which may be as long as 150 residues. The monomeric form of perfringolysin O (PFO), which is a CDC secreted by *Clostridium perfringens*, was the first CDC to be crystallized and its crystal structure (Rossjohn et al. 1997) showed that the monomer is organized in four domains (D1 to D4) (Fig. 11.2). The crystal structures of five other CDCs have been determined and show the same organization of the molecule: intermedilysin (ILY) secreted by *Streptococcus intermedius* (Polekhina et al. 2005), streptolysin O (SLO) secreted by *Streptococcus pyogenes* (Feil et al. 2014), anthrolysin O (ALO) from *Bacillus anthracis* (Bourdeau et al. 2009), suilysin from *Streptococcus suis* (Xu et al. 2010) and listeriolysin O (LLO) from *Listeria monocytogenes* (Köster et al. 2014).

11.2.2 Mechanism of CDC Pore Formation

CDCs are perhaps the best studied family of PFTs with a lot of molecular information on pore assembly (reviewed in Gilbert (2005), Hotze and Tweten (2012)). CDCs coordinate the assembly of about 25 to 40 monomers into an oligomeric structure and inserts into the membrane a β-barrel pore composed of 50 to 80 amphipathic hairpins (Shatursky et al. 1999; Shepard et al. 2000; Hotze et al. 2001;

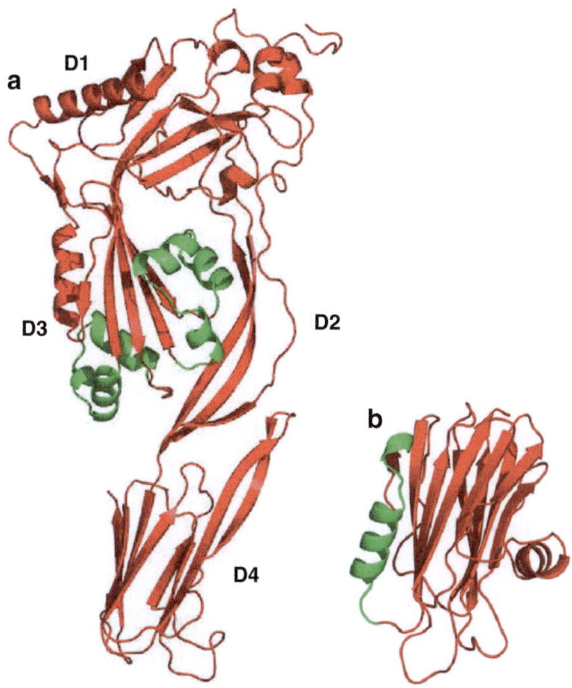

Fig. 11.2 Tridimensional structure of cholesterol-dependent cytolysin and actinoporin. The PFO monomer (PDB-ID 1PFO) (**a**) is organized in four domains (*D1–D4*). Two clusters of α-helices from *D3* that unfold during pore formation and form transmembrane β-hairpins are presented in light color. Equinatoxin II (PDB-ID 1IAZ) (**b**) is a single domain protein that forms a pore by the N-terminal region, which is shown in light color

Czajkowsky et al. 2004; Tilley et al. 2005). Oligomerisation to pore structure at the surface of the membrane is mediated by contacts through D1. The nexus of the β-barrel pore insertion is the disruption of the interface that D3 forms with D1 and D2 (Hotze et al. 2001; Wade et al. 2015). This interface is formed between D1 and D2, and one of the two D3 α-helical bundles. The D3 α-helical structures ultimately unfurl and refold into the two transmembrane β-hairpins (TMH1 and TMH2), which contribute to the formation of the giant β-barrel pore (Shatursky et al. 1999; Shepard et al. 2000; Sato et al. 2013). Pore formation also requires that each monomer undergo a 4 nm vertical collapse that allows the TMHs to span the bilayer when inserted (Czajkowsky et al. 2004; Tilley et al. 2005; Ramachandran et al. 2005).

11.2.3 Lipid Specificity

The binding of CDCs to a target membrane requires the recognition of cholesterol. The exception are CDC members, which require the CD59 membrane-anchored protein in the case of ILY and vaginolysin (VLY) secreted by *Gardnerella vaginalis* (Giddings et al. 2004; Gelber et al. 2008), a glycolipid receptor in the case of pneumolysin (PLY) secreted by *Streptococcus pneumoniae* (Shewell et al. 2014) and carbohydrates in the case of lectinolysin (LLY) secreted by *Streptococcus mitis* (Feil et al. 2012). The presence of cholesterol in the target membrane, instead, is required for pore formation by all CDCs. The depletion of membrane cholesterol from human erythrocytes prevents the insertion of the transmembrane β-barrel and traps the CDCs in an oligomeric inactive complex (Giddings et al. 2003). ILY, for example, will bind only to target membranes containing a specific protein receptor, but cholesterol is required for ILY pore formation.

Membrane binding of CDCs is achieved by D4 (Ramachandran et al. 2004). This structurally conserved domain contains conserved loops, L1-L3, and an undecapeptide region, which was originally thought to be involved in cholesterol recognition (Soltani et al. 2007). Single amino acid modifications in these loops prevented PFO from binding to cholesterol-rich liposomes (Heuck et al. 2010). Farrand and co-workers (Farrand et al. 2010) have shown that only two amino acids, a threonine-leucine pair in loop L1, comprise the cholesterol binding motif and are conserved in all known CDCs. It has now been shown that the conserved undecapeptide is a key element in the allosteric pathway that couples membrane binding to the initiation of structural changes in D3 of the CDC monomer that controls the assembly of the pore complex (Dowd and Tweten 2012).

The phospholipid composition of a cell membrane affects the arrangement of cholesterol within the membrane. For example, PFO will preferentially bind to cholesterol-rich membranes composed mainly of phospholipids containing 18-carbon acyl chains (Ohno-Iwashita et al. 1991). Any modification of the 3β-hydroxy group of cholesterol completely abrogates the activity of LLO, SLO or PFO (Prigent and Alouf 1976; Ohno-Iwashita et al. 1991; Bavdek et al. 2007). These observations suggest that the toxins' specificity is largely confined to the most surface-exposed part of the sterol molecule. The latter conclusion is in line with structural data on PFO and PLY, which indicate that, even after completion of toxin oligomerization and pore formation, only a rather limited part of the domain 4 is actually inserted into the lipid bilayer (Ramachandran et al. 2002; Tilley et al. 2005).

11.2.4 CDC Pore Properties

PLM is a highly sensitive technique that allows the study of single channel openings in real time and was very important for our understanding of the nature of pores formed by CDCs. CDCs punch wide pores, with dimensions much larger than the Debye length, so no influence on the bulk mobility of ions should be expected (Kienker and Lear 1995). In fact LLO pores cannot discriminate between K^+ and Cl^-, and also the current-voltage relationship is linear (Manuela Zanetti and Primož Knap, unpublished result). PLM studies on CDCs have shown a great variety of events occurring in planar lipid membranes. All reports on different CDC single channel recordings show pores with a broad conductance distribution and often three different pore sizes (Blumenthal and Habig 1984; Menestrina et al. 1990; Korchev et al. 1998; Shepard et al. 2000; From et al. 2008; Bavdek et al. 2012; Marchioretto et al. 2013) (Fig. 11.3). Large pores have well-defined conductances, higher than 1 nS, and probably correspond to β-barrel-stave pores, as proposed for PFO (Shepard et al. 2000). Medium and small sized pores are quite frequent, constituting around half of the observed events. In many cases, it is possible to inhibit

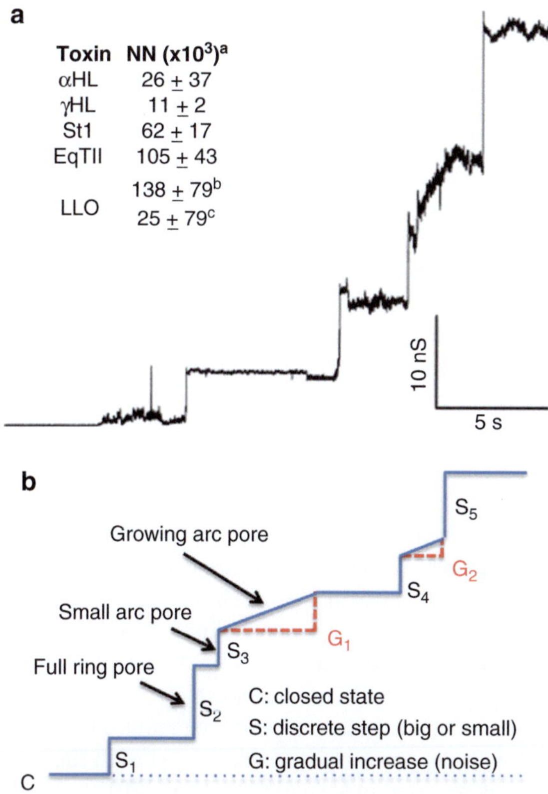

Fig. 11.3 Ionic current of pores formed by LLO in planar lipid membranes. A current trace of LLO in POPC/cholesterol (1/1) membrane is shown at an applied potential of +10 mV (a). The protein concentration was 5 nM and the buffer was 100 mM KCl, 10 mM MES, pH 5.5. The idealized activity in PLM is shown in b: protein activity causes step-wise (S_i), but also unresolved gradual increases (G_i) in the ionic current. These steps have different amplitudes (S_i) and are related to pore opening. Complete rings give rise to the larger steps (S_2, S_5); smaller steps are correlated with the opening of incomplete arcs (S_1, S_3, S_4). The gradual current increases are noisier and could be related to growing arcs, in which the addition of an extra monomer to the partially formed arc correlates with a small current increase. Inset in A: [a] NN is the normalized noise. It is calculated from the standard deviation of the current on a short time interval for the individual current trace. Differences between the standard deviation of the first channel and that of the bare membrane are then normalized by dividing by the average current of the open channel to obtain the normalized noise: $NN = (SD^+ - SD^-)/I$, where SD^- is the noise of the bare membrane, SD^+ is the noise of the first open channel and I is the current of that channel. [b] is the NN of LLO in DOPC/cholesterol 20 % membranes, where proteo-lipid pores are the majority, and [c] is the NN in POPC/cholesterol 50 %, where the ring-shaped pores are predominant. Abbreviations in (a) inset are as follows: *αHL* α-hemolysin, *γHL* γ-hemolysin, *St1* sticholysin I, *EqtII* equinatoxin II, *LLO* listeriolysin O

the activity of small pores with divalent cations, such as Zn^{2+} or Ca^{2+} (Menestrina et al. 1990; Korchev et al. 1998). As reported by Korchev and co-workers, this could reflect either a blocking of pores or a gating phenomenon. However, PLM recordings for CDCs reflect the dynamic nature of pores, which may be in agreement with the lipid involvement in pore structures, as proposed for the arc-shaped pores.

Noise analysis provides additional insights into pore properties. As reported in Praper et al. (2011) the use of normalized noise (NN), obtained by dividing the standard deviation of a short piece of current trace by its average value, permits a comparison among pores with different mean single pore conductance. The NN is used to discriminate between purely protein pores and proteo-lipid (toroidal) pores. In fact, recognized examples of protein pores (i.e. αHL and γHL) have small noise values, in contrast to StI and EqTII which form toroidal pores. Depending on the target membrane lipid composition (and possibly pH), LLO is able to form both kind of pores (i.e. toroidal in DOPC/cholesterol 20 %, or purely protein in POPC/cholesterol 50 %) (Fig. 11.3).

Pore size can be estimated from the single pore conductance (G) defined as the ratio of the amplitude of single step current (I) to the applied voltage (V). The conductance is related to the pore radius (r) by equation 11.1:

$$G=\sigma\left(r^2\pi\right)/\left(L+\pi r/2\right) \tag{11.1}$$

where σ is the conductivity of the salt solution in the chambers and L is the length of the pore. In case of CDCs, the length, estimated by atomic force microscopy (AFM), is 12.5 nm (Czajkowsky et al. 2004). For example, pores formed by PFO in 1-palmitoyl-2-oleoyl-*sn*-glycero-3-phosphocholine (POPC): cholesterol (50 % molar ratio) membrane in 20 mM HEPES, 100 mM KCl, pH 7.4 show a mean conductance value of 14.2 ± 3.7 nS (Marchioretto et al. 2013), that corresponds to a radius value of 10.4 ± 1.6 nm as estimated using equation 11.1. This radius value is quite in agreement with the pore size measured by AFM in the same experimental conditions, 12.4 nm as reported in Marchioretto et al. (2013).

11.2.5 Measuring Pore Activity on Cells

Recently, three published papers show studies on CDC activity in mammalian cells by the patch-clamp technique. Patch clamp is an excellent approach to investigate the electrophysiological properties of pores formed by CDCs. The activity of PLY has been recorded in the membrane of CHO-K1 cells using inside-out patch-clamp configuration (El-Rachkidy et al. 2008). PLY generates in cell membranes pores divided into three conductance classes: small (<200 pS), medium (200–1000 pS) or large (>1 nS). Medium pores are the most frequent, and the smallest ones remain open only for short periods of time. It has also been demonstrated by patch-clamp in whole-cell configuration that recombinant LLO as well as *L. monocytogenes* applied extracellularly induces Ca^{2+} oscillations in HEK293 cells (Repp et al. 2002), which further modulate cellular signaling and gene expression. How IFN-β enhances the ability of LLO to cause permeabilization of plasma membrane and death of macrophages was also studied by perforated patch-clamp analysis (Zwaferink et al. 2008). A broad distribution of currents was observed, and in particular, the larger conductance states were multiples of the smallest conductance. The authors also pointed out the fact that the conductance distribution depends on LLO concentration and on membrane composition. Since the broad CDC pore heterogeneity recorded by PLM is replicated in cells, it is plausible that pores with different sizes may have different functions in physiological systems.

11.2.6 Role of Lipids for CDCs Pore Properties

The cholesterol amount and the fluidity of the bilayer affect the ability of CDCs to bind the membrane and initiate pore formation (as discussed previously), but also the final shape of the structures that can be complete rings or arc-shaped oligomers. PLM experiments with LLO (Bavdek et al. 2012; Marchioretto et al. 2013) clearly identified two different pore-forming characteristics: a well-defined, step-wise increase in current, presumably due to the insertion of completely protein-based ring-shaped pores, and smaller steps that are very noisy and can be related to arc-shaped structures. In this arc configuration, two events could be responsible for the noise in current traces. First, the lipids of both layers have to be rearranged in order to prevent the exposure of the hydrophobic tails to the water environment; and secondly, the contact surface between the monomers at the edge of the arc and the lipids can result in leakage that allows ion passage. Structural studies carried out by electron microscopy (EM) or AFM also demonstrated that both ring and arc configurations are formed by CDC proteins, further lending support to the idea that diverse pore structures exist in membranes (Bhakdi et al. 1985; Palmer et al. 1998; Czajkowsky et al. 2004; Sonnen et al. 2014; Leung et al. 2014).

PLM studies have been carried out in model membrane systems with well-defined lipid composition (Bavdek et al. 2012; Marchioretto et al. 2013). The ordering of acyl chains and the concentration of cholesterol have important effects on the shape of the pores. LLO forms smaller pores showing unresolved current increases when made in membranes composed of lipids with less ordered acyl chains and containing small amount of cholesterol (i.e. 1,2-dioleoyl-*sn*-glycero-3-phosphocholine (DOPC) with 20 % molar ratio of cholesterol). This phenomenon has been assigned to the oligomerization process where sequential addition of monomers to a pre-formed arc structure could occur. Since the new monomer has to transform its α-helices into β-barrels, and the D2 domain has to collapse, the displacement of lipids may not be sharp enough to produce a well-defined increase in current. On membranes composed of more ordered lipids with high amount of cholesterol (i.e. POPC with 50 % molar ratio of cholesterol), LLO preferentially forms larger pores and step-wise, well-resolved increases in current are observed. Thus, the unresolved increases in current could fit well with sequential monomers being added during CDC oligomerization. As already reported for the related protein perforin (Praper et al. 2011), we demonstrated that the lipid composition of target membranes clearly has an effect on the shape of pores. In the case of CDCs, besides the effect of lipids with different degree of acyl chain order, we may also consider the cholesterol effect on the amount of protein bound to the lipid bilayer. Unsaturated lipids (like DOPC) produce a more fluid membrane which could facilitate the prepore to pore transition, allowing it to occur at an earlier stage towards complete ring formation, and therefore permitting the insertion of smaller arc shaped oligomers. These structures are indeed more flexible, as shown by the higher noise into the current traces, and could also tolerate the addition of a new monomer to an already formed and inserted arc-shaped pore. In the case of more ordered lipids (like POPC), the membrane is more rigid and possibly requires higher energy for permitting insertion of the protein oligomer. This free energy could be supplied by the concerted conformational transition of a higher number of monomers involved into the full ring formation.

11.3 Actinoporins

11.3.1 Actinoporins Properties and Pore Formation

Cnidarians are water organisms with radial symmetry. A key characteristic of the phylum is the presence of nematocytes, specialized cells usually associated with venom delivery for preying or defense purposes. Upon receipt of an appropriate chemical and/or mechanical stimulus, a nematocyst, a specialized capsule within nematocytes, discharges and delivers venom to the victim (Tardent 1995). The venom components, however, are spread all over the body. Currently around 250 venom compounds have been identified in cnidarians, which can be divided into two groups: proteinaceous (peptides, proteins, enzymes and proteinase inhibitors) and non-proteinaceous substances (purines, quaternary ammonium compounds, biogenic amines and betaines) (Frazao et al. 2012). The best characterized is the venom delivered by sea anemones (Anthozoa, Actinaria), which contains neurotoxins and cytolytic peptides and proteins (Norton 1991; Anderluh and Maček 2002). Actinoporins are by far the best studied group of sea anemone toxins (Anderluh and Maček 2002; Kristan et al. 2009; Alvarez et al. 2009; Garcia-Ortega et al. 2011). They are produced as soluble monomers and it is believed that their expression is not limited to nematocyst (Frazao et al. 2012; Anderluh and Maček 2002). The most studied representatives of actinoporins are equinatoxin II (EqtII) from the sea anemone *Actinia equina*, and sticholysins I (StI) and II (StII) from the sea anemone *Stichodactyla helianthus*. They share high sequence identity, as well as structural similarity (Athanasiadis et al. 2001; Mancheno et al. 2003). The name "actinoporin" for this group of pore forming toxins was suggested by Kem (1988). Today it is generally accepted that all actinoporins exert their toxicity by forming oligomeric, cation-selective pores within membranes, which results in a colloid-osmotic shock leading to cell death (Anderluh and Maček 2002; Alegre-Cebollada et al. 2007).

The first attempt to find the underlying mechanism by which actinoporins damage cells was conducted with the use of PLM. The addition of partially purified *S. heliantus* toxin to PLM greatly increased membrane permeability without compromising its stability. Because of the high ion flux through the membrane, a carrier-type mechanism was dismissed. Altogether the results implied the formation of transmembrane channels by actinoporins (Michaels 1979). It is well-established that actinoporins bind sphingomyelin (SM) in the target membranes (Bakrač et al. 2008), and form pores by the N-terminal amphiphatic α-helix, which is inserted deep in the membrane (Hong et al. 2002), thus placing actinoporins in the family of α-pore forming toxins. To form a functional cation-selective pore, several monomers have to assemble on the membrane plane (Antonini et al. 2014; Baker et al. 2014).

It is believed that lipid molecules from the target membrane are part of the transmembrane channel (Valcarcel et al. 2001; Anderluh et al. 2003; Antonini et al. 2014); however, due to the structural instability of the pore complex, the molecular details of the final assembly are still not revealed.

11.3.2 Lipid Specificity

Initial studies on lipid specificity of actinoporins have established that SM has an important role for the pore forming mechanism. Treatment of erythrocytes with sphingomyelinase prior to toxin exposure rendered the erythrocytes resistant to lysis by the toxin. Therefore, the authors concluded that SM is a specific receptor for cytolysins present in the sea anemone *S. heliantus* (Bernheimer and Avigad 1976). The lipid specificity of actinoporins was further studied by a PLM approach. Michaels (1979) showed that the toxin from *S. heliantus* readily increased membrane conductance in membranes composed of egg lecithin and cholesterol, devoid of any SM (Michaels 1979). Phosphocholine (PC) membranes are also sensitive to toxin, however, the conductance in the PC:SM mixture is several orders of magnitudes larger than that in the PC membrane alone (Varanda and Finkelstein 1980). The earliest single channel recordings on EqtII from the sea anemone *A. equina* were conducted on POPE planar membranes without the addition of SM (Zorec et al. 1990). Channel formation in PLM was also observed for StI, but a 50-times larger concentration of purified protein had to be added to a voltage clamped PLM in order to achieve similar activity as in the case of a 5 % SM membrane. The lipid specificity of EqtII was studied also by other biochemical and biophysical approaches. Direct binding assays showed that EqtII specifically binds SM but not other lipids (Bakrač et al. 2008). It was also established that interactions with SM are promoted by aromatic amino acid residues placed on one of the broad loops at one end of the molecule (Bakrač and Anderluh 2010). A weak association with PC lipids may be the due to the fact that a phosphorylcholine head group is shared by both SM and phosphatidylcholine (PC). This finding has a strong biological relevance, as sea anemones do not possess SM, but phosphonosphingolipids that have an altered head group, making sea anemone membranes resistant to toxin activity

(Meinardi et al. 1995). The PLM approach was also instructive for understanding actinoporin-membrane association. Upon removal of the toxin from the aqueous compartment in a PLM experiment, the conductance did not decline significantly, pointing to the irreversibility of the membrane-toxin interaction (Varanda and Finkelstein 1980).

11.3.3 Actinoporins Pore Properties

Very low *S. heliantus* toxin concentration (10 ng/mL) at one side of the PLM resulted in discrete step-wise changes in membrane conductance. However, channels were not stable; fluctuations corresponding to the formation and breakdown of individual channels were observed, with the lifetime of a single channel being around two seconds. Fluctuations were also independent of the polarity of applied voltage (Michaels 1979). Similarly, EqtII pores are prone to rapidly fluctuate among various subconductance state levels (Malovrh et al. 2003; Zorec et al. 1990). On the contrary, partially purified *S. heliantus* toxin was reported to form channels which tend to remain open once formed (Varanda and Finkelstein 1980). The steady state probability of the pores to be open was found to be voltage-dependent, as applications of long-lasting large voltage pulses caused the closure of the StI pores (Tejuca et al. 1996). Rojko et al (2014) used droplet interface bilayers (DIB) to directly visualise Ca^{2+} influx through individual EqtII pore by the Fluo8H fluorescent dye and to simultaneously measure the electrical current through the same pore (Rojko et al. 2014). Multiple conductance states of single pores were reported, resulting in different fluorescent intensities from Ca^{2+} influx.

The PLM setup was also used to study other characteristics of actinoporins pores. It was established that actinoporins pores are cation selective (Belmonte et al. 1993; Tejuca et al. 1996; Varanda and Finkelstein 1980). For example, StI pores are approximately 4 times more permeant to Na^+ than Cl^- ions (Tejuca et al. 1996). The current – voltage relationships (I/V) was found to be nonlinear in actinoporin channels, a phenomenon also called rectification (Belmonte et al. 1993; Tejuca et al. 1996; Varanda and Finkelstein 1980; Zorec et al. 1990). In particular, the current flowing at negative voltages is larger than that flowing at positive voltages of the same value. In general, this implies the existence of an asymmetrical potential profile alongside the ion pathway, most probably generated by fixed charges of the toxin which line the lumen of the pore (Tejuca et al. 1996). The rectification did not result from a voltage dependence of channel opening and closing (Varanda and Finkelstein 1980).

PLM also allowed estimating pore radius of actinoporins. The StI pore radius was estimated to be 1.0 nm by analysing single channel conductance in PLM, assuming that the channel is a perfect cylinder, 6 nm long, filled with water, where cations and anions move in a 4:1 ratio, with the same mobility as in water (Tejuca et al. 1996); applying eq. 11.1, we get an estimated radius of 0.90 ± 0.13 nm. A similar value of 0.9 nm was calculated for EqtII pore, although with the assumption of a

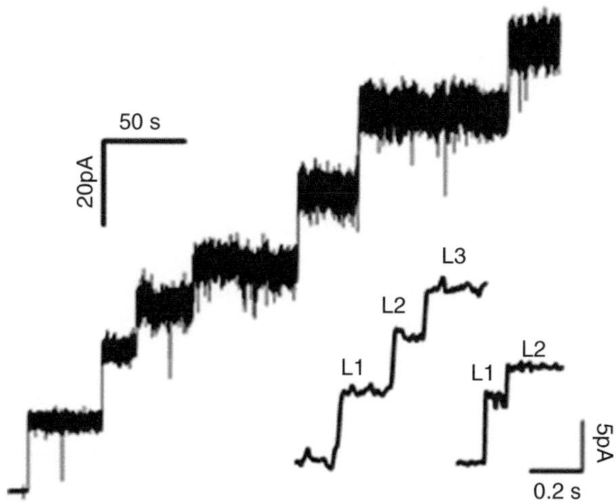

Fig. 11.4 Example of StI pores. Example of StI pore insertion in diphytanoyl-phosphocholine/SM (4:1) membrane. Each step is due to the opening of a stable and structured full pore. The toxin concentration was 5 nM; the buffer was 100 mM KCl, 10 mM Hepes, pH 7.0. In the inset, two representative examples of successive time resolved sublevels are shown; L1, L2, L3 represent three resolved, successive short-lived sub-conductance levels (Adapted from (Antonini et al. 2014))

cylindrical cation-selective channel of 7 nm (Belmonte et al. 1993). These values have to be regarded as approximations; however, they are in agreement with pore size estimation based on haemolysis prevention by osmoprotectant molecules of different sizes (Belmonte et al. 1993; Tejuca et al. 1996; Tejuca et al. 2001; Antonini et al. 2014).

Recently, Antonini et al. (2014) has shown by PLM assay that the stable St1 pore assembles through sequential and transient sub-conductance levels. These 3–4 sub-levels have been assigned to successive incorporation of N-term α-helices, which, together with lipid head groups, constitute a stable tetrameric pore with a toroidal structure (Fig. 11.4).

11.3.4 Molecular Details of Actinoporins Pores

PLM also allowed deriving molecular details of actinoporins pore properties. Actinoporins are single-domain proteins, composed of a β-sandwich core flanked by two α-helices (Athanasiadis et al. 2001; Hinds et al. 2002; Mancheno et al. 2003; Mechaly et al. 2011) (Fig. 11.2). According to helical wheel analysis, the N-terminal α-helix sequence is amphipathic in nature. Closer inspection reveals analogy to some amphipathic polypeptides which are known to interact with lipid membranes, such as the honey-bee venom peptide melittin (Belmonte et al. 1994). Mutating residues of the N-terminal α-helix to cysteine considerably changed the ion selectivity of the pores (Malovrh et al. 2003). After introducing positive or negative charges to the thiol groups of the introduced Cys residues by methanethiosulfonate (MTS) reagents, it was possible to measure the level of exposure of each position in the pore lumen, as newly introduced charges along the ion conductive pathway affect selectivity of the pore. A clear α-helical pattern was observed, whereas no effect on ion selectivity was observed on the hydrophobic side residues. Interestingly, the effect of mutations was fading when moving away from Asp-10, suggesting this is the constriction site of the pore lumen (Malovrh et al. 2003).

The strong asymmetry of the I/V curves of mutants with charge modification and the mutant with the first five amino acids truncated from the N-terminus is the indication that the N-terminus is exposed to the *trans* side of the membrane, opposite to the binding site of the membrane (Kristan et al. 2007). Introducing a His-tag at the N-terminus resulted in channels of lower conductance, asymmetric I/V curve and rapid closures of pores at a negative applied potential. As the selectivity was not modified, it was assumed that the His-tag is translocated to the *trans* side and blocks the channel when negative voltages are applied. In addition, the transfer of the N-terminal α-helix across the membrane was confirmed by the binding of charged MTS reagents to the trans side of the PLM, consequently changing its selectivity (Kristan et al. 2007). It was also established that the side chain of Ser1 is facing the pore lumen, and because the effect on ion selectivity after introducing charged molecules to this position was the same as in case of Asp17, Ser1 is placed at approximately the same position with regard to the center of the pore (Kristan et al. 2007).

One specific amino acid residue was found to be of greater importance in the EqtII oligomerisation process. The K77C mutant formed channels in PLM at much higher concentration than the wild type toxin, but a series of experiments showed that the binding was not affected. The conductance of the wild-type single channel was approximately 400 pS whereas the mutant formed channels of approximately 500 pS conductance. Despite the increase in conductance, the selectivity of the pores remained practically the same and it was concluded that Lys77 may be the residue involved in EqtII oligomerisation on the membrane (Anderluh et al. 2000).

11.3.5 Actinoporins Form Toroidal Pores

Not only protein mutations but also the lipid composition of the target membrane affect actinoporin activity. Valcarcel et al. (2001) observed that, at low doses of either StI or StII, vesicles composed of SM and phosphatidic acid (PA) were permeabilized faster and to a higher extent than vesicles of PC and SM (Valcarcel et al. 2001). It was suggested that the insertion of the toxin channel could induce the formation of a nonlamellar lipid structure, a toroidal lipid pore, where pore walls are composed not only of protein helices, but also lipid molecules. In this case, the presence of lipids favoring a nonlamellar phase like PA, strong inducers of negative curvature in the bilayer, could help in the formation of the pore. This possibility was confirmed by the fact that the formation of toxin pores strongly promotes the rate of transbilayer movement of lipid molecules, which indicates local disruption of the lamellar structure (Valcarcel et al. 2001). To further explore the toroidal pore model, EqtII was added at nanomolar concentrations only to the *cis* side of a PLM composed of SM, PC and variable concentration of PA. Reversal voltages were measured in a 10-fold KCl gradient and translated into a permeability ratio P^+/P^- (where P^+ and P^- refer to cation and anion permeability, respectively) by the Goldmann-Hodgkin-Katz equation. The cation selectivity of the toxin-induced pores increased in the presence of negatively charged PA, indicating the presence of lipid head groups in the conductive channel (Anderluh et al. 2003). Furthermore, the comparison of pore conductance distribution between actinoporins and other pore-forming toxins, such as those of the β-barrel α-hemolysin family, reveals that actinoporin pores are noisier with broader current distribution, which is expected if both protein and lipids form the pore walls (Kristan et al. 2009; Antonini et al. 2014).

11.4 Conclusions

Electrophysiological measurements, specifically the PLM approach, are very important for studies of pores formed by CDCs and actinoporins. The biggest advantage is the possibility to directly observe the properties of pores at the single pore level. PLM measurements in different conditions, such as salt, pH, lipid composition, etc. allowed, therefore, for distinguishing different modes of pore formation by CDCs and actinoporins. It is obvious that these unrelated protein toxin families share some similarities in how pores are formed, namely that the presence of lipids in the conductive pathway cannot be excluded. It is also clear that the lipid composition of the target membrane greatly affects the mechanism of membrane perforation and this needs to be thoroughly studied in the future, if we wish to understand the biological role of these molecules in full.

Acknowledgments We are indebted to Marta Marchioretto and Valeria Antonini for their help with PLM traces reported in Fig. 11.3a. MZ has been supported by the Marie Curie-PAT Cofund project "NanoArtPore". This project has received funding from the European Union's Seventh Framework Programme for research, technological development and demonstration under grant agreement No 226070, project "Trentino". NR and GA were supported by the Slovenian Research Agency.

References

Alegre-Cebollada J, Onaderra M, Gavilanes JG, del Pozo AM (2007) Sea anemone actinoporins: the transition from a folded soluble state to a functionally active membrane-bound oligomeric pore. Curr Protein Pept Sci 8(6):558–572

Alvarez C, Mancheno JM, Martinez D, Tejuca M, Pazos F, Lanio ME (2009) Sticholysins, two pore-forming toxins produced by the Caribbean Sea anemone *Stichodactyla helianthus*: their interaction with membranes. Toxicon 54(8):1135–1147

Anderluh G, Barlič A, Potrich C, Maček P, Menestrina G (2000) Lysine 77 is a key residue in aggregation of equinatoxin II, a pore-forming toxin from sea anemone *Actinia equina*. J Membr Biol 173(1):47–55

Anderluh G, Dalla Serra M, Viero G, Guella G, Maček P, Menestrina G (2003) Pore formation by equinatoxin II, a eukaryotic protein toxin, occurs by induction of nonlamellar lipid structures. J Biol Chem 278(46):45216–45223

Anderluh G, Lakey JH (2008) Disparate proteins use similar architectures to damage membranes. Trends Biochem Sci 33(10):482–490

Anderluh G, Maček P (2002) Cytolytic peptide and protein toxins from sea anemones (Anthozoa: Actiniaria). Toxicon 40(2):111–124

Antonini V, Perez-Barzaga V, Bampi S, Penton D, Martinez D, Dalla Serra M, Tejuca M (2014) Functional characterization of sticholysin I and W111C mutant reveals the sequence of the actinoporin's pore assembly. PLoS One 9(10), e110824

Athanasiadis A, Anderluh G, Maček P, Turk D (2001) Crystal structure of the soluble form of equinatoxin II, a pore-forming toxin from the sea anemone *Actinia equina*. Structure 9(4):341–346

Baker MA, Rojko N, Cronin B, Anderluh G, Wallace MI (2014) Photobleaching reveals heterogeneous stoichiometry for equinatoxin II oligomers. Chembiochem 15(14):2139–2145

Bakrač B, Anderluh G (2010) Molecular mechanism of sphingomyelin-specific membrane binding and pore formation by actinoporins. Adv Exp Med Biol 677:106–115

Bakrač B, Gutierrez-Aguirre I, Podlesek Z, Sonnen AF, Gilbert RJ, Maček P, Lakey JH, Anderluh G (2008) Molecular determinants of sphingomyelin specificity of a eukaryotic pore-forming toxin. J Biol Chem 283(27):18665–18677

Bavdek A, Gekara NO, Priselac D, Gutierrez Aguirre I, Darji A, Chakraborty T, Maček P, Lakey JH, Weiss S, Anderluh G (2007) Sterol and pH interdependence in the binding, oligomerization, and pore formation of Listeriolysin O. Biochemistry 46(14):4425–4437

Bavdek A, Kostanjšek R, Antonini V, Lakey JH, Dalla Serra M, Gilbert RJ, Anderluh G (2012) pH dependence of listeriolysin O aggregation and pore-forming ability. FEBS J 279(1):126–141

Belmonte G, Menestrina G, Pederzolli C, Križaj I, Gubenšek F, Turk T, Maček P (1994) Primary and secondary structure of a pore-forming toxin from the sea anemone, Actinia equina L., and its association with lipid vesicles. Biochim Biophys Acta 1192(2):197–204

Belmonte G, Pederzolli C, Maček P, Menestrina G (1993) Pore formation by the sea anemone cytolysin equinatoxin II in red blood cells and model lipid membranes. J Membr Biol 131(1):11–22

Bernheimer AW, Avigad LS (1976) Properties of a toxin from the sea anemone *Stoichacis helianthus*, including specific binding to sphingomyelin. Proc Natl Acad Sci U S A 73(2):467–471

Bhakdi S, Tranum-Jensen J, Sziegoleit A (1985) Mechanism of membrane damage by streptolysin-O. Infect Immun 47(1):52–60

Blumenthal R, Habig WH (1984) Mechanism of tetanolysin-induced membrane damage: studies with black lipid membranes. J Bacteriol 157(1):321–323

Bourdeau RW, Malito E, Chenal A, Bishop BL, Musch MW, Villereal ML, Chang EB, Mosser EM, Rest RF, Tang WJ (2009) Cellular functions and X-ray structure of anthrolysin O, a cholesterol-dependent cytolysin secreted by *Bacillus anthracis*. J Biol Chem 284:14645–14656

Czajkowsky DM, Hotze EM, Shao Z, Tweten RK (2004) Vertical collapse of a cytolysin prepore moves its transmembrane beta-hairpins to the membrane. EMBO J 23(16):3206–3215

Dalla Serra M, Menestrina G (2000) Characterization of molecular properties of pore-forming toxins with planar lipid bilayers. Methods Mol Biol 145:171–188

Dalla Serra M, Tejuca M (2011) Pore-forming toxins. eLS. doi:10.1002/9780470015902.a0002655.pub2

Dowd KJ, Tweten RK (2012) The cholesterol-dependent cytolysin signature motif: a critical element in the allosteric pathway that couples membrane binding to pore assembly. PLoS Pathog 8(7), e1002787

El-Rachkidy RG, Davies NW, Andrew PW (2008) Pneumolysin generates multiple conductance pores in the membrane of nucleated cells. Biochem Biophys Res Commun 368(3):786–792

Farrand AJ, LaChapelle S, Hotze EM, Johnson AE, Tweten RK (2010) Only two amino acids are essential for cytolytic toxin recognition of cholesterol at the membrane surface. Proc Natl Acad Sci U S A 107(9):4341–4346

Feil SC, Ascher DB, Kuiper MJ, Tweten RK, Parker MW (2014) Structural studies of *Streptococcus pyogenes* streptolysin O provide insights into the early steps of membrane penetration. J Mol Biol 426(4):785–792

Feil SC, Lawrence S, Mulhern TD, Holien JK, Hotze EM, Farrand S, Tweten RK, Parker MW (2012) Structure of the lectin regulatory domain of the cholesterol-dependent cytolysin lectinolysin reveals the basis for its lewis antigen specificity. Structure 20(2):248–258

Frazao B, Vasconcelos V, Antunes A (2012) Sea anemone (Cnidaria, Anthozoa, Actiniaria) toxins: an overview. Mar Drugs 10(8):1812–1851

From C, Granum PE, Hardy SP (2008) Demonstration of a cholesterol-dependent cytolysin in a noninsecticidal *Bacillus sphaericus* strain and evidence for widespread distribution of the toxin within the species. FEMS Microbiol Lett 286(1):85–92

Garcia-Ortega L, Alegre-Cebollada J, Garcia-Linares S, Bruix M, Martinez-Del-Pozo A, Gavilanes JG (2011) The behavior of sea anemone actinoporins at the water-membrane interface. Biochim Biophys Acta 1808(9):2275–2288

Garcia-Saez AJ, Coraiola M, Dalla Serra M, Mingarro I, Muller P, Salgado J (2006) Peptides corresponding to helices 5 and 6 of Bax can independently form large lipid pores. FEBS J 273(5):971–981

Gelber SE, Aguilar JL, Lewis KL, Ratner AJ (2008) Functional and phylogenetic characterization of Vaginolysin, the human-specific cytolysin from *Gardnerella vaginalis*. J Bacteriol 190(11):3896–3903

Giddings KS, Johnson AE, Tweten RK (2003) Redefining cholesterol's role in the mechanism of the cholesterol-dependent cytolysins. Proc Natl Acad Sci U S A 100(20):11315–11320

Giddings KS, Zhao J, Sims PJ, Tweten RK (2004) Human CD59 is a receptor for the cholesterol-dependent cytolysin intermedilysin. Nat Struct Mol Biol 11(12):1173–1178

Gilbert RJ (2005) Inactivation and activity of cholesterol-dependent cytolysins: what structural studies tell us. Structure 13(8):1097–1106

Gilbert RJ, Dalla Serra M, Froelich CJ, Wallace MI, Anderluh G (2014) Membrane pore formation at protein-lipid interfaces. Trends Biochem Sci 39(11):510–516

Gilbert RJ, Mikelj M, Dalla Serra M, Froelich CJ, Anderluh G (2013) Effects of MACPF/CDC proteins on lipid membranes. Cell Mol Life Sci 70(12):2083–2098

Heuck AP, Moe PC, Johnson BB (2010) The cholesterol-dependent cytolysin family of gram-positive bacterial toxins. In: Harris JR (ed) Cholesterol binding and cholesterol transport proteins: structure and function in health and disease. Springer, Dordrecht, pp 551–577

Hinds MG, Zhang W, Anderluh G, Hansen PE, Norton RS (2002) Solution structure of the eukaryotic pore-forming cytolysin equinatoxin II: implications for pore formation. J Mol Biol 315(5):1219–1229

Hong Q, Gutierrez-Aguirre I, Barlič A, Malovrh P, Kristan K, Podlesek Z, Maček P, Turk D, Gonzalez-Manas JM, Lakey JH, Anderluh G (2002) Two-step membrane binding by Equinatoxin II, a pore-forming toxin from the sea anemone, involves an exposed aromatic cluster and a flexible helix. J Biol Chem 277(44):41916–41924

Hotze EM, Le HM, Sieber JR, Bruxvoort C, McInerney MJ, Tweten RK (2013) Identification and characterization of the first cholesterol-dependent cytolysins from Gram-negative bacteria. Infect Immun 81(1):216–225

Hotze EM, Tweten RK (2012) Membrane assembly of the cholesterol-dependent cytolysin pore complex. Biochim Biophys Acta 1818(4):1028–1038

Hotze EM, Wilson-Kubalek EM, Rossjohn J, Parker MW, Johnson AE, Tweten RK (2001) Arresting pore formation of a cholesterol-dependent cytolysin by disulfide trapping synchronizes the insertion of the transmembrane beta-sheet from a prepore intermediate. J Biol Chem 276(11):8261–8268

Iacovache I, van der Goot FG, Pernot L (2008) Pore formation: an ancient yet complex form of attack. Biochim Biophys Acta 1778(7–8):1611–1623

Kem WR (1988) Sea anemone toxins: structure and action. In: Hessinger DA, Lenhoff HM (eds) The biology of nematocysts, vol 375–405. Academic, San Diego

Kienker PK, Lear JD (1995) Charge selectivity of the designed uncharged peptide ion channel Ac-(LSSLLSL)3-CONH2. Biophys J 68(4):1347–1358

Korchev YE, Bashford CL, Pederzolli C, Pasternak CA, Morgan PJ, Andrew PW, Mitchell TJ (1998) A conserved tryptophan in pneumolysin is a determinant of the characteristics of channels formed by pneumolysin in cells and planar lipid bilayers. Biochem J 329(3):571–577

Köster S, van Pee K, Hudel M, Leustik M, Rhinow D, Kuhlbrandt W, Chakraborty T, Yildiz O (2014) Crystal structure of listeriolysin O reveals molecular details of oligomerization and pore formation. Nat Commun 5:3690

Kristan K, Viero G, Maček P, Dalla Serra M, Anderluh G (2007) The equinatoxin N-terminus is transferred across planar lipid membranes and helps to stabilize the transmembrane pore. FEBS J 274(2):539–550

Kristan KC, Viero G, Dalla Serra M, Maček P, Anderluh G (2009) Molecular mechanism of pore formation by actinoporins. Toxicon 54(8):1125–1134

Leung C, Dudkina NV, Lukoyanova N, Hodel AW, Farabella I, Pandurangan AP, Jahan N, Pires Damaso M, Osmanovic D, Reboul CF, Dunstone MA, Andrew PW, Lonnen R, Topf M, Saibil HR, Hoogenboom BW (2014) Stepwise visualization of membrane pore formation by suilysin, a bacterial cholesterol-dependent cytolysin. Elife 3, e04247

Malovrh P, Viero G, Dalla Serra M, Podlesek Z, Lakey JH, Maček P, Menestrina G, Anderluh G (2003) A novel mechanism of pore formation: membrane penetration by the N-terminal amphipathic region of equinatoxin. J Biol Chem 278(25):22678–22685

Mancheno JM, Martin-Benito J, Martinez-Ripoll M, Gavilanes JG, Hermoso JA (2003) Crystal and electron microscopy structures of sticholysin II actinoporin reveal insights into the mechanism of membrane pore formation. Structure 11(11):1319–1328

Marchioretto M, Podobnik M, Dalla Serra M, Anderluh G (2013) What planar lipid membranes tell us about the pore-forming activity of cholesterol-dependent cytolysins. Biophys Chem 182:64–70

Matsuzaki K (1998) Magainins as paradigm for the mode of action of pore forming polypeptides. Biochim Biophys Acta 1376(3):391–400

Mechaly AE, Bellomio A, Gil-Carton D, Morante K, Valle M, Gonzalez-Manas JM, Guerin DM (2011) Structural insights into the oligomerization and architecture of eukaryotic membrane pore-forming toxins. Structure 19(2):181–191

Meinardi E, Florin-Christensen M, Paratcha G, Azcurra JM, Florin-Christensen J (1995) The molecular basis of the self/nonself selectivity of a coelenterate toxin. Biochem Biophys Res Commun 216(1):348–354

Menestrina G, Bashford CL, Pasternak CA (1990) Pore-forming toxins: experiments with *S. aureus* alpha-toxin, *C. perfringens* theta-toxin and *E. coli* haemolysin in lipid bilayers, liposomes and intact cells. Toxicon 28(5):477–491

Menestrina G, Dalla Serra M, Prevost G (2001) Mode of action of beta-barrel pore-forming toxins of the staphylococcal alpha-hemolysin family. Toxicon 39(11):1661–1672

Metkar SS, Marchioretto M, Antonini V, Lunelli L, Wang B, Gilbert RJ, Anderluh G, Roth R, Pooga M, Pardo J, Heuser JE, Dalla Serra M, Froelich CJ (2015) Perforin oligomers form arcs in cellular membranes: a locus for intracellular delivery of granzymes. Cell Death Differ 22(1):74–85

Michaels DW (1979) Membrane damage by a toxin from the sea anemone *Stoichactis helianthus*. I. Formation of transmembrane channels in lipid bilayers. Biochim Biophys Acta 555(1): 67–78

Norton RS (1991) Structure and structure-function relationships of sea anemone proteins that interact with the sodium channel. Toxicon 29(9):1051–1084

Ohno-Iwashita Y, Iwamoto M, Ki M, Ando S, Iwashita S (1991) A cytolysin, {theta}-toxin, preferentially binds to membrane cholesterol surrounded by phospholipids with 18-carbon hydrocarbon chains in cholesterol-rich region. J Biochem 110(3):369–375

Palmer M, Harris R, Freytag C, Kehoe M, Tranum-Jensen J, Bhakdi S (1998) Assembly mechanism of the oligomeric streptolysin O pore: the early membrane lesion is lined by a free edge of the lipid membrane and is extended gradually during oligomerization. EMBO J 17(6): 1598–1605

Polekhina G, Giddings KS, Tweten RK, Parker MW (2005) Insights into the action of the superfamily of cholesterol-dependent cytolysins from studies of intermedilysin. Proc Natl Acad Sci U S A 102(3):600–605

Praper T, Sonnen A, Viero G, Kladnik A, Froelich CJ, Anderluh G, Dalla Serra M, Gilbert RJ (2011) Human perforin employs different avenues to damage membranes. J Biol Chem 286(4):2946–2955

Prigent D, Alouf JE (1976) Interaction of steptolysin O with sterols. Biochim Biophys Acta 443(2):288–300

Ramachandran R, Heuck AP, Tweten RK, Johnson AE (2002) Structural insights into the membrane-anchoring mechanism of a cholesterol-dependent cytolysin. Nat Struct Biol 9(11):823–827

Ramachandran R, Tweten RK, Johnson AE (2004) Membrane-dependent conformational changes initiate cholesterol-dependent cytolysin oligomerization and intersubunit beta-strand alignment. Nat Struct Mol Biol 11(8):697–705

Ramachandran R, Tweten RK, Johnson AE (2005) The domains of a cholesterol-dependent cytolysin undergo a major FRET-detected rearrangement during pore formation. Proc Natl Acad Sci U S A 102(20):7139–7144

Repp H, Pamukci Z, Koschinski A, Domann E, Darji A, Birringer J, Brockmeier D, Chakraborty T, Dreyer F (2002) Listeriolysin of *Listeria monocytogenes* forms Ca2+ – permeable pores leading to intracellular Ca2+ oscillations. Cell Microbiol 4(8):483–491

Rojko N, Cronin B, Danial JS, Baker MA, Anderluh G, Wallace MI (2014) Imaging the lipid-phase-dependent pore formation of equinatoxin II in droplet interface bilayers. Biophys J 106(8):1630–1637

Rossjohn J, Feil SC, McKinstry WJ, Tweten RK, Parker MW (1997) Structure of a cholesterol-binding, thiol-activated cytolysin and a model of its membrane form. Cell 89(5):685–692

Sato TK, Tweten RK, Johnson AE (2013) Disulfide-bond scanning reveals assembly state and beta-strand tilt angle of the PFO beta-barrel. Nat Chem Biol 9(6):383–389

Shatursky O, Heuck AP, Shepard LA, Rossjohn J, Parker MW, Johnson AE, Tweten RK (1999) The mechanism of membrane insertion for a cholesterol-dependent cytolysin: a novel paradigm for pore-forming toxins. Cell 99(3):293–299

Shepard LA, Shatursky O, Johnson AE, Tweten RK (2000) The mechanism of pore assembly for a cholesterol-dependent cytolysin: formation of a large prepore complex precedes the insertion of the transmembrane β-hairpins. Biochemistry 39(33):10284–10293

Shewell LK, Harvey RM, Higgins MA, Day CJ, Hartley-Tassell LE, Chen AY, Gillen CM, James DB, Alonzo F 3rd, Torres VJ, Walker MJ, Paton AW, Paton JC, Jennings MP (2014) The cholesterol-dependent cytolysins pneumolysin and streptolysin O require binding to red blood cell glycans for hemolytic activity. Proc Natl Acad Sci U S A 111(49):E5312–E5320

Soltani CE, Hotze EM, Johnson AE, Tweten RK (2007) Structural elements of the cholesterol-dependent cytolysins that are responsible for their cholesterol-sensitive membrane interactions. Proc Natl Acad Sci U S A 104(51):20226–20231

Sonnen AF, Plitzko JM, Gilbert RJ (2014) Incomplete pneumolysin oligomers form membrane pores. Open Biol 4:140044

Tardent P (1995) The cnidarian cnidocyte, a hightech cellular weaponry. Bioessays 17(4):351–362

Tejuca M, Dalla Serra M, Potrich C, Alvarez C, Menestrina G (2001) Sizing the radius of the pore formed in erythrocytes and lipid vesicles by the toxin sticholysin I from the sea anemone *Stichodactyla helianthus*. J Membr Biol 183(2):125–135

Tejuca M, Dalla Serra M, Ferreras M, Lanio ME, Menestrina G (1996) Mechanism of membrane permeabilization by sticholysin I, a cytolysin isolated from the venom of the sea anemone *Stichodactyla helianthus*. Biochemistry 35(47):14947–14957

Tilley SJ, Orlova EV, Gilbert RJ, Andrew PW, Saibil HR (2005) Structural basis of pore formation by the bacterial toxin pneumolysin. Cell 121(2):247–256

Valcarcel CA, Dalla Serra M, Potrich C, Bernhart I, Tejuca M, Martinez D, Pazos F, Lanio ME, Menestrina G (2001) Effects of lipid composition on membrane permeabilization by sticholysin I and II, two cytolysins of the sea anemone *Stichodactyla helianthus*. Biophys J 80(6):2761–2774

Varanda W, Finkelstein A (1980) Ion and nonelectrolyte permeability properties of channels formed in planar lipid bilayer membranes by the cytolytic toxin from the sea anemone, *Stoichactis helianthus*. J Membr Biol 55(3):203–211

Wade KR, Hotze EM, Kuiper MJ, Morton CJ, Parker MW, Tweten RK (2015) An intermolecular electrostatic interaction controls the prepore-to-pore transition in a cholesterol-dependent cytolysin. Proc Natl Acad Sci U S A. doi:10.1073/pnas.1423754112

Xu L, Huang B, Du H, Zhang XC, Xu J, Li X, Rao Z (2010) Crystal structure of cytotoxin protein suilysin from Streptococcus suis. Protein Cell 1(1):96–105

Xu XP, Zhai D, Kim E, Swift M, Reed JC, Volkmann N, Hanein D (2013) Three-dimensional structure of Bax-mediated pores in membrane bilayers. Cell Death Dis 4:e683

Yang L, Harroun TA, Weiss TM, Ding L, Huang HW (2001) Barrel-stave model or toroidal model? A case study on melittin pores. Biophys J 81(3):1475–1485

Zorec R, Tester M, Maček P, Mason WT (1990) Cytotoxicity of equinatoxin II from the sea anemone *Actinia equina* involves ion channel formation and an increase in intracellular calcium activity. J Membr Biol 118(3):243–249

Zwaferink H, Stockinger S, Hazemi P, Lemmens-Gruber R, Decker T (2008) IFN-beta increases listeriolysin O-induced membrane permeabilization and death of macrophages. J Immunol 180(6):4116–4123

Part IV
Other Unconventional Channels

Chapter 12
Perforins

Robert J.C. Gilbert

Abstract "Perforins" – cytotoxic cell perforin-1, proteins contributing to the complement membrane attack complex (MAC) and other perforin-like proteins – form unconventional pores in biological membranes in the sense that as well as a ring of subunits circumscribing an opening in the lipid bilayer, arcs of subunits induce similar effects. The resulting arc-pore structures are completed by a lipidic edge, conferring distinctive functional characteristics on them. Electrophysiological measurements have played a distinguished role in the discovery and characterisation of this alternative mechanism of pore formation which enables the same protein to form widely-varying sizes of pore and enact a greater diversity of effects than more conventional channels. This review discusses the mechanism of pore formation by the perforins and the way in which it has been illuminated by electrical conductance studies alongside imaging methods such as electron microscopy and atomic force microscopy.

Keywords Perforin • Complement membrane attack complex • Toroidal pore • Proteolipidic pore • Single channel conductance

12.1 Discovery of Perforins

The discovery about the year 1980 that cytotoxic lymphocytes target the antigen-presenting cells they lyse by releasing proteins that generate pore-forming assemblies of multiple subunits revealed the existence of the class of molecules from that time onwards termed *perforins* (Podack and Dennert 1983; Young et al. 1986b) (Fig. 12.1a). The main perforin discussed in this chapter will be perforin-1; though historically just referred to as "perforin" it will be here called perforin-1 to distinguish it from the recently-identified macrophage-specific perforin-2 (see below) (McCormack et al. 2013a, b; Spilsbury et al. 1995; D'Angelo et al. 2012; Wiens et al. 2005). In addition to perforin-1, I will also discuss pore formation by perforin-like proteins (PLPs), of which the most significant are the perforin-homologous

R.J.C. Gilbert
Division of Structural Biology, Wellcome Trust Centre for Human Genetics,
University of Oxford, Roosevelt Drive, Oxford OX3 7BN, UK
e-mail: gilbert@strubi.ox.ac.uk

A.H. Delcour (ed.), *Electrophysiology of Unconventional Channels and Pores*,
Springer Series in Biophysics 18, DOI 10.1007/978-3-319-20149-8_12

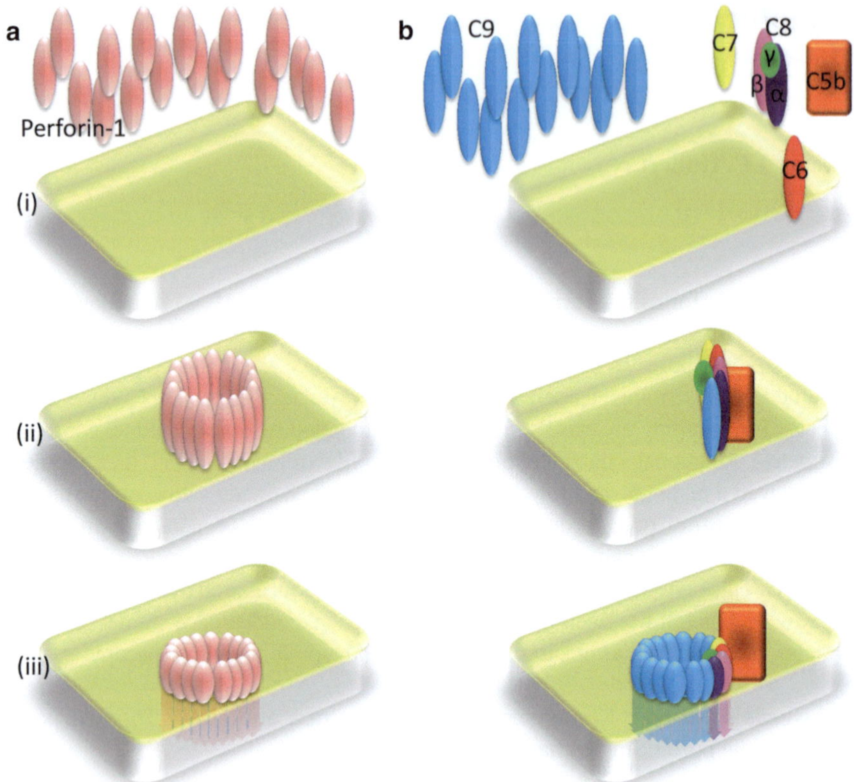

Fig. 12.1 Schematics of pore formation by the perforins perforin-1 and components of the membrane attack complex. (**a**) Pore formation by perforin-1 involves oligomerisation of 18–20 subunits on a membrane, triggered by calcium (Law et al. 2010). (*i*) Like the cholesterol-dependent cytolysins, bacterial proteins which are homologues of the perforins, a prepore assembly (*ii*) converts in a concerted fashion into a pore (*iii*) (Praper et al. 2011b). As described in Fig. 12.3, prepore assemblies can be both complete rings (generating "conventional" pores) and arcs (generating "unconventional" pores) (Gilbert 2002, 2005; Gilbert et al. 2014; Leung et al. 2014; Metkar et al. 2015; Podobnik et al. 2015; Praper et al. 2011b; Sonnen et al. 2014). (**b**) Pore formation by the membrane attack complex is initiated by the binding of an assembly of C5b-8 subunits which is joined by C9. The more copies of C9 bind, the larger the pore that can be generated, up to a complete ring of subunits (Aleshin et al. 2012; Sonnen and Henneke 2014; Lovelace et al. 2011)

subunits together contributing to the formation of the complement membrane attack complex (MAC) (Fig. 12.1b) (Borsos et al. 1964; Michaels et al. 1976; Humphrey and Dourmashkin 1969). Other PLPs mentioned will include pleurotolysin (Lukoyanova et al. 2015; Schlumberger et al. 2014), PLPs found in Apicomplexan parasites such as *Plasmodium* and *Toxoplasma* (Amino et al. 2008; Garg et al. 2013; Kafsack and Carruthers 2010; Kafsack et al. 2009; Tavares et al. 2013) and the human PLPs astrotactins 1 and 2 (Solecki 2012; Wilson et al. 2010). This selection of PLPs will suggest the different kinds of biophysical effects perforins are involved with, in and on living cells, but hardly begins to describe their range of activity (Anderluh and Gilbert 2014).

Perforin-1 itself is produced by cytotoxic T-lymphocytes (CTLs) and natural killer (NK) cells and stored within granules complexed with granzymes (granule-associated enymes) and serglycin (a proteoglycan providing the granule-forming matrix) (Pardo et al. 2009; Froelich et al. 2009). On engagement of the CTL/NK cell with an antigen-presenting cell via a T-cell receptor/major histocompatibility complex protein interaction and immunological synapse formation (Davis and van der Merwe 2011; Dustin and Depoil 2011) the granules are released, whereupon perforin is activated by the binding of calcium ions (Uellner et al. 1997; Voskoboinik et al. 2005). At this point, it is known that perforin-1 mediates the introduction of granzymes within the target cell, though two different kinds of mechanism have been proposed (Fig. 12.2). In one, pore formation at the target cell plasma membrane directly introduces granzymes within its cytosol where they initiate apoptosis (Voskoboinik et al. 2006; Lopez et al. 2013). In the other, perforin-1 and the granzymes are internalised together within endosomes, and perforin-1 then delivers the granzymes into the target cell cytoplasm, again triggering a cell death pathway, by forming pores in the endosomal membrane (Metkar et al. 2015; Praper et al. 2011a; Thiery et al. 2010, 2011). In both cases pore formation is believed to occur as the result of oligomerisation of up to 20 copies of perforin-1 to create a ring of subunits, or an arc of subunits in which the pore is formed at an interface with the lipid membrane (Fig. 12.3) (Gilbert et al. 2013, 2014; Praper et al. 2011b). Rings of subunits represent "conventional pores" whereas arcs interfacing with the lipid bilayer constitute "unconventional pores". In both proposed mechanisms for granzyme delivery (Fig. 12.2) there is a possible role for the unconventional pores formed by arcs of subunits. "Unconventional" pores of this kind (Fig. 12.3b) (Gilbert et al. 2013, 2014) may explain the transience of perforin pores at the cell membrane (Metkar et al. 2015; Thiery et al. 2011; Lopez et al. 2013), and do explain the capacity of perforin-1 to form pores of significantly different functional area, resulting in electrical conductances of differing amplitude (Young et al. 1986a, b; Praper et al. 2011b; Metkar et al. 2015).

Whereas perforin is a single protein which multimerizes to generate pore-forming structures (Fig. 12.1a), complement pores form as the conclusion to a series of protein-protein interactions beginning with the proteolytic activation of C5 to C5b, and then the formation of progressively larger assemblies as first C6, C7 and ultimately C8 join to form a membrane-binding nucleus to the MAC (Borsos et al. 1964; Michaels et al. 1976; Humphrey and Dourmashkin 1969). The addition of a number of copies of C9 then leads to formation of a fully functioning pore (Fig. 12.1b) (Borsos et al. 1964; Michaels et al. 1976; Humphrey and Dourmashkin 1969), though again the pores it generates exist in both conventional (a complete ring of protein subunits) and unconventional (an arc of subunits interfacing with the membrane bilayer) forms (Bhakdi and Tranum-Jensen 1991) (Fig. 12.3). The discovery that not only do perforin-1 and the MAC generate pore-forming oligomers of similar appearance by electron microscopy, but are actually homologues was made via the demonstration of antibody cross-reactivity very soon after perforin-1 had first been discovered (Young et al. 1986b). For this reason perforin-1 and the complement proteins to which it shows homology have been grouped under the family name MACPF (MAC-perforin) and for the purposes of this chapter, both MAC components possessing the canonical MACPF domain and perforin-1 itself

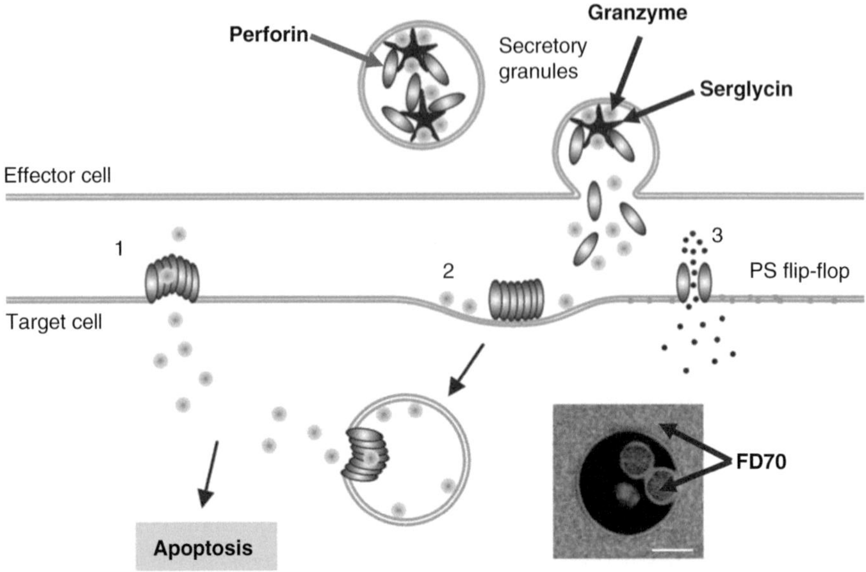

Fig. 12.2 Models for the activity of perforin-1 at the cytotoxic lymphocyte/antigen presenting cell (APC) interface. Granules containing perforin-1 and granzymes in a serglycin matrix released towards the APC deliver perforin-1 to the target cell surface. Pore formation at that surface might result in direct delivery of granzymes into the target cell cytoplasm, initiating apoptosis (*1*) (Lopez et al. 2013; Voskoboinik et al. 2006); or, (*2*) endocytosis of perforin-1 and granzymes together may lead to pore formation in a subsequent endosome leading to granzyme delivery (Lopez et al. 2013; Metkar et al. 2011, 2015; Thiery et al. 2011; Voskoboinik et al. 2006). Perforin is also capable of causing invagination and vesiculation on its own (Praper et al. 2011a) (*inset*, a giant unilamellar vesicle into which perforin has imported fluorescent dextran (FD70) -containing intraluminal vesicles; bar = 20μm). The formation of pores by arcs of perforin-1 completed by a lipid edge ("unconventional" pores; Figs. 12.3 and 12.5) allows flip-flop of lipids such as phosphatidylserine from the inner leaflet to the outer (*3*) (Metkar et al. 2011, 2015). Pores active at the plasma membrane might be full rings of subunits or consist of arcs interfacing with lipids and both would be capable of granzyme delivery (Figure reproduced with permission from (Gilbert et al. 2013) (Fig. 9; © Springer Basel AG, 2012) and subsequently modified for this publication)

will be treated as types of *perforins*. The MACPF domain is found not only in perforins-1 and –2 and the main pore-forming complement component, C9, but also in C8α and β (C8 is an αβγ complex), C7 and C6 (Aleshin et al. 2012; Hadders et al. 2007; Lovelace et al. 2011; Rosado et al. 2007; Rosado et al. 2008); see a recent review for more details (Sonnen and Henneke 2014) (Fig. 12.1b).

Macrophage-specific perforin-2 was first discovered via detection of a novel peforin-1-homologous cDNA expressed in a developmentally-regulated fashion (Spilsbury et al. 1995). Phylogenetic sequence analysis suggests that the gene encoding perforin-2 (Mpeg1) is the evolutionary origin of both perforin-1 and complement C6 – and presumably the rest of the complement MAC components (D'Angelo et al. 2012). In contrast to perforin-1 and the MAC components, however, perforin-2 is a membrane protein (Wiens et al. 2005) suggested to have a single transmembrane helix (McCormack et al. 2013a) which can kill both Gram-positive

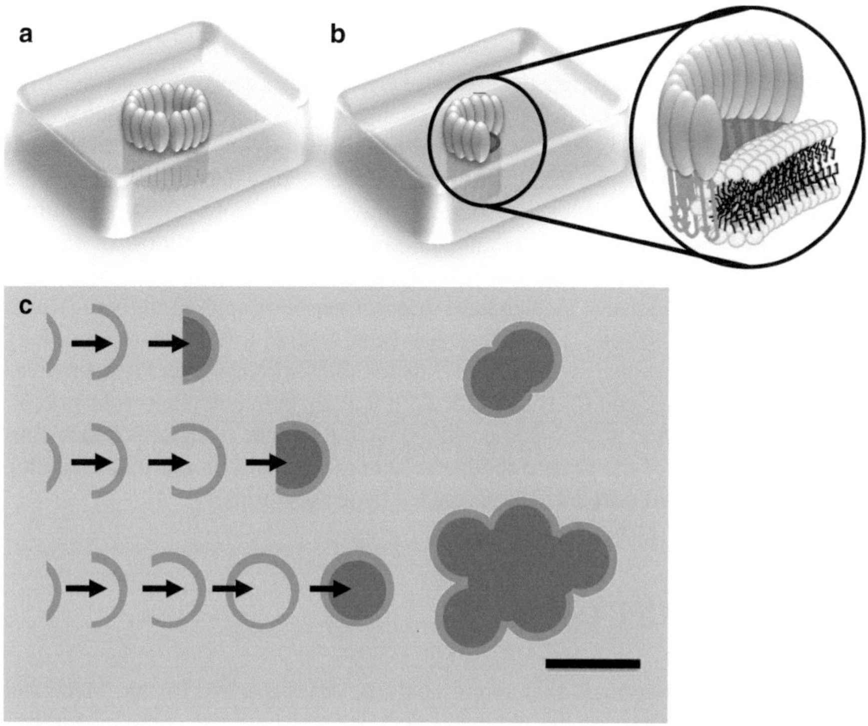

Fig. 12.3 Mechanisms of pore formation by perforin-1. (**a**) Pore formation by a complete ring of subunits, as shown in detail in Fig. 12.1a. (**b**) Pore formation by an arc of subunits, and (*right*) a close-up of the toroidal lipid arrangement believed to form at the interface of an arc and the membrane bilayer. Pore formation is shown by the *darker gray* colouring within the arc. (**c**) Pore formation by perforin-1 occurs via a prepore intermediate, which can however be both an arc of subunits and a ring of subunits. Protein-lipid interface pore formation as in panel (**b**) occurs with subunit arcs. The point of transition from prepore to pore is determined by the concentration of perforin-1 subunits bound to the membrane. This mechanism of pore formation is shared by perforin-1 and the cholesterol-dependent cytolysins (Gilbert 2002, 2005; Gilbert et al. 2014; Leung et al. 2014; Metkar et al. 2015; Podobnik et al. 2015; Praper et al. 2011b; Sonnen et al. 2014)

and Gram-negative bacteria including *Mycobacterium smegmatis*, *Staphylococcus aureus*, *Escherichia coli* and *Salmonella typhimurium* (McCormack et al. 2013b), as well as *Chlamydia* species (Fields et al. 2013). Perforin-2 functions out of a vesicle in which its MACPF domain is within the vesicle lumen such that fusion of vesicles with endosomes/phagosomes containing bacteria exposes them to the perforin-2 ectodomain and results in pore formation within their membranes, and subsequent death (McCormack et al. 2013a). To this extent the activity of perforin-2 is reminiscent of that of astrotactin-2 in humans which controls astrotactin-1-mediated neuron contact with glial fibres and is stored beforehand in vesicles within the neuronal cytoplasm (Solecki 2012; Wilson et al. 2010). Thus, whereas perforin-1 and complement MAC components function in a very similar way to other pore-forming proteins and especially to the cholesterol-dependent cytolysins to which they are related (Gilbert et al. 2013; Hadders et al. 2007; Rosado et al. 2007, 2008), perforin-2 functions more like MACPF proteins involved in developmental regulation, such as the astrotactins (Solecki 2012; Wilson et al. 2010) and *Drosophila* torso-like (Johnson et al. 2013; Stevens et al. 2003). Does its mechanism then involve pore formation? It is thought so (McCormack et al. 2013b), but further studies will be needed to show that this is the case and other mechanisms whereby perforin-2 might afford protection against intracellular bacteria are of course conceivable. There are however no electophysiological data on perforin-2 so it will not be discussed further here. Soluble (i.e. unlike perforin-2, not membrane-inserted) perforin-like proteins are also found in the Apicomplexan parasites *Toxoplasma gondii* and *Plasmodium* spp., however conductance studies have also not yet been applied to them (Roiko and Carruthers 2009). One PLP which has been studied using electrophysiology is the fungal pore-former pleurotolysin A/B combination in which PlyA provides membrane anchorage and PlyB has a MACPF domain structure (Lukoyanova et al. 2015; Schlumberger et al. 2014); this will be included in the discussion below concerning the impact of membrane conductance measurements on our understanding of the mechanisms of pore formation enacted by the perforins.

12.2 The MACPF Domain

From the time it was noted that serum could be cross-reactive for the MAC and perforin-1, and because of the obvious importance of both the complement system and perforin-1 itself for human innate and cellular immunity, it was clear that solving the atomic structure of a MACPF domain would enable a major advance in scientific understanding regarding mechanisms of mammalian defence against pathology. In fact, the determination of the first MACPF domain structures (C8α and a bacterial homologue from *Photorhabdus luminescens* known simply as *Plu-MACPF*) (Hadders et al. 2007; Rosado et al. 2007) was one of the triumphs of structural biology because it revealed a previously wholly unthought homology between the perforins and the large family of bacterial cholesterol-dependent cytolysins (CDCs) (Hadders et al. 2007; Rosado et al. 2007). The CDCs are discussed elsewhere in this volume (see Chap. 11) and the homology between MACPFs and CDCs has been extensively reviewed and analysed (Anderluh and Gilbert 2014; Gilbert et al. 2013; Rosado et al. 2008). This homology is not, however, limited to the polypeptide fold shared by MACPFs and CDCs but extends also to their mechanism of action which appears as much conserved as their structure transparently is (mean RMSD of 1.75 Å over on average 120 residues) (Gilbert et al. 2013, 2014). The extraordinary conservation which the structures show must be concluded at least in part to be explained by the fact that *this* fold *in particular* is especially well adapted to the mechanism of pore formation which is characteristic of it. In turn, it then has to be said that that mechanism appears to be of great selective advantage.

At the heart of the MACPF domain lies the secret to its success, in the form of a set of α helices which undergo refolding to generate a pair of transmembrane hairpins (Fig. 12.4). The discovery of the transmembrane hairpins (TMHs) was initially made in work on the CDC perfringolysin (Shatursky et al. 1999; Shepard et al. 1998)

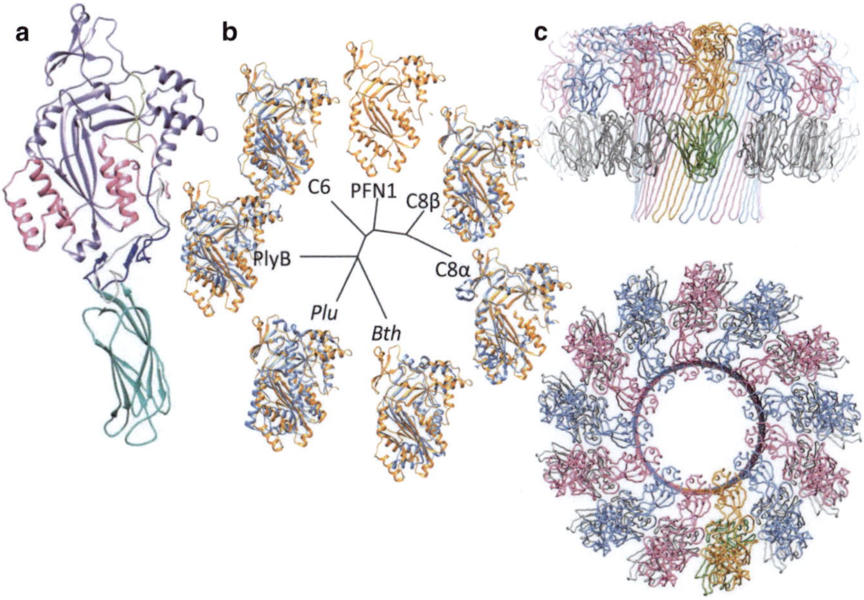

Fig. 12.4 Atomic structures of perforins and their refolding pore-forming transition. (**a**) X-ray crystal structure of murine perforin-1 (Law et al. 2010) with the ribbon backbone for the MACPF domain shown in *violet* with the helical regions refolding into β hairpins during pore formation coloured *pink*. A *blue*-coloured linking domain joins the pore-forming MACPF domain to the C-terminal membrane-binding C2 domain (aquamarine). (**b**) Structure-based phylogenetic analysis of MACPF domains with determined crystal structures (cornflower *blue*), all superimposed for reference on perforin-1 (PFN1, *orange* in each case). RCSB PDB depositions used: *PFN1*, *3NSJ*; *C6*, *3T5O*; *C8α* and *C8β*, 3OJY; *Photorhabdus luminescens* MACPF (*Plu*), 2QP2; *Bacillus thetaiotamicron* MACPF (*Bth*), 3KK7; and pleurotolysin B MACPF (PlyB), 4OEJ. (**c**) Structure of the pore-forming 13meric ring complex formed by pleurotolysin viewed from the side (*upper image*) and as it were from above the membrane surface (*lower image*) (Lukoyanova et al. 2015). Dimers of the PlyA subunit provide a membrane-binding platform for monomers of PlyB. In the *upper panel* and towards the viewer a pair of PlyA subunits coloured in two different shades of *green* are seen supporting an *orange*-coloured PlyB; then, *pink* and cornflower *blue* subunits of PlyB alternate around the ring partnered with dimers of PlyA in tones of *grey*. The β-barrel built from the two pairs of hairpins provided by each PlyB subunit is plainly visible from the side in the *upper image* and forming the central structural ring in the *lower image*

and was then easily mapped onto the MACPF proteins when they were identified as homologues of the CDCs (Rosado et al. 2008). Subsequent studies, first with CDCs (Gilbert et al. 1999; Tilley et al. 2005; Reboul et al. 2014) and now with MACPF proteins themselves (Lukoyanova et al. 2015) have allowed visualisation and modelling of the transmembrane barrel of TMHs formed by rings of MACPF/CDC subunits. As a result it should be acknowledged here that the distinction between MACPFs and CDCs is now inevitably somewhat artificial – or really, non-existent. The transition from a set of α-helices into β-sheets characteristic of the function of the MACPF/CDC fold is a wonderful example of the dependence of protein fold on context, of allostery, and of metastability in protein folds. Interestingly, another kind of pore-forming protein (*E. coli* hemolysin A) was subsequently found to undergo the reverse transition, from a β-stranded structure to an α-helical one, during the process of pore formation (Mueller et al. 2009).

12.3 Triggering a Refold

During pore formation, perforins make use of some very simple steps to facilitate the major change in their polypeptide fold that enables them to penetrate a membrane to generate a pore. The first step is membrane binding, which in the case of perforin itself occurs via a C2-like domain activated by calcium, and in the case of the MAC occurs via proteolytic activation of C5 to C5b and its binding to C6, C7 and C8 before (multiple copies of) C9 associate with the nascent pore-forming assembly (Fig. 12.1a, b). The most critical effect of membrane binding (and in the case of perforin, calcium binding too) is that it potentiates oligomerisation of further protein subunits – in the case of perforin of multiple copies of itself, in the case of the MAC of multiple C9 proteins – by concentrating them on the membrane surface. Even interaction of perforin with calcium in solution, without the presence of a membrane, results in oligomerisation (Metkar et al. 2011), and the resulting arcs or rings, when formed on membranes, undergo a transition to complete membrane insertion and pore formation (Law et al. 2010; Praper et al. 2011b). This mechanism of membrane attack via the formation of a pre-pore complex and then of a pore is common to perforin and the CDCs – and although it was suggested that the orientation of subunits in the perforin pore differs from that of CDCs (Law et al. 2010) this now seems very unlikely (Gilbert et al. 2013; Lukoyanova et al. 2015) (Fig. 12.4). Complement differs slightly in its mechanism as there may not be a pre-pore complex in that case; rather the mechanism of pore formation may be enabled by progressively longer TMH regions ($C7<C6<C8\beta<C8\alpha<C9$) inserting one-by-one gradually to open up a membrane pore and then, with the sequential addition of more copies of C9, an increase in pore size (Aleshin et al. 2012; Lovelace et al. 2011).

12.4 MACPF Proteins as Formers of Unconventional Pores

The mechanism common to perforins and the CDCs is the capacity to form pores using both full rings of subunits and partial rings, or arcs (Fig. 12.3), and these constitute the unconventional pores which MACPF proteins form: pores at the interface between an arciform protein oligomer and a lipid membrane (Gilbert et al. 2014). In my view, one of the chief selective pressures working on MACPF proteins is very likely to have been the capacity of this particular fold to form pores in this way. The advantages of this mechanism are only now beginning to be unpicked (Metkar et al. 2011, 2015; Praper et al. 2011a, b) but probably include the ability of single proteins to form both small and large pores – for example, at different stages in cell invasion or infection. A mechanism like this also enables pores to have flexible perimeters as may be especially useful for perforin-like proteins conferring cell traversal or cell egression phenotypes on parasitic organisms (Amino et al. 2008; Kafsack et al. 2009; Tavares et al. 2013). Other possible advantages are to enable the remodelling of membranes in other ways, such as the invagination and vesiculation which perforin can cause (Praper et al. 2011a).

12.5 Are These Unconventional Pores?

Whether pores formed at protein-lipid interfaces *are* "unconventional" at all is an interesting point to reflect on, since their apparently unusual or unconventional structure may instead just be a feature of the limitations of earlier assumptions framing the ways in which pore forming proteins got studied. If it is assumed that genuine pores in membranes will consist of solid rings of protein subunits then that will colour the ways in which pores are investigated, and this has been the case with MACPF/CDC proteins. The result has been a privileging in recent years of the insights of single particle electron microscopy (Law et al. 2010; Lukoyanova et al. 2015), for example, over those of electrophysiology, which have long indicated the existence of widely varying pore sizes and that lipids may be structural components of the pores themselves (Young et al. 1986a; Marchioretto et al. 2013; Metkar et al. 2015; Praper et al. 2011b; Schlumberger et al. 2014). Indeed, from the time when complement MAC proteins and perforin-1 were defined, onwards, the idea that they might form "unconventional" pores using arcs of subunits was being proposed, inspired by electrical conductance data and also by negatively-stained images of pore-forming complexes obtained using electron microscopy (Amiguet et al. 1985; Borsos et al. 1964; Tschopp 1984; Young et al. 1986a, b). These days, sub-averaging of volumes extracted from cryo-electron tomograms (Sonnen et al. 2014) and atomic force microscopy (AFM) (Leung et al. 2014; Podobnik et al. 2015; Praper et al. 2011b) have been added to the armoury of those investigating the "unconventional" pores constituted by arcs of protein subunits interfacing with a lipid membrane bilayer.

Complement was the first of the "perforins" to be imaged forming pores (Borsos et al. 1964) with both arcs and rings of subunits apparently responsible for pore formation (Young et al. 1986a). Later discussion of "tubular" and "non-tubular" (i.e. arc) pores explicitly proposed the kind of interface (Amiguet et al. 1985) still held today to be the most likely basis on which arc-based pore formation occurs to create lesions of variable size (Gilbert et al. 2014) (Fig. 12.3b). In this understanding, a toroidal arrangement of lipids forms a partial annulus and the lipid component of the structure is stabilised by the reduction in curvature stress at the "lip" of the membrane deriving from their formation by a reduction in line tension (Karatekin et al. 2003; Teissie et al. 2005; Gilbert et al. 2014). "Toroidal pores" (Karatekin et al. 2003; Teissie et al. 2005) – classic examples of unconventional pores – are reckoned to be involved in the formation of lesions by a host of different kinds of pore-forming agents, both specifically adapted pore-forming proteins such as Bax and actinoporins (Garcia-Saez et al. 2007; Valcarcel et al. 2001; Anderluh et al. 2003; Tanaka et al. 2015), and proteins which enable channels to form across membranes by destabilising the bilayer itself, either based on charge-charge effects or mechanical disruption (Bergstrom et al. 2013; Raghava et al. 2013; Steringer et al. 2012). Purely lipidic toroidal pores are also classically associated with electroporation and membrane deformation (Karatekin et al. 2003; Teissie et al. 2005). The presence in toroidal pores of a continuous lipid surface between the two sides of a lipid bilayer enables the lateral movement of lipids between leaflets, or in other words, their flip-flop (Metkar et al. 2011, 2015; Valcarcel et al. 2001) (Fig. 12.2). This is a characteristic taken to provide good evidence for the existence of such pore-forming structures.

Imaging of perforin itself using electron microscopy also from the start identified ring- and arc-shaped protein assemblies forming lesions in target membranes (Podack and Dennert 1983; Young et al. 1986b) and arcs forming pores were even observed in human myocardium which had been damaged during an attack of acute myocarditis (Young et al. 1990). Subsequently arcs of perforin-1 forming pores have been observed directly on the surface of human bladder carcinoma cells (Metkar et al. 2015). So, from the mid-1980s on, some authors were confidently drawing the conclusion that an unconventional mechanism of pore formation involving arcs of subunits and a protein-lipid interface was established as a viable understanding of how perforin and related proteins work – as they asserted, a hole is a hole (Bhakdi and Tranum-Jensen 1991), however you form it. Yet other researchers discounted this possibility, it now seems too fast.

The idea that proteolipidic pore-forming structures are biologically functioning structures has finally become accepted through the use of alternative structural studies to those carried out previously – using sub-averaging of cryo-electron tomograms (Sonnen et al. 2014) and atomic force microscopy (Leung et al. 2014; Podobnik et al. 2015). Although these studies have been limited so far to the CDC arm of the MACPF/CDC family, they will shortly without doubt be complemented by similar work on perforin itself and perhaps other perforin-like proteins. There has therefore been a breakthrough in the general understanding of the viability of such "unconventional" pores, though to the discerning it has been perfectly clear for years that perforins, alongside CDCs, form them (Bhakdi and Tranum-Jensen 1991; Bhakdi et al. 1985; Tschopp 1984). Among the most important data showing this have always been electrophysiological measurements.

12.6 The Distinguished Role of Electrophysiology in Revealing the "Unconventional" Pores Formed by Perforins

Two landmark papers published by Young and colleagues in 1986 identified the homology between perforin-1 and complement proteins, including as previously mentioned the demonstration of antigenic cross-reactivity (Young et al. 1986a, b). For perforin-1, pores of different functional sizes were identified, both on planar lipid membranes and in live cells, using conductance measurements, and a flickering of pore conductance suggested dynamic features to the pore structure – such as is provided by a lipid edge (Young et al. 1986b) – with differing levels of oligomerisation completeness (Young et al. 1986d). A further paper explicitly allied the variable degree of oligomer size to the existence of pores of variable functional size, and showed the discrete opening of pores of differing functional dimensions (Young et al. 1986c). Furthermore it again suggested that incomplete rings might generate pores of smaller functional conductance compared to those formed by full rings of subunits (Young et al. 1986c). It is now known that the sizes of pore-forming oligomer generated by MACPF/CDC proteins are dependent on the concentration of pore-forming subunits available. It seems that smaller pores result from lower local concentrations of monomeric subunits, such that when the prepore-to-pore transition occurs the oligomer is still an arc of subunits and not yet a ring (Gilbert 2002, 2005, 2010). Originally proposed in 2002 (Gilbert 2002), this model has recently been confirmed in general (Leung et al. 2014; Podobnik et al. 2015; Sonnen et al. 2014; Metkar et al. 2015).

Measuring the conductance properties of perforin pores of variable size in living cells provided a further development of the emerging understanding of differently-sized oligomers giving rise to differently-sized functional pores through membranes (Felzen et al. 1994b). Conductance measurements for single guinea pig ventricular myocytes clearly showed variable amplitudes which were assigned to variable sizes of protein oligomer, estimated as being from 3 to 20 subunits in size. This was a beautiful piece of work which matches with striking accuracy the perforin oligomeric size later found by 3D cryo-EM analysis (Law et al. 2010) and also the recent estimate for the smallest size of arciform MACPF/CDC oligomer capable of forming a pore (Leung et al. 2014). Furthermore, an enhanced capacity to generate pores when perforin was delivered as part of granules as opposed to as isolated protein was found, and it was thought this may relate to an independent role for granzme A in pore formation (Felzen et al. 1994b). Indeed, granzymes are cationic proteins which have been suggested to have the capacity to form toroidal pores, in a similar way to the effect shown for cationic cytochrome C (Bergstrom et al. 2013; Gilbert et al. 2014) and analogous to the effect of electrical charge on membranes during electroporation itself (Teissie et al. 2005). Another study took conductance measurements on perforin-1 activity one stage further and made direct electrophysiological measurement of a cytotoxic T lymphocyte-myocyte interaction. A distinct form of electrical activity was identified across the membranes of conjugated target cells (myocytes) compared to that observed when purified perforin-1 or perforin-1-containing CTL granules were added directly to cell membranes (Felzen et al. 1994a). These conductances could not be resolved into the kinds of discrete changes in permeability typical of perforin-1 pores and were more like an increase in conductance noise across the myocyte membrane, though such pores were also observed when granules were added to the still-conjugated myocytes (Felzen et al. 1994a). This strikingly suggests the presence of different kinds of membrane structure changes at the immunological synapse across which CTLs deliver their lethal hit to their targets, compared to effects observed for perforin in isolation (Felzen et al. 1994a, b; Young et al. 1986c). As the authors comment, their electrophysiological measurements suggest that perforin is not involved in the increased conductance noise observed across the membranes of CTL-conjugated myocytes "at least not in the commonly understood fashion" (Felzen et al. 1994a). This most interesting pair of studies are supportive of a mechanism for perforin's delivery of the lethal hit which does not involve the formation of large pores for granzyme delivery at the cell membrane, but instead involves internalization of granzymes within vesicles, possibly associated with the formation of small pores. Larger pores inside vesicles then lead to delivery of granzymes to the cytosol to trigger apoptosis (Metkar et al. 2015; Praper et al. 2011a) (Fig. 12.2).

In reporting ultrastructural and sequence-dependent demonstrations of similarity between perforin and complement pores, Young and colleagues also made use of electrophysiology to probe C9-based pore formation and showed, critically, that lower concentrations of C9 give rise to smaller functional pores (Young et al. 1986a). Indeed, they explicitly linked the variable sizes of functional pores to the presence of "incomplete, half and double-rings" structural assemblies of C9 seen in electron microscopy images (Young et al. 1986a). A paper from another group focusing on complement underscored how the variability in pore size derives from the variable number of C9 subunits incorporated and again showed variable conductances, and opening and closing of pores (Benz et al. 1986). In contrast to the conductances of pores formed by perforin (Fig. 12.5a), complement pores also show some oblique increases in conductance alongside step-wise changes (Benz et al. 1986) (Fig. 12.5b). This capacity of electrophysiology to identify changes in conductance of differing forms maps, we now understand, onto the differing mechanisms of pore formation by complement and perforin – the former is thought to involve, as discussed above, the gradual addition of C9 subunits to already-functional pore-forming complexes (Aleshin et al. 2012; Lovelace et al. 2011) whereas the latter undergoes a prepore-to-pore transition (Praper et al. 2011b) (Fig. 12.1). As Benz and colleagues also show, an increase in the concentration of C9 leads to an increase in the probability of pore formation, up to a maximum level of membrane disruption when the bilayer breaks (Benz et al. 1986).

The capacity of minimal complement assemblies to form pores is underscored by one of the earliest functional studies on them, in which it was shown that a C5b-8 complex alone can cause irreparable damage to targeted cells (Stolfi 1968). Later, it was found using early conductance measurements with low time resolution that C5b-8 addition to targeted planar lipid membranes produces a stepwise increase in conductance which increases with the addition of C9 to a steady state and again with subsequent additions of C9 until the membrane breaks (Michaels et al. 1976). A later study with higher resolution time sampling indicated the presence of transmembrane current fluctuations after C5b-8 addition (reminiscent of the fluctuations observed during CTL-myocyte engagement (Felzen et al. 1994a)) but that the addition of C9 induces larger and discrete pore-forming states of the complement assembly (Shiver et al. 1991). As in earlier work, complement protein C8 was the first subunit to induce electrical conductances of any kind, and then C9 stabilises and allows the expansion of those pores (Shiver et al. 1991). In any case, the mechanism of pore formation is "unconventional" in that, for many individual complement MAC-generated pores, a lipidic region is involved in their structure.

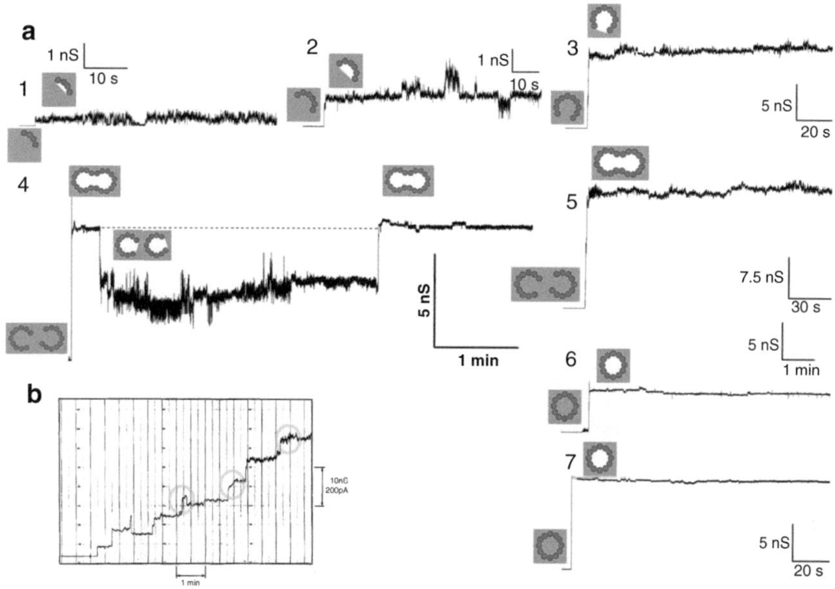

Fig. 12.5 Different grades of pore formed by perforin-1 and complement. (**a**) Electrical conductance traces of perforin-1 pores are shown, as formed by smaller (*1*, *2*) and larger (*3*) arcs of subunits, by pairs of arcs which can separate and rejoin (*4*) or remain together (*5*) and by rings of subunits (*6*, *7*). Pores with lipidic (*gray*) components display enhanced fluctuations in conductance compared to those whose perimeter is formed entirely from protein (Praper et al. 2011b). These traces were all obtained by adding perforin at 0.2–1.1 nM final concentration to preformed planar lipid bilayers with symmetric conditions in a buffer consisting of 100 mM KCl, 20 mM HEPES, 0.1 mM $CaCl_2$, pH 7.4. A constant voltage was applied, generally 40 mV. Perforin was added to the *cis* side of the membrane to which the electrical potential was applied, whereas the *trans* side was grounded (Praper et al. 2011b). (**b**) An image of a conductance trace obtained for complement showing, alongside stepwise increases in conductance, some gradual changes (ringed) (Figure reproduced with permission from (Benz et al. 1986) (Fig. 3; © Springer Basel AG, 1986) and subsequently modified. Here, the symmetrical buffer conditions were 0.15 M KCl, 5 mM HEPES, pH 7.5 and the applied voltage was 20 mV)

12.7 More Recent Use of Electrophysiology Defines Single Perforin Pores and Directly Indicates a Role for Lipids in the Pore Structure

In more recent years a series of papers have been published which have again applied electrophysiological measurements to the activity of mammalian perforin, and which have provided a new and more detailed understanding of its pore-forming mechanism (Metkar et al. 2015; Praper et al. 2011a, b). These findings have been in agreement with the long-held contention that perforin and related proteins form pores using both oligomeric rings and oligomeric arcs of subunits (Fig. 12.3). They have also explicitly demonstrated a role for the bilayer lipids in the actual structure of the pore, as reflected in the "structure" of the conductance traces obtained (Praper et al. 2011b). By using different lipids mixtures it was found that smaller pores in which the current trace was noisier (i.e. showed more random fluctuations) were formed in membranes in which the lipid tails were more flexible/disordered themselves, whereas larger pores with less noise in the current trace were formed in membranes in which the lipid tails were more rigid/ordered (Praper et al. 2011b). The significance of these data are twofold. Firstly, smaller pores having noisier current traces indicates that there are compositional differences to the molecules contributing to pore structure in smaller compared to larger pores. Secondly, the lipids themselves are seen to play a role in the structures of the smaller pores, and the stability of lipid arrangements affects the homogeneity of the pore size over time. Smaller oligomers producing smaller pores will have a greater proportion of their perimeter formed out of lipid, taken to be in a toroidal arrangement (Gilbert et al. 2014) (Fig. 12.3), and this explains their greater noise level (Fig. 12.5a). Comparing the traces numbered 1, 2 and 3 in Fig. 12.5, it would be possible to interpret the larger conductances seen in traces 2 and 3 compared to trace 1 as being due to several pore-forming events triggered by the insertion of arcs of perforin-1 subunits, rather than the insertion of one single but larger arc of subunits. However, the stepwise change in conductance makes it very unlikely that this is the case as the simultaneous insertion of two oligomers is hard to imagine given the stochastic and kinetically-determined transition of prepores to pores (Gilbert 2002, 2005; Gilbert et al. 2014; Leung et al. 2014; Podobnik et al. 2015; Sonnen et al. 2014).

A further implication of the data on effects of lipid bilayer composition on pore conductance is that oligomers forming on more disordered bilayers undergo the prepore-to-pore transition which perforin does undergo (Praper et al. 2011b) more swiftly (or, at an earlier stage in the oligomerisation process) than oligomers forming on bilayers with more ordered structures. Otherwise, more disordered lipids would not result in more, smaller pores (Praper et al. 2011b). This fits exactly with our understanding of the way in which a kinetic mechanism governs pore formation by MACPF/CDC proteins, as argued elsewhere (Gilbert 2002, 2005, 2010) and more recently by others (Leung et al. 2014). It also relates to the fact that the prepore-to-pore transition in MACPF/CDCs can be prevented by lowering the temperature of an experimental system (even just to room temperature) which, clearly, will increase the structure/order in the bilayer (Gilbert 2005; Gilbert et al. 1999; Tilley et al. 2005; Shepard et al. 2000). In other words, the energetic barrier for undergoing the prepore-to-pore transition is increased if the kinetic energy in the system is reduced, which is hardly surprising; more disordered lipids have higher kinetic energy, and higher kinetic energy is conferred by higher temperatures. There will be a greater barrier for the tearing open of the membrane bilayer, the clearance of lipids to form a pore and the conformational changes enabling these steps if the bilayer itself is more ordered. These insights can be referenced back to structural studies of MACPF/CDC pore formation. The first oligomeric form of a MACPF/CDC protein to have its structure determined was a helical form (Gilbert et al. 1999), in which the protein subunits adopted a conformation which later transpired to be the pore-forming conformation (Tilley et al. 2005). This was because they were free to, due to the absence of the constraints imposed by the presence of a lipid bilayer. In contrast, the first membrane-bound oligomeric structure was formed at room temperature and later transpired to be the prepore state (Gilbert et al. 1999). No pores were observed without pre-warming of the protein and liposomes to 37 °C (Gilbert et al. 1999; Tilley et al. 2005) because pore formation could not be completed due to the stability of the membrane and the high thermal (activation energy) barrier it therefore introduced. Therefore, the presence of a membrane bilayer constrains the capacity of a MACPF/CDC oligomer to undergo the process of actual pore formation – a constraint lacking when oligomers form in solution (Gilbert et al. 1999), and reduced at higher temperatures and when the altered dynamics of the lipids in the bilayer renders it a more disordered environment (Praper et al. 2011b).

As noted above, complement C9 was found using electron microscopy and conductance measurements to form pores using "incomplete, half and double-rings" as early as 1986 (Young et al. 1986a). Equivalent insights are now available for perforin-1 (Metkar et al. 2015; Praper et al. 2011b) and, while complete rings of subunits will remain stable and isolated structures, pores formed by arcs of subunits interfacing with a lipid bilayer have the capacity to combine directly to create a larger pore, as discussed above (Fig. 12.5a). Electrophysiology has allowed this to be shown directly and also that these double-arc structures are dynamic in the membrane (Fig. 12.5a) (Praper et al. 2011b). Thus, a pore can form which consists of two arcs together, which having an entirely protein-based edge has a low noise level in the current trace; then, the two arcs can come apart giving rise to two smaller pores with larger noise levels, and when the arcs come together again an identical amplitude to the original is obtained, with lowered noise (Praper et al. 2011b) (Fig. 12.5a, trace 4). The dynamics of pore formation have recently been further explored for listeriolysin, a bacterial MACPF/CDC (Podobnik et al. 2015), and double arc pores were also observed in a recent study using AFM to examine pore formation by the CDC suilysin (Leung et al. 2014).

The application of electrophysiological measurements to perforin pore formation has been further developed through the use of an antibody (Pf80) which does not prevent pore formation but limits oligomer size – thus promoting pore formation by arcs of oligomerised subunits rather than rings of them (Metkar et al. 2015). In line with the understanding of pore formation which is being advanced here, shaped by electrical conductance measurements and electron microscopy, this then results in pores whose conductance traces show enhanced levels of noise (Metkar et al. 2015).

12.8 Other Perforins: Perforin-like Proteins

The class of perforin-like proteins is likely to prove one of the more interesting and important groups of proteins in biology, because they are responsible for the achievement of feats of membrane remodelling critical, it seems, for the growth and migration of cells, the laying down of patterns of body formation, infection by parasites and bacteria, defence against infection and a host of other phenomena

(Anderluh and Gilbert 2014). A recent structure of the fungal PLP pleurotolysin, a 13meric pore-forming ring of subunits, provided a higher-resolution structure for a perforin pore than previously obtained, and defines in better detail than previously the mechanism of pore formation as being very similar to that of the CDCs (Lukoyanova et al. 2015). To some, the structural snapshot provided will seem to define entirely how pleurotolysin operates, but for others who look at the functional characteristics of its pores there will be more to be said. Electrophysiology has a key role here – and has shown the formation by pleurotolysin of pores of variable size on planar lipid membranes, which are characteristic of the use of arcs of subunits as well as rings to generate pore-forming structures, as found in the other forms of perforin-like protein already discussed (Schlumberger et al. 2014). It will certainly be the case that single-channel conductance measurements have a role to play in discovering the different ways in which the different perforins form pores as we come to understand more and more about their different mechanisms of action – whether in the cell traversal and egress of Apicomplexan parasites (Amino et al. 2008; Garg et al. 2013; Kafsack et al. 2009; Tavares et al. 2013), or the migration of neurons and structuring of mammalian tissue (Solecki 2012; Wilson et al. 2010).

12.9 Electrical Conductance Traces as a Measure of Membrane Area

The best-understood role of perforin-1 is pore formation, but its activity seems not to be limited to this effect on membranes. In experiments with lowered concentrations, perforin was found to induce invaginations of target membranes and the formation of intraluminal vesicles (ILVs) (Praper et al. 2011a). This process appears similar to the membrane engulfment associated with endocytosis, but to be independent of its classical mechanisms, being neither clathrin- nor receptor- nor ATP-dependent (Praper et al. 2011a). For example, invagination and vesiculation can be demonstrated on large and giant unilamellar vesicles, and on cells in which ATP production has been inhibited; it also appears to be based on a mechanism other than pore formation itself since neither the pore-forming protein lysenin nor the MACPF/CDC pneumolysin show the same effect (Praper et al. 2011a). Invagination of membranes involves an increase in local membrane area and electrophysiological measurements have been used to observe this process directly, by measurement of the capacitance of the membrane (that is, its capacity to store charge) which in the absence of changes in membrane thickness is directly related to membrane area (Praper et al. 2011a). Given the role of some perforin-like proteins in a range of very different membrane-remodelling activities (such as cell traversal of the malarial parasite (Amino et al. 2008; Tavares et al. 2013) and recycling of neuronal contacts with glial cells via vesicle formation and trafficking (Solecki 2012; Wilson et al. 2010)) the use of capacitance to measure the ways in which perforins including PLPs affect membrane area is likely to be of greater significance in the years ahead.

12.10 A Cautionary Tale?

The assessment provided so far in this chapter of the significance of electrophysiology in the study of perforins has been positive in tone – it seems that conductance measurements have provided a very important contribution to the emergence of an understanding of perforin pore-forming mechanism which was once controversial but is now accepted (Gilbert et al. 2014). This understanding – in short, that perforins can form unconventional pores (Borsos et al. 1964; Young et al. 1986a, b; Michaels et al. 1976) – was an early and accurate insight which should have been given more credit than it was (Dunstone and Tweten 2012; Law et al. 2010). However another strand of conductance measurements using perforin is harder to reconcile with the understanding of its mechanism of action that we now have, including the similarity of action for proteins like perforin-1 and the cholesterol-dependent cytolysins (Gilbert et al. 2013).

In 1991 it was shown that a 34-residue amino-terminal fragment of perforin can form pores in membranes, and the pores were characterised using electrical conductance measurements (Ojcius et al. 1991). The experiments were quite convincing, they clearly do show channel formation in lipid bilayers, including the formation of channels associated with discrete stepwise increments in conductance, and should be taken at face value (Ojcius et al. 1991). Yet it is hard to see how this peptide could in fact have a significant role in pore formation by perforin-1 in a biological context (Law et al. 2010; Lukoyanova et al. 2015). Still, a later study showed that the pore-forming capacity of the perforin amino-terminus could be further defined as associated with the first 19 residues, again with ample data to indicate pore formation – and, almost as importantly, pore formation was shown not to occur with some of the amino-terminal perforin-1 peptides tested and with some kinds of membrane (Persechini et al. 1992). What is one to make of this?

Context is most likely the answer. In fact, while one approach would be to dismiss work on the pore-forming role of perforin's amino terminal residues as demonstrating the lack of value in electrophysological experiments, that would almost certainly be the wrong approach. Instead this work reminds us about *context*, the context of a fragment of polypeptide as being critical to an understanding of its role – and it also reminds us that there has been limited understanding about the roles of lipids as collaborators in pore formation by proteins and peptides (Gilbert 2010). So, is it likely that the amino-terminal 34 (or 19) residues of perforin are directly involved in its native pore structure? No – but this reminds us that amphipathic peptides have remarkable effects on lipid bilayer membranes and can induce stable rearrangements of lipids to produce pore-forming structures – such as the toroidal lipidic pores discussed previously (Gilbert et al. 2014) (Fig. 12.3a). It also indirectly reminds us that the actual pore-forming regions of perforin are hidden within its MACPF domain in the form of a series of α-helices which will refold on membrane insertion to generate a set of β-strands – and who could have guessed that? That too reminds us of the importance of context. And an alternative hypothesis is that the channel-forming activity observed for the amino-terminus of perforin-1 relates not to pore formation as such, but its capacity to induce invagination and vesiculation.

Other forms of unconventional pore have been studied using similar approaches to that which argued that perforin's amino terminus has a role in pore formation. For example, the α5 and α6 peptides of the pro-apoptotic pore-forming protein Bax have been implicated strongly in pore formation and it was thought that they may form a transmembrane hairpin partially lining a pore which otherwise also had structural lipid components in a toroidal arrangement (Fuertes et al. 2010, 2011; Qian et al. 2008; Garcia-Saez et al. 2004). A new understanding of the mechanism of pore formation by Bax and related proteins is now emerging; there are important roles in it for helices 5 and 6, but not of the kind originally envisaged (Bleicken et al. 2014; Westphal et al. 2014). Studies on isolated regions of Bax which are involved in pore formation told just part of the story (Fuertes et al. 2010, 2011; Qian et al. 2008; Garcia-Saez et al. 2004).

Other cautionary elements to the interpretation of electrophysiological data relate to the effects of ions on pore function or structure. For example, in one of the leading conductance-based papers on complement it was concluded that zinc ions accelerate pore formation by enhancing the oligomerisation of C9, when in fact the most likely explanation for the great increase in conductance associated with the presence of zinc is its destabilisation of the negatively-charged surface of a pore-forming toroidal arrangement of lipids interfacing with a MAC arc (Young et al. 1986a).

12.11 Conclusion: Discretion Is the Better Part of Valour

Electrophysiology has played a very important part in the discovery of some fundamental characteristics of pores formed by perforins. In particular, it proved ahead of the game compared to high-resolution structural methods in identifying an important mechanism by which they act, that is, the formation of unconventional pores. Combined with an open-minded appraisal of negatively-stained electron microscopy images, single-channel conductance measurements indicated a novel and controversial mechanism for the formation of an unconventional kind of pore which has proved correct (Gilbert 2002, 2005; Gilbert et al. 2014; Leung et al. 2014; Podobnik et al. 2015; Sonnen et al. 2014).

These unconventional pores, generated at the interface between an oligomeric arc of pore-forming protein subunits and a lipid bilayer, confer characteristics to the channel generated which are likely to turn out in the years ahead to be extremely adaptively advantageous. For example, the pores formed have an enhanced capacity to prove transitory compared to complete rings of subunits, they can be affected by the fluidity of the bilayer, and they allow the transport of material not only through the lumen of the pore but also over its surface which, in its lipidic regions, possesses inherent fluidity. For the Apicomplexan parasites such pores might transpire to be especially useful for the way they fight their way out of infected cells, or in another mode traverse them. Another major – and perhaps the most basic – characteristic such structures confer is the capacity for a single protein to form pores of widely varying size (Figs. 12.3c and 12.5a). In this sense the discretion of a pore generated at a protein-lipid interface – smaller and less mechanically damaging to the cell membrane, less likely for example to trigger necrosis or the loss of cytoplasmic components – may be the better part of valour, as the cell is more efficiently and discretely targeted or remodelled in the attack. Among the CDCs, *Listeria* seem to use this mechanism – small pores to enable neutralisation of endosomes, larger pores to allow bacteria to escape into the cytoplasm – and for perforin-1 itself it remains the case that this may be a critical aspect of the way it arms cytotoxic cells to silently eliminate their infected or mitogenically-transformed prey.

References

Aleshin AE, Schraufstatter IU, Stec B, Bankston LA, Liddington RC, Discipio RG (2012) Structure of complement C6 suggests a mechanism for initiation and unidirectional, sequential assembly of the Membrane Attack Complex (MAC). J Biol Chem 287:10210–10222

Amiguet P, Brunner J, Tschopp J (1985) The membrane attack complex of complement: lipid insertion of tubular and nontubular polymerized C9. Biochemistry 24(25):7328–7334

Amino R, Giovannini D, Thiberge S, Gueirard P, Boisson B, Dubremetz JF, Prevost MC, Ishino T, Yuda M, Menard R (2008) Host cell traversal is important for progression of the malaria parasite through the dermis to the liver. Cell Host Microbe 3(2):88–96

Anderluh G, Gilbert RJC (2014) MACPF/CDC proteins – agents of defence, attack and invasion. In: Subcellular biochemistry, vol 80. Springer, Dordrecht

Anderluh G, Dalla Serra M, Viero G, Guella G, Macek P, Menestrina G (2003) Pore formation by equinatoxin II, a eukaryotic protein toxin, occurs by induction of nonlamellar lipid structures. J Biol Chem 278(46):45216–45223. doi:10.1074/jbc.M305916200

Benz R, Schmid A, Wiedmer T, Sims PJ (1986) Single-channel analysis of the conductance fluctuations induced in lipid bilayer membranes by complement proteins C5b-9. J Membr Biol 94(1):37–45

Bergstrom CL, Beales PA, Lv Y, Vanderlick TK, Groves JT (2013) Cytochrome c causes pore formation in cardiolipin-containing membranes. Proc Natl Acad Sci U S A 110(16):6269–6274. doi:10.1073/pnas.1303819110

Bhakdi S, Tranum-Jensen J (1991) Complement lysis: a hole is a hole. Immunol Today 12(9):318–320; discussion 321. doi: 10.1016/0167-5699(91)90007-G

Bhakdi S, Tranum-Jensen J, Sziegoleit A (1985) Mechanism of membrane damage by streptolysin-O. Infect Immun 47(1):52–60

Bleicken S, Jeschke G, Stegmueller C, Salvador-Gallego R, Garcia-Saez AJ, Bordignon E (2014) Structural model of active Bax at the membrane. Mol Cell 56(4):496–505. doi:10.1016/j.molcel.2014.09.022

Borsos T, Dourmashkin RR, Humphrey JH (1964) Lesions in erythrocyte membranes caused by immune haemolysis. Nature 202:251–252

D'Angelo ME, Dunstone MA, Whisstock JC, Trapani JA, Bird PI (2012) Perforin evolved from a gene duplication of MPEG1, followed by a complex pattern of gene gain and loss within Euteleostomi. BMC Evol Biol 12:59. doi:10.1186/1471-2148-12-59

Davis SJ, van der Merwe PA (2011) Lck and the nature of the T cell receptor trigger. Trends Immunol 32(1):1–5. doi:10.1016/j.it.2010.11.003

Dunstone MA, Tweten RK (2012) Packing a punch: the mechanism of pore formation by cholesterol dependent cytolysins and membrane attack complex/perforin-like proteins. Curr Opin Struct Biol 22(3):342–349. doi:10.1016/j.sbi.2012.04.008

Dustin ML, Depoil D (2011) New insights into the T cell synapse from single molecule techniques. Nat Rev Immunol 11(10):672–684. doi:10.1038/nri3066

Felzen B, Berke G, Rosen D, Binah O (1994a) Mechanisms whereby cytotoxic T lymphocytes damage guinea-pig ventricular myocytes in vitro. Pflugers Arch 427(5–6):422–431

Felzen B, Berke G, Rosen D, Coleman R, Tschopp J, Young JD, Binah O (1994b) Effects of purified perforin and granzyme A from cytotoxic T lymphocytes on guinea pig ventricular myocytes. Cardiovasc Res 28(5):643–649

Fields KA, McCormack R, de Armas LR, Podack ER (2013) Perforin-2 restricts growth of Chlamydia trachomatis in macrophages. Infect Immun 81(8):3045–3054. doi:10.1128/IAI.00497-13

Froelich CJ, Pardo J, Simon MM (2009) Granule-associated serine proteases: granzymes might not just be killer proteases. Trends Immunol 30(3):117–123. doi:10.1016/j.it.2009.01.002

Fuertes G, Garcia-Saez AJ, Esteban-Martin S, Gimenez D, Sanchez-Munoz OL, Schwille P, Salgado J (2010) Pores formed by Baxalpha5 relax to a smaller size and keep at equilibrium. Biophys J 99(9):2917–2925. doi:10.1016/j.bpj.2010.08.068

Fuertes G, Gimenez D, Esteban-Martin S, Sanchez-Munoz OL, Salgado J (2011) A lipocentric view of peptide-induced pores. Eur Biophys J 40(4):399–415. doi:10.1007/s00249-011-0693-4

Garcia-Saez AJ, Mingarro I, Perez-Paya E, Salgado J (2004) Membrane-insertion fragments of Bcl-xL, Bax, and Bid. Biochemistry 43(34):10930–10943. doi:10.1021/bi036044c

Garcia-Saez AJ, Chiantia S, Salgado J, Schwille P (2007) Pore formation by a Bax-derived peptide: effect on the line tension of the membrane probed by AFM. Biophys J 93(1):103–112. doi:10.1529/biophysj.106.100370

Garg S, Agarwal S, Kumar S, Yazdani SS, Chitnis CE, Singh S (2013) Calcium-dependent permeabilization of erythrocytes by a perforin-like protein during egress of malaria parasites. Nat Commun 4:1736. doi:10.1038/ncomms2725

Gilbert RJ (2002) Pore-forming toxins. Cell Mol Life Sci 59(5):832–844

Gilbert RJ (2005) Inactivation and activity of cholesterol-dependent cytolysins: what structural studies tell us. Structure 13(8):1097–1106

Gilbert RJ (2010) Cholesterol-dependent cytolysins. Adv Exp Med Biol 677:56–66

Gilbert RJ, Jimenez JL, Chen S, Tickle IJ, Rossjohn J, Parker M, Andrew PW, Saibil HR (1999) Two structural transitions in membrane pore formation by pneumolysin, the pore-forming toxin of Streptococcus pneumoniae. Cell 97(5):647–655

Gilbert RJ, Mikelj M, Dalla Serra M, Froelich CJ, Anderluh G (2013) Effects of MACPF/CDC proteins on lipid membranes. Cell Mol Life Sci 70(12):2083–2098. doi:10.1007/s00018-012-1153-8

Gilbert RJ, Dalla Serra M, Froelich CJ, Wallace MI, Anderluh G (2014) Membrane pore formation at protein-lipid interfaces. Trends Biochem Sci 39:510–516

Hadders MA, Beringer DX, Gros P (2007) Structure of C8alpha-MACPF reveals mechanism of membrane attack in complement immune defense. Science 317(5844):1552–1554

Humphrey JH, Dourmashkin RR (1969) The lesions in cell membranes caused by complement. Adv Immunol 11:75–115

Johnson TK, Crossman T, Foote KA, Henstridge MA, Saligari MJ, Forbes Beadle L, Herr A, Whisstock JC, Warr CG (2013) Torso-like functions independently of Torso to regulate Drosophila growth and developmental timing. Proc Natl Acad Sci U S A 110(36):14688–14692. doi:10.1073/pnas.1309780110

Kafsack BF, Carruthers VB (2010) Apicomplexan perforin-like proteins. Commun Integr Biol 3(1):18–23

Kafsack BF, Pena JD, Coppens I, Ravindran S, Boothroyd JC, Carruthers VB (2009) Rapid membrane disruption by a perforin-like protein facilitates parasite exit from host cells. Science 323(5913):530–533

Karatekin E, Sandre O, Guitouni H, Borghi N, Puech PH, Brochard-Wyart F (2003) Cascades of transient pores in giant vesicles: line tension and transport. Biophys J 84(3):1734–1749. doi:10.1016/S0006-3495(03)74981-9

Law RH, Lukoyanova N, Voskoboinik I, Caradoc-Davies TT, Baran K, Dunstone MA, D'Angelo ME, Orlova EV, Coulibaly F, Verschoor S, Browne KA, Ciccone A, Kuiper MJ, Bird PI,

Trapani JA, Saibil HR, Whisstock JC (2010) The structural basis for membrane binding and pore formation by lymphocyte perforin. Nature 468(7322):447–451
Leung C, Dudkina NV, Lukoyanova N, Hodel AW, Farabella I, Pandurangan AP, Jahan N, Pires Damaso M, Osmanovic D, Reboul CF, Dunstone MA, Andrew PW, Lonnen R, Topf M, Saibil HR, Hoogenboom BW (2014) Stepwise visualization of membrane pore formation by suilysin, a bacterial cholesterol-dependent cytolysin. eLife 3. doi:10.7554/eLife.04247
Lopez JA, Susanto O, Jenkins MR, Lukoyanova N, Sutton VR, Law RH, Johnston A, Bird CH, Bird PI, Whisstock JC, Trapani JA, Saibil HR, Voskoboinik I (2013) Perforin forms transient pores on the target cell plasma membrane to facilitate rapid access of granzymes during killer cell attack. Blood 121(14):2659–2668. doi:10.1182/blood-2012-07-446146
Lovelace LL, Cooper CL, Sodetz JM, Lebioda L (2011) Structure of human C8 protein provides mechanistic insight into membrane pore formation by complement. J Biol Chem 286(20):17585–17592. doi:10.1074/jbc.M111.219766
Lukoyanova N, Kondos SC, Farabella I, Law RH, Reboul CF, Caradoc-Davies TT, Spicer BA, Kleifeld O, Traore DA, Ekkel SM, Voskoboinik I, Trapani JA, Hatfaludi T, Oliver K, Hotze EM, Tweten RK, Whisstock JC, Topf M, Saibil HR, Dunstone MA (2015) Conformational changes during pore formation by the perforin-related protein pleurotolysin. PLoS Biol 13(2):e1002049. doi:10.1371/journal.pbio.1002049
Marchioretto M, Podobnik M, Dalla Serra M, Anderluh G (2013) What planar lipid membranes tell us about the pore-forming activity of cholesterol-dependent cytolysins. Biophys Chem 182:64–70. doi:10.1016/j.bpc.2013.06.015
McCormack R, de Armas L, Shiratsuchi M, Podack ER (2013a) Killing machines: three pore-forming proteins of the immune system. Immunol Res 57(1–3):268–278. doi:10.1007/s12026-013-8469-9
McCormack R, de Armas LR, Shiratsuchi M, Ramos JE, Podack ER (2013b) Inhibition of intracellular bacterial replication in fibroblasts is dependent on the perforin-like protein (perforin-2) encoded by macrophage-expressed gene 1. J Innate Immun 5(2):185–194. doi:10.1159/000345249
Metkar SS, Wang B, Catalan E, Anderluh G, Gilbert RJ, Pardo J, Froelich CJ (2011) Perforin rapidly induces plasma membrane phospholipid flip-flop. PLoS One 6(9):e24286
Metkar S, Marchioretto M, Antonini V, Lunelli L, Wang B, Gilbert RJC, Anderluh G, Roth R, Pooga M, Pardo J, Heuser JE, Dalla Serra M, Froelich CJ (2015) Perforin oligomers form arcs in cellular membranes: a locus for intracellular delivery of granzymes. Cell Death Differ 22:78–85
Michaels DW, Abramovitz AS, Hammer CH, Mayer MM (1976) Increased ion permeability of planar lipid bilayer membranes after treatment with the C5b-9 cytolytic attack mechanism of complement. Proc Natl Acad Sci U S A 73(8):2852–2856
Mueller M, Grauschopf U, Maier T, Glockshuber R, Ban N (2009) The structure of a cytolytic alpha-helical toxin pore reveals its assembly mechanism. Nature 459(7247):726–730. doi:10.1038/nature08026
Ojcius DM, Persechini PM, Zheng LM, Notaroberto PC, Adeodato SC, Young JD (1991) Cytolytic and ion channel-forming properties of the N terminus of lymphocyte perforin. Proc Natl Acad Sci U S A 88(11):4621–4625
Pardo J, Aguilo JI, Anel A, Martin P, Joeckel L, Borner C, Wallich R, Mullbacher A, Froelich CJ, Simon MM (2009) The biology of cytotoxic cell granule exocytosis pathway: granzymes have evolved to induce cell death and inflammation. Microbes Infect 11(4):452–459. doi:10.1016/j.micinf.2009.02.004
Persechini PM, Ojcius DM, Adeodato SC, Notaroberto PC, Daniel CB, Young JD (1992) Channel-forming activity of the perforin N-terminus and a putative alpha-helical region homologous with complement C9. Biochemistry 31(21):5017–5021
Podack ER, Dennert G (1983) Assembly of two types of tubules with putative cytolytic function by cloned natural killer cells. Nature 302(5907):442–445
Podobnik M, Marchioretto M, Zanetti M, Bavdek A, Kisovec M, Cajnko MM, Lunelli L, Dalla Serra M, Anderluh G (2015) Plasticity of listeriolysin O pores and its regulation by pH and a unique histidine. Sci Rep 5:9623

Praper T, Sonnen AF, Kladnik A, Andrighetti AO, Viero G, Morris KJ, Volpi E, Lunelli L, Dalla Serra M, Froelich CJ, Gilbert RJ, Anderluh G (2011a) Perforin activity at membranes leads to invaginations and vesicle formation. Proc Natl Acad Sci U S A 108(52):21016–21021

Praper T, Sonnen AF, Viero G, Kladnik A, Froelich CJ, Anderluh G, Dalla Serra M, Gilbert RJ (2011b) Human perforin employs different avenues to damage membranes. J Biol Chem 286:2946–2955

Qian S, Wang W, Yang L, Huang HW (2008) Structure of transmembrane pore induced by Bax-derived peptide: evidence for lipidic pores. Proc Natl Acad Sci U S A 105(45):17379–17383. doi:10.1073/pnas.0807764105

Raghava S, Giorda KM, Romano FB, Heuck AP, Hebert DN (2013) SV40 late protein VP4 forms toroidal pores to disrupt membranes for viral release. Biochemistry 52(22):3939–3948. doi:10.1021/bi400036z

Reboul CF, Whisstock JC, Dunstone MA (2014) A new model for pore formation by cholesterol-dependent cytolysins. PLoS Comput Biol 10(8):e1003791. doi:10.1371/journal.pcbi.1003791

Roiko MS, Carruthers VB (2009) New roles for perforins and proteases in apicomplexan egress. Cell Microbiol 11(10):1444–1452

Rosado CJ, Buckle AM, Law RH, Butcher RE, Kan WT, Bird CH, Ung K, Browne KA, Baran K, Bashtannyk-Puhalovich TA, Faux NG, Wong W, Porter CJ, Pike RN, Ellisdon AM, Pearce MC, Bottomley SP, Emsley J, Smith AI, Rossjohn J, Hartland EL, Voskoboinik I, Trapani JA, Bird PI, Dunstone MA, Whisstock JC (2007) A common fold mediates vertebrate defense and bacterial attack. Science 317(5844):1548–1551. doi:10.1126/science.1144706

Rosado CJ, Kondos S, Bull TE, Kuiper MJ, Law RH, Buckle AM, Voskoboinik I, Bird PI, Trapani JA, Whisstock JC, Dunstone MA (2008) The MACPF/CDC family of pore-forming toxins. Cell Microbiol 10(9):1765–1774. doi:10.1111/j.1462-5822.2008.01191.x

Schlumberger S, Kristan KC, Ota K, Frangez R, Molgomicron J, Sepcic K, Benoit E, Macek P (2014) Permeability characteristics of cell-membrane pores induced by ostreolysin A/pleurotolysin B, binary pore-forming proteins from the oyster mushroom. FEBS Lett 588(1):35–40. doi:10.1016/j.febslet.2013.10.038

Shatursky O, Heuck AP, Shepard LA, Rossjohn J, Parker MW, Johnson AE, Tweten RK (1999) The mechanism of membrane insertion for a cholesterol-dependent cytolysin: a novel paradigm for pore-forming toxins. Cell 99(3):293–299

Shepard LA, Heuck AP, Hamman BD, Rossjohn J, Parker MW, Ryan KR, Johnson AE, Tweten RK (1998) Identification of a membrane-spanning domain of the thiol-activated pore-forming toxin Clostridium perfringens perfringolysin O: an alpha-helical to beta-sheet transition identified by fluorescence spectroscopy. Biochemistry 37(41):14563–14574. doi:10.1021/bi981452f

Shepard LA, Shatursky O, Johnson AE, Tweten RK (2000) The mechanism of pore assembly for a cholesterol-dependent cytolysin: formation of a large prepore complex precedes the insertion of the transmembrane beta-hairpins. Biochemistry 39(33):10284–10293

Shiver JW, Dankert JR, Esser AF (1991) Formation of ion-conducting channels by the membrane attack complex proteins of complement. Biophys J 60(4):761–769. doi:10.1016/S0006-3495(91)82110-5

Solecki DJ (2012) Sticky situations: recent advances in control of cell adhesion during neuronal migration. Curr Opin Neurobiol 22(5):791–798. doi:10.1016/j.conb.2012.04.010

Sonnen AF, Henneke P (2014) Structural biology of the membrane attack complex. Subcell Biochem 80:83–116. doi:10.1007/978-94-017-8881-6_6

Sonnen AF, Plitzko J, Gilbert RJC (2014) Incomplete pneumolysin oligomers form membrane pores. Roy Soc Open Biol 4:140044

Spilsbury K, O'Mara MA, Wu WM, Rowe PB, Symonds G, Takayama Y (1995) Isolation of a novel macrophage-specific gene by differential cDNA analysis. Blood 85(6):1620–1629

Steringer JP, Bleicken S, Andreas H, Zacherl S, Laussmann M, Temmerman K, Contreras FX, Bharat TA, Lechner J, Muller HM, Briggs JA, Garcia-Saez AJ, Nickel W (2012) Phosphatidylinositol 4,5-bisphosphate (PI(4,5)P2)-dependent oligomerization of fibroblast growth factor 2 (FGF2) triggers the formation of a lipidic membrane pore implicated in unconventional secretion. J Biol Chem 287(33):27659–27669. doi:10.1074/jbc.M112.381939

Stevens LM, Beuchle D, Jurcsak J, Tong X, Stein D (2003) The Drosophila embryonic patterning determinant torsolike is a component of the eggshell. Curr Biol 13(12):1058–1063

Stolfi RL (1968) Immune lytic transformation: a state of irreversible damage generated as a result of the reaction of the eighth component in the guinea pig complement system. J Immunol 100(1):46–54

Tanaka K, Caaveiro JM, Morante K, Gonzalez-Manas JM, Tsumoto K (2015) Structural basis for self-assembly of a cytolytic pore lined by protein and lipid. Nat Commun 6:6337. doi:10.1038/ncomms7337

Tavares J, Formaglio P, Thiberge S, Mordelet E, Van Rooijen N, Medvinsky A, Menard R, Amino R (2013) Role of host cell traversal by the malaria sporozoite during liver infection. J Exp Med 210(5):905–915. doi:10.1084/jem.20121130

Teissie J, Golzio M, Rols MP (2005) Mechanisms of cell membrane electropermeabilization: a minireview of our present (lack of ?) knowledge. Biochim Biophys Acta 1724(3):270–280. doi:10.1016/j.bbagen.2005.05.006

Thiery J, Keefe D, Saffarian S, Martinvalet D, Walch M, Boucrot E, Kirchhausen T, Lieberman J (2010) Perforin activates clathrin- and dynamin-dependent endocytosis, which is required for plasma membrane repair and delivery of granzyme B for granzyme-mediated apoptosis. Blood 115(8):1582–1593

Thiery J, Keefe D, Boulant S, Boucrot E, Walch M, Martinvalet D, Goping IS, Bleackley RC, Kirchhausen T, Lieberman J (2011) Perforin pores in the endosomal membrane trigger the release of endocytosed granzyme B into the cytosol of target cells. Nat Immunol 12(8):770–777

Tilley SJ, Orlova EV, Gilbert RJ, Andrew PW, Saibil HR (2005) Structural basis of pore formation by the bacterial toxin pneumolysin. Cell 121(2):247–256

Tschopp J (1984) Ultrastructure of the membrane attack complex of complement. Heterogeneity of the complex caused by different degree of C9 polymerization. J Biol Chem 259(12):7857–7863

Uellner R, Zvelebil MJ, Hopkins J, Jones J, MacDougall LK, Morgan BP, Podack E, Waterfield MD, Griffiths GM (1997) Perforin is activated by a proteolytic cleavage during biosynthesis which reveals a phospholipid-binding C2 domain. EMBO J 16(24):7287–7296. doi:10.1093/emboj/16.24.7287

Valcarcel CA, Dalla Serra M, Potrich C, Bernhart I, Tejuca M, Martinez D, Pazos F, Lanio ME, Menestrina G (2001) Effects of lipid composition on membrane permeabilization by sticholysin I and II, two cytolysins of the sea anemone Stichodactyla helianthus. Biophys J 80(6):2761–2774. doi:10.1016/S0006-3495(01)76244-3

Voskoboinik I, Thia MC, Fletcher J, Ciccone A, Browne K, Smyth MJ, Trapani JA (2005) Calcium-dependent plasma membrane binding and cell lysis by perforin are mediated through its C2 domain: A critical role for aspartate residues 429, 435, 483, and 485 but not 491. J Biol Chem 280(9):8426–8434. doi:10.1074/jbc.M413303200

Voskoboinik I, Smyth MJ, Trapani JA (2006) Perforin-mediated target-cell death and immune homeostasis. Nat Rev Immunol 6(12):940–952

Westphal D, Dewson G, Menard M, Frederick P, Iyer S, Bartolo R, Gibson L, Czabotar PE, Smith BJ, Adams JM, Kluck RM (2014) Apoptotic pore formation is associated with in-plane insertion of Bak or Bax central helices into the mitochondrial outer membrane. Proc Natl Acad Sci U S A 111(39):E4076–E4085. doi:10.1073/pnas.1415142111

Wiens M, Korzhev M, Krasko A, Thakur NL, Perovic-Ottstadt S, Breter HJ, Ushijima H, Diehl-Seifert B, Muller IM, Muller WE (2005) Innate immune defense of the sponge Suberites domuncula against bacteria involves a MyD88-dependent signaling pathway. Induction of a perforin-like molecule. J Biol Chem 280(30):27949–27959. doi:10.1074/jbc.M504049200

Wilson PM, Fryer RH, Fang Y, Hatten ME (2010) Astn2, a novel member of the astrotactin gene family, regulates the trafficking of ASTN1 during glial-guided neuronal migration. J Neurosci 30(25):8529–8540. doi:10.1523/JNEUROSCI.0032-10.2010

Young JD, Cohn ZA, Podack ER (1986a) The ninth component of complement and the pore-forming protein (perforin 1) from cytotoxic T cells: structural, immunological, and functional similarities. Science 233(4760):184–190

Young JD, Hengartner H, Podack ER, Cohn ZA (1986b) Purification and characterization of a cytolytic pore-forming protein from granules of cloned lymphocytes with natural killer activity. Cell 44(6):849–859

Young JD, Nathan CF, Podack ER, Palladino MA, Cohn ZA (1986c) Functional channel formation associated with cytotoxic T-cell granules. Proc Natl Acad Sci U S A 83(1):150–154

Young JD, Podack ER, Cohn ZA (1986d) Properties of a purified pore-forming protein (perforin 1) isolated from H-2-restricted cytotoxic T cell granules. J Exp Med 164(1):144–155

Young LH, Joag SV, Zheng LM, Lee CP, Lee YS, Young JD (1990) Perforin-mediated myocardial damage in acute myocarditis. Lancet 336(8722):1019–1021

Chapter 13
Gap Junction Channels: The Electrical Conduit of the Intercellular World

Richard D. Veenstra

Abstract The concept of an intercellular channel that mediates the direct intracellular passage of ions, second messengers, and metabolites between adjacent cells was born from the observations of electrical coupling within cellular excitable and nonexcitable tissues in the 1950s and 1960s. Originally called the "nexus", the term gap junction was applied to this membrane structure after the observation of 1–2 nm gap between adjacent cell membranes populated with hexameric protein bridges containing a central aqueous pore in 1967. Molecular structural models of the gap junction channel evolved in 1977 and approximately every 20 years since with increasing resolution. The first junction channel current records and the cloning of the connexin subunits occurred in the mid-1980s, ushering in the molecular era of gap junction channel research. This atypical intracellular double membrane channel is now known to be formed by 20 different mammalian connexins with 60 % of them linked to inherited human diseases. Once thought to be a large static channel, gap junction channels exhibit with conductances that vary from 10 to 300 pS with ionic and molecular permeabilities that defy the simple principles of a nonselective large aqueous pore. Their activity is modulated by numerous mechanisms including calcium signaling, intracellular pH, post-translation modifications (PTMs) like protein phosphorylation and acetylation, and a variety of lipophiles. Despite knowledge of the connexin membrane topology and primary amino acid sequences for three decades, precise identification of the molecular pore composition of this unconventional ion channel remains poorly understood. Recent development of a 3.5 Å three-dimensional structure and molecular dynamic simulations have facilitated development of precise hypotheses for the origin of cationic and anionic pore permeability barriers amenable to experimental verification.

Keywords Gap junction • Nexus • Connexin • Hemichannel • Conductance • Permeability • Disease

R.D. Veenstra
Department of Pharmacology, SUNY Upstate Medical University,
750 East Adams Street, Syracuse, NY 13210, USA
e-mail: veenstrr@upstate.edu

A.H. Delcour (ed.), *Electrophysiology of Unconventional Channels and Pores*,
Springer Series in Biophysics 18, DOI 10.1007/978-3-319-20149-8_13

13.1 Introduction: The Nexus – Structural Evidence for a "Gap Junction" Channel

The typical ion channel mediates some form of electrochemical signaling between the external and internal milieu of a living cell. As you will read about in this volume, there are many types of channel forming molecules (e.g. polyene antibiotics) and proteins that do not conform to this paradigm, from intracellular organellar channels to cytotoxic channels. Since 1985 during my NIH postdoctoral fellowship at Emory University, my scientific curiosity has been devoted to studying the ionic basis for electrical coupling between cardiomyocytes, namely the cardiac gap junction channel. The saga of the cardiac gap junction channel, as if there is only one (not true), began with the observation by Silvio Weidmann in 1952 that when current was injected into one cell of a cardiac Purkinje fiber, the resulting voltage decayed exponentially from the injection site over several cell lengths (Weidmann 1952; Fig. 13.1). The exponential decay constant, the space constant (λ), was equivalent to the square root of the ratio of membrane and intracellular resistances, $\lambda = \sqrt{r_m / r_i}$. The phenomenon of intercellular electrical coupling was later extended to other excitable cells, from the crayfish giant axon to goldfish Mauthner cells (Furshpan and Potter 1959; Furshpan and Furukawa 1962). The concept of electrically coupled cells jumped from muscle cells and neurons to non-excitable secretory gland and Malpighian tubule cells with the work of Loewenstein et al. (1965), demonstrating for the first time that passive spread of cellular membrane voltage was a universal concept of multicellular tissues, not just nerve and muscle (Fig. 13.1b). The site of this intercellular electrical communication was hypothesized to be the "nexus", a specialized region where the cell membranes of adjacent

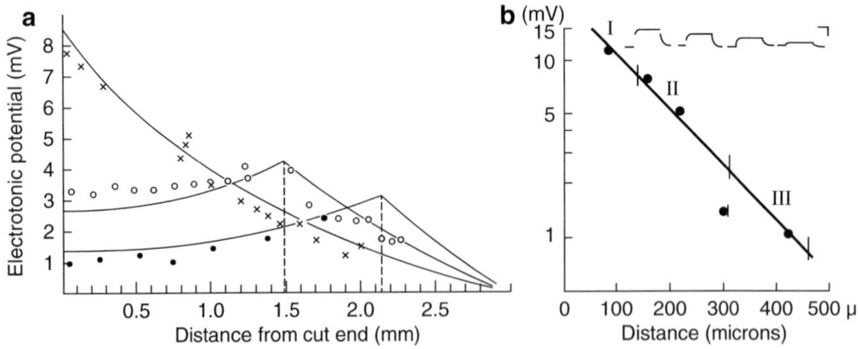

Fig. 13.1 (**a**) Passive spread of electrotonic potential (*mV*) from three different sites (*left end* (x), center (○), *mid-right center* (•)) of intracellular current injection in a cut Purkinje fiber. The change in membrane voltage decreased exponentially with distance in excess of cell length (0.1 mm) (From Weidmann (1952) with permission). (**b**) Voltage recordings and change in membrane potential as a function of distance from the site of current injection in a *Chironomus* Malpighian tubule (From Loewenstein et al. (1965) with permission)

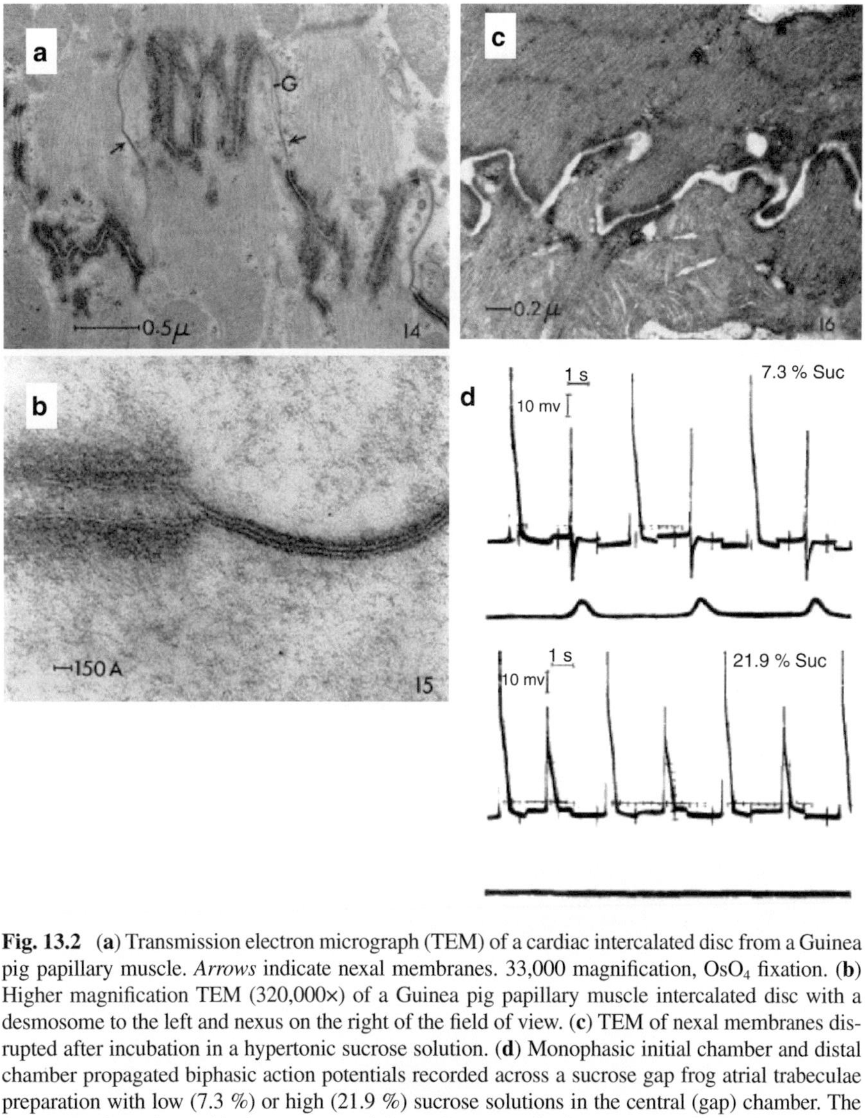

Fig. 13.2 (**a**) Transmission electron micrograph (TEM) of a cardiac intercalated disc from a Guinea pig papillary muscle. *Arrows* indicate nexal membranes. 33,000 magnification, OsO_4 fixation. (**b**) Higher magnification TEM (320,000×) of a Guinea pig papillary muscle intercalated disc with a desmosome to the left and nexus on the right of the field of view. (**c**) TEM of nexal membranes disrupted after incubation in a hypertonic sucrose solution. (**d**) Monophasic initial chamber and distal chamber propagated biphasic action potentials recorded across a sucrose gap frog atrial trabeculae preparation with low (7.3 %) or high (21.9 %) sucrose solutions in the central (gap) chamber. The *lower trace* in each panel is the post-gap atrial muscle contraction in response to the (biphasic) action potential in the distal chamber, which was abolished by the hypertonic sucrose solution (From Barr et al. ©1965 Rockefeller University Press. *Journal of General Physiology*. 48:797–823)

cells appeared to fuse (Dewey and Barr 1962). Demonstration that the nexus was the site of intercellular propagation of the cardiac action potential was achieved by hypertonic solution disruption of the nexal structure and cardiac action potential transfer (Barr et al. 1965; Fig. 13.2). The nexus was a defined anatomical structure and the precise basis for the transmission of electric current in the form of net ionic flux was not known at the time.

13.2 The Gap Junction Channel in the Pre-molecular Era (Before Cloning) and Today

The demonstration of intercellular hydrophilic fluorescent dye coupling with an apparent molecular size permeability limit of approximately 1000 Da without leakage to the extracellular environment led Werner Loewenstein (1966, 1981) to hypothesize that the nexal junctional pathway was composed of "channel units" where "Each unit consists of a pair of permeable membrane elements, one from each cell, aligned and tightly joined, thus permitting a central aqueous passageway insulated peripherally from the extracellular space. Many such units in parallel were considered to make up the entire junctional pathway, but no two units were thought to be mediated by a common compartment. The junctional units proposed were thus discrete, continuous, and direct cell-to-cell passageways or channels" (Fig. 13.3a). Definitive electrophysiological proof of this intercellular channel concept would have to wait until the development of the whole cell patch clamp technique (Hamill et al. 1981), but further anatomical studies helped refine the structure of the nexus to the level of a molecular model within the next 10 years. The pioneering work of Revel and Karnovsky (1967) defined, for the first time, a hexagonal packing of protein bridges traversing the narrow 2 nm intercellular gap with an apparent central aqueous pore that culminated in the modern day name for this intercellular junction, the "gap junction" (Fig. 13.3b–d). X-ray diffraction studies of mouse liver gap junction membranes resulted in the classical molecular model for the gap junction where a hexamer of proteins in each cell membrane formed a 60 Å diameter "connexon" with a 20 Å axial pore (Makowski et al. 1977; Fig. 13.3e). Each protein hexamer

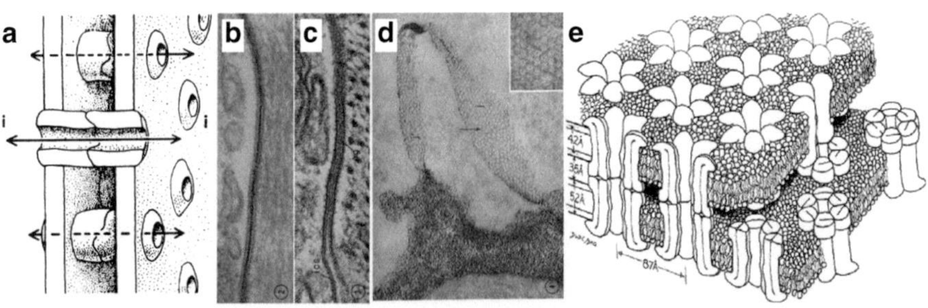

Fig. 13.3 (**a**) Diagram of the channel hypothesis for the intercellular gap junction (From Loewenstein (1981) with permission). (**b**) TEM (200,000×) of a uranium stained gap junction membrane preparation from mouse heart with the extracellular gap filled with lanthanum to illustrate the 2 nm gap between adjacent cell membranes. (**c**) TEM (250,000×) of a mouse heart gap junction without La^{3+} illustrating the parallel railroad track appearance of the gap junction membranes. (**d**) Tangential TEM section (*arrows*) through a lanthanum filled mouse heart gap junction membrane illustrating the hexameric pillars with a central aqueous pore spanning the gap between the intercellular junction membranes (From Revel and Karnovsky ©1967 Rockefeller University Press. *Journal of Cell Biology*. 33:C7–C12). (**e**) Structural model of a gap junction from X-ray diffraction studies of mouse liver gap junctions (From Makowski et al. ©1977 Rockefeller University Press. *Journal of Cell Biology*. 74:629–645)

was approximately 80 Å long and 20 Å wide. The salient features of this molecular gap junction model remain true to this day with only minor adjustments to the "actual" pore diameter and the discovery of the molecular architecture of the hexameric protein subunits of the connexon, appropriately named the "connexin" (Beyer et al. 1990).

One could say that the mid-1980s was the beginning of the "molecular era" for the study of gap junctions. This time period was defined by the first direct gap junction channel recordings and the cloning of the first two, liver and heart, gap junction proteins (Neyton and Trautmann 1985; Veenstra and DeHaan 1986; Paul 1986; Beyer et al 1987). Prior to these events, the closest anyone had been to proving the channel hypothesis for gap junctions was the occurrence of quantal increases in junctional conductance between two *Xenopus laevis* blastocysts during the formation of a nascent gap junction by mechanical manipulation of cell-to-cell contact (Loewenstein et al 1978; Fig. 13.4a, b). The electrical and temporal resolution of the voltage clamp – voltage follower apparatus was limited to seconds, microvolts (μV), and nanoamperes (nA), thus precluding full resolution of the previously hypothesized 100 picosiemen (100 pS or 10 gigaΩ^{-1}) discrete channel events (Loewenstein 1976). These cell-cell channels were described as "quite stable in their open state… for at least several minutes; closures were not seen unless Ca^{2+} was injected", although channel openings of < 1 s duration were not ruled out owing to the slow temporal response of the detector (Loewenstein et al. 1978). The astonishing accuracy of the hypothesized 100 pS gap junction channel conductance was illustrated by the independent observations of 120 pS rat lacrimal gland and 165 pS embryonic chick heart cell gap junction channel events (Neyton and Trautmann 1985; Veenstra

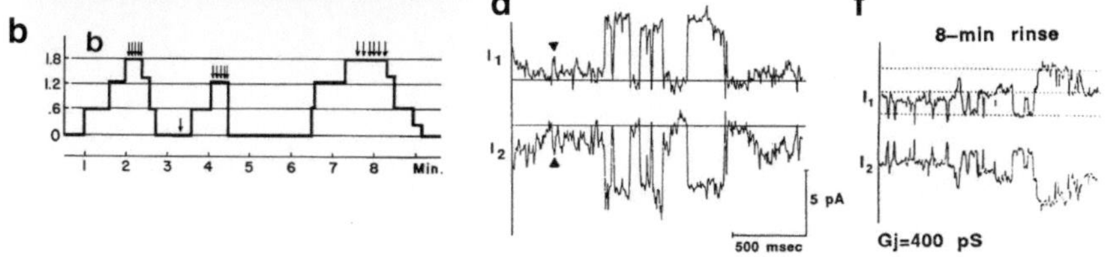

Fig. 13.4 (**a**) Quantum jumps of membrane voltage during nascent gap junction formation between two voltage-clamped *Xenopus* embryo blastocysts. (**b**) Diagram of idealized 0.6 μV quantal steps of membrane voltage in cell 2 from time point *b* in panel **a** after iontophoretic Ca^{2+} injections (*arrows*) (From Loewenstein (1981) with permission). (**c**) Whole-cell patch clamp diagram of paired embryonic chick ventricular heart cells. (**d**) Simultaneous whole cell currents of 160 pS gap junction channel events illustrating the "equal amplitude and opposite polarity" appearance of junctional currents from a patch-clamped cell pair. (**e, f**) Absence and reappearance of cardiac gap junction channel activity in the presence of and during washout of 120 μM 1-octanol (From Veenstra and DeHaan (1986) with permission)

and DeHaan 1986). These first gap junction channel records demonstrated another manner in which gap junction channels are atypical of other, more conventional, ion channels: their order of magnitude higher channel conductance and slower channel gating kinetics. These images also define the criteria by which an intercellular gap junction channel current may be distinguished from nonjunctional plasmalemmal ion channels, by the "equal magnitude and opposite polarity" appearance of the spontaneous whole cell current fluctuations recorded simultaneous from two apposed cells (Fig. 13.4c, d). These "mirror image" tandem whole cell channel recordings were also reversibly blocked by 1-octanol (Fig. 13.4e, f), the gold standard gap junction uncoupling agent of its time, and downregulated by elevated internal [Ca^{2+}] (Johnston et al. 1980; Veenstra and DeHaan 1988).

13.3 Functional Diversity of Connexin-Specific Gap Junctions

13.3.1 Channel Conductance and Selectivity

From the initial debate over whether heart and liver possessed two different gap junction proteins to today's knowledge of 21 human connexin genes and 20 proteins, the concept of the gap junction channel has increased in complexity owing to the distinct channel conductance, gating, and selective permeability properties of the some 20 different connexin-specific gap junction channels that possess a common membrane topology (Manjunath et al. 1985; Beyer and Berthoud 2009, Fig. 13.5a).

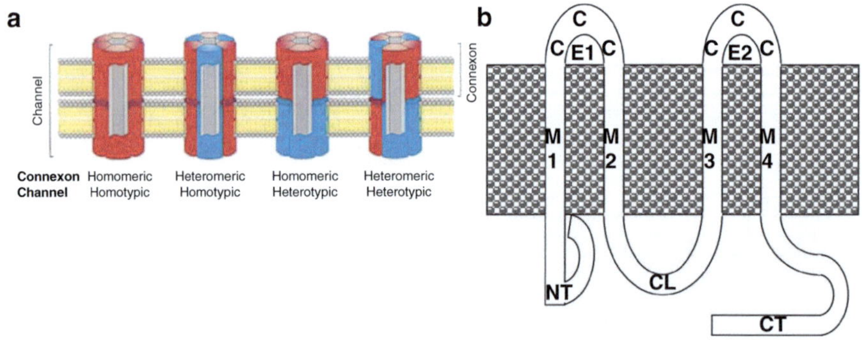

Fig. 13.5 (**a**) Illustration of the possible heteromeric and heterotypic connexin subunit arrangements in the formation of a gap junction channel (From Kumar and Gilula (1996) with permission). (**b**) Conserved membrane topology of the connexin proteins beginning with the cytoplasmic amino-terminal (*NT*) and ending with the carboxy-terminal (*CT*) domains. There are four transmembrane domains (*M1–M4*) linked by one cytoplasmic loop (*CL*) domain and two extracellular loop (*E1 & E2*) domains, each containing three conserved cysteine (*C*) residues that form inter-loop, intramolecular, anti-parallel disulfide bonds required for functional gap junction channel formation (Foote et al. 1998). The CL and CT domains possess the most diverse sequences

The identification of additional connexins beyond the "liver" connexin32 (Cx32) and "heart" connexin43 (Cx43) evolved from further cloning studies and the electrophysiological characterization of these homotypic gap junctions formed by a single connexin necessarily followed with the exogenous expression of these newly cloned connexins in communication-deficient cell lines (Veenstra et al. 1992). The hypothesized gap junction channel conductance of ≈ 100 pS is now known to vary from 12 to 300 pS. Table 13.1 summarizes the single gap junction channel conductances (γ_j) for the known human or mammalian connexin-specific gap junctions (see also Harris and Locke 2009).

It is apparent from Table 13.1 that the gap junction channel conductance actually exhibits a 30-fold variation in γ_j. The concept that the gap junction channel is a large aqueous pore exhibiting poor selectivity among solutes up to 1 kDa in molecular weight was also proven to be an oversimplification. Fluorescent dye permeability and cation/anion conductance ratio measurements quickly revealed that many of the higher γ_j connexin-specific gap junction channels were more-cation-selective and

Table 13.1 Channel conductances of connexin-specific gap junctions

Connexin (*Gene*)	Channel conductance (γ_j)[a]	Reference
Cx43 (*GJA1*)	21, 40-45, **70** pS 40, **60** pS 60–70, **90–100** pS	Takens-Kwak and Jongsma (1992), Traub et al. (1994), Moreno et al. (1994)
Cx46 (*GJA3*)	10–60, **140** pS	Hopperstad et al. (2000)
Cx37 (*GJA4*)	53, 123, 165, **219** pS 63, **300** pS	Reed et al. (1993), Veenstra et al. (1994a)
Cx40 (*GJA5*)	121, **153** pS **158** pS	Traub et al. (1994), Beblo et al. (1995)
Cx50 (*GJA8*)	45, 90, **290** pS 43, **220** pS	Donaldson and Kistler (1992) Srinivas et al. (1999a)
Cx62 (*GJA10*) {mCx57}	**27** pS	Manthey et al. (1999)
Cx32 (*GJB1*)	25–35, **120–130** pS **53** pS **68** pS 9, **31** pS	Moreno et al. (1991), Suchyna et al. (1999), Oh et al. (1999), Valiunas et al. (1999)
Cx26 (*GJB2*)	**135** pS 15–30, **140** pS 17, 102 pS	Suchyna et al. (1999), Oh et al. (1999), Valiunas et al. (1999)
Cx31 (*GJB3*)	**85** pS	Abrams et al. (2006)
Cx30 (*GJB6*)	48, **179** pS	Valiunas et al. (1999)
Cx45 (*GJC1*)	19, **26** pS	Veenstra et al. (1994b)
Cx47 (*GJC2*)	8, **55** pS	Teubner et al. (2001)
Cx36 (*GJD2*)	**10–15** pS **14** pS	Srinivas et al. (1999b), Teubner et al. (2000)
Cx31.9 (*GJD3*)	**12–15** pS	White et al. (2002)

[a]Fully open (main state) channel conductance is indicated by **bold** font

less dye permeable than their 50–100 pS Cx32 and Cx43 counterparts (Beblo et al. 1995; Elfgang et al. 1995; Veenstra et al. 1994a, b, 1995; Veenstra 1996; Suchyna et al. 1999; reviewed in Harris 2007). Thus, the concept that the gap junction permeability barrier consists basically of a passive physical spatial constraint, i.e. steric hindrance or molecular sieve, is not valid. In fact, the earliest evidence for charge selectivity of gap junction channels dates back to the early fluorescent tracer studies that demonstrated reduced permeability relative to their molecular size for negatively charged molecules like 6-carboxyfluorescein (Flagg-Newton et al. 1979; Brink and Dewey 1980). Hence, the premise that the gap junction channel contains fixed anionic charges was born. The current concepts for the molecular basis of the gap junction channel permeability barriers will be discussed later in the Sect. 13.6 of this review.

Another level of complexity was added to the subject of gap junction channel conductance and selectivity as the capability of distinct connexins to intermix to form heteromeric connexons of "mixed" connexin composition (Koval et al. 2014). Connexons of homomeric or heteromeric connexin composition are also capable of combining with connexins of unlike connexin composition to form heterotypic gap junction channels. Thus, there are four possible gap junction conformations based on the expression of two distinct connexins as illustrated in Fig. 13.5a (Kumar and Gilula, 1996). The cloning of multiple cardiac connexins (e.g. Cx40, Cx45) and a second liver connexin, Cx26, revealed that multiple connexin expression was the norm (Zhang and Nicholson 1989; Kanter et al. 1992). Soon thereafter, it was revealed that coexpressed connexins in liver and lens readily heteromerize and that the formation of heteromeric connexons affects the molecular permeability limit of the resultant gap junction channel (Stauffer 1995; Jiang and Goodenough 1996; Bevans et al. 1998; Bevans and Harris 1999). These initial observations with Cx26 and Cx32 have been extended to numerous other heteromeric and/or heterotypic connexin gap junction channels including Cx43 with Cx37, Cx40, and Cx45; mouse Cx30.2 (human Cx31.9) with Cx40 and Cx43; Cx26 with Cx30; and Cx46 with Cx50 to mention some of the most prominent examples (Brink et al. 1997; Hopperstad et al. 2000; Valiunas et al. 2000, 2001; Cottrell and Burt 2001; Cottrell et al. 2002; Sun et al. 2005; Yum et al. 2007; Gemel et al. 2008). Heteromeric connexin compatibility is influenced by a cytoplasmic loop – third transmembrane (CL-M3) domain motif (see Fig. 13.5b) that subdivides the majority of the connexins into "R" or "W" type groups (Smith et al. 2012; reviewed by Koval et al. 2014). Heterotypic connexin compatibility correlates closely with a second extracellular loop (E2) motif, thus subdividing most of the connexins into "K-N" or "H" type groups (Gong et al. 2013; Koval et al. 2014). These heterotypic and heteromeric connexin interactions affect not only channel permeability, but also cause rectification of gap junction channel conductance and transjunctional voltage gating. Additional consequences of mutant and wild-type connexin heteromerization will be discussed in the Sect. 13.4 of this article.

13.3.2 Gating and Modulation of Gap Junction Channels

13.3.2.1 Voltage

Gap junction channels also differ from typical plasmalemmal ion channels in their response to voltage. Whereas many ion channels modulate their open probability (P_o) in response to the cellular membrane voltage (V_m or $V_{i\text{–}o}$ (for $V_{inside\text{-}outside}$) in gap junction terminology) relative to the extracellular environment, mammalian gap junction channels respond predominantly to the transjunctional voltage (V_j), i.e. the difference in V_m between adjacent coupled cells (Spray et al. 1981a; Harris et al. 1981). Some mammalian gap junctions are V_m-sensitive, though the modulation of macroscopic gap junction conductance (g_j) by V_m is usually less than the V_j-dependent mechanism (Barrio et al. 1991; Revilla et al. 2000; reviewed in González et al. 2007). Another atypical feature of gap junction channels is that the connexins do not contain a highly charged transmembrane voltage sensor domain analogous to the S4 domain of voltage-gated ion channels. Instead, the gap junction V_j gating mechanism principally involves charged amino acid residues on the connexin amino terminus (NT) and the first transmembrane (M1) – first extracellular loop (E1) border (Verselis et al. 1994; reviewed in Bargiello and Brink 2009). Each connexon is proposed to possess a V_j gate with a gating polarity (+, −) determined by the charged NT and M1-E1 domains of each composite connexin (Fig. 13.6a–c) and only one connexin V_j gate is required to translocate to close the gap junction channel to the subconductance state (Oh et al. 2000; Fig. 13.6d). Other connexin domains have been implicated in the V_j gating process beginning with a conserved second transmembrane (M2) domain proline residue (P87/88) hypothesized to translocate between open and closed channel configurations (Suchyna et al. 1993; Ri et al. 1999). The V_j gating process became more complicated with the frequent observation that gap junction channels rapidly transitioned to a subconductance state and that CO_2-mediated pH-dependent gating resulted in slow transitions to the fully closed state (Bukauskas and Peracchia 1997). Thus each connexon is now considered to possess two gates in series, a "fast" V_j gate and a "slow" chemical or "loop" gate. The carboxyl terminal (CT) domain has been implicated in the fast V_j gating mechanism of Cx43 (Revilla et al. 1999; Moreno et al. 2002), but a recent re-examination of this concept has concluded that the CT of Cx32 is not involved in fast V_j gating. Connexin-specific gap junctions possess disparate V_j gating properties, a feature that is frequently examined during the functional characterization of a newly cloned connexin (e.g. Table 13.1). The half-inactivation voltages ($V_{½}$ or V_0) vary from 10–20 mV (e.g. Cx45) to nearly 100 mV (Cx26, Cx36) with no V_j or V_m sensitivity detected for Cx31.9 (mCx30.2) (Barrio et al. 1991, 1997; White et al. 2002; reviewed in González et al. 2007).

Another atypical feature of a homotypic gap junction channel is that, by necessity, this channel is symmetrical, thus accounting for the bell-shaped appearance of the normalized g_j (G_j) – V_j Boltzmann curve (Fig. 13.6a, b). Heterotypic gap junction

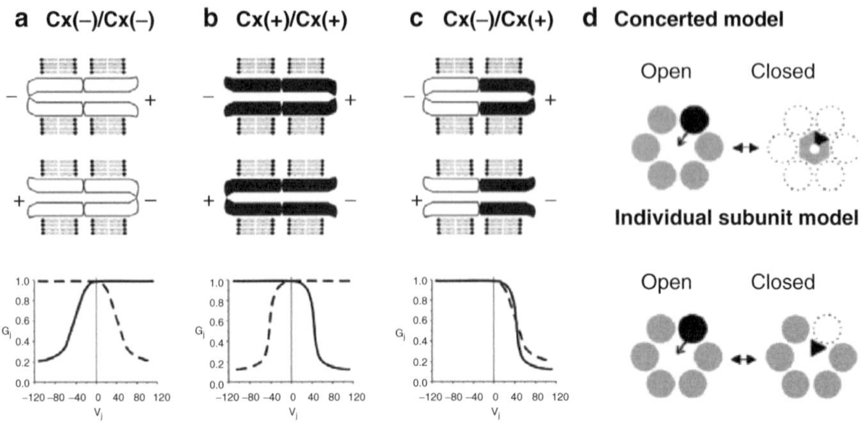

Fig. 13.6 (**a–c**) V_j gating diagrams for homomeric connexin gap junctions with a negative (**a**) or positive (**b**) gating polarity and a heterotypic gap junction channel (**c**) with opposite gating polarities. The *solid line* is the Boltzmann equation curve for the G_j–V_j relationship of the negative polarity (–) hemichannel in panel **a**, the positive polarity (+) hemichannel in panels **b** and **c**, and the *dashed line* represents the Boltzmann curve for the other hemichannel in panels **a** and **b** and the negative polarity hemichannel in panel **c** (From Bargiello and Brink (2009) with permission). (**d**) Diagram of alternative connexin stoichiometric models for V_j gating from the open state to the partially closed subconductance state of one hemichannel of a gap junction channel wherein all six subunits transit to the "closed" conformation (concerted (dependent) model) or only one connexin subunit is required to gate to the closed conformation (individual (independent) model). Experimental data supported the independent model (Modified from Oh et al. ©2000 Rockefeller University Press. *Journal of General Physiology*. 116:13–31)

channels, by contrast, are asymmetric and the resultant gap junction frequently exhibits asymmetric G_j –V_j curves (Fig. 13.6c) and non-linear, i.e. rectifying, $\gamma_j - V_j$ relationships. The first such heterotypic gap junction to be described was the Cx26/Cx32 gap junctions formed by the two liver connexins (Barrio et al. 1991). A chimeric Cx26 with the NT and M1 domains of Cx32 (Cx26*32NTM1) resulted in a gap junction that gated in only one direction when paired with Cx26 and a nearly symmetrical $G_j - V_j$ curve when paired with Cx32 (Verselis et al. 1994; Fig. 13.7a, b). Rectification occurs not only at the macroscopic G_j level, but the heterotypic Cx26/Cx32 gap junction channel also displayed an asymmetric $\gamma_j - V_j$ relationship (Bukauskas et al, 1995; Suchyna et al. 1999, Fig. 13.7c). Thus, the rectification of the heterotypic Cx26/Cx32 gap junction results from not only from the asymmetric steady-state $G_j - V_j$ relationship, but also from the asymmetric channel conductance and permeability properties of this heterotypic gap junction channel. The functionality of the heterotypic and heteromeric combinations of Cx40 and Cx43, the two major cardiac connexins of the atria and ventricular conduction system, has been highly controversial with conflicting reports in the literature of the heterotypic incompatibility or compatibility of Cx40 and Cx43 to form functional gap junctions (Bruzzone et al. 1993; Elfgang et al. 1995; Rackauskas et al. 2007; Valiunas et al. 2000, 2002; Cottrell and Burt 2001; Cottrell et al. 2002). The functional heteromeric mixing of Cx40 and Cx43 is also controversial with reports in favor of and opposed

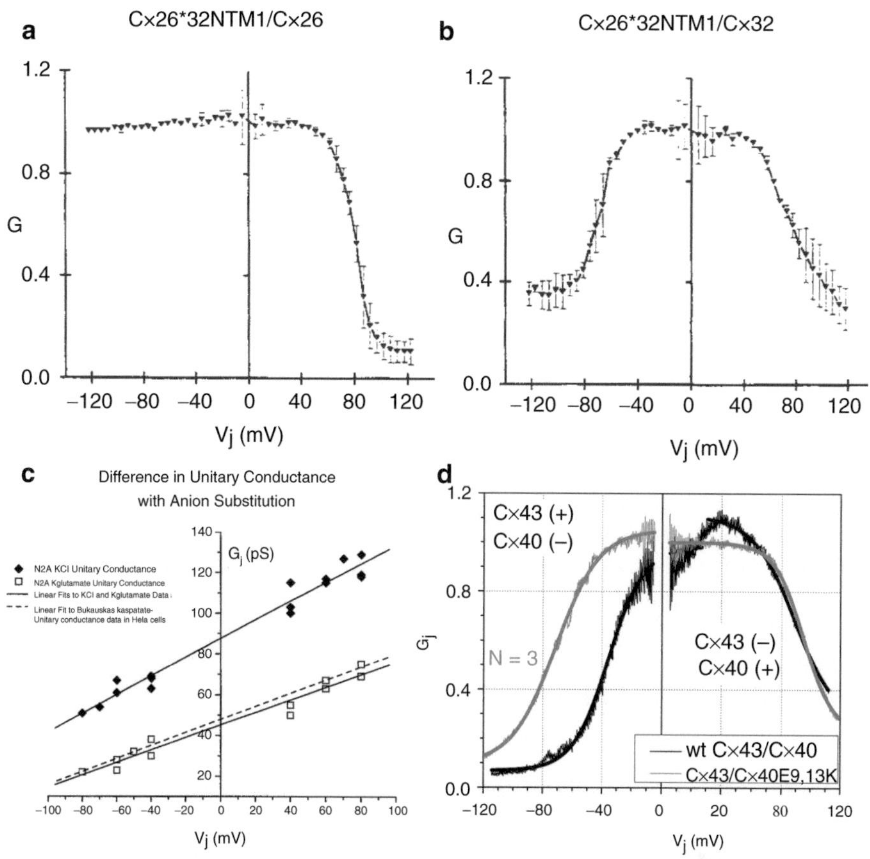

Fig. 13.7 (**a**) Normalized gap junction conductance (G) – V_j curve for a Cx26*32NTM1 chimeric protein coupled with wild-type Cx26 illustrating V_j gating only in the Cx26(+) and Cx32NT(−) direction. (**b**) The same chimeric protein coupled with wild-type Cx32 exhibits nearly symmetrical V_j gating, demonstrating that V_j gating polarity is primarily determined by the connexin NT domain (From Verselis et al. (1994) with permission). (**c**) Single channel conductance – V_j relationships for the heterotypic Cx26/Cx32 gap junction channel under symmetrical KCl (◆) or Kglutamate (□) conditions illustrate the rectification of an asymmetric heterotypic gap junction channel. The *dashed line* represents the Bukauskas et al. (1995) γ_j-V_j relationship (From Suchyna et al. (1999) with permission). (**d**) The heterotypic G_j–V_j relationship for a wild-type heterotypic Cx43/Cx40 gap junction and the nearly symmetrical G_j–V_j curve resulting from heterotypically pairing wild-type Cx43 with the Cx40 E9K, E13K (E9, 13K) double mutant protein (From Lin et al. (2014) with permission)

to their heteromeric compatibility (He et al. 1999; Burt et al. 2001; Valiunas et al. 2001). Two recent studies suggest that the functionality of the heteromeric Cx40/Cx43 gap junctions may be impaired by a specific interaction between the oppositely charged NT domains of these two connexins (rat Cx43 K9 and K13, rat Cx40 E9 and E13; Beyer et al. 2013). Mutation of these same negatively charged NT domain sites at positions 9 and 13 of Cx40 to the Cx43 K9 and K13 sequence nearly eliminated the rectification of the heterotypic Cx40/Cx43 gap junctions, thus illustrating that charge differences on the NT domains of the α-subfamily of connexins also determines the gating polarity of these connexins (Lin et al. 2014, Fig. 13.7d).

Precisely how the gap junction channel pore is occluded, or partially occluded, by the V_j gating mechanism is not fully understood. Structural studies of the NT domain of Cx32 indicated that a "glycine-hinge" motif in the middle of the predominantly α-helical NT segment bends the first half of this domain into the pore where it translocates in a V_j-dependent manner towards the center of the pore (Purnick et al. 2000). Similarly, the connexin M2 domain with its conserved proline-kink is thought to transit into and out of the pore in a V_j-dependent manner (Ri et al. 1999). Oshima et al. (2007, 2011) have modeled the NT domain of Cx26 as forming a central "plug" that occludes the gap junction channel pore in a V_j-dependent manner. However, the glycine-hinge motif is specific to the β-subfamily connexins (e.g. Cx32, Cx26, etc., *GJB#*; Table 13.1; see Beyer and Berthoud 2009) and substitution of this motif into the α-subfamily connexins (e.g. Cx43, Cx40, etc., *GJA#*) rendered Cx43 and Cx40 gap junctions insensitive to V_j (Gemel et al. 2006). Nonetheless, numerous structure-function studies have implicated charged amino acids on the NT domain of α-subfamily connexin gap junctions in the V_j gating and γ_j properties of the resultant gap junctions (Musa et al. 2004; Tong et al. 2004; Lin et al. 2006; Xin et al. 2010). These results are consistent with the NT domain of α- and β-subfamily connexins residing within the pore (ion permeation pathway) and V_j field of the assembled gap junction channel.

13.3.2.2 Calcium/Calmodulin and pH

Historically, heart muscle was known to recover its resting membrane potential within minutes of excision, a process called "healing-over" that was demonstrated to be dependent on the presence of extracellular calcium (Délèze 1970). In the absence of external Ca^{2+}, intracellular acidification to pH 4–5.0 was required to effect complete healing over (De Mello 1983). Thus, the concept that intracellular Ca^{2+} and pH ($[Ca^{2+}]_i$ and pH_i) regulates cellular electrical coupling was developed. There are conflicting reports in the gap junction literature about the relative importance of low pH_i or elevated $[Ca^{2+}]_i$ to induce closure of gap junctions, an interpretation that is complicated by the interdependence of pH_i and $[Ca^{2+}]_i$ (Rose and Rick 1978; Turin and Warner 1980; Spray et al. 1981b; Peracchia 1990; Vaughan-Jones et al. 1983). Calmodulin (CaM), a cytoplasmic Ca^{2+}-binding protein that modulates the activity of numerous ion channels (Kovalevskaya et al. 2013; Ben-Johny and Yue 2014), was reported to bind to lens gap junction membranes and later to regulate gap junction coupling even under conditions of respiratory acidosis (100 % CO_2), a process used to uncouple *Xenopus* blastomeres by lowering pH_i (Welsh et al. 1982; Peracchia 1984; reviewed in Peracchia 2004). A molecular pH gating model was developed for Cx43 involving trans-domain CL-CT interactions initially discovered by truncation of the Cx43 CT domain at residue position 257 and subsequent demonstration of pH-dependent binding to residues 119–144 of the CL domain (Morley et al 1996; Duffy et al. 2002).

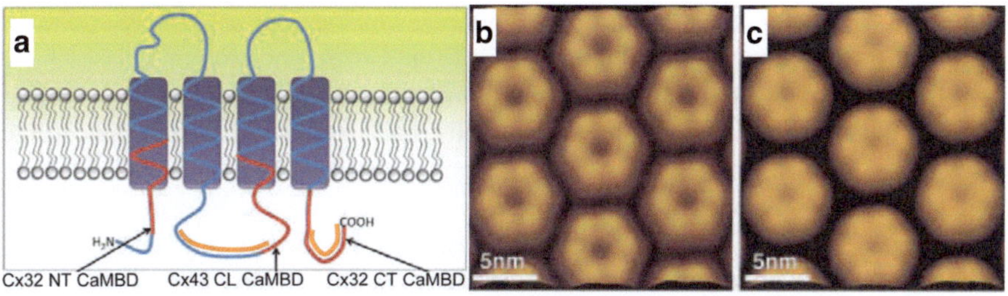

Fig. 13.8 (**a**) Diagram of the connexin membrane topology with the Cx32 NT (^{1}MNWTGLYTLLSGVNRHSTAIGRVWLSVIFIFRIM^{34}) and Cx32 CT (^{206}VAEVVYLIIRACARRAQRRSNP^{227}) calmodulin binding domains (CaMBD) and Cx43, Cx44, Cx50 CL CaMBDs indicated in *red*. The CL-CT cytoplasmic pH gate of Cx43 is indicated by *orange lines*. The Cx43 and lens connexins CL CaMBDs conform to a 1-5-10 hydrophobic CaM binding motif (e.g. Cx43 ^{136}KYGIEEHGKVKMRGGLLRTYIISILFKS^{163}) (Modified from Zou et al. (2014) with permission). (**b**) Atomic force microscope (AFM) image of the open conformation of a Cx26 hemichannel in the absence of external Ca^{2+}. (**c**) AFM image of the external Cx26 hemichannel surface in the presence of 0.5 mM Ca^{2+} (From Müller et al. (2002) with permission)

Multiple basic residues located on the proximal end of the Cx32 CT influence CO_2-sensitivity (Wang and Peracchia 1997), but pH gating of Cx40 was not affected by deletion of the CT domain (Peracchia et al. 2004), thus negating a generalized model for the pH gating of connexin-specific gap junctions. Peracchia and colleagues have developed a trans-domain gating model for the Ca^{2+}/CaM gating of Cx32 gap junctions (Dodd et al. 2008). Recently, the Ca^{2+}/CaM gating hypothesis has undergone a resurrection with the identification of a 1-5-10 hydrophobic CaM binding domain (CaMBD) motif on the CL domain of numerous connexins beginning with Cx43 (Zhou et al. 2007; reviewed in Zou et al. 2014; Fig. 13.8a). Electrophysiological experiments have shown that an influx of Ca^{2+} ions induces closure of Cx43 gap junction channels in a CaM-dependent manner, but the precise molecular mechanism for the closure of the cytoplasmic pore remains poorly understood. Theories vary from an “iris” model involving tilt of all six connexin subunits, a “plug” model where Ca^{2+}/CaM may bind and plug the pore at the CL-M3 interface, to a particle-receptor model where CaM binding induces a closed conformation of the CL domain (Unwin and Ennis 1984; Xu et al. 2012). Atomic force microscopic images of Cx26 and Cx40 connexons reveals narrowing of the extracellular and cytoplasmic pore openings in the presence of 0.5–3.6 mM Ca^{2+}, which may represent the irreversible (condensed particle) closed (iris) state of the channel (Müller et al. 2002; Allen et al 2011; Fig. 13.8b, c).

13.3.2.3 Post-translational Modifications

Protein phosphorylation is a common mechanism for modulating the activity of ion channels, and connexin gap junctions are no exception to this rule (Levitan 1994; Solan and Lampe 2005). Early on, pp60v-src was linked to the downregulation of gap junction communication by Cx43 but not Cx32, once again demonstrating the connexin-specificity of gap junction regulation (Swenson et al. 1990). No connexin has been more extensively studied than Cx43 and different protein kinases have been linked to Cx43 gap junction assembly, degradation, reduction of Cx43 gap junction channel conductance and permeability, and lateralization from the cardiac intercalated disc by dephosphorylation (Takens-Kwak and Jongsma 1992; Moreno et al. 1994; Warn-Cramer et al. 1996; Beardslee et al. 2000; Ek-Vitorin et al. 2006; Solan et al. 2007; Dunn and Lampe 2014; reviewed in Solan and Lampe 2014). Recently, another post-translational modification has been show to reduce cardiac expression and function of Cx43 gap junctions and induce their lateralization not unlike Cx43 dephosphorylation (Colussi et al. 2011; Xu et al. 2013). Protein acetylation is becoming recognized as a modulatory post-translational modification akin to phosphorylation beyond its long recognized role of epigenetic modulation of gene expression (Kouzarides 2000; Haberland et al. 2009). Gap junctions, once thought to be a static electrical conduit for electrical and biochemical intercellular communication, are now known to be dynamic structures that are modulated by many cellular biochemical mechanisms.

13.3.2.4 Gap Junction Agonists and Antagonists

As previously mentioned, the first identified gap junction uncoupling agent, or antagonist, was long chain alkanols like 1-octanol (Johnston et al. 1980; see Fig. 13.4). Many additional pharmacological blockers of gap junctions have been identified over the past three decades. Such gap junction antagonists include unsaturated free fatty acids, glycyrrhetinic acid and carbenoxolone (CBX), volatile anesthetics, connexin-mimetic peptides, fenamates, 2-aminoethoxydiphenyl borate (2-APB), and spermine (Aylsworth et al. 1986; Davidson et al. 1986; Burt and Spray 1989; Warner et al. 1995, Evans and Boitano 2001; Harks et al. 2001, 2003; Musa and Veenstra 2003; Wang et al. 2013). However, these connexin-mimetic peptides don't exhibit connexin sequence-specific effects as anticipated and also inhibit pannexin channel activity (Dahl 2007; Wang et al 2007). CBX may be the most specific chemical gap junction antagonist in terms of not affecting cardiac ionic currents, but CBX is reported to inhibit retinal Ca^{2+} channels and pannexin 1 channels (de Groot et al. 2003; Vessey et al. 2004; Dahl et al. 2013). Gap junction agonists are less common. Derivatives of the atrial antiarrhythmic peptide such as rotigaptide (ZP123) and GAP-134 (ZP1609, danegaptide) improves cardiac conduction velocity and inhibits atrial fibrillation, apparently by enhancing gap junction communication (Kjølbye et al. 2003; Lin et al. 2008; Rossman et al. 2009; Skyschally et al. 2013). RXP-E, a different sort of connexin-binding peptide directed against the CT domain of Cx43 inhibits octanol-induced and low pH-induced uncoupling (Shibayama et al. 2007). So far, no gap junction agonists have received approval for human therapeutic use as an antiarrhythmic drug or other indication.

13.4 Connexin-linked Diseases

Channelopathies is a term given to describe the human pathological condition or disease associated with a mutational malfunction of an ion channel (Hoffman 1995). Human diseases linked to genetic mutations in ion channel genes have been fundamental in demonstrating the functional importance of ion channels to the normal physiological human condition and gap junction channels are no exception to this rule. The first mutant connexin linked to a human disease was Cx32 and X-linked Charcot Marie Tooth Syndrome (CMTX; Bergoffen et al. 1993). Since this initial discovery, the list of connexin-associated inherited human diseases has expanded to 12 of the 20 human connexin proteins, as summarized in Table 13.2. The molecular mechanisms by which human connexin mutations produce a human pathological condition vary from dominant-negative functional effects (Omori et al. 1996), abnormal connexin protein trafficking and absent or reduced gap junction assembly (Bone et al 1997; Deschênes et al. 1997), reduced connexin expression (Kabzińska et al. 2011), and accelerated connexin protein degradation (Gemel et al. 2014). Cx32 and Cx47 mutations cause peripheral and central nervous system demyelinating neuropathies presumably by the loss of "reflexive" gap junctions between the layers of the Schwann cell myelin sheath and oligodendrite-astrocyte coupling (Bone et al. 1997; Menichella et al. 2003). The Cx37 C1019T polymorphism (P319S mutation) associated with atherosclerotic artery disease apparently alters ATP-dependent cell adhesion (Boerma et al. 1999; Derouette et al. 2009). The most common result of disease-associated connexin mutations is a loss-of-function that disrupts cellular-tissue homeostasis leading to cell death and the phenotypic expression of the disease symptoms is based on the organ where the particular connexin is expressed. There is one type of connexin mutation that does not result in loss of function, increased function and/or permeability of connexin hemichannels.

The undocked hexamer of connexin subunits that forms the historically named "connexon" was discovered to function as a "half-gap junction channel" or hemichannel quite paradoxically when expression of the newly cloned Cx46 in *Xenopus* oocytes caused their lysis (Paul et al. 1991). Aberrant plasmalemmal connexin hemichannel activity may create a "leaky" cell that shunts the normal V_m, resulting in membrane depolarization, and/or induce excessive Ca^{2+} influx into the cell (Abrams and Bennett 2002). Both effects will result in a loss of cellular homeostasis and result in cell death. Metabolic inhibition activates Cx43 hemichannel activity, presumably by s-nitrosylation of intracellular cysteine residues, also culminating in loss of cellular homeostasis and death (John et al. 1999; Retamal et al. 2006). Connexin hemichannels are normally gated closed by physiological mM extracellular $[Ca^{2+}]$ and human connexin disease-linked mutations that affect this extracellular Ca^{2+} gate or increase hemichannel Ca^{2+} permeability recapitulate known disease symptoms, thus confirming the pathophysiological ramifications of aberrant mutant connexin hemichannel activity (Ebihara and Steiner 1993; Pfahnl and Dahl 1999; Gerido et al. 2007; Sánchez et al. 2010; Mese et al. 2011).

Table 13.2 Connexin-linked diseases

Connexin (*Gene*)	Human disease	Reference
Cx32 (*GJB1*)	X-linked Charcot-Marie-Tooth (CMTX) [>300 references] Hepatocarcinogenesis (chemical-induced)	Bergoffen et al. (1993), Fairweather et al. (1994), Ionasescu et al. (1994), Dagli et al. (2004)
Cx26 (*GJB2*)	Hereditary deafness [>700 references] Keratitis-ichthyosis-deafness (KID)	Kelsell et al. (1997), Richard et al. (2002)
Cx31 (*GJB3*)	Erythrokeratodermia variabilis (EKV)	Wilgoss et al. (1999)
	Palmoplantar keratoderma	Kelsell et al. (2000)
	Hereditary deafness	Kelsell et al. (2000)
	Peripheral neuropathy	López-Bigas et al. (2001)
Cx30.3 (*GJB4*)	Erythrokeratodermia variabilis EKV	Macari et al. (2000)
Cx30 (*GJB6*)	Hereditary deafness keratitis-ichthyosis-deafness (KID)	Grifa et al. (1999), Jan et al. (2004)
Cx43 (*GJA1*)	Developmental heart defects Hereditary deafness Oculodentodigital dysplasia (ODDD) Atrial fibrillation	Britz-Cunningham et al. (1995), Liu et al. (2001), Paznekas et al. (2003, 2009), Thibodeau et al. (2010)
Cx46 (*GJA3*)	Congenital zonular pulverulent cataract (CZP1)	Mackay et al. (1999)
Cx37 (*GJA4*)	Atherosclerosis Acute myocardial infarction	Boerma et al. (1999), Listì et al. (2005)
Cx40 (*GJA5*)	Atrial fibrillation	Firouzi et al. (2004), Gollob et al. (2006), Yang et al. (2010), Shi et al. (2013), Makita et al. (2005), Firouzi et al. (2006), Wagner et al. (2007), Krattinger et al. (2007), Makita et al. (2012)
	Atrial standstill Hypertension	
	Familial heart block	
Cx50 (*GJA8*)	Congenital zonular pulverulent cataract (CZP1)	Shiels et al. (1998)
Cx47 (*GJC2*)	Pelizaeus-Merzbacher-like disease (PMD)	Salviati et al. (2007)
Cx36 (*GJD2*)	Schizophrenia	Meyer et al. (2002)
	Diabetes	Ravier et al. (2005)

13.5 Connexin Hemichannels: A More Typical Ion Channel?

The study of connexin hemichannel function has been fundamental to our understanding of the conductance, gating, and permeability properties of the gap junction channels formed by the distinct connexins. Inside-out patch clamp recordings revealed that Cx46 hemichannels exhibit both fast V_j-gating and slow loop-gating between the fully open, subconductance, and fully closed states of the channel with a maximum conductance (γ_{hc}) of 300 pS, twice the γ_j of Cx46 gap junction channels

(Trexler et al. 1996). Development of an artificial chimeric connexin Cx32E143 hemichannel, a Cx32 connexin with the E1 domain replaced by Cx43, led to the identification of the first putative pore-lining residues (Pfahnl et al. 1997; Zhou et al. 1997). Subsequent chimeric connexin domain swaps and cysteine scanning accessibility method (SCAM) mutagenesis plus methanethiolsulfonate (MTS) reagent studies further implicated NT, M1, and E1 connexin domains as pore lining domains (Oh et al. 1999; Kronengold et al 2003; Hu et al. 2006, Ma and Dahl 2006). The connexin M3 domain was originally predicted to form the transmembrane pore, owing to its amphipathic nature, and one SCAM-MTS study supported this interpretation (Bennett et al. 1991; Skerrett et al. 2002). Thus, unlike the identification of the selective pore-forming domains of ligand-gated and voltage-gated ion channels within the first decade of their cloning (Imoto et al. 1988; Yellen et al. 1991; Heinemann et al. 1992; Yang et al. 1993), a consensus as to the pore-forming domains of connexin channels remained unresolved more than two decades after their molecular identification (reviewed in Verselis 2009).

13.6 The Molecular Model for the Gap Junction Channel

Two decades later, the three-dimensional resolution (3D) image of a gap junction was improved to 7.5 Å with the electron crystallographic image of a CT truncated, oleamide-treated, Cx43 gap junction (Unger et al. 1999). This 3D image confirmed the presence of 24 (6 subunits × 4) transmembrane domains consistent with α-helical conformations (Fig. 13.9a). Cytoplasmic domain and secondary structures were not resolved and specific assignment of M1-M4 domains was not ascertained. Specific assignment of M1-M4 α-helical domains as M1 = B, M2 = A, M3 = C, and M4 = D was achieved by mapping conserved connexin sequences and known mutations onto the 3D Cx43 crystal structure to generate a Calpha model (Fleishman et al. 2004). This structural model was based on the premise that conserved sequences preferentially pack into helix-helix interfaces and supported the view that M3 was the pore-lining helix. Hence, the electrophysiological data and structural models for the gap junction channel remained in dispute over the assignment of the gap junction channel pore lining residues. This discrepancy was apparently resolved with the resolution of the crystal structure of the Cx26 gap junction channel to 3.5 Å (Fig. 13.9c, d; Maeda et al. 2009). The salient features of the Cx26 gap junction channel 3D structural model include an NT domain that folds into the cytoplasmic pore opening (Purnick et al. 2000), an M1 pore-lining helix, and an E1 3–10 pore-lining helix extending into the extracellular space. This model also incorporated the E1-E2 anti-parallel disulfide (C-C) bonds based on the findings of Foote et al. (1998). The Maeda et al (2009) model further predicted NT-M1 hydrophobic W3-M34 and NT-NT hydrophilic D2-T5 interdomain interactions between adjacent connexin subunits that stabilize the NT domain within the pore. This 3D model became the accepted model of an open gap junction channel that accounts for the majority of predicted structural features and electrophysiological connexin

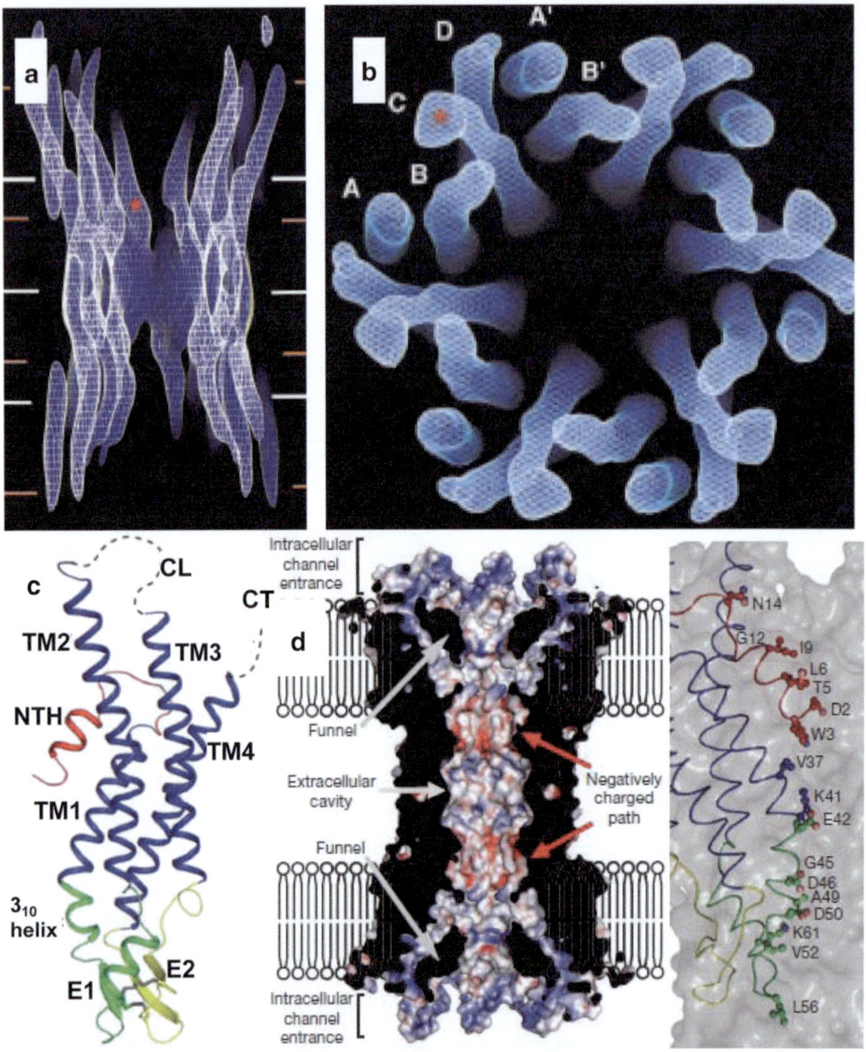

Fig. 13.9 (**a**) Image of the interior of the Cx43 gap junction channel with a central extracellular compartment that is narrower than the cytoplasmic vestibule in each cell. The *red asterisk* marks the location of the narrowest diameter (15 Å) of the pore within the membrane. (**b**) End on view looking towards the intracellular opening of the channel with the transmembrane helicies M1–M4 of one connexin labeled *A–D* (A' and B' indicates adjacent connexin subunit helicies). The *red asterisk* in this image indicates the helical density with the furthest extension from the membrane (From Unger et al. (1999) with permission). (**c**) Side view of Cx26 subunit with the NT domain (*red*) folding into the pore lined by M1 (*blue*) and E1 (*green*) helicies with the non-pore lining E2 domain indicated by *yellow*. CL and CT structures were not resolved. (**d**) Cross section of the Cx26 gap junction pore illustrating the negatively charged extracellular chamber and intracellular funnel into the transmembrane pore (*left panel*) with amino acid residues lining the ion permeation pathway from one subunit indicated by numbers (*right panel*) (From Maeda et al. (2009) with permission)

hemichannel and gap junction channel pore data despite the inability to resolve the specific positioning of connexin protein amino acid side chains. Nonetheless, this model provides a working framework for gap junction channel-structure function experiments and has been utilized to explain the structural basis for the disparate functional effects of Cx43 ODDD NT domain mutations and the heterotypic docking of Cx26 and Cx32 connexons (Nakagawa et al. 2011; Shao et al. 2012). However, many of the structural and functional implications of the Maeda et al. (2009) model have not been verified experimentally.

13.7 Future Directions for Gap Junction Channel Research, Unresolved Questions

Questions remain about the open structure of the gap junction channel. Short of direct electrophysiological verification of the model predictions of Maeda et al. (2009), some investigators have performed molecular dynamic (MD) simulations of the Cx26 hemichannel structure. Two initial MD studies did not support the hypothesized W3-M34 hydrophobic NT-M1 interaction proposed to stabilize the NT domain within the pore (Kwon et al. 2011; Zonta et al. 2012). Additionally, these *in silico* MD simulation studies provided new insight as to the potential identity of the pore permeability barriers to cations and anions. Kwon et al. (2011) agreed with the Cx26 crystal structure that the Met1 position and the reduced pore volume of the NH_2-terminus of the protein affects ion permeation, but also that major energy barriers to anion (Cl^-) permeation are formed by D2 and E42 of the Cx26 hemichannel while K41 presents the largest energy barrier to permeating cations (K^+). They also concluded that acetylation of multiple cytoplasmic lysine (K15, K102, K103, K105, K108, K112, K116) residues and γ-carboxyglutamation of three glutamic acid (E42, E47, E114) residues, detected in a mass spectrographic study of Cx26 post-translational modifications (Locke et al. 2009), contribute to the selectivity of the Cx26 channel. One final conclusion of this study was that the MD simulations suggest that the Cx26 crystal structure represents the closed state of the channel. By homology mapping the sequence of Cx30 onto the Cx26 crystal structure, Zonta et al. (2012) added the interpretation that E49 creates the narrowest pore diameter and the highest energy ion permeability barrier in this connexin channel and that Cx26 E42, D46, E47, and E50 residues interact with external Ca^{2+} ions near the pore restriction and may serve as the Cx26 hemichannel calcium gate. These *in silico* studies lend themselves to experimental verification by site-directed mutagenesis and patch clamp electrophysiological studies and provide a useable framework for future studies into the structure-function relationships of gap junction channels and connexin hemichannels. A combination of structural, computational, and electrophysiological approaches to the study of connexin hemichannels and gap junction channels may finally provide a solution to the question of how connexins assemble to connect the cellular world to form an electrochemical conduit that creates a synergy of anatomical form and physiological function that is essential to the life of the organism.

References

Abrams CK, Bennett MV (2002) Voltage opens unopposed gap junction hemichannels formed by a connexin 32 mutant associated with X-linked Charcot-Marie-Tooth disease. Proc Natl Acad Sci U S A 99(6):3980–3984

Abrams CK, Freidin MM et al (2006) Properties of human connexin 31, which is implicated in hereditary dermatological disease and deafness. Proc Natl Acad Sci U S A 103(13):5213–5218

Allen MJ, Gemel J et al (2011) Atomic force microscopy of Connexin40 gap junction hemichannels reveals calcium-dependent three-dimensional molecular topography and open-closed conformations of both the extracellular and cytoplasmic faces. J Biol Chem 286(25):22139–22146

Aylsworth CF, Trosko JE et al (1986) Influence of lipids on gap-junction-mediated intercellular communication between Chinese hamster cells in vitro. Cancer Res 46(9):4527–4533

Bargiello T, Brink P (2009) Voltage gating mechanisms of connexin channels. In: Harris A, Locke D (eds) Connexins: a guide. Humana Press, Totowa, pp 103–128

Barr L, Dewey MM et al (1965) Propagation of action potentials and the structure of the nexus in cardiac muscle. J Gen Physiol 48:797–823

Barrio LC, Suchyna T et al (1991) Gap junctions formed by connexins 26 and 32 alone and in combination are differently affected by applied voltage. Proc Natl Acad Sci U S A 88(19):8410–8414

Barrio LC, Capel J et al (1997) Species-specific voltage-gating properties of connexin-45 junctions expressed in Xenopus oocytes. Biophys J 73(2):757–769

Beardslee MA, Lerner DL et al (2000) Dephosphorylation and intracellular redistribution of ventricular connexin43 during electrical uncoupling induced by ischemia. Circ Res 87(8):656–662

Beblo DA, Wang HZ et al (1995) Unique conductance, gating, and selective permeability properties of gap junction channels formed by connexin40. Circ Res 77(4):813–822

Ben-Johny M, Yue DT (2014) Calmodulin regulation (calmodulation) of voltage-gated calcium channels. J Gen Physiol 143(6):679–692

Bennett MV, Barrio LC et al (1991) Gap junctions: new tools, new answers, new questions. Neuron 6(3):305–320

Bergoffen J, Scherer SS et al (1993) Connexin mutations in X-linked Charcot-Marie-Tooth disease. Science 262(5142):2039–2042

Bevans CG, Harris AL (1999) Direct high affinity modulation of connexin channel activity by cyclic nucleotides. J Biol Chem 274(6):3720–3725

Bevans CG, Kordel M et al (1998) Isoform composition of connexin channels determines selectivity among second messengers and uncharged molecules. J Biol Chem 273(5):2808–2816

Beyer EC, Berthoud VM (2009) The family of connexin genes. In: Harris AL, Locke D (eds) Connexins, a guide. Humana Press, New York, pp 3–26

Beyer EC, Paul DL et al (1987) Connexin43: a protein from rat heart homologous to a gap junction protein from liver. J Cell Biol 105(6):2621–2629

Beyer EC, Paul DL et al (1990) Connexin family of gap junction proteins. J Membr Biol 116(3):187–194

Beyer EC, Lin X et al (2013) Interfering amino terminal peptides and functional implications for heteromeric gap junction formation. Front Pharmacol 4:67

Boerma M, Forsberg L et al (1999) A genetic polymorphism in connexin 37 as a prognostic marker for atherosclerotic plaque development. J Intern Med 246(2):211–218

Bone LJ, Deschênes SM et al (1997) Connexin32 and X-linked Charcot-Marie-Tooth disease. Neurobiol Dis 4(3–4):221–230
Brink PR, Dewey MM (1980) Evidence for fixed charge in the nexus. Nature 285(5760):101–102
Brink PR, Cronin K et al (1997) Evidence for heteromeric gap junction channels formed from rat connexin43 and human connexin37. Am J Physiol 273(4):C1386–C1396
Britz-Cunningham SH, Shah MM et al (1995) Mutations of the Connexin43 gap-junction gene in patients with heart malformations and defects of laterality. N Engl J Med 332(20):1323–1329
Bruzzone R, Haefliger JA et al (1993) Connexin40, a component of gap junctions in vascular endothelium, is restricted in its ability to interact with other connexins. Mol Biol Cell 4(1):7–20
Bukauskas FF, Peracchia C (1997) Two distinct gating mechanisms in gap junction channels: CO2-sensitive and voltage-sensitive. Biophys J 72(5):2137–2142
Bukauskas FF, Elfgang C et al (1995) Heterotypic gap junction channels (connexin26-connexin32) violate the paradigm of unitary conductance. Pflugers Arch 429(6):870–872
Burt JM, Spray DC (1989) Volatile anesthetics block intercellular communication between neonatal rat myocardial cells. Circ Res 65(3):829–837
Burt JM, Fletcher AM et al (2001) Alteration of Cx43:Cx40 expression ratio in A7r5 cells. Am J Physiol Cell Physiol 280(3):C500–C508
Colussi C, Rosati J et al (2011) N^{ε}-lysine acetylation determines dissociation from GAP junctions and lateralization of connexin 43 in normal and dystrophic heart. Proc Natl Acad Sci U S A 108(7):2795–2800
Cottrell GT, Burt JM (2001) Heterotypic gap junction channel formation between heteromeric and homomeric Cx40 and Cx43 connexons. Am J Physiol Cell Physiol 281(5):C1559–C1567
Cottrell GT, Wu Y et al (2002) Cx40 and Cx43 expression ratio influences heteromeric/ heterotypic gap junction channel properties. Am J Physiol Cell Physiol 282(6):C1469–C1482
Dagli ML, Yamasaki H et al (2004) Delayed liver regeneration and increased susceptibility to chemical hepatocarcinogenesis in transgenic mice expressing a dominant-negative mutant of connexin32 only in the liver. Carcinogenesis 25(4):483–492
Dahl G (2007) Gap junction-mimetic peptides do work, but in unexpected ways. Cell Commun Adhes 14(6):259–264
Dahl G, Qiu F et al (2013) The bizarre pharmacology of the ATP release channel pannexin1. Neuropharmacology 75:583–593
Davidson JS, Baumgarten IM et al (1986) Reversible inhibition of intercellular junctional communication by glycyrrhetinic acid. Biochem Biophys Res Commun 134(1):29–36
de Groot JR, Veenstra T et al (2003) Conduction slowing by the gap junctional uncoupler carbenoxolone. Cardiovasc Res 60(2):288–297
De Mello WC (1983) The influence of pH on the healing-over of mammalian cardiac muscle. J Physiol 339:299–307
Délèze J (1970) The recovery of resting potential and input resistance in sheep heart injured by knife or laser. J Physiol 208(3):547–562
Derouette JP, Desplantez T et al (2009) Functional differences between human Cx37 polymorphic hemichannels. J Mol Cell Cardiol 46(4):499–507
Deschênes SM, Walcott JL et al (1997) Altered trafficking of mutant connexin32. J Neurosci 17(23):9077–9084
Dewey MM, Barr L (1962) Intercellular connection between smooth muscle cells: the nexus. Science 137(3531):670–672
Dodd R, Peracchia C et al (2008) Calmodulin association with connexin32-derived peptides suggests trans-domain interaction in chemical gating of gap junction channels. J Biol Chem 283(40):26911–26920
Donaldson P, Kistler J (1992) Reconstitution of channels from preparations enriched in lens gap junction protein MP70. J Membr Biol 129(2):155–165
Duffy HS, Sorgen PL et al (2002) pH-dependent intramolecular binding and structure involving Cx43 cytoplasmic domains. J Biol Chem 277(39):36706–36714

Dunn CA, Lampe PD (2014) Injury-triggered Akt phosphorylation of Cx43: a ZO-1-driven molecular switch that regulates gap junction size. J Cell Sci 127(2):455–464
Ebihara L, Steiner E (1993) Properties of a nonjunctional current expressed from a rat connexin46 cDNA in Xenopus oocytes. J Gen Physiol 102(1):59–74
Ek-Vitorin JF, King TJ et al (2006) Selectivity of connexin 43 channels is regulated through protein kinase C dependent phosphorylation. Circ Res 98(12):1498–1505
Elfgang C, Eckert R et al (1995) Specific permeability and selective formation of gap junction channels in connexin-transfected HeLa cells. J Cell Biol 129(3):805–817
Evans WH, Boitano S (2001) Connexin mimetic peptides: specific inhibitors of gap-junctional intercellular communication. Biochem Soc Trans 29(Pt 4):606–612
Fairweather N, Bell C et al (1994) Mutations in the connexin 32 gene in X-linked dominant Charcot-Marie-Tooth disease (CMTX1). Hum Mol Genet 3(1):29–34
Firouzi M, Ramanna H et al (2004) Association of human connexin40 gene polymorphisms with atrial vulnerability as a risk factor for idiopathic atrial fibrillation. Circ Res 95(4):e29–e33
Firouzi M, Kok B et al (2006) Polymorphisms in human connexin40 gene promoter are associated with increased risk of hypertension in men. J Hypertens 24(2):325–330
Flagg-Newton J, Simpson I et al (1979) Permeability of the cell-to-cell membrane channels in mammalian cell junction. Science 205(4404):404–407
Fleishman SJ, Unger VM et al (2004) A Calpha model for the transmembrane alpha helices of gap junction intercellular channels. Mol Cell 15(6):879–888
Foote CI, Zhou L et al (1998) The pattern of disulfide linkages in the extracellular loop regions of connexin 32 suggests a model for the docking interface of gap junctions. J Cell Biol 140(5):1187–1197
Furshpan EJ, Furukawa T (1962) Intracellular and extracellular responses of the several regions of the Mauthner cell of the goldfish. J Neurophysiol 25:732–771
Furshpan EJ, Potter DD (1959) Transmission at the giant motor synapses of the crayfish. J Physiol 145(2):289–325
Gemel J, Lin X et al (2006) N-terminal residues in Cx43 and Cx40 determine physiological properties of gap junction channels, but do not influence heteromeric assembly with each other or with Cx26. J Cell Sci 119(Pt 11):2258–2268
Gemel J, Lin X et al (2008) Cx30.2 can form heteromeric gap junction channels with other cardiac connexins. Biochem Biophys Res Commun 369(2):388–394
Gemel J, Simon AR et al (2014) Degradation of a connexin40 mutant linked to atrial fibrillation is accelerated. J Mol Cell Cardiol 74:330–339
Gerido DA, DeRosa AM et al (2007) Aberrant hemichannel properties of Cx26 mutations causing skin disease and deafness. Am J Physiol Cell Physiol 293(1):C337–C345
Gollob MH, Jones DL et al (2006) Somatic mutations in the connexin 40 gene (GJA5) in atrial fibrillation. N Engl J Med 354(25):2677–2688
Gong XQ, Nakagawa S et al (2013) A mechanism of gap junction docking revealed by functional rescue of a human-disease-linked connexin mutant. J Cell Sci 126(14):3113–3120
González D, Gómez-Hernández JM et al (2007) Molecular basis of voltage dependence of connexin channels: an integrative appraisal. Prog Biophys Mol Biol 94(1-2):66–106
Grifa A, Wagner CA et al (1999) Mutations in GJB6 cause nonsyndromic autosomal dominant deafness at DFNA3 locus. Nat Genet 23(1):16–18
Haberland M, Montgomery RL et al (2009) The many roles of histone deacetylases in development and physiology: implications for disease and therapy. Nat Rev Genet 10(1):32–42
Hamill OP, Marty A et al (1981) Improved patch-clamp techniques for high-resolution current recording from cells and cell-free membrane patches. Pflugers Arch 391:85–100
Harks EG, de Roos AD et al (2001) Fenamates: a novel class of reversible gap junction channel blockers. J Pharmacol Exp Ther 298(3):1031–1041
Harks EG, Camiña JP et al (2003) Besides affecting intracellular calcium signaling, 2-APB reversibly blocks gap junctional coupling in confluent monolayers, thereby allowing measurement of single-cell membrane currents in undissociated cells. FASEB J 17(8):941–943

Harris AL (2007) Connexin channel permeability to cytoplasmic molecules. Prog Biophys Mol Biol 94(1-2):120–143

Harris AL, Locke D (2009) Permeability of connexin channels. In: Harris AL, Locke D (eds) Connexins, a guide. Humana Press, NY, pp 165–206

Harris AL, Spray DC et al (1981) Kinetic properties of a voltage-dependent junctional conductance. J Gen Physiol 77(1):95–117

He DS, Jiang JX et al (1999) Formation of heteromeric gap junction channels by connexins 40 and 43 in vascular smooth muscle cells. Proc Natl Acad Sci U S A 96(11):6495–6500

Heinemann SH, Terlau H et al (1992) Calcium channel characteristics conferred on the sodium channel by single mutations. Nature 356(6368):441–443

Hoffman EP (1995) Voltage-gated ion channelopathies: inherited disorders caused by abnormal sodium, chloride, and calcium regulation in skeletal muscle. Annu Rev Med 46:431–441

Hopperstad MG, Srinivas M et al (2000) Properties of gap junction channels formed by Cx46 alone and in combination with Cx50. Biophys J 79(4):1954–1966

Hu X, Ma M et al (2006) Conductance of connexin hemichannels segregates with the first transmembrane segment. Biophys J 90(1):140–150

Imoto K, Busch C et al (1988) Rings of negatively charged amino acids determine the acetylcholine receptor channel conductance. Nature 335(6191):645–648

Ionasescu V, Searby C et al (1994) Point mutations of the connexin32 (GJB1) gene in X-linked dominant Charcot-Marie-Tooth neuropathy. Hum Mol Genet 3(2):355–358

Jan AY, Amin S et al (2004) Genetic heterogeneity of KID syndrome: identification of a Cx30 gene (GJB6) mutation in a patient with KID syndrome and congenital atrichia. J Invest Dermatol 122(5):1108–1113

Jiang JX, Goodenough DA (1996) Heteromeric connexons in lens gap junction channels. Proc Natl Acad Sci U S A 93(3):1287–1291

John SA, Kondo R et al (1999) Connexin-43 hemichannels opened by metabolic inhibition. J Biol Chem 274(1):236–240

Johnston MF, Simon SA et al (1980) Interaction of anesthetics with electrical synapses. Nature 286(5772):498–500

Kabzińska D, Kotruchow K et al (2011) Two pathogenic mutations located within the 5′-regulatory sequence of the GJB1 gene affecting initiation of transcription and translation. Acta Biochim Pol 58(3):359–363

Kanter HL, Saffitz JE et al (1992) Cardiac myocytes express multiple gap junction proteins. Circ Res 70(2):438–444

Kelsell DP, Dunlop J et al (1997) Connexin 26 mutations in hereditary non-syndromic sensorineural deafness. Nature 387(6628):80–83

Kelsell DP, Wilgoss AL et al (2000) Connexin mutations associated with palmoplantar keratoderma and profound deafness in a single family. Eur J Hum Genet 8(2):141–144

Kjølbye AL, Knudsen CB et al (2003) Pharmacological characterization of the new stable antiarrhythmic peptide analog Ac-D-Tyr-D-Pro-D-Hyp-Gly-D-Ala-Gly-NH2 (ZP123): in vivo and in vitro studies. J Pharmacol Exp Ther 306(3):1191–1199

Kouzarides T (2000) Acetylation: a regulatory modification to rival phosphorylation? EMBO J 19:1176–1179

Koval M, Molina SA et al (2014) Mix and match: investigating heteromeric and heterotypic gap junction channels in model systems and native tissues. FEBS Lett 588(8):1193–1204

Kovalevskaya NV, van de Waterbeemd M et al (2013) Structural analysis of calmodulin binding to ion channels demonstrates the role of its plasticity in regulation. Pflugers Arch 465(11):1507–1519

Krattinger N, Capponi A et al (2007) Connexin40 regulates renin production and blood pressure. Kidney Int 72(7):814–822

Kronengold J, Trexler EB et al (2003) Single-channel SCAM identifies pore-lining residues in the first extracellular loop and first transmembrane domains of Cx46 hemichannels. J Gen Physiol 122(4):389–405

Kumar NM, Gilula NB (1996) The gap junction communication channel. Cell 84(3):381–388
Kwon T, Harris AL et al (2011) Molecular dynamics simulations of the Cx26 hemichannel: evaluation of structural models with Brownian dynamics. J Gen Physiol 138(5):475–493
Levitan IB (1994) Modulation of ion channels by protein phosphorylation and dephosphorylation. Annu Rev Physiol 56:193–212
Lin X, Fenn E et al (2006) An amino-terminal lysine residue of rat connexin40 that is required for spermine block. J Physiol 570(2):251–269
Lin X, Zemlin C et al (2008) Enhancement of ventricular gap-junction coupling by rotigaptide. Cardiovasc Res 79(3):416–426
Lin X, Xu Q et al (2014) Functional formation of heterotypic gap junction channels by connexins-40 and -43. Channels 8(5):433–443
Listì F, Candore G et al (2005) Association between C1019T polymorphism of connexin37 and acute myocardial infarction: a study in patients from Sicily. Int J Cardiol 102(2):269–271
Liu XZ, Xia XJ et al (2001) Mutations in GJA1 (connexin 43) are associated with non-syndromic autosomal recessive deafness. Hum Mol Genet 10(25):2945–2951
Locke D, Bian S et al (2009) Post-translational modifications of connexin26 revealed by mass spectrometry. Biochem J 424(3):385–398
Loewenstein WR (1966) Permeability of membrane junctions. Ann N Y Acad Sci 137(2):441–472
Loewenstein WR (1976) Permeable junctions. Cold Spring Harb Symp Quant Biol 40:49–63
Loewenstein WR (1981) Junctional intercellular communication: the cell-to-cell membrane channel. Physiol Rev 61(4):829–913
Loewenstein WR, Socolar SJ et al (1965) Intercellular communication: renal, urinary bladder, sensory, and salivary gland cells. Science 149(3681):295–298
Loewenstein WR, Kanno Y et al (1978) Quantum jumps of conductance during formation of membrane channels at cell-cell junction. Nature 274(5667):133–136
López-Bigas N, Olivé M et al (2001) Connexin 31 (GJB3) is expressed in the peripheral and auditory nerves and causes neuropathy and hearing impairment. Hum Mol Genet 10(9): 947–952
Ma M, Dahl G (2006) Cosegregation of permeability and single-channel conductance in chimeric connexins. Biophys J 90(1):151–163
Macari F, Landau M et al (2000) Mutation in the gene for connexin 30.3 in a family with erythrokeratodermia variabilis. Am J Hum Genet 67(5):1296–1301
Mackay D, Ionides A et al (1999) Connexin46 mutations in autosomal dominant congenital cataract. Am J Hum Genet 64(5):1357–1364
Maeda S, Nakagawa S et al (2009) Structure of the connexin 26 gap junction channel at 3.5 a resolution. Nature 458(7238):597–602
Makita N, Sasaki K et al (2005) Congenital atrial standstill associated with coinheritance of a novel SCN5A mutation and connexin 40 polymorphisms. Heart Rhythm 2(10):1128–1134
Makita N, Seki A et al (2012) A connexin40 mutation associated with a malignant variant of progressive familial heart block type I. Circ Arrhythm Electrophysiol 5(1):163–172
Makowski L, Caspar DL et al (1977) Gap junction structures. II. Analysis of the x-ray diffraction data. J Cell Biol 74(2):629–645
Manjunath CK, Goings GE et al (1985) Proteolysis of cardiac gap junctions during their isolation from rat hearts. J Membr Biol 85(2):159–168
Manthey D, Bukauskas F et al (1999) Molecular cloning and functional expression of the mouse gap junction gene connexin-57 in human HeLa cells. J Biol Chem 274(21):14716–14723
Menichella DM, Goodenough DA et al (2003) Connexins are critical for normal myelination in the CNS. J Neurosci 23(13):5963–5973
Mese G, Sellitto C et al (2011) The Cx26-G45E mutation displays increased hemichannel activity in a mouse model of the lethal form of keratitis-ichthyosis-deafness syndrome. Mol Biol Cell 22(24):4776–4786

Meyer J, Mai M et al (2002) Mutational analysis of the connexin 36 gene (CX36) and exclusion of the coding sequence as a candidate region for catatonic schizophrenia in a large pedigree. Schizophr Res 58(1):87–91

Moreno AP, de Carvalho AC et al (1991) Voltage-dependent gap junction channels are formed by connexin32, the major gap junction protein of rat liver. Biophys J 59(4):920–925

Moreno AP, Sáez JC et al (1994) Human connexin43 gap junction channels. Regulation of unitary conductances by phosphorylation. Circ Res 74(6):1050–1057

Moreno AP, Chanson M et al (2002) Role of the carboxyl terminal of connexin43 in transjunctional fast voltage gating. Circ Res 90(4):450–457

Morley GE, Taffet SM et al (1996) Intramolecular interactions mediate pH regulation of connexin43 channels. Biophys J 70(3):1294–1302

Müller DJ, Hand GM et al (2002) Conformational changes in surface structures of isolated connexin 26 gap junctions. EMBO J 21(14):3598–3607

Musa H, Veenstra RD (2003) Voltage-dependent blockade of connexin40 gap junctions by spermine. Biophys J 84(1):205–219

Musa H, Fenn E et al (2004) Amino terminal glutamate residues confer spermine sensitivity and affect voltage gating and channel conductance of rat connexin40 gap junctions. J Physiol 557(3):863–878

Nakagawa S, Gong XQ et al (2011) Asparagine 175 of connexin32 is a critical residue for docking and forming functional heterotypic gap junction channels with connexin26. J Biol Chem 286(22):19672–19681

Neyton J, Trautmann A (1985) Single-channel currents of an intercellular junction. Nature 317(6035):331–335

Oh S, Rubin JB et al (1999) Molecular determinants of electrical rectification of single channel conductance in gap junctions formed by connexins 26 and 32. J Gen Physiol 114(3): 339–364

Oh S, Abrams CK et al (2000) Stoichiometry of transjunctional voltage-gating polarity reversal by a negative charge substitution in the amino terminus of a connexin32 chimera. J Gen Physiol 116(1):13–31

Omori Y, Mesnil M et al (1996) Connexin 32 mutations from X-linked Charcot-Marie-Tooth disease patients: functional defects and dominant negative effects. Mol Biol Cell 7(6):907–916

Oshima A, Tani K et al (2007) Three-dimensional structure of a human connexin26 gap junction channel reveals a plug in the vestibule. Proc Natl Acad Sci U S A 104(24):10034–10039

Oshima A, Tani K et al (2011) Asymmetric configurations and N-terminal rearrangements in connexin26 gap junction channels. J Mol Biol 405(3):724–735

Paul DL (1986) Molecular cloning of cDNA for rat liver gap junction protein. J Cell Biol 103(1):123–134

Paul DL, Ebihara L et al (1991) Connexin46, a novel lens gap junction protein, induces voltage-gated currents in nonjunctional plasma membrane of Xenopus oocytes. J Cell Biol 115(4):1077–1089

Paznekas WA, Boyadjiev SA et al (2003) Connexin 43 (GJA1) mutations cause the pleiotropic phenotype of oculodentodigital dysplasia. Am J Hum Genet 72(2):408–418

Paznekas WA, Karczeski B et al (2009) GJA1 mutations, variants, and connexin 43 dysfunction as it relates to the oculodentodigital dysplasia phenotype. Hum Mutat 30(5):724–733

Peracchia C (1984) Communicating junctions and calmodulin: inhibition of electrical uncoupling in Xenopus embryo by calmidazolium. J Membr Biol 81(1):49–58

Peracchia C (1990) Increase in gap junction resistance with acidification in crayfish septate axons is closely related to changes in intracellular calcium but not hydrogen ion concentration. J Membr Biol 113(1):75–92

Peracchia C (2004) Chemical gating of gap junction channels; roles of calcium, pH and calmodulin. Biochim Biophys Acta 1662(1-2):61–80

Peracchia C, Chen JT et al (2004) CO(2) sensitivity of voltage gating and gating polarity of gapjunction channels–connexin40 and its COOH-terminus-truncated mutant. J Membr Biol 200(2):105–113
Pfahnl A, Dahl G (1999) Gating of Cx46 gap junction hemichannels by calcium and voltage. Pflugers Arch 437(3):345–353
Pfahnl A, Zhou XW et al (1997) A chimeric connexin forming gap junction hemichannels. Pflugers Arch 433(6):773–779
Purnick PE, Benjamin DC et al (2000) Structure of the amino terminus of a gap junction protein. Arch Biochem Biophys 381(2):181–190
Rackauskas M, Kreuzberg MM et al (2007) Gating properties of heterotypic gap junction channels formed of connexins 40, 43, and 45. Biophys J 92(6):1952–1965
Ravier MA, Güldenagel M et al (2005) Loss of connexin36 channels alters beta-cell coupling, islet synchronization of glucose-induced Ca2+ and insulin oscillations, and basal insulin release. Diabetes 54(6):1798–1807
Reed KE, Westphale EM et al (1993) Molecular cloning and functional expression of human connexin37, an endothelial cell gap junction protein. J Clin Invest 91(3):997–1004
Retamal MA, Cortés CJ et al (2006) S-nitrosylation and permeation through connexin 43 hemichannels in astrocytes: induction by oxidant stress and reversal by reducing agents. Proc Natl Acad Sci U S A 103(12):4475–4480
Revel JP, Karnovsky MJ (1967) Hexagonal array of subunits in intercellular junctions of the mouse heart and liver. J Cell Biol 33(3):C7–C12
Revilla A, Castro C et al (1999) Molecular dissection of transjunctional voltage dependence in the connexin-32 and connexin-43 junctions. Biophys J 77(3):1374–1383
Revilla A, Bennett MV et al (2000) Molecular determinants of membrane potential dependence in vertebrate gap junction channels. Proc Natl Acad Sci U S A 97(26):14760–14765
Ri Y, Ballesteros JA, Abrams CK et al (1999) The role of a conserved proline residue in mediating conformational changes associated with voltage gating of Cx32 gap junctions. Biophys J 76(6):2887–2898
Richard G, Rouan F et al (2002) Missense mutations in GJB2 encoding connexin-26 cause the ectodermal dysplasia keratitis-ichthyosis-deafness syndrome. Am J Hum Genet 70(5):1341–1348
Rose B, Rick R (1978) Intracellular pH, intracellular free Ca, and junctional cell-cell coupling. J Membr Biol 44(3–4):377–415
Rossman EI, Liu K et al (2009) The gap junction modifier, GAP-134 [(2S,4R)-1-(2-aminoacetyl)-4-benzamido-pyrrolidine-2-carboxylic acid], improves conduction and reduces atrial fibrillation/flutter in the canine sterile pericarditis model. J Pharmacol Exp Ther 329(3):1127–1133
Salviati L, Trevisson E et al (2007) A novel deletion in the GJA12 gene causes Pelizaeus-Merzbacher-like disease. Neurogenetics 8(1):57–60
Sánchez HA, Mese G et al (2010) Differentially altered Ca^{2+} regulation and Ca^{2+} permeability in Cx26 hemichannels formed by the A40V and G45E mutations that cause keratitis ichthyosis deafness syndrome. J Gen Physiol 136(1):47–62
Shao Q, Liu Q et al (2012) Structure and functional studies of N-terminal Cx43 mutants linked to oculodentodigital dysplasia. Mol Biol Cell 23(17):3312–3321
Shi HF, Yang JF et al (2013) Prevalence and spectrum of GJA5 mutations associated with lone atrial fibrillation. Mol Med Rep 7(3):767–774
Shibayama J, Lewandowski R et al (2007) Identification of a novel peptide that interferes with the chemical regulation of connexin43. Circ Res 98(11):1365–1372
Shiels A, Mackay D et al (1998) A missense mutation in the human connexin50 gene (GJA8) underlies autosomal dominant "zonular pulverulent" cataract, on chromosome 1q. Am J Hum Genet 62(3):526–532
Skerrett IM, Aronowitz J et al (2002) Identification of amino acid residues lining the pore of a gap junction channel. J Cell Biol 159(2):349–360
Skyschally A, Walter B et al (2013) The antiarrhythmic dipeptide ZP1609 (danegaptide) when given at reperfusion reduces myocardial infarct size in pigs. Naunyn Schmiedebergs Arch Pharmacol 386(5):383–391

Smith TD, Mohankumar A et al (2012) Cytoplasmic amino acids within the membrane interface region influence connexin oligomerization. J Membr Biol 245(5–6):221–230
Solan JL, Lampe PD (2005) Connexin phosphorylation as a regulatory event linked to gap junction channel assembly. Biochim Biophys Acta 1711(2):154–163
Solan JL, Lampe PD (2014) Specific Cx43 phosphorylation events regulate gap junction turnover in vivo. FEBS Lett 588(8):1423–1439
Solan JL, Marquez-Rosado L et al (2007) Phosphorylation at S365 is a gatekeeper event that changes the structure of Cx43 and prevents down-regulation by PKC. J Cell Biol 179(6):1301–1309
Spray DC, Harris AL et al (1981a) Equilibrium properties of a voltage-dependent junctional conductance. J Gen Physiol 77(1):77–93
Spray DC, Harris AL et al (1981b) Gap junctional conductance is a simple and sensitive function of intracellular pH. Science 211(4483):712–715
Srinivas M, Costa M et al (1999a) Voltage dependence of macroscoic and unitary currents of gap junction channels formed by mouse Cx50 expressed in rat neuroblastoma cells. J Physiol 517(3):673–689
Srinivas M, Rozental R et al (1999b) Functional properties of channels formed by the neuronal gap junction protein connexin36. J Neurosci 19(22):9848–9855
Stauffer KA (1995) The gap junction proteins beta 1-connexin (connexin-32) and beta-2 connexin (connexin-26) can form heteromeric hemichannels. J Biol Chem 270(12):6768–6772
Suchyna TM, Xu LX et al (1993) Identification of a proline residue as a transduction element involved in voltage gating of gap junctions. Nature 365(6449):847–849
Suchyna TM, Nitsche JM et al (1999) Different ionic selectivities for connexins 26 and 32 produce rectifying gap junction channels. Biophys J 77(6):2968–2987
Sun J, Ahmad S et al (2005) Cochlear gap junctions coassembled from Cx26 and 30 show faster intercellular Ca2+ signaling than homomeric counterparts. Am J Physiol Cell Physiol 288(3):C613–C623
Swenson KI, Piwnica-Worms H et al (1990) Tyrosine phosphorylation of the gap junction protein connexin43 is required for the pp60v-src-induced inhibition of communication. Cell Regul 1(13):989–1002
Takens-Kwak BR, Jongsma HJ (1992) Cardiac gap junctions: three distinct single channel conductances and their modulation by phosphorylating treatments. Pflugers Arch 422(2):198–200
Teubner B, Degen J et al (2000) Functional expression of the murine connexin 36 gene coding for a neuron-specific gap junctional protein. J Membr Biol 176(3):249–262
Teubner B, Odermatt B et al (2001) Functional expression of the new gap junction gene connexin47 transcribed in mouse brain and spinal cord neurons. J Neurosci 21(4):1117–1126
Thibodeau IL, Xu J et al (2010) Paradigm of genetic mosaicism and lone atrial fibrillation: physiological characterization of a connexin 43-deletion mutant identified from atrial tissue. Circulation 122(3):236–244
Tong JJ, Liu X et al (2004) Exchange of gating properties between rat cx46 and chicken cx45.6. Biophys J 87(4):2397–2406
Traub O, Eckert R et al (1994) Immunochemical and electrophysiological characterization of murine connexin40 and -43 in mouse tissues and transfected human cells. Eur J Cell Biol 64(1):101–112
Trexler EB, Bennett MV et al (1996) Voltage gating and permeation in a gap junction hemichannel. Proc Natl Acad Sci U S A 93(12):5836–5841
Turin L, Warner AE (1980) Intracellular pH in early Xenopus embryos: its effect on current flow between blastomeres. J Physiol 300:489–504
Unger VM, Kumar NM et al (1999) Three-dimensional structure of a recombinant gap junction membrane channel. Science 283(5405):1176–1180
Unwin PN, Ennis PD (1984) Two configurations of a channel-forming membrane protein. Nature 307(5952):609–613
Valiunas V, Niessen H et al (1999) Electrophysiological properties of gap junction channels in hepatocytes isolated from connexin32-deficient and wild-type mice. Pflugers Arch 437(6): 846–856

Valiunas V, Weingart R et al (2000) Formation of heterotypic gap junction channels by connexins 40 and 43. Circ Res 86(2):E42–E49
Valiunas V, Gemel J et al (2001) Gap junction channels formed by coexpressed connexin40 and connexin43. Am J Physiol Heart Circ Physiol 281(4):H1675–H1689
Valiunas V, Beyer EC et al (2002) Cardiac gap junction channels show quantitative differences in selectivity. Circ Res 91(2):104–111
Vaughan-Jones RD, Lederer WJ et al (1983) Ca^{2+} ions can affect intracellular pH in mammalian cardiac muscle. Nature 301(5900):522–524
Veenstra RD (1996) Size and selectivity of gap junction channels formed from different connexins. J Bioenerg Biomembr 28(4):327–337
Veenstra RD, DeHaan RL (1986) Measurement of single channel currents from cardiac gap junctions. Science 233(4767):972–974
Veenstra RD, DeHaan RL (1988) Cardiac gap junction channel activity in embryonic chick ventricle cells. Am J Physiol 254(1):H170–H180
Veenstra RD, Wang HZ et al (1992) Multiple connexins confer distinct regulatory and conductance properties of gap junctions in developing heart. Circ Res 71(5):1277–1283
Veenstra RD, Wang HZ et al (1994a) Connexin37 forms high conductance gap junction channels with subconductance state activity and selective dye and ionic permeabilities. Biophys J 66(6):1915–1928
Veenstra RD, Wang HZ et al (1994b) Selective dye and ionic permeability of gap junction channels formed by connexin45. Circ Res 75(3):483–490
Veenstra RD, Wang HZ et al (1995) Selectivity of connexin-specific gap junctions does not correlate with channel conductance. Circ Res 77(6):1156–1165
Verselis VK (2009) The connexin channel pore: pore-lining segments and residues. In: Harris AL, Locke D (eds) Connexins, a guide. Humana Press, New York, pp 77–102
Verselis VK, Ginter CS et al (1994) Opposite voltage gating polarities of two closely related connexins. Nature 368(6469):348–351
Vessey JP, Lalonde MR et al (2004) Carbenoxolone inhibition of voltage-gated Ca channels and synaptic transmission in the retina. J Neurophysiol 92(2):1252–1256
Wagner C, de Wit C et al (2007) Connexin40 is essential for the pressure control of renin synthesis and secretion. Circ Res 100(4):556–563
Wang XG, Peracchia C (1997) Positive charges of the initial C-terminus domain of Cx32 inhibit gap junction gating sensitivity to CO_2. Biophys J 73(2):798–806
Wang J, Ma M et al (2007) Modulation of membrane channel currents by gap junction protein mimetic peptides: size matters. Am J Physiol Cell Physiol 293(3):C1112–C1119
Wang N, De Bock M et al (2013) Connexin targeting peptides as inhibitors of voltage- and intracellular Ca^{2+}-triggered Cx43 hemichannel opening. Neuropharmacology 75:506–516
Warn-Cramer BJ, Lampe PD et al (1996) Characterization of the MAP kinase phosphorylation sites on the connexin43 gap junction protein. J Biol Chem 271(7):3779–3786
Warner A, Clements DK et al (1995) Specific motifs in the external loops of connexin proteins can determine gap junction formation between chick heart myocytes. J Physiol 488(3):721–728
Weidmann S (1952) The electrical constant of Purkinje fibers. J Physiol 118(3):348–360
Welsh MJ, Aster JC et al (1982) Calmodulin binds to chick lens gap junction protein in a calcium-independent manner. Science 216(4546):642–644
White TW, Srinivas M et al (2002) Virtual cloning, functional expression, and gating analysis of human connexin31.9. Am J Physiol Cell Physiol 283(3):C960
Wilgoss A, Leigh IM et al (1999) Identification of a novel mutation R42P in the gap junction protein beta-3 associated with autosomal dominant erythrokeratoderma variabilis. J Invest Dermatol 113(6):1119–1122
Xin L, Gong XQ et al (2010) The role of amino terminus of mouse Cx50 in determining transjunctional voltage-dependent gating and unitary conductance. Biophys J 99(7):2077–2086
Xu Q, Lin X et al (2012) Gating of connexin 43 gap junctions by a cytoplasmic loop calmodulin binding domain. Am J Physiol Cell Physiol 302(10):C1548–C1556

Xu Q, Lin X et al (2013) Histone deacetylase inhibition reduces cardiac connexin43 expression and gap junction communication. Front Pharmacol 4:44

Yang J, Ellinor PT et al (1993) Molecular determinants of Ca^{2+} selectivity and ion permeation in L-type Ca^{2+} channels. Nature 366(6451):158–161

Yang YQ, Liu X et al (2010) Novel connexin40 missense mutations in patients with familial atrial fibrillation. Europace 12(10):1421–1427

Yellen G, Jurman ME et al (1991) Mutations affecting internal TEA blockade identify the probable pore-forming region of a K^+ channel. Science 251(4996):939–942

Yum SW, Zhang J et al (2007) Human connexin26 and connexin30 form functional heteromeric and heterotypic channels. Am J Physiol Cell Physiol 293(3):C1032–C1048

Zhang JT, Nicholson BJ (1989) Sequence and tissue distribution of a second protein of hepatic gap junctions, Cx26, as deduced from its cDNA. J Cell Biol 109(6):3391–3401

Zhou XW, Pfahnl A et al (1997) Identification of a pore lining segment in gap junction hemichannels. Biophys J 72(5):1946–1953

Zhou Y, Yang W et al (2007) Identification of the calmodulin binding domain of connexin 43. J Biol Chem 282(48):35005–35017

Zonta F, Polles G et al (2012) Permeation pathway of homomeric connexin 26 and connexin 30 channels investigated by molecular dynamics. J Biomol Struct Dyn 29(5):985–998

Zou J, Salarian M et al (2014) Gap junction regulation by calmodulin. FEBS Lett 588(8):1430–1438

Chapter 14
Amyloid Peptide Channels

Rustam Azimov and Bruce L. Kagan

Abstract Amyloid peptides and proteins appear to play a pathophysiologic role in amyloid diseases. Many amyloid peptides have been found to form ion channels under physiologic conditions. The channels from various diseases share common properties including heterodispersity, irreversibility, weak ionic selectivity, voltage-independent inhibition by Congo red and blockade by Zn^{+2}. These features would make these channels likely to depolarize target cells and mitochondria, disrupt Ca^{+2} regulation, and deplete cellular energy stores leading to cell dysfunction and death. The failure of many anti-amyloid drugs in human clinical trials may be related to the membrane location of amyloid channels.

Keywords Amyloid • Channels • Pores • Membranes • Toxicity

14.1 Introduction

Amyloid deposits were first described a century ago by the brilliant scientist and physician, Rudolph Virchow. He observed amorphous deposits that stained with iodine under light microscopy and thought they were primarily composed of starch due to their appearance. Later studies showed dyes such as Congo red exhibited a characteristic green birefringence when seen under polarized light. Extended amyloid "fibrils" have been exhaustively studied and even in several varieties (Sipe and Cohen 2000). "Amyloid" now refers to more than two dozen clinical syndromes where these classic amyloid deposits can be observed (see Table 14.1). The role of these deposits in causing disease has been hotly debated for many years. In the mid-20th Century, the "tombstone" hypothesis held sway, arguing that amyloid deposits were a "scar" or marker of some earlier damaging event. In 1992 Hardy and Higgins proposed the "amyloid cascade hypothesis" of Alzheimer's disease (AD) which implicated the beta-amyloid peptide (Abeta) as the pathological agent in AD. Further

R. Azimov • B.L. Kagan (✉)
Department of Psychiatry and Biobehavioral Sciences,
Semel Institute for Neuroscience and Human Behavior,
David Geffen School of Medicine at UCLA, Los Angeles, CA, USA
e-mail: bkagan@mednet.ucla.edu

A.H. Delcour (ed.), *Electrophysiology of Unconventional Channels and Pores*,
Springer Series in Biophysics 18, DOI 10.1007/978-3-319-20149-8_14

Table 14.1 Neurodegenerative protein folding diseases

Disease	Protein	Abbreviation
Alzheimer's disease	Amyloid precursor protein (Abeta 1–42)	APP (Abeta 1–42)
Down's Syndrome (Trisomy 21)		
Heredity cerebral angiopathy (Dutch)		
Kuru	Prion protein	PrP^c/PrP^{sc}
Gerstmann-Straussler Syndrome (GSS)		
Creutzfeld-Jacob Disease		
Scrapie (sheep)		
Bovine spongiform encephalopathy ("mad cow").		
Type II Diabetes mellitus (adult onset)	Islet amyloid polypeptide (Amylin)	IAPP
Dialysis associated amyloidosis	β2-Microglobulin	β2M
Senile cardiac amyloidosis	Atrial natriuretic factor	ANF
Familial amyloid Polyneuropathy	Transthyretin	TTR
Reactive amyloidosis familial mediterranean fever	Serum amyloid A	SAA
Familial amyloid polyneuropathy (Finnish)	Gelsolin	Agel
Macroglobulinemia	Gamma-1 heavy chain	AH
Primary systemic amyloidoses	Ig-lambda, Ig-kappa	AL
Familial Polyneuropathy – Iowa (Irish)	Apolipoprotein A1	ApoA1
Hereditary cerebral myopathy- Iceland	Cystatin C	Acys
Nonneuropathic herediatary amyloid with renal disease	Fibrinogen α	AFibA
Nonneuropathic herediatary amyloid with renal disease	Lysozyme	Alys
Familial British dementia	FBDP	A Bri
Familial Danish dementia	FDDP	A Dan
Diffuse Lewy body disease	Alpha-synuclein	AS
Parkinson's disease		
Fronto-temporal dementia	tau	tau
Amyotrophic lateral sclerosis	Superoxide Dismutase-1	SoD-1
Triplet-Repeat Diseases (Huntington's, Spinocerebellar ataxias, Spinal & bulbar muscular atrophy, Spinocerebellar ataxia17 etc.)	Polyglutamine tracts in: Hantingtin, Androgen receptor, Ataxins, TATA box-binding protein	PG

studies identified Abeta oligomers (aggregates of 2–100 Abeta molecules) as the most likely candidate for causing tissue damage and destruction. Which oligomers are most toxic (dimers, tetramer, 100 mers) has been a source of controversy, but evidence has accumulated that suggest that monomers and fibrils cause little, if any tissue damage. Indeed, some studies show that amyloid deposits which are sequestered in inclusion borders may reflect a defensive tactic on the part of the host to a wall off toxic material. (Arrasate et al. 2004).

Although deposits of Abeta and other amyloid peptides have been repeatedly demonstrated to be toxic to cells, the mechanism of cellular dysfunction and death remains elusive. Two decades ago, Arispe et al. (1994) proposed the channel hypothesis of amyloid disease. They demonstrated the ability of Abeta to form large, relatively non-selective voltage-independent ion-permeable channels in planar phospholipid bilayer membranes. They proposed that channel formation in cell membranes could cause cell depolarization, calcium dysregulation, depletion of cellular energy stores and other cellular dysfunction. Although this hypothesis has been hard to prove conclusively, it has stood the test of time, outlasting most of the other theories of amyloid peptide damage. In this review, we will summarize the case for and against the channel hypothesis after 20 years of experiments. Although most of the experiments have focused on the Abeta peptide from AD, there is ample evidence to imply that other amyloid peptides behave in a similar fashion in their respective disease processes.

14.2 The Beta-Sheet Conformation

Although amyloid is defined by its ability to bind certain dyes such as Congo red and exhibits characteristic staining properties, it is now clear that these are properties of beta-sheet peptides. Although amyloid peptides and proteins exhibit no primary sequence homology, they are all able to readily adopt beta-sheet conformation in at least some part of the peptide. Beta-sheet peptides tend to aggregate in aqueous solution and this aggregation can be markedly enhanced by the presence of membranes (Kagan and Thundimadathil 2010). Membranes enhance the tendency of peptides to form beta-sheet structures, although in amyloid disease, there is usually another trigger to beta-sheet formation. This trigger can be a mutation, pH change, high concentration, protolytic cleavage, metal ion presence or other physical chemical disruption that destabilizes native protein structure. Beta-sheet peptides have a naturally high affinity for membranes and can spontaneously insert into them, sometimes after aggregating. This transition creates new hydrogen binding possibilities which can help drive the process energetically (Fernandez and Berry 2003). Peptide aggregation within membranes is facilitated by the "hydrophobic effect" which reflects the tendency of hydrophobic protein elements to avoid exposure to water (Tanford 1980). As native protein structures destabilize and unfold into beta-sheet, these regions tend to aggregate and insert into membranes. Proteins with hydrogen bonding defects are especially likely to bind to lipid membranes (Fernandez and Berry 2003). Thus the tendency of amyloid peptides to aggregate, seek out and insert into membranes is no mere coincidence, but a direct result of the changes in physical properties which accompany the transition to beta-sheet conformation.

14.3 Toxicity of Ion Channels

There are ample precedents for cytotoxicity caused by ion channel formation in the world of bacteria (Schein et al. 1978). Prokaryotes use their inner cell membrane for active transport, respiration, and energy generation. For this reason, the inner membrane is relatively impermeable to common physiologic ions such as Na^+, K^+, Ca^{+2}.

Breaches of this membrane can cause major deficits in cellular function and lead rather quickly to cell death. Because bacterial cells are relatively small volumes compared to eukarotic cells, they are also vulnerable to rapid depletion of vital intracellular constituents such as K^+ and Mg^{+2} when the inner cell membrane is punctured. Channel formation by colicins, defensins, and protegins has been clearly recognized as the mechanism of toxicity of these proteins and peptides (Jang et al. 2010; Kayed et al. 2003; Kagan et al. 1990; Sokolov et al. 1999).

While eukaryotic cells are less vulnerable to channel forming toxins due to their larger volume and stiffer, sterol-containing membranes, they are not invulnerable. Yeast killer toxins are able to kill their targets by the formation of very large, highly stable pores in the plasma membrane of yeast (Kagan 1983). Some channel forming drugs, such as nystatin and amphotericin B actually bind to sterols and use this property to insert into fungal membranes to kill cells (Ng et al. 2003). This propensity is likely the cause of their relatively high toxicity for human cells which also contain sterols.

Bacteria which prey on eukaryotic hosts have developed a group of cholesterol binding cytolysins. Perfringolysin of Clostridia is a typical member of this group. Channels formed are gigantic by bacterial standards and channel formation requires the transition of alpha-helical regions to beta-sheet. Two beta-sheet hairpins from each toxin molecule contribute to forming a large beta-barrel structure in the host membrane. Aggregates can contain 50 monomers and have a diameter of as much as 300A. Alpha toxins from *Staphylococcus* form a similar toxic channel with a well characterized structure containing 14 beta-sheet strands from 7 alpha toxin molecules (Song et al. 1996). Beta barrel channel structures have also been observed for the outer membrane channels of bacteria known as porins (Cowan et al. 1992). Thus the beta barrel is a well-established channel forming structure.

Although eukaryotic cells enjoy some protection from bacterial toxins due to their volume and membrane stiffness, certain cells are more vulnerable. Neurons, for example, require the maintenance of a stable membrane potential for signal transmission to occur reliably. This necessitates the maintenance of ionic gradents of Na^+, K^+ and Ca^{+2}. These gradients are maintained by the energetically costly actions of the sodium, calcium and other pumps. Insertion of an Abeta size nonspecific cation leak in the plasma membrane of a neuron can rapidly alter cellular Na^+ concentration (Arispe et al. 1993a). This would lead to energetic stress at least, and to failure of signaling, Ca^{+2} influx, and cell death at worst.

14.4 Abeta Channels

The discovery of Abeta channels led to a wide range of experiments on amyloid peptides. More than a dozen channel forming amyloid peptides have been reported (Table 14.2) and they exhibit remarkably similar properties.

The properties of Abeta described above have mostly been known for some time. However, investigation of other amyloid peptides has revealed that almost all of

Table 14.2 Channel properties of amyloid peptides from neurodegenerative diseases

Peptide	Single channel conductance	Ion selectivity (permeability ratio)	Blockade by zinc	Blockade by copper	Inhibition by congo red
Aβ25-35	10–400 pS	Cation ($P_K/P_{Cl}=1.6$)	+	+	+
Aβ1-40	10–2000 pS	Cation ($P_K/P_{Cl}=1.8$)	+		N.D.
Aβ1-40	50–4000 pS	Cation ($P_K/P_{Cl}=11.1$)	+		N.D.
Aβ1-42	10–2000 pS	Cation ($P_K/P_{Cl}=1.8$)	+		+
Aβ1-40	Variable	Cation	+		
Aβ1-40/42	Variable	Cation	+		
Aβ17-42 (p3)	Variable	Cation	+		+
CT105 (C-terminal fragment of Amyloid Precursor Protein (APP)	120 pS	Cation	+		+
Islet amyloid polypeptide (Amylin)	7.5 pS	Cation ($P_K/P_{Cl}=1.9$)	+		+
PrP106-126	10–400 pS	Cation ($P_K/P_{Cl}=2.5$)	+		+
PrP106-126	140,900, 1444 pS	Cation (PK/PCl >10)	N.D.		N.D.
PrP 82-146		Cation (variable)	N.D.		N.D.
PrP106-126 (deamdated)				+	
Serum amyloid A	10–1000 pS	Cation ($P_K/P_{Cl}=2.9$)	+		+
C-type natriuretic peptide	21, 63 pS	Cation ($P_K/P_{Cl>10}$)	+		+
b2-Microglobulin	0.5–120 pS	Non-selective	+		+
K3 fragment of b2M	Variable	Cation ($P_K/P_{Cl\sim5}$)	+		
Transthyretin	Variable	Cation (variable)	+		+
Polyglutamine(AVG MV=6000)	19–220 pS	Non-selective	–		–
Polyglutamine	17 pS	Cation	N.D.		N.D.
NAC (Alpha-Synuclein 65–95)	10–300 pS	Variable	+		+
Alpha synuclein	50–1300 pS	Cation	N.D.		N.D.
Alpha synuclein	50–250 pS	Cation	N.D.		N.D.
Abri	Variable	N.D.	+		+
Adan	Variable	N.D.	+		+

these properties are conserved amongst amyloid channel formers (Kagan et al. 2004). These properties, which distinguish amyloid channels from the classic ion channels of nerve and muscle, explain the toxicity of amyloid peptides. Amyloid channels constitute an unregulated ionic leakage pathway in cellular membranes, leading to cell depolarization and unregulated fluxes of ions such as Na^+, K^+ and Ca^{+2}. This disrupts cellular processes and runs down vital energy stores as cellular pumps try to keep up with the leakage.

Atomic force microscopy (AFM) has revealed Abeta pore structures in membranes (Jang et al. 2010). Pores with an outer diameter of 40 Å and inner diameter of 20 Å have been observed and appear to be composed of several subunits arranged in a ring. These structures are consistent with the ionic conductances seen in electrophysiological studies. Molecular modeling has suggested that highly mobile subunits of Abeta oligomers may aggregate to form large channel structures (Jang et al. 2009). This molecular fluidity is consistent with the multiple conductance states observed in the electrophysiolgic experiments. The physical pore sizes predicted by molecular modeling are also consistent with the AFM and electrophysiologic data.

14.4.1 Properties of Abeta Channels

1. Channels are large – Conductances up to 5 nS have been reported for Abeta channels under physiologic conditions (Arispe et al. 1993b). This represents a 1000-fold greater conductance than the native channels of nerve and muscle cells.
2. Channels have long lifetimes (open times). The Abeta channels have lifetimes of minutes to hours, vastly longer than the millisecond lifetimes of native channels.
3. Channels are irreversibly associated with the membrane. In the time course of usual experiments, channels rarely leave the membrane. Even in *in vivo* experiments with Abeta in cells, channels persist for very long times in the membrane (Fraser et al. 1997).
4. Channels exhibit heterodisprese conductances (Arispe et al. 1993b; Fraser et al. 1997; Hirakura et al. 1999; Anekonda et al. 2010). Initially reported by Arispe et al. (1993b), this property of amyloid channels strongly suggests that multiple species of oligomers representing aggregates of varying molecular size can form ion channels. Hirakura et al. (1999) demonstrated that exposure to solvents which changed the aggregation state of the Abeta peptide could alter the distribution of single channel conductances. Inhibitors of aggregation such as Congo red have also been shown to inhibit channel formation, thus strongly confirming the link between peptide aggregation and channel formation (Hirakura et al. 1999).
5. Channels form in many different kinds of membranes. Reports have confirmed the ability of Abeta to form ion channels in planar lipid bilayers, liposomes, neurons, oocytes and fibroblasts (Fraser et al. 1996). Evidence also exists for

Abeta channel formation in mitochondrial membranes (Camilleri et al. 2013), and mitochondria are known to be depolarized in AD, Parkinson's Disease (PD) and Huntington's disease (HD) (Panov et al. 1999; Orr et al. 2008). The fact that the Abeta channels exhibit similar properties in these different membrane environments suggests that similar structures are forming, and that these are likely proteinacious pore structures rather than toroidal lipidic pores.

6. Channels are weakly cation selective: Although Abeta channels exhibit some preference for cations over anions, there is evidence that Na^+, K^+, and Ca^{2+} all permeate the channels. This lack of selectivity is very different from the high degree of ion selectivity exhibited *in vivo* by K^+, Na^+ or Ca^{2+} permeable channels. Some reports (Hirakura et al. 1999) suggest that anions such as Cl^- may also have some permeability to the Abeta channels,
7. Channels are voltage- independent (mostly). While the classical channels of nerve and muscle are often sharply dependent on membrane voltage for opening and closing, Abeta channels seem to insert in an open state and to remain open regardless of the membrane voltage. Certain peptide fragments of Abeta such as Abeta 25–35 exhibit a degree of voltage dependent opening and closing, but these species are not found *in vivo* (Mirzabekov et al. 1994).
8. Channels are blocked by Zn^{+2}, NA7 (a specifically designed pore binding peptide) (Diaz et al. 2006) and amines (Diaz et al. 2009).

   Zn^{2+} has proved to be a general blocker of nearly all amyloid channels. The block is reversible and rescues and protects cells from Abeta toxicity (Zhu et al. 2000). This result strongly implicates channel formation in the toxicity of amyloid peptides. Zn^{+2} blockade also prevents Abeta mediated transport of Ca^{2+} into liposomes and cells, implicating Ca^{+2} in the toxicity induced by Abeta (Lin et al. 1999; Rhee et al. 1998). NA7 can also rescue cells from Abeta cytoxicity (Diaz et al. 2006).

   Other studies have implicated Ca^{+2} dysregulation in the pathogenesis of AD and other amyloid diseases (Garwood et al. 2013). Abeta may disrupt Ca^{+2} regulation by at least three mechanisms. First, Abeta may directly allow Ca^{+2} to enter cells from the extracellular space by passing through the Abeta pore (Lin et al. 1999). Second, depolarization of cells by insertion of non-specific Abeta pores may lead to opening of voltage-dependent Ca^{+2} channels and Ca^{+2} influx (Anekonda et al. 2010), Third, mitochondrial depolarization mediated by Abeta pores may lead to Ca^{+2} efflux from mitochondria into the cytosol (Kim et al. 2002). In a transgenic mouse model of AD, diazoxide, a K^+ channel opener which would counter the depolarization effects of Abeta, has been shown to improve memory as well as decreasing tau and Abeta levels (Liu et al. 2010). Voltage-dependent Ca^{+2} channel blockers have been shown to protect neurons in culture from Abeta toxicity (Anekonda et al. 2010). Taken together, these results imply that Abeta pore formation is a critical mechanism of Abeta toxicity.
9. Channels can be inhibited by Congo red (Hirakura et al. 1999; Mirzabekov et al. 1994). The binding of Congo red to Abeta can prevent aggregation of Abeta into oligomers, a necessary step for channel formation. This is further

confirmation of the necessary role played by the beta-sheet conformation of Abeta in aggregation and toxicity. Congo red binds tightly to beta-sheet structures, making aggregation impossible (Kagan and Thundimadathil 2010).

10. Abeta can have direct effects on mitochondria including decrease of membrane potential, opening of the permeability transition pore, and release of cytochrome c (Kim et al. 2002; Parks et al. 2001). Mitochondria from AD patients exhibit deceased membrane potential and membrane bound Abeta (Panov et al. 1999). Abeta can also cause Ca^{+2} dependent depolarization of mitochondria leading to an overall collapse of membrane potential (Abramov et al. 2004).

Although plasma membranes are a vulnerable target, increasing evidence points to mitochondrial membranes as a locus of action of amyloid channels. Mitochondria evolved from bacteria and possess a similar dual membrane structure. The inner membrane of mitochondria must maintain relative impermeability to accomplish the coupling of respiration to oxidative phosphorylation via a proton gradient. Non-specific leakage channels would uncouple these events, depolarize mitochondria and potentially lead to escape of Ca^{+2}, cytochrome c, and programmed cell death. Curiously, a number of cell death related proteins such as bAX and Bcl-2 are channel formers with some homology to the channel forming bacterial toxins, colicins (Lazebnik 2001).

14.5 Prion Channels

Prion diseases are infectious, genetic, or sporadic neurodegenerative illnesses caused by a proteinaceous agent, the scrapie prion protein or PrPsc. Mutations in the normal cellular version of the prion protein PrPc can cause illness but are not required. The infectious prion protein conformation (PrPsc) can induce PrPc to change its conformation. Thus, prion diseases can spread via a transmitted conformational change from PrPsc to PrPc (Cobb and Surewicz 2009). In this transition, alpha-helical protein regions convert to beta-sheet (Pan et al. 1993). Prion diseases are also characterized by amyloid deposition in the brain, with fibrils of PrPsc composing the bulk of the deposits. Although the toxic mechanism of prion induced neurodegeneration remains uncertain, the channel hypothesis of prion diseases (Lin et al. 1997) remains a viable explanation. The prior protein fragment PrP 106-126 was shown to be both neurotoxic (Yankner et al. 1989; Forloni et al. 1993) and channel forming (Lin et al. 1997) at comparable concentrations. PrP 106-126 aggregation was necessary for channel formation, and PrP 106-126 channels possessed all of the properties of Abeta channels. Additionally, it was reported that exposure to acidic pH, which enhanced PrP 106-126 aggregation and neurotoxicity, also enhanced channel activity and altered the distribution of single channel conductances to larger conductances. These results were confirmed and extended by Kourie and Culverson (2000) and channel conductances from 40 to 1500 pS were reported. Quinacine was reported to be a PrP 106-126 channel blocker (Farrelly et al. 2003).

A larger fragment, PrP 82-146, was later identified as a channel former with very similar properties to PrP106-126 (Bahadi et al. 2003). This fragment is identical to the protein found in the brains of patients with Gerstrman-Straussler-Schenker disease. It was also demonstrated that the 106–126 region was critical for channel formation. Recently the full length prion protein has been reported to form channels with similar properties (Solomon et al. 2012).

There is other evidence to implicate prion channels in pathogenesis. In one prion disease in which amyloid deposits are not seen, the mutant prion protein is found embedded in the cell membrane, perhaps as a leakage pathway (Hegde et al. 1998).

Expression of PrP deletion mutants induces large ionic current fluctuations (Solomon et al. 2010). The most toxic deletion is 105–125, the region responsible for pore formation. Expression of wild type PrPc can inhibit these current fluctuations, as can a sulfated glycosaminoglycan, normally a component of amyloid deposits. Thus deletion or mutation of the protective peptide form allows the disease form PrPc to make toxic channels. Expression of prion proteins with point mutations corresponding to familial prion disease led to the observance of abnormal ionic current fluctuations in cells. It was proposed that PrPc has a channel inhibition function which is disrupted by mutations.

Another portion of the prion protein, PrP 170-175, has been reported to form channels when mutated (N171S). This mutation results in a form of schizoaffective disorder (Berest et al. 2003), although not an amyloid disease.

14.6 Alpha-Synuclein Channels

alpha-Synuclein (alphaS) is a synaptic vesicle associated protein that forms amyloid deposits in the Lewy bodies of neurons in patients with Parkinson's disease (PD). Although alphaS is 140 residues long, a fragment known as NAC (non-amyloid component) comprising residues 66–95 is typically found in the amyloid deposits. Mutations in alphaS are responsible for some forms of familial PD (Bekris et al. 2010). Both the alphaS and NAC can form channels in membranes (Volles and Lansbury 2002) and annular pore structures as viewed by electron microscopy (Kourie et al. 2003). alphaS is an intrinsically disordered protein, although monomeric alphaS can adopt an alpha-helical conformation in membranes to form voltage-dependent pores (Zakharov et al. 2007). NAC channels closely resemble other amyloid channels in their membrane properties (Azimova and Kagan 2003). The NAC peptide adopts beta-sheet conformation similar to other amyloid pore forms. NAC most likely represents the ultimate processed form of alpha S that is responsible for channel formation and cellular toxicity. Oligomer formation by alphaS can be induced by organic solvents and iron. These oligomers can also form pores in membranes recognized by the A11 anti-body developed against oligomeric forms of amyloid (Kostka et al. 2008; Kim et al. 2009). In dopamine neurons, alphaS mutants can cause increased membrane permeability, elevated levels of Ca^{+2} and oxidative stress. Ca^{+2} chelators can protect neurons from this toxicity, again suggesting a key role for Ca^{+2} disregulation in amyloid pathophysiology. alphaS channels are Ca^{+2} permeable, and likely cause a direct leak of Ca^{+2} into neurons.

14.7 Models of Amyloid Pores

Early molecular models of amyloid pores were based on secondary structure prediction rules. Durell et al. (1994) modelled Abeta channels as a beta-sheet followed by alpha helix, beta turn and alpha helix. The pore was lined by beta hairpins and Arispe and Doh (2002) used this prediction to synthesize a pore blocking peptide NA7. The remarkable success of this peptide as a blocker suggests that *beta* structure is a key component of the pore model. Other classes of model for the pore proposed by Arispe et al. have not been supported by experimental evidence. The pore blocking peptide also protects neurons from Abeta toxicity, again supporting the idea that pore formation is a key part of pathogenesis.

Molecular dynamics simulations (Jang et al. 2008) have led to a new class of models for the pore. These models incorporate more recent experimental data suggesting that the Abeta peptide adapts a u-shaped beta-strand-beta-turn-beta strand structure in the membrane (Fig. 14.1). These models contain more peptides which is consistent with other data suggesting that larger oligomers of amyloid peptides are the toxic species. Jang et al. described an annular structure of 24 peptides with

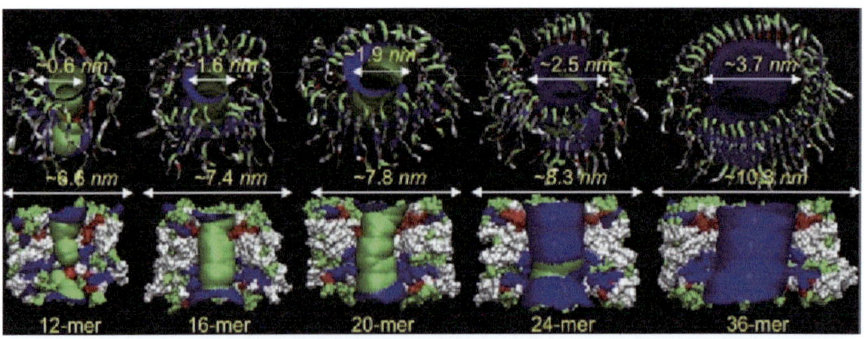

Fig. 14.1 Averaged pore structures calculated by the HOLE program embedded in the averaged channel conformations during the simulations for the 12-, 16-, 20-, 24-, and 36-mer $A\beta_{9-42}$ channels. The 12-, 16-, 20-, and 36-mer structures were obtained from the simulations in the zwitterionic DOPC bilayer. The 24-mer structure was obtained from the anionic bilayer containing POPC and POPG with a molar ratio of 4:1. In the angle views of the pore structure (*upper row*), whole channel structures are shown with the ribbon representation. In the lateral views of the pore structure (*lower row*), cross-sectioned channels are given in the surface representation. For the pore structures in the surface representation, the degree of the pore diameter is indicated by the color codes in the order of *red*<*green*<*blue*, but the scale of these colors is relative to each channel. In the channel structures, hydrophobic residues are shown in *white*, polar and Gly residues are shown in *green*, positively charged residues are shown in *blue*, and negatively charged residues are shown in *red* (Reproduced from Jang et al. (2009), with permission)

both beta strands traversing the lipid bilayer. The diameter of the inner pore ranged from 22 to 27 Å, consistent with data from atomic force microscopy measurements. This is also consistent with electrophysiological data suggesting that the Abeta pore is much larger electrically than the pores of nerve and muscle. While electrical conductance measurements cannot be translated into physical pore diameters in a linear fashion, the pore dimensions of the model proposed by Jang et al. is consistent with the typical conductances of pores of this physical size. The molecular dynamic simulations suggest that subinits of the pore are highly mobile in the membrane. This is compatible with electrophysiological data showing current fluctuations suggesting different size channels in the membrane.

14.8 Amyloid Channels and the Beta Sheet

The beta-sheet conformation plays a central role in amyloid peptide aggregation, membrane insertion and channel formation (Kagan and Thundimadathil 2010). Many pathways can lead mature protein structures to alter their conformation to beta-sheet. Once this occurs, the propensity of beta-sheets to aggregation, especially in the presence of lipid membranes, is quite strong. While the classical channels of nerve and muscle have been primarily formed from alpha helical structures, there are now a number of channels known to be formed by beta-sheet peptides including porins (Nikaido et al. 1983), staphyloccal alpha-hemolysin (Krasilnikov et al. 1988), anthrax toxin (Blaustein et al. 1989), and clostridium perfringolysin (Shepard et al. 2000). Bacterial pore forming toxins and amyloid channels both require aggregation as a prerequisite for channel formation. Channels are relatively large and non-selective (Fig. 14.2). Both kinds of channels cause leakage pathways that damage critical cellular functions (Caughey and Lansbury 2003). Both channels represent the conversion of soluble cellular proteins into transmembrane proteins. The pore diameters of bacterial toxins (15–35 A) are comparable to those hypothesized for the amyloid channels by molecular dynamics and observed by atomic force microscopy (Fig. 14.3).

While amyloid peptide channels are not "designed" to be cytotoxic, it is noteworthy that they exhibit so many similarities to the highly evolved lethal toxins of bacteria. One might consider amyloid peptide channels as a kind of "autotoxin" that damages cells from within (Fig. 14.4). The beta sheets (red) of the Abeta peptide are shown assembling in solution into oligomers, protofibrils, and fibrils, as well as into a channel that spans the cell membrane and permits ionic flux. The kinetic relationships amongst these various Abeta species is complex and poorly understood. Altering the equilibrium may be key to controlling Alzheimer's disease.

Model beta-sheet peptides (xSxG) have been shown to form voltage-dependent channels in planar lipid bilayers (Thundimadathil et al. 2006). The physiologic properties of these peptide channels resemble those of the bacterial outer membrane porins. The high conductance and weak ion selectivity of these channels are strikingly reminiscent of amyloid peptide channels, suggesting that these may be generic properties of beta-sheet channels.

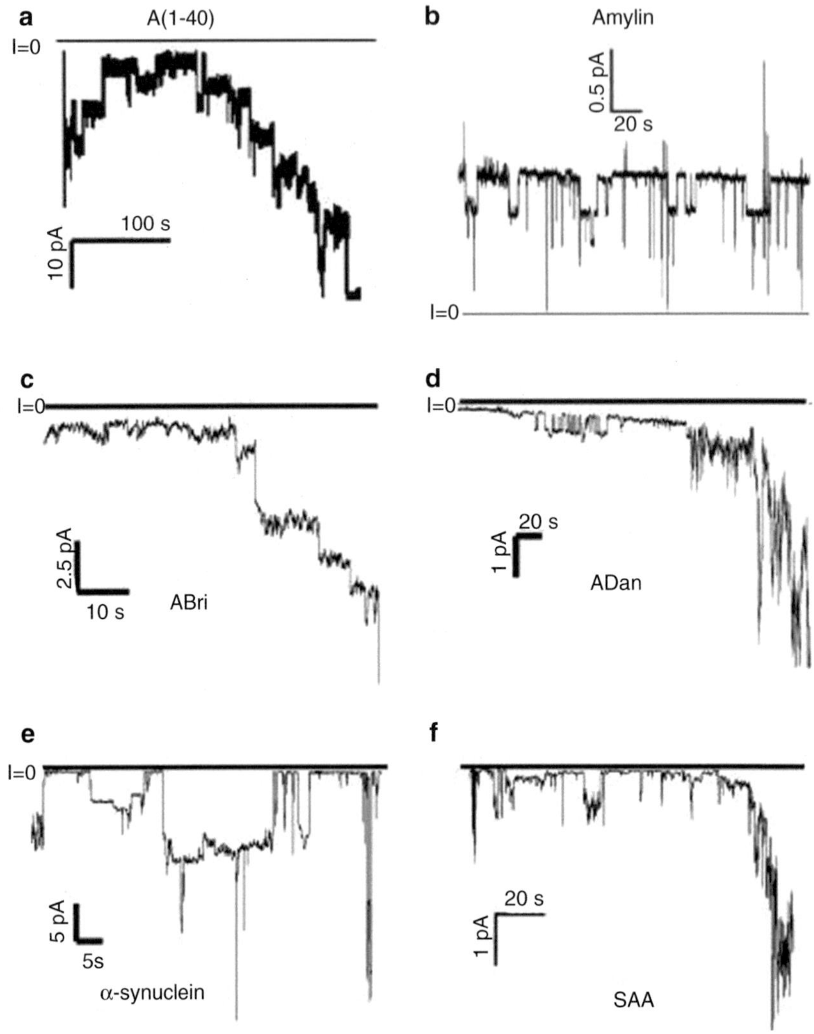

Fig. 14.2 Single-channel records of amyloid channels in planar lipid bilayers. Current traces as a function of time under voltage-clamp conditions are shown. Current jumps corresponding to the opening or closing of individual ion channels can be observed for all six amyloid peptides. Solutions contained 100 mM KCL (except *B*, which contained 10 mM KCl, and *F*, which contained 1 M KCl), buffered to pH 7.4. Peptide concentrations were as follows. (**a**) Ab(1–40): 21 ug/ml, *V*=−30 mV, (**b**) Amylin: 3 ug/ml, *V*=+50 mV, (**c**) ABri: 50 ug/ml, *V*=−50 mV, (**d**) ADan: 100 ug/ml, *V*=−50 mV, (**e**) NAC: 15 nM, *V*=−50 mV, (**f**) SAA: 1 ug/ml, *V*=−60 mV (Reproduced from Quist et al. (2005), with permission)

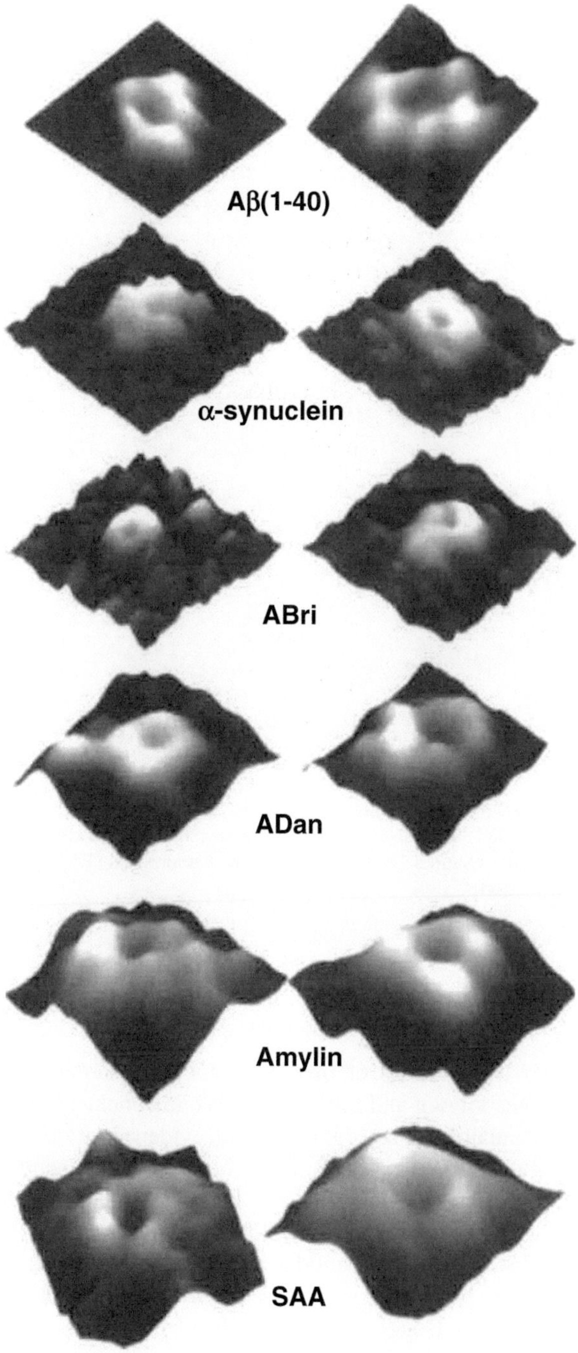

Fig. 14.3 Individual channel-like structures observed by atomic force microscopy. Two examples are shown for each kind of molecule, in which a central pore can be observed. The number of subunits observed protruding from the surface varies from four to eight subunits. Resolution of AFM images is not enough to resolve individual subunit structures. (Image sizes are 25 nm for Ab(1–40), 25 nm for a-synuclein, 35 nm for ABri, 20 nm for ADan, 25 nm for amylin, and 20 nm for SAA) (Reproduced from Quist et al. (2005), with permission)

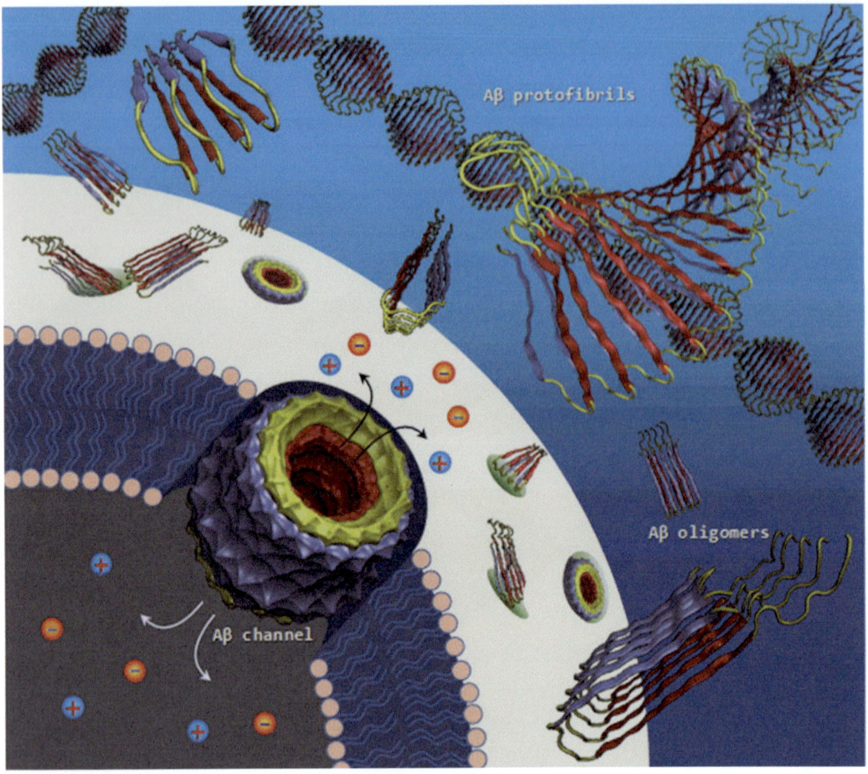

Fig. 14.4 Schematic diagram of cell toxicity by Abeta oligomers. Small fibril-like oligomers with the parallel beta-sheet structures and an exposed hydrophobic surface are believed to be toxic through a channel formation in the cell membranes (Reproduced from Jang et al. (2013), with permission)

14.9 Conclusions

After two decades of investigations, the channel hypothesis of amyloid disease remains vibrant and provocative. While conclusive proof of the relevance of channels remains elusive, circumstantial evidence for the role of channels in pathophysiology continues to accumulate. The non-specific nature of these channels continues to be an obstacle to finding specific blockers that might prove the hypothesis *in vivo*. We will summarize the evidence for the channel hypothesis below.

First, all amyloid peptides and proteins tested seem to form channels. The channels are large, heterodisperse, non-selective, voltage-independent, irreversible, inhibited by Congo red, and blocked by Zn^{+2}. Second, channels are toxic to target cells in vitro. Zn^{+2} and NA7 can rescue cells from the toxic channel effects. Third, Ca^{2+} dysregulation appears to be a general feature of amyloid pathophysiology. Channel formation easily explains alterations in Ca^{+2} levels. Fourth, mitochondria in AD, PD and Huntington's disease are depolarized and coated with amyloid peptides. In vitro, amyloid peptides can depolarize mitochondria by binding to VDAC and stimulating formation of the PTP (Permeability Transition Pore) leading to Ca^{+2} efflux, cytochrome c efflux and apoptosis (Camilleri et al. 2013). Abeta pore structures have also been identified in AD brain (Inoue 2008). Current fluctuations consistent with channel formation have been identified in a PD mouse model (Wu et al. 2010). Fifth, amyloid peptides seem to lack an enzymatic activity or specific cellular binding site. Sixth, amyloid aggregation into oligomers is necessary for both toxicity and for channel formation. Variable aggregation of amyloid into different size oligomers could account for the heterogeneous distribution of single channel conductances observed in electrophysiological experiments (Fig. 14.2). Seventh, the failure of several anti-amyloid drugs in human clinical trials may be explained by the channel hypothesis. As demonstrated by our experiments with Congo red, once inserted into the membrane, amyloid channels are resistant to the effects of *beta sheet* binding molecules (Hirakura et al. 2000; Mirzabekov et al. 1994; Lin et al. 1997).

New drug candidates may need to block amyloid channels or bind to them prior to membrane insertion Specific channel blockers could lead to the development of new treatments for amyloid diseases.

Acknowledgments We thank Ms. Doris Finck for editorial assistance.

References

Abramov AY, Canevari L, Duchen MR (2004) Beta-amyloid peptides induce mitochondrial dysfunction and oxidative stress in astrocytes and death of neurons through activation of NADPH oxidase. J Neurosci 24(2):565–575

Anekonda TS, Quinn JF, Harris C, Frahler K, Wadsworth TL, Woltier RL (2010) L-type voltage-Au10 gated calcium channel blockade with isradipine as a therapeutic strategy for Alzheimer's disease. Neurobiol Dis 41:62–70. 21

Arispe N, Doh M (2002) Plasma membrane cholesterol controls the cytotoxicity of Alzheimer's disease AbetaP(1–40) and(1–42) peptides. FASEB J 16:1526–1536

Arispe N, Rojas E, Pollard HB (1993a) Alzheimer disease amyloid beta protein forms calcium channels in bilayer membranes: blockade by tromethamine and aluminum. Proc Natl Acad Sci U S A 90(2):567–571

Arispe N, Pollard HB, Rojas E (1993b) Giant multilevel cation channels formed by Alzheimer disease amyloid beta-protein [A beta P-(1–40)] in bilayer membranes. Proc Natl Acad Sci U S A 90(22):10573–10577

Arispe N, Pollard HB, Rojas E (1994) beta-Amyloid Ca(2+)-channel hypothesis for neuronal death in Alzheimer disease. Mol Cell Biochem 140(2):119–125

Arrasate M, Mitra S, Schweitzer ES, Segal MR, Finkbeiner S (2004) Inclusion body formation reduces levels of mutant huntingtin and the risk of neuronal death. Nature 431(7010):805–810

Azimova RK, Kagan BL (2003) Ion channels formed by a fragment of alpha-synuclein (NAC) in lipid membranes. Biophys J 84(2):53a

Bahadi R, Farrelly PV, Kenna BL, Kourie JI, Tagliavini F, Forloni G et al (2003) Channels formed with a mutant prion protein PrP(82–146) homologous to a 7-kDa fragment in diseased brain of GSS patients. Am J Physiol Cell Physiol 285(4):C862–C872

Bekris LM, Mata IF, Zabetian CP (2010) The genetics of Parkinson disease. J Geriatr Psychiatry Neurol 23(4):228–242. Epub 2010 Oct 11. 23

Berest V, Rutkowski M, Rolka K, Łegowska A, Debska G, Stepkowski D et al (2003) The prion peptide forms ion channels in planar lipid bilayers. Cell Mol Biol Lett 8(2):353–362

Blaustein RO, Koehler TM, Collier RJ, Finkelstein A (1989) Anthrax toxin: channel-forming activity of protective antigen in planar phospholipid bilayers. Proc Natl Acad Sci U S A 86(7):2209–2213

Camilleri A, Zarb C, Caruana M, Ostermeier U, Ghio S, Högen T, Schmidt F, Giese A, Vassallo N (2013) Mitochondrial membrane permeabilisation by amyloid aggregates and protection by polyphenols. Biochim Biophys Acta 1828(11):2532–2543

Caughey B, Lansbury PT Jr (2003) Protofibrils, pores, fibrils, and neurodegeneration: separating the responsible protein aggregates from the innocent bystanders. Annu Rev Neurosci 26:267–298

Cobb NJ, Surewicz WK (2009) Prion diseases and their biochemical mechanisms. Biochemistry 48(12):2574–2585

Cowan SW, Schirmer T, Rummel G, Steiert M, Ghosh R, Pauptit RA et al (1992) Crystal structures explain functional properties of two E coli porins. Nature 358(6389):727–733

Díaz JC, Linnehan J, Pollard H, Arispe N (2006) Histidines 13 and 14 in the Abeta sequence are targets for inhibition of Alzheimer's disease Abeta ion channel and cytotoxicity. Biol Res 39(3):447–460

Diaz JC, Simakova O, Jacobson KA, Arispe N, Pollard HB (2009) Small molecule blockers of the Alzheimer Abeta calcium channel potently protect neurons from Abeta cytotoxicity. Proc Natl Acad Sci U S A 106(9):3348–3353

Durell SR, Guy HR, Arispe N, Rojas E, Pollard HB (1994) Theoretical models of the ion channel structure of amyloid beta-protein. Biophys J 67(6):2137–2145

Farrelly PV, Kenna BL, Laohachai KL, Bahadi R, Salmona M, Forloni G et al (2003) Quinacrine blocks PrP(106–126) formed channels. J Neurosci Res 74(6):934–941

Fernandez A, Berry RS (2003) Proteins with H-bond packing defects are highly interactive with lipid bilayers: implications for amyloidogenesis. Proc Natl Acad Sci U S A 100:2391–2396

Forloni G, Angeretti N, Chiesa R, Monzani E, Salmona M, Bugiani O et al (1993) Neurotoxicity of a prion protein fragment. Nature 362(6420):543–546

Fraser SP, Suh YH, Chong YH, Djamgoz MB (1996) Membrane currents induced in Xenopus oocytes by the C-terminal fragment of the amyloid protein (APP). J Neurochem 66(5):2034–2040

Fraser SP, Suh YH, Djamgoz MB (1997) Ionic effects of the Alzheimer's disease beta-amyloid precursor protein and its metabolic fragments. Trends J Neurosci 20:67–72

Garwood C, Faizullabhoy A, Wharton SB, Ince PG, Heath PR, Shaw PJ, Baxter L, Gelsthorpe C, Forster G, Matthews FE, Brayne C, Simpson JE, MRC Cognitive Function and Ageing Neuropathology Study Group (2013) Calcium dysregulation in relation to Alzheimer-type pathology in the ageing brain. Neuropathol Appl Neurobiol 39(7):788–799

Hardy JA, Higgins GA (1992) Alzheimer's disease: the amyloid cascade hypothesis. Science 256:184–185

Hegde RS, Mastrianni JA, Scott MR, DeFea KA, Tremblay P, Torchia M et al (1998) A transmembrane form of the prion protein in neurodegenerative disease. Science 279:827–834

Hirakura Y, Lin MC, Kagan BL (1999) Alzheimer amyloid abeta1-42 channels: effects of solvent, pH, and Congo red. J Neurosci Res 57:458–466

Hirakura Y, Azimov R, Azimova R, Kagan BL (2000) Polyglutamine-induced ion channels: a possible mechanism for the neurotoxicity of Huntington and other CAG repeat diseases. J Neurosci Res 60:490–494

Inoue S (2008) In situ Abeta pores in AD brain are cylindrical assembly of Abeta protofilaments. Amyloid 15(4):223–233

Jang H, Zheng J, Lal R, Nussinov R (2008) New structures help the modeling of toxic amyloidbeta ion channels. Trends Biochem Sci 33:91–100. 26

Jang H, Arce FT, Capone R, Ramachandran S, Lal R, Nussinov R (2009) Misfolded amyloid ion channels present mobile beta-sheet subunits in contrast to conventional ion channels. Biophys J 97(11):3029–3037

Jang H, Teran AF, Ramachandran S, Capone R, Lal R, Nussinov R (2010) Structural convergence among diverse, toxic beta-sheet ion channels. J Phys Chem B 114(29):9445–9451

Jang H, Connelly L, Arce FT, Ramachandran S, Lal R, Kagan BL (2013) R Nussinov Alzheimer's disease: which type of amyloid-preventing drug agents to employ? Phys Chem Chem Phys 15:8868–8877

Kagan BL (1983) Mode of action of yeast killer toxins: channel formation in lipid bilayer membranes. Nature 302(5910):709–711

Kagan BL, Thundimadathil J (2010) Amyloid peptide pores and the beta sheet conformation. Adv Exp Med Biol 677:150–167

Kagan BL, Selsted ME, Ganz T, Lehrer RI (1990) Antimicrobial defensin peptides form voltage dependent ion-permeable channels in planar lipid bilayer membranes. Proc Natl Acad Sci U S A 87:210–214

Kagan BL, Azimov R, Azimova R (2004) Amyloid peptide channels. J Membr Biol 202(1):1–10

Kayed R, Head E, Thompson JL, McIntire TM, Milton SC, Cotman CW et al (2003) Common structure of soluble amyloid oligomers implies common mechanism of pathogenesis. Science 300(5618):486–489

Kim HS, Lee JH, Lee JP, Kim EM, Chang KA, Park CH et al (2002) Amyloid beta peptide induces cytochrome C release from isolated mitochondria. Neuroreport 13:1989–1993

Kim HY, Cho MK, Kumar A, Maier E, Siebenhaar C, Becker S et al (2009) Structural properties of pore-forming oligomers of alpha-synuclein. J Am Chem Soc 131(47):17482–17489

Kostka M, Högen T, Danzer KM, Levin J, Habeck M, Wirth A et al (2008) Single particle characterization of iron-induced pore-forming alpha-synuclein oligomers. J Biol Chem 283(16):10992–11003

Kourie JI, Culverson A (2000) Prion peptide fragment PrP[106–126] forms distinct cation channel types. J Neurosci Res 62:120–133. 22

Kourie JI, Kenna BL, Tew D, Jobling MF, Curtain CC, Masters CL et al (2003) Copper modulation of ion channels of PrP[106–126] mutant prion peptide fragments. J Membr Biol 193(1):35–45

Krasilnikov OV, Sabirov RZ, Ternovsky VI, Merzliak PG, Tashmukhamedov BA (1988) The structure of Staphylococcus aureus alpha-toxin-induced ionic channel. Gen Physiol Biophys 7(5):467–473

Lazebnik Y (2001) Why do regulators of apoptosis look like bacterial toxins? Curr Biol 11(19): R767–R768

Lin MC, Mirzabekov T, Kagan BL (1997) Channel formation by a neurotoxic prion protein fragment. J Biol Chem 272:44–47

Lin H, Zhu YJ, Lal R (1999) Amyloid beta protein (1–40) forms calcium-permeable, Zn^{2+}-sensitive channel in reconstituted lipid vesicles. Biochemistry 38:11189–11196

Liu D, Pitta M, Lee JH, Ray B, Lahiri D, Furukawa K et al (2010) The KATP channel activator Au10 diazoxide ameliorates amyloid-b and tau pathologies and improves memory in the 3xTgAD mouse model of Alzheimer's disease. J Alzheimers Dis 22:443–457

Mirzabekov T, Lin MC, Yuan WL, Marshall PJ, Carman M, Tomaselli K et al (1994) Channel formation in planar lipid bilayers by a neurotoxic fragment of the beta-amyloid peptide. Biochem Biophys Res Commun 202:1142–1148

Ng AW, Wasan KM, Lopez-Berestein G (2003) Development of liposomal polyene antibiotics: an historical perspective. J Pharm Pharm Sci 6(1):67–83

Nikaido H, Rosenberg EY, Foulds J (1983) Porin channels in Escherichia coli: studies with β-lactams in intact cells. J Bacteriol 153:232–240

Orr AL, Li S, Wang CE, Li H, Wang J, Rong J, Xu X, Mastroberardino PG, Greenamyre JT, Li XJ (2008) N-terminal mutant huntingtin associates with mitochondria and impairs mitochondrial trafficking. J Neurosci 28(11):2783–2792

Pan KM, Baldwin M, Nguyen J, Gasset M, Serban A, Groth D et al (1993) Conversion of alpha-helices into beta-sheets features in the formation of the scrapie prion proteins. Proc Natl Acad Sci U S A 90:10962–10966

Panov A, Obertone T, Bennett-Desmelik J, Greenamyre JT (1999) Ca(2+)-dependent permeability transition and complex I activity in lymphoblast mitochondria from normal individuals and patients with Huntington's or Alzheimer's disease. Ann N Y Acad Sci 893:365–368

Parks JK, Smith TS, Trimmer PA, Bennett JP Jr, Parker WD Jr (2001) Neurotoxic Abeta peptides increase oxidative stress *in vivo* through NMDA-receptor and nitric-oxide-synthase mechanisms, and inhibit complex IV activity and induce a mitochondrial permeability transition in vitro. J Neurochem 76:1050–1056

Quist A, Doudevski I, Lin H, Azimova R, Ng D, Frangione B et al (2005) Amyloid ion channels: a common structural link for protein-misfolding disease. Proc Natl Acad Sci U S A 102(30): 10427–10432

Rhee SK, Quist AP, Lal R (1998) Amyloid beta protein-(1–42) forms calcium-permeable, Zn^{2+}sensitive channel. J Biol Chem 273:13379–13382

Schein SJ, Kagan BL, Finkelstein A (1978) Colicin K acts by forming voltage-dependent channels in phospholipid bilayer membranes. Nature 276:159–163

Shepard LA, Shatursky O, Johnson AE, Tweten RK (2000) The mechanism of pore assembly for a cholesterol-dependent cytolysin: formation of a large prepore complex precedes the insertion of the transmembrane beta-hairpins. Biochemistry 39(33):10284–10293

Sipe JD, Cohen AS (2000) Review: history of the amyloid fibril. J Struct Biol 130:88–98

Sokolov Y, Mirzabekov T, Martin DW, Lehrer RI, Kagan BL (1999) Membrane channel formation by antimicrobial protegrins. Biochim Biophys Acta 1420(1):23–29. 27

Solomon IH, Huettner JE, Harris DA (2010) Neurotoxic mutants of the prion protein induce spontaneous ionic currents in cultured cells. J Biol Chem 285(34):26719–26726

Solomon IH, Biasini E, Harris DA (2012) Ion channels induced by the prion protein: mediators of neurotoxicity. Prion 6(1):40–45

Song L, Hobaugh MR, Shustak C, Cheley S, Bayley H, Gouaux JE (1996) Structure of staphylococcal alpha-hemolysin, a heptameric transmembrane pore. Science 274:1859–1866

Tanford C (1980) The hydrophobic effect: formation of micelles and biological membranes. Wiley-Interscience, New York. ISBN 10: 0471048933 ISBN 13: 9780471048930

Thundimadathil J, Roeske RW, Guo L (2006) Effect of membrane mimicking environment on the conformation of a pore-forming (xSxG)6 peptide. Biopolymers 84:317–328

Volles MJ, Lansbury PT Jr (2002) Vesicle permeabilization by protofibrillar alpha-synuclein is sensitive to Parkinson's disease-linked mutations and occurs by a pore-like mechanism. Biochemistry 41(14):4595–4602

Wu N, Joshi PR, Cepeda C, Masliah E, Levine MS (2010) Alpha-synuclein overexpression in mice alters synaptic communication in the corticostriatal pathway. J Neurosci Res 88(8):1764–1776

Yankner BA, Dawes LR, Fisher S, Villa-Komaroff L, Oster-Granite ML, Neve RL (1989) Neurotoxicity of a fragment of the amyloid precursor associated with Alzheimer's disease. Science 245(4916):417–420

Zakharov SD, Hulleman JD, Dutseva EA, Antonenko YN, Rochet JC, Cramer WA (2007) Helical alpha-synuclein forms highly conductive ion channels. Biochemistry 46:14369–14379

Zhu YJ, Lin H, Lal R (2000) Fresh and nonfibrillar amyloid beta protein(1–40) induces rapid cellular degeneration in aged human fibroblasts: evidence for AbetaP-channel-mediated cellular toxicity. FASEB J 14(9):1244–1254

Chapter 15
From Phototaxis to Biomedical Applications: Investigating the Molecular Mechanism of Channelrhodopsins

Ryan Richards and Robert E. Dempski

Abstract Phototactic response in the green algae *Chlamydomonas reinhardtii* is mediated through two rhodopsin proteins, channelrhodopsin-1 (ChR1) and channelrhodopsin-2 (ChR2). Similar to other microbial-type rhodopsins, ChRs have a seven transmembrane motif with a retinal moiety bound to a highly conserved lysine residue that activates the protein upon illumination with blue light. While most bacterial and algal homologues function as an ion pump, ChR2 is an inwardly rectified non-selective cation channel. Although ChR2 shares structural similarity to the proton pump bacteriorhodopsin (bR), ChR2 and bR have distinct functionality. The molecular determinants which define the differing protein functionalities can be elucidated through structure/function experiments. Moreover, the unique properties of ChR2 have paved the way for the emerging field of optogenetics. Cell-specific expression of ChR2 *in vivo* can invoke action potentials upon light activation and trigger downstream behavioral responses. However, the wild type properties of ChR2 limit the utility of this protein in biomedical applications. Therefore, understanding the underlying structural changes and mechanism of ion translocation for ChRs will be crucial for targeted engineering of ChR mutants with favorable optogenetic properties.

Keywords Membrane protein • Ion channel • Ion pump • Optogenetics • Bioenergetics • Phototaxis

15.1 Introduction

The ability of unicellular organisms to optimize their light environment is a central component in the marine food chain. As part of the requirement for light optimization, all photosynthetic organisms have evolved a mechanism by which they can swim towards or away from light (phototaxis) (Foster and Smyth 1980). Phototaxis

R. Richards • R.E. Dempski (✉)
Department of Chemistry and Biochemistry, Worcester Polytechnic Institute,
100 Institute Road, Worcester, MA 01609, USA
e-mail: rdempski@wpi.edu

A.H. Delcour (ed.), *Electrophysiology of Unconventional Channels and Pores*,
Springer Series in Biophysics 18, DOI 10.1007/978-3-319-20149-8_15

is essential not only for viability of a single organism, but for entire food chains. For example, in the marine food web, producers use light energy to orient themselves in an optimal light environment to use the energy of photons to create complex substances, such as sugars and amino acids. These producers include diatoms (ex. plankton and algae) as well as dinoflagellates. Producers are then eaten by first order consumers such as cleaner shrimp, copepodis and pteropoids. Consumers are ingested by subsequently larger organisms until reaching the top of the food chain. Therefore, conversion of light energy to complex substances is a requirement for homeostasis of the marine food web.

The physiology of light detection has been studied for nearly 200 years. Studies on phototaxis originated in 1817 through the study of the algae *Draparnaldia* (Treviranus 1817). Research in this field has ebbed and flowed since then. However the first molecular identification of the proteins which initiate the process of phototaxis within the past 12 years has precipitated an explosion of research in this field, both for basic science and as a tool for biomedical applications (Foster and Smyth 1980; Lenci and Colombetti 1978; Nagel et al. 2002, 2003).

Chlamydomonas reinhardtii is a single cell green alga which has been used as a model organism to study evolutionary biology, produce biopharmaceuticals as well as investigate phototaxis (Colegrave 2002; Demurtas et al. 2013). *C. reinhardtii* can thrive in fresh water as well as soil. It contains an eyespot region which is a photoreceptive organelle and enables the algae to sense light direction and intensity. Equally, *C. reinhardtii* has two flagella. These flagella swim asymmetrically upon light exposure to change the direction of the alga. In the absence of light, *C. reinhardtii* can thrive when provided with complex substances. In comparison, *C. reinhardtii* is viable in media lacking organic carbon and chemical energy sources when illuminated. Therefore, *C. reinhardtii* is able to transform light energy into a driving force to make complex substances. This response system to light, which enables synthesis of complex compounds, likely includes a receptor which responds to the light stimulus, a transducer as well as a mechanism for controlling cell movement. Together, this complex behavior enables algae to swim towards (positive phototaxis) or away (negative phototaxis) from light to optimize their environment. In 2002 and 2003, it was shown that channelrhodopsin-1 (ChR1) and channelrhodopsin-2 (ChR2) are two receptors which mediate the first committed step in photosynthesis which respond to light stimulus (Nagel et al. 2002, 2003).

15.2 The Channelrhodopsin-2 Photocycle

Although both ChR1 and ChR2 are important for phototactic response in *C. reinhardtii*, this review focuses on ChR2 as this protein has been most extensively studied. ChR2 is activated through the photoisomerization of all-trans retinal at C13 with blue light (~470 nm). Similar to other microbial-rhodopsins, ChR2 is comprised of seven transmembrane spanning domains with a retinal moiety bound to a single lysine residue on helix 7 (K257 in ChR2). However, unlike other microbial rhodopsins, ChR2 is a light-activated cation channel, and not a pump (Fig. 15.1a).

ChR2 can conduct a large array of both monovalent and divalent cations, with the highest affinity for protons. Large cationic molecules such as methylated amines can also pass through the channel.

Conceptually, ChRs differ from retinylidene-containing ion pumps as ChRs facilitate passive downhill ion transport. In contrast, ion pumps are capable of transporting ions against the thermodynamic gradient. Recently, it has been suggested that this mechanistic difference between ion channels and ion pumps is due to the fact that channels have one or more gates which are open at the same time (Gadsby 2009). When the gate (gates) is (are) open, ions are conducted down the chemoelectric gradient. As pumps are performing uphill transport

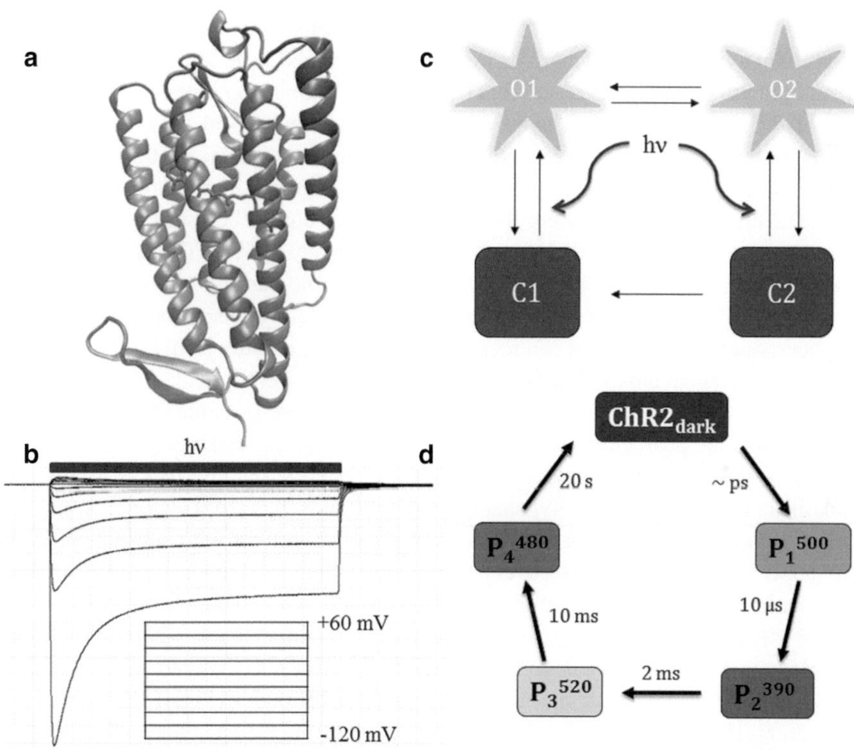

Fig. 15.1 The overall structure of ChR2 and continuous illumination/single turnover photocycles. (**a**) Homology model of ChR2 adapted from the C1C2 crystal structure. All images were created using Visual Molecular Dynamics Software (Humphrey et al. 1996). (**b**) Representative two-electrode voltage clamp recording of ChR2 current in oocytes under continuous illumination. Each trace is recorded at the corresponding membrane potential shown below ($\Delta V_m = 20$ mV). (**c**) Continuous illumination photocycle with two open (O1 and O2) and two closed states (C1 and C2). Blue light initiates the photocycle to the O1 state. O1 transitions to O2 under constant illumination until the light is turned off, where ChR2 molecules enter the desensitized C2 state. With sufficient recovery in the dark, C2 thermally relaxes back to the ground state (C1). (**d**) Single turnover photocycle. Kinetic transition values represent the $\tau_{1/2}$ values. The large conformational changes of the backbone and hydration of the helices occurs prior to and persists through the P_3^{520} conducting state

against the chemoelectric gradient, at least two gates are required to sequester ions from the thermodynamic gradient. These gates cannot be open at the same time as ions would flow down the thermodynamic gradient and opposite of the desired direction. Therefore, it is likely that retinal isomerization induces a distinct conformational change in ChRs when compared to retinylidene-containing ion pumps which enables cations to be conducted down the chemoelectric gradient. However, at this time, the molecular underpinnings which differentiate ion channels from ion pumps, especially as this relates to the opsin family of proteins, are not understood.

Photoisomerization of retinal initiates the photocycle reaction and ion conductance in ChR2. Under continuous illumination conditions, ChR2 photocurrent decays from an initial peak current to a steady state. Turning the light off causes the steady state photocurrent to decay bi-exponentially to the baseline (Fig. 15.1b). The rates of these transitions vary with voltage and light intensity (Nagel et al. 2003; Nikolic et al. 2009). Interestingly, ChR2 exhibits highly reduced peak currents with subsequent light pulses. This suggested that ChR2 can accumulate in a desensitized closed state after light illumination. The initial proposed photocycle used a three-state model to qualitatively describe ChR2 photocurrents (Nagel et al. 2003). In this model, the closed state (C) transitions into an open state (O) during illumination with light ($\tau_{CO}=0.2$ ms, $pH_i=7.3$). Turning the light off causes channel closure and transition from O to the dark desensitized state (D) ($\tau_{OD}=20$ ms, $pH_i=7.3$). The D state relaxes back to the C state after prolonged recovery time in the dark ($\tau=\sim 2$ s at $pH_i=7.3$) (Nagel et al. 2003). The possibility for a four state model was also proposed that used two open states with different conductances to explain the decay from peak to stationary current (Hegemann et al. 2005; Nikolic et al. 2009; Berndt et al. 2010). Channel inactivation, typically described as the ratio of I_{ss}/I_p, where I_{SS} is the steady state current and I_P is the peak current, and the bi-exponential decay for this process is most accurately described by a model containing two open states and two closed states (Nikolic et al. 2009). For clarity, channel inactivation differs from channel desensitization. Desensitization refers to the reduction in photocurrent with subsequent light activation without adequate recovery in the dark whereas channel inactivation is a measure in the loss of photocurrent during prolonged light exposure (Fig. 15.1b).

The current four-state photocycle model uses two open states (O1 and O2) and two closed states (C1 and C2) to quantitatively describe ChR2 photocurrents which previously could not be explained by the three-state model (Fig. 15.1c) (Nikolic et al. 2009; Hegemann et al. 2005; Berndt et al. 2010). Upon light activation, the C1 to O1 transition occurs rapidly. Immediately following light excitation, O1 transitions to the O2 state. Here it is important to note that the peak current is not an accurate representation of O1 as the peak current is composed of a mixture of O1 and O2 states (Berndt et al. 2010). Rather, the peak current describes the moment at which the decrease in O1 and increase in O2 is equal, resulting in no change in current with respect to time. O1 can be determined by extrapolating the peak current to $t=0$. When the light is turned off, I_{ss} decays bi-exponentially to baseline, representing the O2 $\rightarrow$ C2 transition. C2 thermally relaxes back to C1 after a recovery period in the dark. Without sufficient recovery between repetitive illumination cycles, the intensity of subsequent peak currents is reduced, while the stationary currents remain unchanged (Nagel et al. 2003). The reduced I_p is indicative of a higher population of the O2 state as C2 can be directly converted to the lesser conducting O2 state by blue light, resulting in a lower population of the high conducting O1 state (Fig. 15.1c). Analysis of computational studies have concluded that the experimental current requires a branched four-state photocycle (Fig. 15.1c) to quantitatively model ChR2 photocurrent (Stefanescu et al. 2013; Williams et al. 2013; Nikolic et al. 2009; Grossman et al. 2011).

Single turnover conditions offer a different view of a unidirectional ChR2 photocycle, where several states and intermediates have been elucidated by UV–vis and time-resolved infrared spectroscopic measurements (Fig. 15.1d) (Bamann et al. 2008; Ernst et al. 2008; Radu et al. 2009). Although it has been suggested that 6–8 exponentials are required to describe ChR2 kinetics, only four states have been spectroscopically identified under single turnover conditions (Bamann et al. 2008; Verhoefen et al. 2010; Lórenz-Fonfría and Heberle 2014; Lórenz-Fonfría et al. 2013; Ernst et al. 2008). The photocycle described here only includes these identified states. Upon excitation, ChR2 transitions to the P_1^{500} non-conducting state in the picosecond range and undergoes a large conformational change in the backbone structure (Ritter et al. 2008; Radu et al. 2009). The conformational changes are likely driven by rearrangement of the hydrogen-bonding networks that govern interhelical interactions (Lórenz-Fonfría et al. 2013). The prominent movement of the backbone structure precedes the conducting state of ChR2 and suggests that gating for cation translocation involves much smaller structural changes (Radu et al. 2009). Such fast and large conformational changes are an unusual property of ChR2 not seen in other microbial-type rhodopsins.

The photocycle of channelrhodopsin can be described as the following process. Deprotonation of the Schiff base by D253, the internal proton acceptor, leads to P_2^{390}, a non-conducting intermediate (Fig. 15.2) (Bamann et al. 2008). Transition to the P_3^{520} occurs after reprotonation of the Schiff base by the internal proton donor D156 (Fig. 15.2, green arrows). Furthermore, rise of P_3^{520} occurs simultaneously with the release of a proton to the extracellular side from an unidentified release group. Interestingly, the release of this proton is not tied to the deprotonation of the Schiff base as D253 remains protonated throughout the conducting state (Lórenz-Fonfría et al. 2013). The rise of the P_4^{480} non-conducting state is concomitant with channel closing kinetics. This transition involves the reprotonation of D156 and deprotonation of E90. E90 resides in the putative pore region and deprotanation of this residue suggests an important step for gating ChR2 (Radu et al. 2009; Eisenhauer et al. 2012). P_4^{480} represents the desensitized channel state which decays back to ground state after ~10 s (Bamann et al. 2008). The transition back to the ground state can be accelerated at low pH values, where it is coupled to proton uptake (Fig. 15.2). This correlates with reprotonation of E90 that occurs during the $P_4^{480} \rightarrow$ ground state transition (Radu et al. 2009). There is also evidence that the P_4^{480} state is actually part of a branching photocycle, where only ~25 % of ChR2 molecules enter from P_3^{520} (Lórenz-Fonfría et al. 2013). The remaining ChR2 molecules relax directly to the ground state. It is important to note that the large structural changes that occur in the initial excitation of the channel persist through the $P_3^{520} \rightarrow P_4^{480}$ transition. These structural changes do not reverse until the channel enters P_4^{480}, but the process is much slower than the onset (Radu et al. 2009).

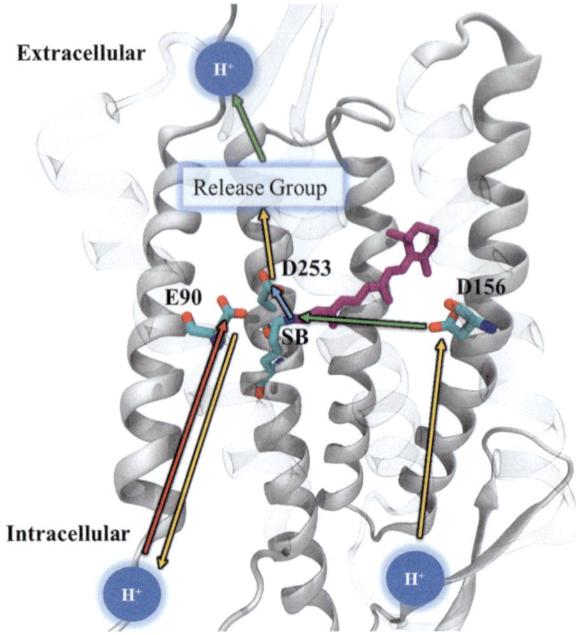

Fig. 15.2 Proton relay pathway in ChR2 adapted from Lorenz-Fonfria et al. *Blue arrow* – Deprotonation of the SB representing the $P_1^{500} \rightarrow P_2^{390}$ transition. *Green arrows* – Reprotonation of the Schiff base from D156 and release of proton from an unknown proton release group occurring during the $P_2^{390} \rightarrow P_3^{520}$ transition. *Orange arrows* – Deprotonation of E90 and reprotonation of D156 during the $P_3^{520} \rightarrow P_4^{480}$ transition. *Red Arrow* – Reprotonation of E90 during recovery from P_4^{480} to the ground state

Originally, it was thought that the internal proton acceptor from the Schiff base was E123. E123 in ChR2 is the homologous residue to the proton acceptor in bR (D85). (Bamann et al. 2008; Luecke et al. 1998; Berndt et al. 2009; Sineshchekov et al. 2013). However, the crystal structure of the C1C2 chimera (ChR1 – helices 1–5; ChR2 – helices 6–7) revealed that the distance between E123 and the Schiff base was 3.4 Å, but only 3.0 Å for D253. Therefore, it was proposed that D253, not E123, was the internal proton acceptor (Kato et al. 2012). This was later experimentally determined by monitoring protonation changes of specific residues via time-resolved FTIR spectroscopy in the P_2^{390} state (Lórenz-Fonfría et al. 2013). This also provided explanation as to why the E123T mutation (ChETA variant) results in little change for the rise and decay kinetics of P_2^{390} and P_3^{520} and a similar current-voltage relationship to WT ChR2 (Lórenz-Fonfría et al. 2013; Gunaydin et al. 2010). Moreover, the D253E mutation results in nearly a 50-fold increase in kinetics for the deprotonation of the Schiff base ($P_1^{500} \rightarrow P_2^{390}$ transition). This is analogous to the ~30-fold increase in M-state formation of the D85E mutant in bR (Heberle et al. 1993; Lórenz-Fonfría et al. 2013). An acidic residue at D253 in known ChRs is completely conserved, but not at E123, providing strong evidence of D253 as the proton acceptor.

Understanding the photocycle kinetics is important for elucidating the mechanism of ion transport and ion selectivity. Several kinetic models have been applied in tandem with electrophysiological measurements to describe the photocurrents and ionic composition (Gradmann et al. 2011; Berndt et al. 2010; Schneider et al. 2013). These models have an advantage over the Goldman-Hodgkin-Katz (GHK) equation because ChR2 is non-selective for various cations and these ions do not move independently through the channel. Additionally, ChR2 is a multi-valent channel which is not accounted for in the standard GHK equation (Hille 2001). A three state and four-state kinetic model was described by Nikolic et al. that were able to fit ChR2 photocurrent measurements under short (2 ms) and prolonged light exposure. Although this three state model reproduced photocurrents qualitatively, it could not accurately describe the light-off currents. Instead, the four-state model with an added open state term was able to accurately fit ChR2 photocurrents. Here, it is important to note that O1 and O2 have differing ion selectivity and that kinetic models need to account for these differences. Therefore, models were extended to reveal the selectivity between ions and selectivity between conducting states. Using differences in reversal potentials in the presence and absence of Na^+, it was found that the selectivity for H^+ over monovalent metals (Na^+ and/or K^+) was ~10^6 (Berndt et al. 2010). Furthermore, the preference for divalent metals (Ca^{2+} and/or Mg^{2+}) over monovalent metals was ~0.01. Lastly, during the transition to O2, the divalent metal selectivity decreases 2-fold. These results showed that ChR2 specificity increases temporally under prolonged light pulses. Based on these results, it was suggested that, with respect to *C. reinhardtii* physiology, this mechanism appears to be a way for the cell to save energy during prolonged light exposure (Berndt et al. 2010).

In comparison, Hegemann and colleagues utilized an enzyme model to elucidate the ionic composition of initial and steady-state photocurrents under conditions where there is a large inward driving force (Schneider et al. 2013). This model could accurately recreate the magnitude of photocurrents of wild-type ChR2, the ChR1 and *Volvox carteri* ChR1 chimera (C1V1), and the highly calcium permeable L132C mutant (CatCh) and also predict the ionic compositions of the currents. Another enzyme kinetic model was applied to fit the non-linear current-voltage relationship of ChR2 (Gradmann et al. 2011). Determination of the kinetic rate constants of the ChR2 photocycle reaction showed the parameters that had the largest effect on the non-linear I-V relationship. Moreover, it was found that the empty binding site had an apparent charge of about −0.3 and rectification of ChR2 is caused by non-linear transport and competition between cations permeating the channel. Elucidation of the underlying mechanisms and kinetics of the photocycle reaction are imperative for the implementation of ChR2 as an optogenetic tool.

15.3 The 'DC Gate'

It is well established that inter- and intramolecular non-covalent forces present in membrane proteins are critical for proper folding, insertion in the membrane, and transport functions (Popot and Engelman 1990; Dawson et al. 2002; Adamian et al. 2003; Luecke et al. 1998; Jayasinghe et al. 2001; Fleming et al. 1997). An integral force that has been of great relevance to ion channels is the hydrogen bonding network within transmembrane domains. The importance of hydrogen bonding in ChR2 has also been observed to be crucial for proper function (Bamann et al. 2010; Stehfest et al. 2010). In bR two residues, T90 and D115, form an interhelical hydrogen bond between TMs 3 and 4 which is critical for proton pumping and photocycle kinetics (Peralvarez-Marin et al. 2004). Sequencing alignments reveal that residues

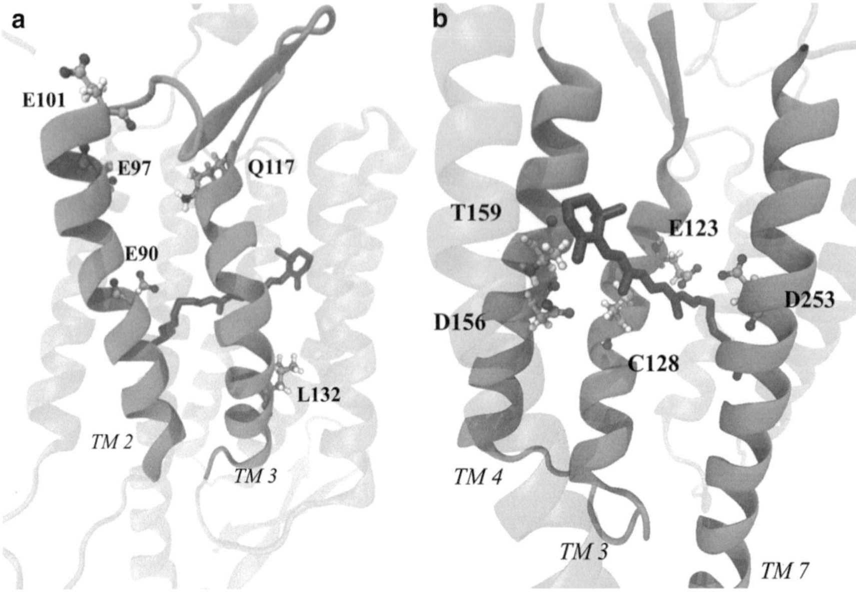

Fig. 15.3 (**a**) Homology model of ChR2 highlighting residues that affect cation selectivity. (**b**) ChR2 homology model highlighting residues that affect photocycle kinetics. TM 6 is not shown for clarity

at these two positions are highly conserved among archael rhodopsins and ChR2, where the homologous residues are C128 and D156 (Ihara et al. 1999). Therefore, it was proposed that D156 on TM 4 and C128 on TM 3 formed a putative hydrogen-bond that was closely tied to the photoreaction and the conducting state lifetime (Radu et al. 2009; Nack et al. 2010). The hydrogen bond between C128 and D156 was named the "DC Gate." To investigate the role of these residues in ChR2 function, mutations at C128 (C128X series) with threonine, serine or alanine caused a 100–10,000 times increase in the conducting state lifetime through retardation of closing kinetics after light is turned off (Fig. 15.3b) (Berndt et al. 2009). Mutations at C128 also affect the ground state spectrum due to the proximity to the retinal chromophore. A similar phenotype was observed for D156A (Fig. 15.3b) (Bamann et al. 2010). Analysis of FTIR experiments suggested that a hydrogen-bond was present between the carbonyl oxygen of D156 and the thiol group on C128. Furthermore, this hydrogen bond was present at different strengths in the P_1^{500} (stronger) and P_4^{480} (weaker) photointermediates (Nack et al. 2010). Replacing D156 with the larger glutamic acid removed the interaction with C128 in both states, while the C128T mutant retained the interaction in the P_1^{500} state. Additionally, the C=O vibrational stretch of D156 is protonated, sensitive to the presence of C128, and hydrogen-bonded to an undetermined residue. Therefore, it was hypothesized that the interaction between D156 and C128 was responsible for the small structural changes that gate ChR2 conductance after the large backbone movement induced by light excitation.

However, recent evidence has suggested that the distance between D156 and C128 is too far for a hydrogen bond to be formed and does not function similarly to the analogous gate in bR (Kato et al. 2012; Watanabe et al. 2012, 2013). The crystal structure of the C1C2 chimera calculated the distance between D156 and C128 to be 4.4 Å (Kato et al. 2012). Here, the distances in the crystal structure correspond to the ChR1 structure. Analysis of molecular dynamic simulations of the C1C2 chimera hinted at the instability of the conformation of the two residues (Watanabe et al. 2013). Interestingly, MD simulations performed by Watanabe and colleagues on a ChR2 model suggested that a water mediated hydrogen-bond forms between C128 and D156. Further simulations utilizing the C128T mutation showed the absence of water in the same region, and may explain the slow photocycle kinetics of this mutant. More recently, analysis of time-resolved FTIR experiments elucidated D156 as the internal proton donor for retinal (Lórenz-Fonfría et al. 2013). This explains why the conducting state lifetime is increased for the D156A as reprotonation of the Schiff base becomes the rate limiting step. Additionally, the recent D156C mutation (ChR2-XXL) has a long open state, significantly increased light sensitivity and expression, and the largest photocurrent of all known ChR variants thus far (38 ± 10 μA at −100 mV compared to 2 ± 0.5 uA for WT) (Dawydow et al. 2014). The functional role of C128 is unclear; however it is proposed that it lowers the pK_a of D156 for reprotonation of the Schiff base. A similar result has also been observed for the T90A mutation in bR (Peralvarez-Marin et al. 2004).

15.4 Channel Selectivity, Conductance, and Function

One of the most interesting features of ChR2 is that, despite the sequence homology to other microbial rhodopsins, it functions as an ion channel *in vivo* and not an ion pump. The ClC family of channels and transporters are also known to include transporters and channels with structural homology (Cohen and Schulten 2004). Substitution of a single glutamate residue can transform the Cl^-/H^+ exchanger into a channel for the passive transport of Cl^- (Accardi and Miller 2004; Feng et al. 2012; Lisal and Maduke 2008). Although transporters and channels function very differently, structurally they are similar. To this end, microbial rhodopsins and particularly ChR2 can be used to tease out the molecular determinants that govern channel/transporter behavior. Even more unique to ChR2 is that in the absence of an electrochemical gradient, it can act as an outwardly driven proton pump (Feldbauer et al. 2009). ChR2 has been shown to function as an inefficient proton pump; pumping only 0.3 H^+/ChR2 molecule (Nack et al. 2012). Additionally, proton release for ChR2 occurs with the reprotonation of the Schiff base in contrast to bR, where proton release is coupled with deprotonation of the Schiff base (Fig. 15.2) (Nack et al. 2012; Grzesiek and Dencher 1986).

The ion conductance properties of ChR2 have been most often measured using electrophysiology. ChR2 can be expressed heterologously into light-insensitive cells, usually *Xenopus laevis* oocytes or in mammalian cells such as human embryonic kidney (HEK) cells. This allows for the precise control of the extracellular buffer conditions and, under patch clamp conditions, control of the intracellular buffer. It also enables a facile method to examine changes in ion conductance upon mutagenesis. For example, replacement of L132 with cysteine (CatCh mutant) resulted in a construct which exhibited a change in the peak to stationary current ratio as well as a change in the on and off kinetics (Fig. 15.4a, b) (Kleinlogel et al. 2011). Interestingly, photocurrents in the presence of 80 mM calcium also resulted in a substantially higher photocurrent when compared to WT ChR2 (Fig. 15.4c). Consistent with this result, it was observed that there was an increase in calcium permeability (as well as protons), but no change in the permeabilities for sodium or potassium (Fig. 15.4d). The relative Ca^{2+} permeability (P_{Ca}/P_{Na}) of the L132C mutant when compared to WT ChR2 was increased from 0.15 to 0.24. Photocurrents are measured under specific ionic conditions at different holding potentials which can be used to generate current-voltage curves and elucidate the reversal potential of the channel (Fig. 15.4e). Reversal potentials can be used to determine the relative permeability of ions flowing through the channel. Here, changing the extracellular buffer from 140 mM Na^+ to 90 mM Ca^{2+} causes a negative shift in the reversal potential of WT ChR2 and CatCh. The less negative shift for CatCh compared to the WT indicates a higher Ca^{2+} permeability. The increase in calcium permeability of the L132C mutant when compared to WT ChR2 was validated when Fura-2, a fluorescent Ca^{2+} indicator, was used to quantify calcium update into HEK293 cells (Fig. 15.4f).

ChR2 can conduct a wide array of cations with differing selectivity (Nagel et al. 2003). ChR2 has the highest affinity for protons, which is about 10^6 fold higher than that of the alkali metals. Likewise, alkali metals are ~10 to 100-fold more selective over alkaline metals. The selectivity among periodic groups decreases with increasing atomic radius, consistent with the hydration shell being stripped during translocation (Nagel et al. 2003). Even large cationic compounds can permeate the channel, such as methyl- and dimethyl-amine. This is suggestive of a pore diameter larger than voltage-gated sodium channels, but smaller than the nicotinic acetylcholine receptor (Nagel et al. 2003; Richards and Dempski 2012). The dependence of reversal potentials on alkali metals was used to determine the minimum pore diameter of ChR2 to be ~6.2 A, within reason with previous estimates (Richards and Dempski 2012).

The first structural information on ChR2 came from projection map images using cryo-electron microscopy (Mueller et al. 2011). It was not until the crystal structure of the ChR1-ChR2 chimera (C1C2; ChR1 TM 1–5, ChR2 TM 6–7) was solved that the putative pore of ChR2 was found to form between transmembrane helices 1–3 and 7 (Kato et al. 2012). An open vestibule in this region on the extracellular side is lined with polar residues (Fig. 15.5). This water accessible pathway is occluded on the cytoplasmic side by two constriction sites. The first site is composed of S63, E90, and N258 which are highly conserved across ChRs. Mutations made at E90 (A or Q) and N258 (D) alter cation selectivity while E90Q, N258D, and S63D affect kinetics (Kato et al. 2012). This central restriction site provides a hydrophobic barrier through a network of interconnected hydrogen-bonds during the dark state that prevents water from entering the pore (Kato et al. 2012; Wietek et al. 2014). The

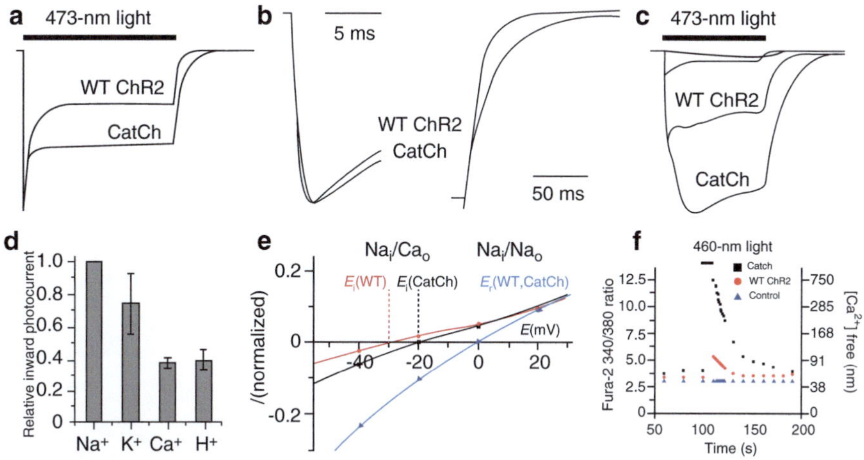

Fig. 15.4 Comparison of the Permeability Properties between WT and CatCh (L132C) mutant. (**a**) Photocurrents in response to a 1-s blue-light pulse. Traces are normalized to the peak photocurrent amplitude to illustrate the increase in the steady state–to–peak current ratio in CatCh compared with the wild type. (**b**) Comparison of on kinetics (*left*) and off kinetics (*right*) of photocurrents normalized to peak and steady-state currents, respectively. (**c**) Typical responses to a 600-ms blue-light pulse of CatCh- and wild type ChR2–expressing *Xenopus* oocytes in 80 mM extracellular Ca^{2+} (pH 9) at −120 mV (*continuous lower traces*). Injection of 1 mM BAPTA (Ca^{2+} chelator) abolished the superimposed currents of the intrinsic Ca^{2+}-activated chloride channels, while residual channelrhodopsin Ca^{2+} currents remained (*dashed upper traces*). Currents were normalized to the wild-type ChR2 peak current. (**d**) Ion flux characteristics of CatCh in HEK293 cells at −80 mV (mean ± s.d., $n=6$). (**e**) Shift of the reversal potential (E_r) for wild-type ChR2 (*red circles*) and CatCh (*black squares*) when exchanging the extracellular solution from 140 mM Na^+ (blue triangles, wild-type and CatCh are superimposed) to 90 mM Ca^{2+}. The intracellular solution is kept constant with 140 mM Na^+ inside that leads to a negative E_r in the presence of Ca^{2+} (mean ± s.d., $n=5$). (**f**) Fura-2 measurements of Ca^{2+} influx in HEK293 cells expressing wild-type ChR2 (*red circles*) and CatCh (*black squares*) to 10 s of 460-nm light in the presence of 90 mM extracellular Ca^{2+} ($n=10$). The fluorescence ratios at 340/380 are normalized to the YFP fluorescence. Control untransfected HEK293 cells are also shown (*blue triangles*) (Kleinlogel et al. 2011)

inner occlusion site is composed of the sterically bulky Y70 on TM 1 (Kato et al. 2012). Experiments utilizing spin-labeled ChR2 and EPR spectroscopy revealed large movement of TM 2 after retinal isomerization with light excitation (Sattig et al. 2013). This contrasts with the opening mechanism for bR, where movement of TM 7, induced by photoisomerization of retinal, is sensed by TM 6 (Hirai and Subramaniam 2009). Another set of experiments using stop-scan FTIR spectroscopy in tandem with MD simulations discovered that E90 is responsible for the movement of TM 2 and provided insight on how a continuous pore forms between the two occlusion sites (Kuhne et al. 2014). E90 faces towards the intracellular side of the membrane and forms two hydrogen-bonds with N258 in the dark state. Retinal photoisomerization disturbs the position of N258, and one of the hydrogen-bonds with E90 is broken. This causes E90 to flip toward the extracellular side and tilts TM2, opening a fully connected pore between the two occlusion sites (Kuhne

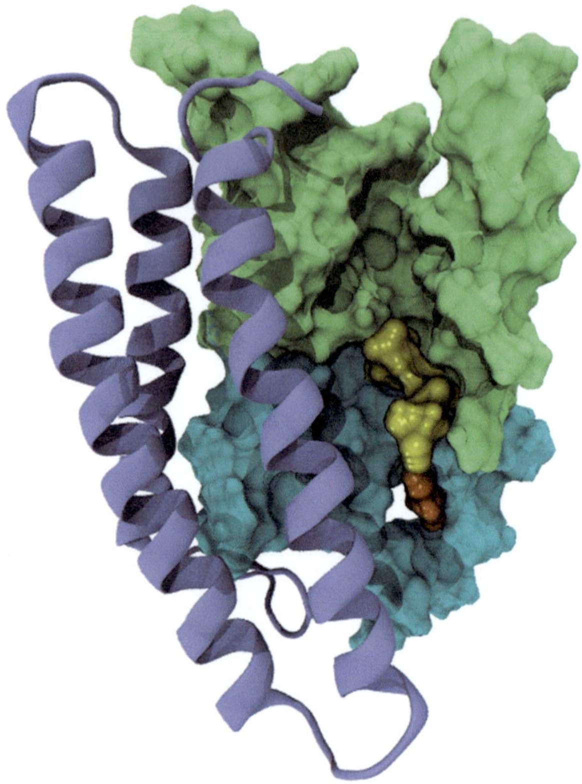

Fig. 15.5 Homology model of the conducting pore region of ChR2. The extracellular side of the channel contains an open vestibule (*lime color*) where cations enter the pore. The pore is constricted by two occlusion sites. The first site of S63, E90, and N258 (*yellow*) blocks the extracellular side while the second site is formed by Y70 (*orange*) and blocks the intracellular side. The bottom half of the pore is colored *cyan* and helices 4–6 are colored *blue*. For clarity, TM 7 is not shown but also lines the conducting pore

et al. 2014; Sattig et al. 2013). This is consistent with previous experiments that demonstrated that the TM domains become saturated with water upon cation flux (Lórenz-Fonfría et al. 2013).

The large movement of TM 2 also explains the importance of this helix to cation selectivity (Fig. 15.3a). TM2 contains multiple charged polar residues facing the pore that contribute to permeability and kinetics (Lórenz-Fonfría et al. 2013; Kato et al. 2012; Ruffert et al. 2011). Three charged glutamate residues (E90, E97, and E101) have the largest effect on the selectivity of Na^+ and H^+ (Ruffert et al. 2011; Schneider et al. 2013). Charge inversion mutations introduced at these positions significantly reduced Na^+ and H^+ photocurrents at pH 4, and produced negligible currents at alkaline and neutral pH (Ruffert et al. 2011; Schneider et al. 2013; Eisenhauer et al. 2012). Interestingly, the E90K mutation has no Na^+ conductance but large photocurrents at pH 4, which were originally attributed to H^+ flux. However, no change in reversal potentials at different pH values was observed (Eisenhauer et al. 2012). It was later discovered that this mutation, along with E90R, turned ChR2 into an outward rectified Cl^- channel with minimal conductance for cations (Wietek et al. 2014).

TM3 has also been found to be important for ion conductance and selectivity. Five cysteine mutations located on TM3 (R115C, Q117C, E123C, L132C, and

I133C), when treated with the cysteine-specific reagents 2-aminomethanthiosulfonate (MTSEA) and 2-sulfonatoethyl-methanethiosulfonate (MTSES), exhibited significant increased or reduced photocurrents compared to untreated cells (Gaiko and Dempski 2013). Analysis of these experimental results suggested these residues line part of the conducting pathway. Mutation of Q117 to glutamate showed reduced permeability ratios for alkali metals when compared to WT ChR2. Q117 sits at the interface of extracellular and transmembrane domains and it was reasoned that a hydrogen bond (broken in the Q117E mutant) plays a role in the dehydration of cations entering the pore (Fig. 15.3a). ChR2 lacks several TM serine residues that are present in bR and important for both structure and function (Richards and Dempski 2012; Marti et al. 1991). Re-introduction of these serine mutations at homologous positions (C87, G181, I197, G224, P234, M255, and V269) reduced the minimum pore diameter of ChR2 and altered the relative permeability of alkali metal ions (Richards and Dempski 2012). The largest affect was seen for V269S, which reduced the pore diameter to 4.6 Å and had highly reduced permeability ratios. Moreover, removal of an endogenous serine (S136A) increased the pore diameter from 6.2 to 7.0 Å. This change in pore diameter and selectivity was attributed to the high propensity for TM serine residues to form inter- or intrahelical hydrogen-bonds, thus restricting the conformational flexibility of the protein (Luecke et al. 1998; Gray and Matthews 1984; Dawson et al. 2002).

15.5 Channelrhodopsin Diversity and Functional Relevance

The discovery of new ChRs has proven beneficial in examining the molecular determinants that contribute to channel activity and selectivity. Initially, the closest structural and sequence homologs were those of microbial-type rhodopsins such as bR. However, homology cloning performed on *C. yellowstonensis*, *C. augustae*, and *C. raudensis*, algal species that exhibit similar light-mediated photocurrents to *C. reinhardtii*, produced several new ChR variants (Hou et al. 2012). These new ChRs had variable kinetic parameters and activation wavelengths when compared to ChR1 although they shared similar and conserved domains and motifs. Additional ChRs were discovered by transcriptome sequencing of 127 algal species (Klapoetke et al. 2014). From these sequencing experiments, 61 algal opsins were discovered and characterized by their function in different wavelengths of light, kinetics, and recovery.

The activation wavelength of ChR2 has become of particular interest to researchers for multi-wavelength activation of cells containing distinct populations of engineered ChR2 mutants. Several factors affect the activation wavelength and wavelength shifts within retinylidene proteins including the counterions near the Schiff base, the electrostatic environment surrounding retinal, and the hydrogen-bonding environment (Lasogga et al. 2010; Govorunova et al. 2013). The C1C2 crystal structure revealed that the retinal binding pocket is lined by W124, D156, F178, W223, F226, and F230 (Fig. 15.6) (Kato et al. 2012). It was also suggested

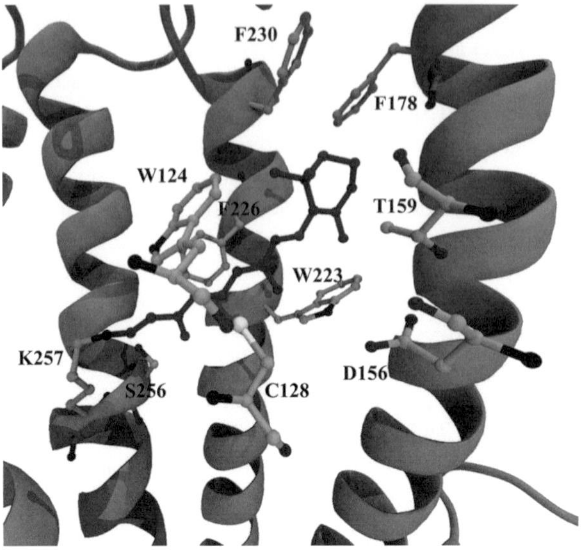

Fig. 15.6 Retinal binding pocket of ChR2 homology model. Residues that line the binding pocket are labeled

that three residues forming a pocket around the β-ionone ring (C128, T159, and S256) may also contribute to the color shift of ChR2 compared to bR (556 nm). Known ChRs contain a highly conserved glutamate (E123 in ChR2) and completely conserved aspartate (D253 in ChR2) that are both in close proximity to the Schiff base. D253 has been identified as the proton acceptor of the Schiff base in the photocycle reaction (Lórenz-Fonfría et al. 2013), while E123 was identified as the original internal acceptor (Fig. 15.2) (Bamann et al. 2008; Berndt et al. 2009; Lórenz-Fonfría et al. 2013). Both residues are located in close proximity to the Schiff base. Action spectrum recordings of the blue-shifted PsChR (found in *Platymonas subcordiformis*) found that mutations at the homologous positions E106 and D236 to Q and N caused a red-shift in the spectrum. The same mutations made at E123 and D253 had a similar effect on the ChR2 spectrum (Govorunova et al. 2013). Absorption experiments with the E123T (neutral) and E123D (negative) mutations revealed a red-shift in the ChR2 spectrum (Scholz et al. 2012). Although it is counter-intuitive that a conserved charge mutation (E123D) would have the same effect as a charge inversion (E123Q), the red-shifted spectrum likely occurs from the distortion of the local geometry by the extra methylene or from alteration of the hydrogen-bonding network (Scholz et al. 2012). Interestingly, the homolgous E123T mutation in a chimera ChR comprised of ChR1 and VChR1 produced a blue-shifted spectrum, suggesting structural differences at the counterion site between ChR1 and ChR2 (Prigge et al. 2012).

The discovery of new ChRs has revealed interesting insights on some of the molecular determinants that influence spectral shifts in channel activation. Correlations can be made from newly discovered ChRs, the structure and composition of the retinal binding pocket, and spectral characteristics. Understanding the structural elements that govern retinal absorbance is of significant importance to the optical control of cells. Blue light activated ChRs are disadvantageous because of biological tissue scattering lower wavelengths of light (Tromberg et al. 2000). Therefore, there is a concerted effort to discover and engineer new ChRs that are activated by red light for greater tissue penetration. The recently discovered Chrimson ChR, found in the algal species *Chlamydomonas noctigama*, has the farthest red activated spectrum yet discovered at 660 nm (Klapoetke et al. 2014). The action spectrum for this protein is 45 nm more red-shifted than the engineered ReaChR variant, which used a chimeric approach with ChR1, VChR1, and ChR2 (Lin et al. 2013). Compared to ChR2, Chrimson has several differences in the residues lining the retinal binding pocket. S256 and T159, which surround the ring structure, are alanine and methionine in Chrimson, respectively. Furthermore, residues that line the pore differ between ChR2 and Chrimson. D156 is a cysteine while the two hydrophobic aromatics F178 and F226 are tyrosine residues. These changes to the environment surrounding the chromophore may account for the large shift in absorbance.

15.6 Application to the Biomedical Field

Recognition that channelopsins could be used as a tool to control excitable membranes was acknowledged upon the first molecular characterization of ChR2 where Bamberg and colleagues wrote “expression of ChR2…may be used as a powerful tool to increase cytoplasmic Ca^{2+} concentration or to depolarize the cell membrane, simply by illumination (Nagel et al. 2003).” The first experiments fulfilling this idea came only 2 years later when it was shown that excitable membranes could be manipulated in a light-dependent manner using ChR2 (Boyden et al. 2005). Here mammalian hippocampal neurons were transformed with ChR2-YFP (yellow fluorescent protein) using a lentivirus. Light-activation of ChR2 was sufficient to generate a threshold potential which resulted in neuronal spiking. Equally, neuronal spiking could be resolved on the millisecond time scale. In these experiments, both excitatory and inhibitory synaptic transmission could be controlled upon the light-induced activation of channelrhodopsin. A glimpse of the impending wide utility of these proteins to control neuronal cells was soon established as it was also shown that ChR2 could also be used to control the membrane potential of PC-12 cells (Ishizuka et al. 2006).

While using ChR2 to control neuronal spiking in cell-based experiments was an exciting development, it was subsequently shown that ChR2 could be used as a tool to manipulate whole animal preparations. Here, Herlitze and colleagues electroporated a GFP-ChR2 gene, under the cytomegalovirus (CMV) promoter into the spinal cords for stage 16 (embryonic day 2–3) chick embryos (Li et al. 2005). At embryonic day 4.5–5, the spinal cord and hind limbs were separated and the GFP-ChR2 fusion protein was found distributed in motor and interneurons. Equally, when exposed to continuous light, it was observed that electroporation did not have a deleterious effect on the development of cord circuits, but did increase the frequency of neuronal firing using suction electrode recordings.

The first successful *in vivo* utilization of ChR2 as a tool to control the behavior of an organism was achieved by the Gottschalk group where ChR2 was functionally expressed in excitable cells of *Caenorhabditis elegans* (Nagel et al. 2005). When ChR2 was expressed and activated upon light stimulation in muscle cells, muscle contractions were observed. Equally, following functional expression of ChR2 in mechanosensory neurons light-induced activation of ChR2 induced withdrawal behaviors which are normally associated with mechanical stimulation. This work was an important benchmark in establishing ChR2 as a tool to study excitable cells in living organisms. Of course, a point which cannot be under-emphasized in this and other studies is the observation that no externally applied retinal is required for the functional expression of channelrhodopsins. If delivery of retinal was required for channelrhodopsin function, then the technical challenges to successfully use this protein as a tool in tissues and organisms would greatly increase. However, it is apparent that even in intact brain tissue, enough retinal is available to generate a sufficient population of ChR2 to induce threshold activation upon light stimulation.

The development of ChR2 functional expression in intact organisms opened a new wave of research strategies to apply ChR2 as a tool to investigate cellular processes. These included using ChR2 as a tool to treat retinal degenerative diseases by injecting ChR2 into mouse and rat eyes, triggering appetitive or aversive learning in *Drosophila* larvae, investigating long-term potentiation in rat hippocampal slice cultures, learning behaviors, sleep, as well as mapping circuits between presynaptic and postsynaptic neurons (Bi et al. 2006; Schroll et al. 2006; Zhang and Oertner 2007; Huber et al. 2008; Petreanu et al. 2007; Adamantidis et al. 2007).

While ChR2 has become a powerful tool to investigate excitable membranes, limitations arise due to the excitation wavelength, ion conductance and kinetic properties and the absence of a mechanism to inhibit action potentials. Therefore, there has been a sustained effort to identify new constructs either through discovery of new channelrhodopsin homologs or manipulation of known genes. Considering that phototaxis is a well-established biological process through multiple organsisms, it is no surprise that related microbes have light activated cation channels with differing properties. One of the first of these novel proteins to be identified was isolated from *Volvox carteri* and named channelrhodopsin (VChR1) (Zhang et al. 2008). This protein has a red-shifted excitation maximum at 589 nm when compared to ChR2 (470 nm). Identification of this protein, for the first time, enabled simultaneous investigation of the interaction of different cell types in circuit computation or behavior. Equally, light at this longer wavelength can penetrate tissue further enable investigation of more distal cells in intact organisms.

In a parallel avenue of research, there has been an emphasis to identify ChR mutants with novel properties. For instance, mutation of L132 to cysteine in ChR2 also revealed a construct with enhanced calcium permeability (CatCh mutant, Figs. 15.3a and 15.4) (Kleinlogel et al. 2011). Considering that some excitable membranes, most notably heart tissue, are controlled by changing the cytosolic concentration of calcium, identification of this construct has facilitated a new method to control heart tissue (Williams et al. 2013; Karathanos et al. 2014; Park et al. 2014; Ambrosi et al. 2014). Finally, it has been demonstrated that the D156C mutation results in a construct, named ChR2-XXL, which has extra high expression and a longer open-state lifetime (Fig. 15.3b) (Dawydow et al. 2014). This construct was successfully used in *Drosophila melanogaster* to investigate innate male courtship and olfactory memories.

A new construct named ChEF demonstrated decreased photoactivation. This protein is a chimeric protein resulting from transmembrane domains 1–5 of ChR1 and 6 and 7 of ChR2 as well as site directed mutations (Lin et al. 2009).

15.7 Summary

Starting with its central role in phototaxis, the behavior and molecular mechanism of channelrhodopsins has been studied for nearly 200 years. While the outline of the molecular mechanism of cation conduction induced by light activation has been elucidated, it is clear that there is much to learn about the photocycle of channelrhodopsins. Equally, identification of mutants and recent sequencing efforts has uncovered a growing array of proteins with novel functionality. These discoveries have opened up new lines of studies in fields as diverse as learning behaviors, vision and cardiac physiology. Thus, the future should bring an exciting array of new developments both to our basic understanding of channelopsins as well as applications in the fields of optogenetics.

Acknowledgements The Dempski laboratory gratefully acknowledges the WPI Research Foundation for support.

References

Accardi A, Miller C (2004) Secondary active transport mediated by a prokaryotic homologue of ClC Cl- channels. Nature 427(6977):803–807. doi:10.1038/nature02314

Adamantidis A, Zhang F, Aravanis AM, Deisseroth K, de Lecea L (2007) Neural substrates of awakening probed with optogenetic control of hypocretin neurons. Nature 450:420–424

Adamian L, Jackups R, Binkowski TA, Liang J (2003) Higher-order interhelical spatial interactions in membrane proteins. J Mol Biol 327(1):251–272. doi:10.1016/s0022-2836(03)00041-x

Ambrosi CM, Klimas A, Yu J, Entcheva E (2014) Cardiac applications of optogenetics. Prog Biophys Mol Biol 115(2–3):294–304. doi:10.1016/j.pbiomolbio.2014.07.001

Bamann C, Kirsch T, Nagel G, Bamberg E (2008) Spectral characteristics of the photocycle of channelrhodopsin-2 and its implication for channel function. J Mol Biol 375(3):686–694. doi:10.1016/j.jmb.2007.10.072

Bamann C, Gueta R, Kleinlogel S, Nagel G, Bamberg E (2010) Structural guidance of the photocycle of channelrhodopsin-2 by an interhelical hydrogen bond. Biochemistry 49(2):267–278. doi:10.1021/bi901634p

Berndt A, Yizhar O, Gunaydin LA, Hegemann P, Deisseroth K (2009) Bi-stable neural state switches. Nat Neurosci 12(2):229–234. doi:10.1038/nn.2247

Berndt A, Prigge M, Gradmann D, Hegemann P (2010) Two open states with progressive proton selectivities in the branched channelrhodopsin-2 photocycle. Biophys J 98(5):753–761. doi:10.1016/j.bpj.2009.10.052

Bi A, Cui J, Ma YP, Olshevskaya E, Pu M, Dizhoor AM, Pan ZH (2006) Ectopic expression of a microbial-type rhodopsin restores visual responses in mice with photoreceptor degeneration. Neuron 50:23–33

Boyden ES, Zhang F, Bamberg E, Nagel G, Deisseroth K (2005) Millisecond-timescale, genetically-targeted optical control of neural activity. Nat Neurosci 8:1263–1268

Cohen J, Schulten K (2004) Mechanism of anionic conduction across ClC. Biophys J 86(2):836–845. doi:10.1016/s0006-3495(04)74159-4

Colegrave N (2002) Sex releases the speed limit on evolution. Nature 420(6916):664–666. doi:10.1038/nature01191

Dawson JP, Weinger JS, Engelman DM (2002) Motifs of serine and threonine can drive association of transmembrane helices. J Mol Biol 316(3):799–805. doi:10.1006/jmbi.2001.5353

Dawydow A, Gueta R, Ljaschenko D, Ullrich S, Hermann M, Ehmann N, Gao S, Fiala A, Langenhan T, Nagel G, Kittel RJ (2014) Channelrhodopsin-2-XXL, a powerful optogenetic tool for low-light applications. Proc Natl Acad Sci U S A 111(38):13972–13977. doi:10.1073/pnas.1408269111

Demurtas OC, Massa S, Ferrante P, Venuti A, Franconi R, Giuliano G (2013) A Chlamydomonas-derived human papillomavirus 16 E7 vaccine induces specific tumor protection. PLoS One 8(4):e61473. doi:10.1371/journal.pone.0061473

Eisenhauer K, Kuhne J, Ritter E, Berndt A, Wolf S, Freier E, Bartl F, Hegemann P, Gerwert K (2012) In channelrhodopsin-2 Glu-90 is crucial for ion selectivity and is deprotonated during the photocycle. J Biol Chem 287(9):6904–6911. doi:10.1074/jbc.M111.327700

Ernst OP, Sanchez Murcia PA, Daldrop P, Tsunoda SP, Kateriya S, Hegemann P (2008) Photoactivation of channelrhodopsin. J Biol Chem 283(3):1637–1643. doi:10.1074/jbc.M708039200

Feldbauer K, Zimmermann D, Pintschovius V, Spitz J, Bamann C, Bamberg E (2009) Channelrhodopsin-2 is a leaky proton pump. Proc Natl Acad Sci U S A 106(30):12317–12322. doi:10.1073/pnas.0905852106

Feng L, Campbell EB, MacKinnon R (2012) Molecular mechanism of proton transport in CLC Cl-/H+ exchange transporters. Proc Natl Acad Sci U S A 109(29):11699–11704. doi:10.1073/pnas.1205764109

Fleming KG, Ackerman AL, Engelman DM (1997) The effect of point mutations on the free energy of transmembrane alpha-helix dimerization. J Mol Biol 272(2):266–275. doi:10.1006/jmbi.1997.1236

Foster KW, Smyth RD (1980) Light antennas in phototactic algae. Microbiol Rev 44(4):572–630

Gadsby DC (2009) Ion channels versus ion pumps: the principal difference, in principle. Nat Rev Mol Cell Biol 10(5):344–352. doi:10.1038/nrm2668

Gaiko O, Dempski RE (2013) Transmembrane domain three contributes to the ion conductance pathway of channelrhodopsin-2. Biophys J 104(6):1230–1237. doi:10.1016/j.bpj.2013.02.013

Govorunova EG, Sineshchekov OA, Li H, Janz R, Spudich JL (2013) Characterization of a highly efficient blue-shifted channelrhodopsin from the marine alga Platymonas subcordiformis. J Biol Chem 288(41):29911–29922. doi:10.1074/jbc.M113.505495

Gradmann D, Berndt A, Schneider F, Hegemann P (2011) Rectification of the channelrhodopsin early conductance. Biophys J 101(5):1057–1068. doi:10.1016/j.bpj.2011.07.040

Gray TM, Matthews BW (1984) Intrahelical hydrogen bonding of serine, threonine and cysteine residues within alpha-helices and its relevance to membrane-bound proteins. J Mol Biol 175(1):75–81. doi:10.1016/0022-2836(84)90446-7

Grossman N, Nikolic K, Toumazou C, Degenaar P (2011) Modeling study of the light stimulation of a neuron cell with channelrhodopsin-2 mutants. IEEE Trans Biomed Eng 58(6):1742–1751. doi:10.1109/tbme.2011.2114883

Grzesiek S, Dencher NA (1986) Time-course and stoichiometry of light-induced proton release and uptake during the photocycle of bacteriorhodopsin. FEBS Lett 208(2):337–342. doi:10.1016/0014-5793(86)81045-6

Gunaydin LA, Yizhar O, Berndt A, Sohal VS, Deisseroth K, Hegemann P (2010) Ultrafast optogenetic control. Nat Neurosci 13(3):387–392

Heberle J, Oesterhelt D, Dencher NA (1993) Decoupling of photo- and proton cycle in the Asp85-Glu mutant of bacteriorhodopsin. EMBO J 12(10):3721–3727

Hegemann P, Ehlenbeck S, Gradmann D (2005) Multiple photocycles of channelrhodopsin. Biophys J 89(6):3911–3918. doi:10.1529/biophysj.105.069716

Hille B (2001) Ion channels of excitable membranes. Sinauer, Sunderland, 3rd Edition

Hirai T, Subramaniam S (2009) Protein conformational changes in the bacteriorhodopsin photocycle: comparison of findings from electron and X-Ray crystallographic analyses. PLoS One 4(6):16. doi:10.1371/journal.pone.0005769

Hou S-Y, Govorunova EG, Ntefidou M, Lane CE, Spudich EN, Sineshchekov OA, Spudich JL (2012) Diversity of Chlamydomonas channelrhodopsins. Photochem Photobiol 88(1):119–128. doi:10.1111/j.1751-1097.2011.01027.x

Huber D, Petreanu L, Ghitani N, Ranade S, Hromadka T, Mainen Z, Svoboda K (2008) Sparse optical microstimulation in barrel cortex drives learned behaviour in freely moving mice. Nature 451(7174):61–64. doi:10.1038/nature06445

Humphrey W, Dalke A, Schulten K (1996) VMD: visual molecular dynamics. J Mol Graph Model 14(1):33–38

Ihara K, Umemura T, Katagiri I, Kitajima-Ihara T, Sugiyama Y, Kimura Y, Mukohata Y (1999) Evolution of the archaeal rhodopsins: evolution rate changes by gene duplication and functional differentiation. J Mol Biol 285(1):163–174

Ishizuka T, Kakuda M, Araki R, Yawo H (2006) Kinetic evaluation of photosensitivity in genetically engineered neurons expressing green algae light-gated channels. Neurosci Res 54:85–94

Jayasinghe S, Hristova K, White SH (2001) Energetics, stability, and prediction of transmembrane helices. J Mol Biol 312(5):927–934. doi:10.1006/jmbi.2001.5008

Karathanos TV, Boyle PM, Trayanova NA (2014) Optogenetics-enabled dynamic modulation of action potential duration in atrial tissue: feasibility of a novel therapeutic approach. Europace 16(Suppl 4):iv69–iv76. doi:10.1093/europace/euu250

Kato HE, Zhang F, Yizhar O, Ramakrishnan C, Nishizawa T, Hirata K, Ito J, Aita Y, Tsukazaki T, Hayashi S, Hegemann P, Maturana AD, Ishitani R, Deisseroth K, Nureki O (2012) Crystal structure of the channelrhodopsin light-gated cation channel. Nature 482(7385):369–374. doi:10.1038/nature10870

Klapoetke NC, Murata Y, Kim SS, Pulver SR, Birdsey Benson A, Cho YK, Morimoto TK, Chuong AS, Carpenter EJ, Tian Z, Wang J, Xie Y, Yan Z, Zhang Y, Chow BY, Surek B, Melkonian M, Jayaraman V, Constantine-Paton M, Wong GK-S, Boyden ES (2014) Independent optical excitation of distinct neural populations. Nat Methods 11(3):338–346. doi:10.1038/nmeth.2836

Kleinlogel S, Feldbauer K, Dempski RE, Fotis H, Wood PG, Bamann C, Bamberg E (2011) Ultra light-sensitive and fast neuronal activation with the Ca(2)+–permeable channelrhodopsin CatCh. Nat Neurosci 14(4):513–518. doi:10.1038/nn.2776

Kuhne J, Eisenhauer K, Ritter E, Hegemann P, Gerwert K, Bartl F (2014) Early formation of the ion-conducting pore in channelrhodopsin-2. Angew Chem Int Ed Engl 54(16):4953–4957. doi:10.1002/anie.201410180

Lasogga L, Rettig W, Otto H, Wallat I, Bricks J (2010) Model systems for the investigation of the opsin shift in bacteriorhodopsin. J Phys Chem A 114(5):2179–2188. doi:10.1021/jp904132f

Lenci F, Colombetti G (1978) Photobehavior of microorganisms: a biophysical approach. Annu Rev Biophys Bioeng 7:341–361. doi:10.1146/annurev.bb.07.060178.002013

Li X, Gutierrez DV, Hanson MG, Han J, Mark MD, Chiel H, Hegemann P, Landmesser LT, Herlitze S (2005) Fast noninvasive activation and inhibition of neural and network activity by vertebrate rhodopsin and green algae channelrhodopsin. Proc Natl Acad Sci U S A 102(49):17816–17821

Lin JY, Lin MZ, Steinbach P, Tsien RY (2009) Characterization of engineered channelrodopsin variants with improved properties and kinetics. Biophys J 96(5):1803–1814

Lin JY, Knutsen PM, Muller A, Kleinfeld D, Tsien RY (2013) ReaChR: a red-shifted variant of channelrhodopsin enables deep transcranial optogenetic excitation. Nat Neurosci 16(10):1499–1508. doi:10.1038/nn.3502

Lisal J, Maduke M (2008) The ClC-0 chloride channel is a 'broken' Cl(−)/H(+) antiporter. Nat Struct Mol Biol 15(8):805–810. doi:10.1038/nsmb.1466
Lórenz-Fonfría VA, Heberle J (2014) Channelrhodopsin unchained: structure and mechanism of a light-gated cation channel. Biochim Biophys Acta 1837(5):626–642
Lórenz-Fonfría VA, Resler T, Krause N, Nack M, Gossing M, Fischer von Mollard G, Bamann C, Bamberg E, Schlesinger R, Heberle J (2013) Transient protonation changes in channelrhodopsin-2 and their relevance to channel gating. Proc Natl Acad Sci U S A 110(14):E1273–E1281. doi:10.1073/pnas.1219502110
Luecke H, Richter HT, Lanyi JK (1998) Proton transfer pathways in bacteriorhodopsin at 2.3 Angstrom resolution. Science 280(5371):1934–1937. doi:10.1126/science.280.5371.1934
Marti T, Otto H, Mogi T, Rösselet SJ, Heyn MP, Khorana HG (1991) Bacteriorhodopsin mutants containing single substitutions of serine or threonine residues are all active in proton translocation. J Biol Chem 266(11):6919–6927
Mueller M, Bamann C, Bamberg E, Kuehlbrandt W (2011) Projection structure of channelrhodopsin-2 at 6 angstrom resolution by electron crystallography. J Mol Biol 414(1):86–95. doi:10.1016/j.jmb.2011.09.049
Nack M, Radu I, Gossing M, Bamann C, Bamberg E, von Mollard GF, Heberle J (2010) The DC gate in Channelrhodopsin-2: crucial hydrogen bonding interaction between C128 and D156. Photochem Photobiol Sci 9(2):194–198. doi:10.1039/b9pp00157c
Nack M, Radu I, Schultz B-J, Resler T, Schlesinger R, Bondar A-N, del Val C, Abbruzzetti S, Viappiani C, Bamann C, Bamberg E, Heberle J (2012) Kinetics of proton release and uptake by channelrhodopsin-2. FEBS Lett 586(9):1344–1348. doi:10.1605/01.301-0019506574.2012
Nagel G, Ollig D, Fuhrmann M, Kateriya S, Musti AM, Bamberg E, Hegemann P (2002) Channelrhodopsin-1: a light-gated proton channel in green algae. Science 296(5577):2395–2398
Nagel G, Szellas T, Huhn W, Kateriya S, Adeishvili N, Berthold P, Ollig D, Hegemann P, Bamberg E (2003) Channelrhodopsin-2, a directly light-gated cation-selective membrane channel. Proc Natl Acad Sci U S A 100(24):13940–13945. doi:10.1073/pnas.1936192100
Nagel G, Brauner M, Liewald JF, Adeishvili N, Bamberg E, Gottschalk A (2005) Light activation of channelrhodopsin-2 in excitable cells of Caenorhabditis elegans triggers rapid behavioral responses. Curr Biol 15(21):2279–2284
Nikolic K, Grossman N, Grubb MS, Burrone J, Toumazou C, Degenaar P (2009) Photocycles of Channelrhodopsin-2. Photochem Photobiol 85(1):400–411. doi:10.1111/j.1751-1097.2008.00460.x
Park SA, Lee SR, Tung L, Yue DT (2014) Optical mapping of optogenetically shaped cardiac action potentials. Sci Rep 4:6125. doi:10.1038/srep06125
Peralvarez-Marin A, Marquez M, Bourdelande JL, Querol E, Padros E (2004) Thr-90 plays a vital role in the structure and function of bacteriorhodopsin. J Biol Chem 279(16):16403–16409. doi:10.1074/jbc.M313988200
Petreanu L, Huber D, Sobczyk A, Svoboda K (2007) Channelrhodopsin-2-assisted circuit mapping of long range callosal projections. Nat Neurosci 10(5):663–668
Popot JL, Engelman DM (1990) Membrane protein folding and oligomerization: the two-stage model. Biochemistry 29(17):4031–4037. doi:10.1021/bi00469a001
Prigge M, Schneider F, Tsunoda SP, Shilyansky C, Wietek J, Deisseroth K, Hegemann P (2012) Color-tuned channelrhodopsins for multiwavelength optogenetics. J Biol Chem 287(38):31804–31812. doi:10.1074/jbc.M112.391185
Radu I, Bamann C, Nack M, Nagel G, Bamberg E, Heberle J (2009) Conformational changes of channelrhodopsin-2. J Am Chem Soc 131(21):7313–7319. doi:10.1021/ja8084274
Richards R, Dempski RE (2012) Re-introduction of transmembrane serine residues reduce the minimum pore diameter of channelrhodopsin-2. PLoS One 7(11), e50018. doi:10.1371/journal.pone.0050018
Ritter E, Stehfest K, Berndt A, Hegemann P, Bartl FJ (2008) Monitoring light-induced structural changes of channelrhodopsin-2 by UV-visible and fourier transform infrared spectroscopy. J Biol Chem 283(50):35033–35041. doi:10.1074/jbc.M806353200

Ruffert K, Himmel B, Lall D, Bamann C, Bamberg E, Betz H, Eulenburg V (2011) Glutamate residue 90 in the predicted transmembrane domain 2 is crucial for cation flux through channelrhodopsin 2. Biochem Biophys Res Commun 410(4):737–743. doi:10.1016/j.bbrc.2011.06.024

Sattig T, Rickert C, Bamberg E, Steinhoff HJ, Bamann C (2013) Light-induced movement of the transmembrane Helix B in channelrhodopsin-2. Angew Chem Int Ed 52(37):9705–9708. doi:10.1002/anie.201301698

Schneider F, Gradmann D, Hegemann P (2013) Ion selectivity and competition in channelrhodopsins. Biophys J 105(1):91–100. doi:10.1016/j.bpj.2013.05.042

Scholz F, Bamberg E, Bamann C, Wachtveitl J (2012) Tuning the primary reaction of channelrhodopsin-2 by imidazole, pH, and site-specific mutations. Biophys J 102(11):2649–2657. doi:10.1016/j.bpj.2012.04.034

Schroll C, Riemensperger T, Bucher D, Ehmer J, Voller T, Erguth K, Gerber B, Hendel T, Nagel G, Buchner E, Fiala A (2006) Light-induced activation of distinct modulatory neurons triggers appetitive or adversive learning in drosophila larvae. Curr Biol 16(17):1741–1747

Sineshchekov OA, Govorunova EG, Wang J, Li H, Spudich JL (2013) Intramolecular proton transfer in channelrhodopsins. Biophys J 104(4):807–817. doi:10.1016/j.bpj.2013.01.002

Stefanescu RA, Shivakeshavan RG, Khargonekar PP, Talathi SS (2013) Computational modeling of channelrhodopsin-2 photocurrent characteristics in relation to neural signaling. Bull Math Biol 75(11):2208–2240. doi:10.1007/s11538-013-9888-4

Stehfest K, Ritter E, Berndt A, Bartl F, Hegemann P (2010) The branched photocycle of the slow-cycling channelrhodopsin-2 mutant C128T. J Mol Biol 398(5):690–702. doi:10.1016/j.jmb.2010.03.031

Treviranus LC (1817) Vermischte Schriften anatomischen und physiologischen Inhalts, vol 2, Fernere Beobachtungen uber die Bewegung der grunen Materie im Pflanzenreiche. J.G. Heyse, Bremen

Tromberg BJ, Shah N, Lanning R, Cerussi A, Espinoza J, Pham T, Svaasand L, Butler J (2000) Non-invasive *in vivo* characterization of breast tumors using photon migration spectroscopy. Neoplasia 2(1–2):26–40. doi:10.1038/sj.neo.7900082

Verhoefen M-K, Bamann C, Blöcher R, Förster U, Bamberg E, Wachtveitl J (2010) The photocycle of channelrhodopsin-2: ultrafast reaction dynamics and subsequent reaction steps. Chemphyschem 11(14):3113–3122. doi:10.1002/cphc.201000181

Watanabe HC, Welke K, Schneider F, Tsunoda S, Zhang F, Deisseroth K, Hegemann P, Elstner M (2012) Structural model of channelrhodopsin. J Biol Chem 287(10):7456–7466. doi:10.1074/jbc.M111.320309

Watanabe HC, Welke K, Sindhikara DJ, Hegemann P, Elstner M (2013) Towards an understanding of channelrhodopsin function: simulations lead to novel insights of the channel mechanism. J Mol Biol 425(10):1795–1814. doi:10.1016/j.jmb.2013.01.033

Wietek J, Wiegert JS, Adeishvili N, Schneider F, Watanabe H, Tsunoda SP, Vogt A, Elstner M, Oertner TG, Hegemann P (2014) Conversion of channelrhodopsin into a light-gated chloride channel. Science 344(6182):409–412. doi:10.1126/science.1249375

Williams JC, Xu J, Lu Z, Klimas A, Chen X, Ambrosi CM, Cohen IS, Entcheva E (2013) Computational optogenetics: empirically-derived voltage- and light-sensitive channelrhodopsin-2 model. PLoS Comput Biol 9(9), e1003220. doi:10.1371/journal.pcbi.1003220

Zhang YP, Oertner TG (2007) Optical induction of synaptic plasticity using a light-sensitive channel. Nat Methods 4(2):139–141. doi:10.1038/nmeth988

Zhang F, Prigge M, Beyriere F, Tsunoda SP, Mattis J, Yizhar O, Hegemann P, Deisseroth K (2008) Red-shifted optogenetic excitation: a tool for fast neural control derived from Volvox carteri. Nat Neurosci 11(6):631–633

Index

A

Actinoporins, 185, 267–281, 297
Amphipathic alpha (α)-helix, 154, 155, 169, 171, 174, 279
Amyloid, 343–357
Anthrax toxin, 146, 209–233, 243, 353
Apoptosis, 4, 24, 34, 59–71, 76, 78, 85, 89, 92, 96, 163, 170, 171, 203, 291, 292, 300, 357
ATP transport, 55–101, 20

B

Bacteria, 48–50, 101–118, 123–146, 174, 185–190, 196, 197, 210, 242, 243, 268–270, 290, 293, 294, 303, 304, 345, 346, 350, 353
β-Barrel structure, 103, 117, 125, 145, 242, 346
Bax, 60, 62, 64–71, 86–89, 92, 190, 297, 306, 350
Bcl2 family, 65, 67–68, 70, 78, 85, 86, 88, 350
Bcl-xL, 65, 66, 70, 85–89, 92, 187, 190
Beta (β)-barrel channels, 346
Bi-component toxin protein, 243, 260, 261
Bioelectrochemistry,
Bioenergetics, 43, 83, 345

C

CED-9, 85, 86
Cell membrane permeability, 128, 138, 243, 254, 255, 257, 261, 270, 272, 274, 291, 299–300, 314, 316, 345, 346, 351, 353, 356, 375
Ceramide, 75–96
Ceramide analogs, 85, 88–91
Channels
  dynamics, 14, 78, 79, 83, 91, 136, 139, 140, 145, 193, 195, 221, 226, 242, 248, 303, 326
  gating, 12, 13, 49, 126, 140, 160, 318
  regulation, 195–203
Chemical modification, 10, 104, 244, 257, 258
Cholesterol-dependent cytolysins, 267–281, 290, 293, 294, 305, 307
Colicin immunity protein, 195–203
Complement membrane attack complex, 290
Conductance, 7, 36, 60, 79, 103, 126, 154, 192, 216, 247, 269, 291, 317, 347, 364
Connexin, 317–332
Cysteine scanning mutagenesis, 169, 246

D

Disease, 4, 26, 65, 69–71, 138, 155, 171, 172, 175, 185, 209–211, 327–328, 343–345, 347, 349–351, 356, 357, 376

E

Electrophysiology, 5, 13, 17, 35–50, 59–71, 78, 80–84, 102, 103, 113, 115, 116, 118, 123–146, 154–163, 166, 168, 172–175, 189, 216, 220–224, 228, 232, 246, 247, 253, 255–256, 268–270, 274, 281, 294, 297–306, 316, 319, 329, 331, 348, 353, 357, 366, 369
Electrostatics, 101, 13, 20, 21, 38, 40, 41, 53, 103–111, 114, 142, 189, 212, 228–230, 261, 373

A.H. Delcour (ed.), *Electrophysiology of Unconventional Channels and Pores*, Springer Series in Biophysics 18, DOI 10.1007/978-3-319-20149-8

G
Gap junction, 69, 313–332
Gram-negative bacteria, 101, 102, 123, 124, 127, 133, 142, 186, 188, 189, 270, 293

H
Hemichannel, 322, 325, 327–329, 331

I
Inhibitors MAC (iMAC), 64, 70, 71
Inner membrane (IM), 4, 9, 25, 34–36, 44, 45, 48, 60–62, 66, 71, 77, 124, 125, 128, 133, 144, 188, 190, 195–198, 201, 202, 345, 350
Ion channel, 5, 7, 20, 36–38, 48, 52, 60, 66, 104, 115, 154, 156, 157, 159–169, 171–175, 185–203, 314, 318, 321, 324, 325, 327–329, 345–346, 348, 354, 363, 364, 367, 369
Ion pump, 117, 363, 364, 369
Ion transport, 103, 107, 110, 113, 253, 363, 366

L
Lanthanum, 316
Lipid, 5, 36, 66, 76, 103, 124, 154, 188, 215, 242, 268, 291, 345
Lipid membranes, 5, 7, 12, 14, 15, 21, 154, 242, 245, 248, 254, 255, 261, 268, 270, 272, 273, 279, 291, 296–298, 300, 304, 345, 353

M
Mega-channel, 63, 71
Membranes
  biogenesis, 48, 142–146
  protein, 5, 12, 20, 35, 41–43, 45, 49, 50, 124, 127, 131, 143, 144, 155, 157, 161, 164, 169, 172, 189, 191, 196, 203, 242, 247, 258, 292, 367
Mitochondria, 3–26, 33–53, 60–69, 71, 77, 82, 85, 87, 90, 92, 144, 146, 161, 163, 171, 188, 227, 349, 350, 356, 357
Mitochondrial apoptosis-induced channel (MAC), 59–71, 290–294, 296, 297, 300, 306
Mitochondrion, 61, 62, 64, 227
Modulation, 9, 15–19, 24–26, 116, 146, 163, 321–326

N
Nexus, 271, 314–316
Non-selective cation channels, 173, 345, 353, 356, 366

O
Optogenetics, 367, 377
Outer membrane (OM), 4, 34–36, 45, 49, 50, 60, 61, 64, 68–69, 71, 77, 85, 88, 101, 102, 111, 124, 125, 127, 131–134, 138, 139, 144–146, 186–192, 346, 353

P
Patch clamp, 60–64, 66, 68, 70, 71, 103, 124, 136, 137, 139, 140, 157–158, 160, 162, 167, 170, 172, 173, 274, 275, 316, 317, 328, 331, 370
Perforin, 268, 276, 289–307
Permeability, 4–7, 9, 14, 20, 24, 25, 37, 60, 62, 63, 65, 66, 76–78, 80, 85–93, 106, 140, 154, 156, 162, 163, 165, 171, 247, 277, 280, 281, 316, 318–320, 322, 326–328, 345, 347, 349–352, 357, 367, 370–373, 376
Phototaxis, 361–377
Planar bilayer, 6, 12, 22, 36, 42, 48, 60, 102, 103, 189, 222, 228, 232
Planar lipid bilayer (PLB), 6, 36, 42, 49, 124, 125, 127–134, 136, 137, 139, 140, 143, 144, 157–172, 193–195, 215, 216, 222–224, 246, 247, 251, 301, 348, 353, 354
Planar lipid membrane (PLM), 5, 7, 12, 14, 15, 248, 255, 267–269, 272–275, 277–281, 298, 300, 304
Planar membrane, 12, 22, 81, 82, 88, 93, 277
Polymer partitioning, 20, 248–250
Pore-forming colicin, 185–203, 232, 275, 292, 297, 300, 302, 305
Pores, 4, 35, 60, 79, 102, 124, 154, 185, 214, 242, 268, 289, 316, 346, 365
Porin, 50, 51, 60, 101–118, 132, 145, 189, 346, 353
Protein-import channels, 33–53, 60, 61, 227
Protein translocation, 43, 46, 49, 50, 124, 138, 143, 144, 214, 216, 220, 222, 223, 227, 229
Proteolipidic pore, 270, 298
Proton channel, 156–162, 164, 170, 171, 174
Proton gradient, 144, 214, 350

R

Rational protein design, 12, 14, 154, 159, 227

S

Secretion, 69, 123–125, 127–139, 143, 215, 242, 243, 261, 268, 270, 271, 294
Single channel conductance, 77, 112, 42, 45, 50, 51, 104, 128, 143, 154, 158–160, 165, 169, 173, 247, 255, 256, 277, 278, 304, 306, 323, 347, 348, 350, 357
Sphingolipid, 85
Subunit stoichiometry, 256, 258

T

Toroidal pore, 131, 195, 268, 269, 274, 279–281, 293, 297–299, 302, 305, 306, 349
Toxicity, 186–188, 195, 197, 198, 345–346, 348–352, 356, 357
Translocation, 77, 36, 66, 93, 124, 186, 214, 247, 280, 321, 365
Tubulin, 66, 12, 15–26

U

Unfolding, 35, 36, 46, 48, 49, 124, 127–129, 135, 143, 189–192, 194, 209–233, 248, 259, 271, 345
Unitary conductance, 174, 249, 250, 253, 255–257, 259

V

Viral ion channel, 154, 157, 159, 163, 171, 174
Virulence, 123, 128, 131, 134, 135, 146, 168, 171, 210–215, 243, 268
Voltage-dependent anion channel (VDAC), 44–26, 60, 62, 65–67, 81, 357
Voltage gating, 5, 7, 9–12, 14, 46, 196, 203, 320, 321, 329, 370

X

Xenopus oocytes, 160, 161, 164, 167, 169, 172, 173, 327, 371

MIX
Papier aus verantwortungsvollen Quellen
Paper from responsible sources
FSC® C105338

If you have any concerns about our products,
you can contact us on
ProductSafety@springernature.com

In case Publisher is established outside the EU,
the EU authorized representative is:
Springer Nature Customer Service Center GmbH
Europaplatz 3, 69115 Heidelberg, Germany

Printed by Libri Plureos GmbH
in Hamburg, Germany